Recycled Aggregate Concrete and Alternative Binders for Sustainable Building Engineering

Recycled Aggregate Concrete and Alternative Binders for Sustainable Building Engineering

Miguel Bravo
Rui Vasco Silva
António P.C. Duarte

MDPI

Basel • Beijing • Wuhan • Barcelona • Belgrade • Novi Sad • Cluj • Manchester

Editors
Miguel Bravo
CERIS, IST-ID, Department of Civil Engineering, Architecture and Georresources
Universidade de Lisboa
Lisbon
Portugal

Rui Vasco Silva
CERIS, IST-ID, Department of Civil Engineering, Architecture and Georresources
Universidade de Lisboa
Lisbon
Portugal

António P.C. Duarte
CERIS, IST-ID, Department of Civil Engineering, Architecture and Georresources
Universidade de Lisboa
Lisbon
Portugal

Editorial Office
MDPI AG
Grosspeteranlage 5
4052 Basel, Switzerland

This is a reprint of articles from the Special Issue published online in the open access journal *Materials* (ISSN 1996-1944) (available at: www.mdpi.com/journal/materials/special_issues/Recycle_Aggregate_Concrete_Alternative_Binders_Sustainable).

For citation purposes, cite each article independently as indicated on the article page online and as indicated below:

Lastname, A.A.; Lastname, B.B. Article Title. *Journal Name* **Year**, *Volume Number*, Page Range.

ISBN 978-3-7258-2328-4 (Hbk)
ISBN 978-3-7258-2327-7 (PDF)
doi.org/10.3390/books978-3-7258-2327-7

Contents

About the Editors

Miguel Bravo

Miguel Bravo is an invited assistant professor in civil engineering at Instituto Superior Técnico and CERIS Integrated Researcher. He completed his Ph.D. on the use of recycled aggregates derived from construction and demolition waste in the production of structural concrete. He conducts his research in the area of concrete with recycled aggregates from different origins and made with cementitious materials from several industrial by-products. He is the author of several publications concerning sustainable construction, including three Highly Cited Papers (as first author). He participated in four FCT projects on concrete with recycled aggregates and on replacing cement with more sustainable binders (magnesium oxide, alkali-activated solid municipal waste, and alkali-activated CDW). In two of these four research projects, he was the managing director of the project. He is currently participating in another three research projects on sustainable concrete, being the principal investigator in one of these projects. He is also currently participating in two international research projects on sustainable construction. Productivity: A book chapter; 40 papers (20 as first author; 34 in high-level international journals; 32 in Q1 or Q2 journals in relevant categories); H-index of 19/21 (Scopus/Google Scholar); Total citations of 2003/2809 (Scopus/Google Scholar). He has several collaborations with different universities around the world (Spain, Italy, and Brazil). He is Guest Editor of four Special Issues (three were in Q1 or Q2 journals), and he is Editorial Board Member of journals *Sustainability* (Q1 journal), *Materials* (Q1 journal), and *Frontiers in Built Environment* (Q2 journal).

Rui Vasco Silva

Rui Vasco Silva is an invited assistant professor of Civil Engineering in the Department of Civil Engineering, Architecture, and Georesources at the IST, University of Lisbon, Portugal. He completed his Ph.D. on the use of recycled aggregates derived from construction, demolition and excavation waste in the production of structural concrete. He is a member of the CERIS Research and Development Unit, hosted by IST-ID, RILEM, and an honorary member of the CIB Student Chapter at IST. He has notable experience in several state-funded research projects. He is now the Principal Investigator of the research project EXPL/ECI-EGC/0288/2021: "Carbon Capture: Performance Enhancing CO_2-Based Curing of Fully Recycled Alkali-Activated Concrete". He is the author of several publications concerning sustainable construction and is currently researching other approaches that can further decrease the environmental impact of concrete, namely reactive magnesium oxide, alkali-activated materials, municipal solid waste incinerator bottom ashes, and electric arc furnace slag. He is the author of 50+ publications in WoS/Scopus databases, 20+ in conferences, 15 book chapters, and 2 international books (H-index of 32/26 in Google Scholar/Scopus with total citations of 5688/4028, respectively).

António P.C. Duarte

António Pedro Carones Duarte, Portuguese, born in Lisbon in 1987, has a PhD in Civil Engineering (Structures) and an MSc in Civil Engineering (Structures) from Instituto Superior Técnico (IST), University of Lisbon (UL), Portugal. Currently he is an integrated researcher of CERIS-IST-UL (in 2019 he was granted a 6-year research contract funded by FCT under the Scientific Employment Stimulus) and Assistant Professor at Escola Superior de Tecnologia do Barreiro—Polytechnic Institute of Setúbal, where he teaches two courses in the area of Materials Science and Technology. He is also Co-Editor of Materials' Special Issue entitled 'Recycled Aggregate Concrete and Alternative Binders for Sustainable Building Engineering' and a reviewer in some of the top journals in civil engineering. He develops his research activities in the areas of structural and computational mechanics and materials science in the scope of both civil and aerospace engineering. In his research activity, he interacted with 30+ collaborators, being the author and co-author of 23 papers in international peer-review journals, of 18 technical-scientific reports, and of 26 papers in national and international conference proceedings, having also participated in the organization of 2 international and 1 national conference. His publications have an H-index of 10 and 524 citations according to Scopus (and an h-index of 10 and 615 citations according to Google Scholar). António participates (ed) as a co-principal investigator in 1 project, as a researcher in 5 projects, and as a Ph.D. fellow in 1 project and supervises(ed) or co-supervises(ed) 2 Ph.D. theses and 16 MSc dissertations in the areas of civil engineering and aerospace engineering. He received an honor diploma in the Innovation Valorpneu Award in 2014. António is a member of the Portuguese Steelwork Association (CMM) and an honorary member of the CIB Student Chapter.

Preface

The construction industry is considered one of the most important sectors of socio-economic development. However, its dynamism gives rise to two major environmental issues: (i) a significant increase of construction and demolition waste (CDW), which represents more than 30% of the solid waste generated by all economic sectors; and (ii) an increase in the production of Portland cement concrete, which is the most widely used construction material in the world and, through the production of cement alone, constitutes about 7% of all greenhouse gas emissions. Hence, it is essential to find, on the one hand, solutions for discarded CDW in addition to their use as recycled aggregates and, on the other hand, alternative sustainable binders that can replace Portland cement.

This reprint focuses on the development of sustainable cementitious composites, either by the replacement of natural aggregates with recycled aggregates or by the use of sustainable binding technologies that constitute a viable alternative to Portland cement.

The main topics covered in this reprint are, but are not limited to, the following:

- Experimental characterization (rheological, physical, mechanical, durability-related, thermal, etc.) of cementitious composites containing recycled aggregates (from all origins).
- Experimental characterization of new sustainable alternative binders and cementitious composites containing them.
- Development and implementation of analytical methods, numerical models, and optimization algorithms applied to the characterization of sustainable materials and structures produced with them.
- Development of inspection and diagnostic methodologies and techniques for evaluating the performance of sustainable materials and structures containing them.
- Life cycle assessment (LCA) and life cycle cost analysis (LCC) of sustainable materials and structures with them and structures made with them.
- Studies on subjects other than the aforementioned that contribute to advancing knowledge of sustainable materials and their applications were also welcome in this reprint.

Miguel Bravo, Rui Vasco Silva, and António P.C. Duarte
Editors

materials

MDPI

rticle

nfluence of Bentonite on Mechanical and Durability Properties f High-Calcium Fly Ash Geopolymer Concrete with Natural nd Recycled Aggregates

ana Muhammad Waqas [1,*], Faheem Butt [1,*], Aamar Danish [2], Muwaffaq Alqurashi [3], lohammad Ali Mosaberpanah [4], Bilal Masood [5] and Enas E. Hussein [6,*]

1 Civil Engineering Department, University of Engineering & Technology, Taxila 47080, Pakistan
2 Civil Engineering Department, Faculty of Engineering Sciences, National University of Modern Languages, Rawalpindi 04631, Pakistan; aamar.danish@numl.edu.pk
3 Department of Civil Engineering, College of Engineering, Taif University, Taif 21944, Saudi Arabia; m.gourashi@tu.edu.sa
4 Civil Engineering Department, Cyprus International University, Nicosia 99258, Turkey; mmosaberpanah@ciu.edu.tr
5 Civil Engineering Department, Wah Engineering College, University of Wah, Wah Cantt 47040, Pakistan; bilal.masood@wecuw.edu.pk
6 National Water Research Center, P.O. Box 74, Shubra El-Kheima 13411, Egypt
* Correspondence: rana.waqas@uettaxila.edu.pk (R.M.W.); faheem.butt@uettaxila.edu.pk (F.B.); enas_el-sayed@nwrc.gov.eg (E.E.H.)

tation: Waqas, R.M.; Butt, F.; anish, A.; Alqurashi, M.; osaberpanah, M.A.; Masood, B.; ussein, E.E. Influence of Bentonite Mechanical and Durability operties of High-Calcium Fly Ash eopolymer Concrete with Natural d Recycled Aggregates. *Materials* 21, *14*, 7790. https://doi.org/.3390/ma14247790

cademic Editors: Miguel Bravo, Rui sco Silva, António P.C. Duarte and rio De Domenico

eceived: 12 October 2021
ccepted: 13 December 2021
ublished: 16 December 2021

Abstract: In this study, bentonite (a naturally occurring pozzolana) was incorporated as a partial replacement (up to 20%) for high-calcium fly ash (HCFA)-based geopolymeric natural aggregate concrete (GNAC) and geopolymeric recycled aggregate concrete (GRAC). The mechanical (compressive strength and splitting tensile strength), durability (chloride migration coefficient, water absorption, and acid attack resistance), and rheological properties (slump test, fresh density, and workability) were investigated. The results revealed that incorporation of bentonite (10 wt % with ordinary Portland cement) showed appreciable improvement in the strength and durability of both the GNAC and GRAC, though its effect is more significant for GRAC than the GNAC.

Keywords: geopolymer concrete; bentonite; recycled aggregates; GRAC; GNAC; mechanical properties; durability; rheological properties

1. Introduction

Concrete is an abundantly used material in the construction industry due to its distinctive properties such as cost-effectiveness, high strength, durability, easy availability of its constituent materials, and ease of formability into desired shapes. Ordinary Portland cement (OPC) is the main constituent of concrete necessary to bind aggregates together [1]. However, the production of OPC requires an appreciable amount of energy and generates approximately 10% of total carbon dioxide (CO_2) emissions worldwide, which is one of the main reasons for global warming [2]. According to an estimate, about 2.8 billion tons of cement are produced each year and will reach 4 billion tons by 2050 [3]. Over the past two decades, constructional activities have grown rapidly around the globe due to increased infrastructure demand arising from the increase in global population and urbanization. Accordingly, the production of cement is also increasing to meet the increased demand for infrastructure. The production of cement causes serious damage to the environment due to greenhouse gas (GHG) emissions. Therefore, serious environmental concerns have been raised regarding the production of OPC [4]. To address such concerns, researchers are suggesting partial or full replacement of OPC in concrete construction with supplementary cementitious materials (SCMs) [2]. Due to urbanization and infrastructure developments, there is a significant increase in the generation of construction and demolition was (C&DW) materials resulting from the demolition of old infrastructures [5]. Moreover, t extraction of natural coarse aggregates (NCA) involves a large amount of energy ar depletion of natural resources, which is affecting the eco-system. Therefore, to dispose and manage these wastes (C&DW) and develop alternatives to traditional OPC has becon an essential requirement for sustainability. Consequently, researchers are trying to find ne ways of recycling C&DW to save natural aggregates reserves and reduce environment pollution [6–8]. The fine and coarse aggregates constitute 75–80% of concrete's tot volume and C&DW can be used as recycled coarse aggregates (RCA) in new concrete minimize the consumption of natural resources that are used to produce NCA [9]. Howev the studies have found that concrete incorporating RCA produces inferior mechanic and durability properties than concrete containing NCA [10–17]. It was reported Etxeberria et al. [8], Sagoe Crenstil et al. [9], and Tabsh and Abdelfatah [11] that OP concrete containing 100% RCA has 10–40% lower compressive strength than the concre containing NCA. Moreover, Marinkovic et al. [12] reported that the incorporation of RC in OPC concrete leads to an increase in the absorption capacity and drying shrinkage th the concrete with NCA. This is due to the porous nature, low density, and high-wat absorption of old cement mortar attached to the surface of RCA [13–15]. Similarly, the u of electric arc furnace (EAF) slag and ground granulated blast furnace slag (GGBFS) partial or total substitution of natural aggregates (NAs) in structural concrete elements ha also been studied by different researchers. This material is a byproduct of steel productic in electric arc furnace plants [18,19]. Domenico et al. studied the use of EAF as parti replacement of coarse aggregates in reinforced concrete (RC) beams [18]. They found th the presence of EAF led to a higher flexural and shear capacity than the correspondir traditional RC beams, crack widths are reduced, and the overall ductility is increased. Tl use of waste materials such as C&DW, EAF, and GGBFS as aggregates in concrete will lea to a sustainable and economical practice in the construction industry.

Geopolymer is an emerging and innovative binder that can provide an environme tally friendly replacement of OPC. The materials with geological origin, i.e., metakaoli and by-products such as fly ash (FA), ground granulated blast furnace slag (GGBS), and ri husk ash (RHA), chemically activated by alkaline solution (sodium silicate and sodium h droxide) are generally used to synthesize geopolymer. It has been estimated that geopol mer binder generates 80% less CO_2 emissions and requires 60% less energy than th manufacturing OPC [20]. Extensive research has been carried out on the development geopolymer as a sustainable binder [21–24]. The incorporation of RCA in the geopolym concrete (GPC) is a step to further improve environmental sustainability. Some recent stu ies have revealed that the negative effects of RCA on mechanical and durability properti of GPC are similar to that of the conventional concrete with RCA [25,26].

To improve properties of the geopolymeric recycled aggregate concrete (GRAC) va ious methods are suggested, e.g., increasing the fineness of fly ash in geopolymers improve the degree of geopolymerization [27,28], and incorporation of calcium-rich mat rials such as OPC and GGBS to enhance strength (early age) and setting time [29,30]. Son studies have suggested incorporating SiO_2-rich materials (rice husk ash, nano SiO_2) th improve the strength and durability of the resulting GRAC [31,32]. Nuaklong et al. [3 reported the use of metakaolin, a mineral admixture, to improve the mechanical an durability properties of high-calcium fly ash-based geopolymer concrete containing RC

The low-calcium bentonite is a naturally found pozzolana, rich in silica and alumin that has been extensively used for many years to improve the mechanical and durabilit properties of conventional concrete [34,35]. Bentonite is an alumino-phyllosilicate clay, cor prising mainly montmorillonite minerals and usually formed by the chemical weatherin of volcanic ash in the presence of water. The use of bentonite clay as a partial replacemer of cement in conventional concrete containing both natural and recycled coarse aggr gates has been widely reported [36–43]. The industrial bentonites are generally either th sodium or calcium variants. Bentonites have excellent rheological and absorbent prope ties. Among various types of bentonites, calcium bentonite has a much lower swelling capacity [36]. Bentonite is found in many regions of Pakistan, including Attock, Jhelum, Nowshera, and Karak. Memon et al. [42] and Mirza et al. [43] have reported the use of low-calcium bentonite as a partial replacement of OPC. In previous studies, bentonite was used for mass replacement of OPC up to 21%. Bentonite mixes outperformed conventional/OPC concrete at 56- and 90-day strength, and resistance improvement against acid attack was also observed. Masood et al. [41] reported the use of low-calcium bentonite as a partial replacement of cement in OPC concrete containing RCA [43] and found that the incorporation of bentonite in concrete with RCA resulted in improved mechanical and durability properties compared to the concrete with NCA. The improved mechanical and durability properties of concrete are attributed to the high pozzolanic reactivity and filling effects of bentonite. In terms of geopolymer concrete (GPC), Ekaputri and Antoni [44] evaluated the effect of bentonite on the GPC containing geothermal silica. They cured the specimens for a specific amount of time at a temperature between 80–120 °C and then kept them at room temperature for 7 days. The results revealed that the addition of bentonite and higher initial curing temperature improved the mechanical properties of GPC. In another study by Yang et al. [45], porosity of geopolymer prepared with pre-swelled bentonite slurry was analyzed. They concluded that incorporation of bentonite reduces pore size in the matrix, leading to improved compressive strength and thermal insulation performance. Similarly, Huang [46] reported the influence of calcium bentonite on the compressive strength, aesthetics, and drying shrinkage of fly ash-based geopolymer mortar. They observed that incorporation of calcium bentonite improves the compressive strength of specimen owing to the increased ratio of silica/alumina (Si/Al). They also concluded that the addition of calcium bentonite reduced drying shrinkage of the specimen, attributed to the peculiar properties of calcium bentonite, i.e., volume expansion and water adsorption. Moreover, the study reported the elimination of efflorescence in calcium bentonite-modified geopolymer mortar due to the adsorption of sodium ion by calcium bentonite via ion exchange.

GPC has emerged as an innovative construction material with environmental benefits and better engineering properties. The addition of RCA in geopolymer concrete can be associated with a further increase in environmental sustainability. The geopolymeric recycled aggregate concrete (GRAC) has benefits of reducing the CO_2 emissions owing to the production of cement and preservation of natural aggregate resources. Furthermore, GRAC can resolve issues associated with the disposal of large volumes of construction and demolishing wastes (C&DW). Despite several environmental benefits of GRAC, it has inferior engineering properties due to weak internal structure as compared to GNAC. The specific objective of the present study is to improve the mechanical and durability properties of GRAC mixes by incorporating SCM. The review of the existing literature indicates that low-calcium bentonite has promising effects on the mechanical and durability properties of conventional concrete [37,42,43,47–50]. However, the studies on the inclusion of bentonite to improve the performance of geopolymer concrete with NCA and RCA are rather scarce. The present study, therefore, attempts to present the effect of bentonite on various properties of GNAC and GRAC mixes, namely, workability, air content, fresh density, compressive strength, split tensile strength, water absorption, chloride migration coefficient, and sulfuric acid attack resistance. The high-calcium fly ash (HCFA) was used as the primary source material to produce the geopolymer concrete; therefore, it can set and harden at ambient temperature. Bentonite was used as a partial replacement (0%, 10%, 15%, and 20%) for fly ash in both the GNAC and GRAC. A comparison between the behavior of both GNAC and GRAC mixes with different levels of bentonite is also presented to highlight their significant features.

Research Significance

Despite several environmental benefits of GRAC, it has inferior engineering properties due to weak internal structure as compared to geopolymer natural aggregate concrete

(GNAC). Low-calcium bentonite has been extensively used in the literature to improve t properties of conventional cementitious composites. However, the studies highlighti the effect of bentonite on the mechanical and durability properties of geopolymer concre with RCA are rather scarce. In this study, an effort has been made to present the effe of bentonite on various properties of both GNAC and GRAC mixes, namely, workabili air content, fresh density, compressive strength, split tensile strength, water absorptio chloride migration coefficient, and sulfuric acid attack resistance.

2. Experimental Program

In the experimental program, a total of 10 mixes were designed to study the influen of bentonite clay content on the rheological, mechanical, and durability properties geopolymer concrete containing natural and recycled coarse aggregates. The HCFA w used as the primary source material to produce the geopolymer binder. Low-calciu bentonite clay was obtained from its natural deposits in Jahangira (Nowshera), Pakista for partial replacement of HCFA in geopolymer concrete. The mixes were divided into tw groups, GNAC and GRAC, based on the aggregates used, i.e., NCA and RCA, respective For studying the influence of bentonite clay, HCFA was replaced by bentonite clay b 0%, 5%, 10%, 15%, and 20% (by weight) in both the GNAC and GRAC mixes. A seri of experiments were conducted to study the effect of bentonite content on rheologic mechanical, and durability properties of both GNAC and GRAC mixes. The fresh properti of the concrete mixes such as workability, fresh density, and air content were measure soon after completion of the mixing process. The mechanical properties of the mixes we evaluated by compressive and split tensile strength tests. The durability of the GNA and GRAC mixes was evaluated by water absorption, chloride migration coefficient, an sulfuric acid resistance tests of specimens.

2.1. Materials and Methods

2.1.1. Fly Ash and Bentonite

In this study, HCFA was used as the primary source material to produce geopolym concrete. The chemical composition of HCFA is given in Table 1. For partial replaceme of HCFA, low-calcium bentonite clay was obtained from its natural deposits in Jahangi (Nowshera), Pakistan. The feasibility of this bentonite as a partial replacement of OPC (u to 25% by weight) in conventional concrete has been verified by the previous studies [41,4

Table 1. Chemical composition for Ordinary Portland cement, bentonite, and fly ash.

Oxides	OPC (% wt)	Bentonite (% wt)	Fly Ash (% wt)
SiO_2	54.55	52.8	35.8
Al_2O_3	31.93	16.4	20.2
Fe_2O_3	3.12	5.8	11.4
MgO	1.42	1.4	1.80
CaO	4.64	4.6	14.3
Na_2O	0.3	0.62	1.20
K_2O	1.3	0.7	2.2
Relative density (g/cm^3)	2.10	2.64	1.8
Specific surface area (cm^2/gm)	3207	4900	3250
Loss on ignition (%)	1.30	9.6	0.57

The raw bentonite obtained from the natural deposits was processed by wet grindin to make it pass through a 325 μm sieve. The XRF analysis confirmed that the bentonite use is low-calcium pozzolana as per ASTM C618-17a [51]. Moreover, scanning electron micr scopic (SEM) (VEGA3 TE model, TESCAN, Brno, Czech Republic) imaging of bentonit revealed spherical and thick flake-shaped particles, as shown in Figure 1. The chemica composition and physical properties of bentonite are shown in Table 1. It can be seen fro the table that the specific surface area of bentonite used in this study is higher than the OPC. This is due to the wet grinding process performed on the bentonite, as mentioned above. The comparison of the surface area of bentonite and OPC (Table 1) depicts more fineness of the bentonite than the OPC. The particle size of bentonite varies between 0.2 and 200 μm while the particle size range for cement is 0.2 to 100 μm. Fly ash consists of silt-sized particles which are generally spherical, typically ranging in size between 10 and 100 microns.

Figure 1. SEM image of low-calcium bentonite.

2.1.2. Alkaline Activation Solution

The solutions of sodium silicate (SS) and sodium hydroxide (SH) were mixed in a ratio of 1:2 (SS/SH = 2.0) to make the alkaline activation solution. This solution was kept as 40% of the total binder quantity. The solution of SH of 12 M molarity was prepared by mixing 98–99% pure pellets of SH (with a specific gravity of 2.10) in water. It was prepared 24 h before the application due to the heat development following the exothermic reaction. The SS solution with a ratio of $SiO_2/Na_2O = 2.0$ was obtained from a local company (Sika, Rawalpindi, Pakistan). It was composed of Na_2O, SiO_2, and H_2O in 15.06%, 29.95%, and 54.99% by mass, respectively. The properties of the SS solution are shown in Table 2.

Table 2. Properties of SS solution.

Entity	Specification
Color	Colorless
Density, kg/m^3	1458
Total solid content	48 (% mass)

2.1.3. Aggregates

The natural river sand from Lawrencepur (Attock) Pakistan was used as fine aggregates in surface dry condition (SSD) conforming to ASTM C128-15 [52] for both GNAC and GRAC. The gradation of the fine aggregates falls under the acceptable limits, as shown in Figure 2a. The NCA with a maximum size of 19 mm were obtained from the Margallah quarry source in Pakistan to produce GNAC. The RCA were prepared by crushing laboratory tested concrete specimens with a compressive strength of 30–35 MPa. The gradation curve of both NCA and RCA is kept the same, as shown in Figure 2b. The properties of fine and coarse aggregates are shown in Table 3.

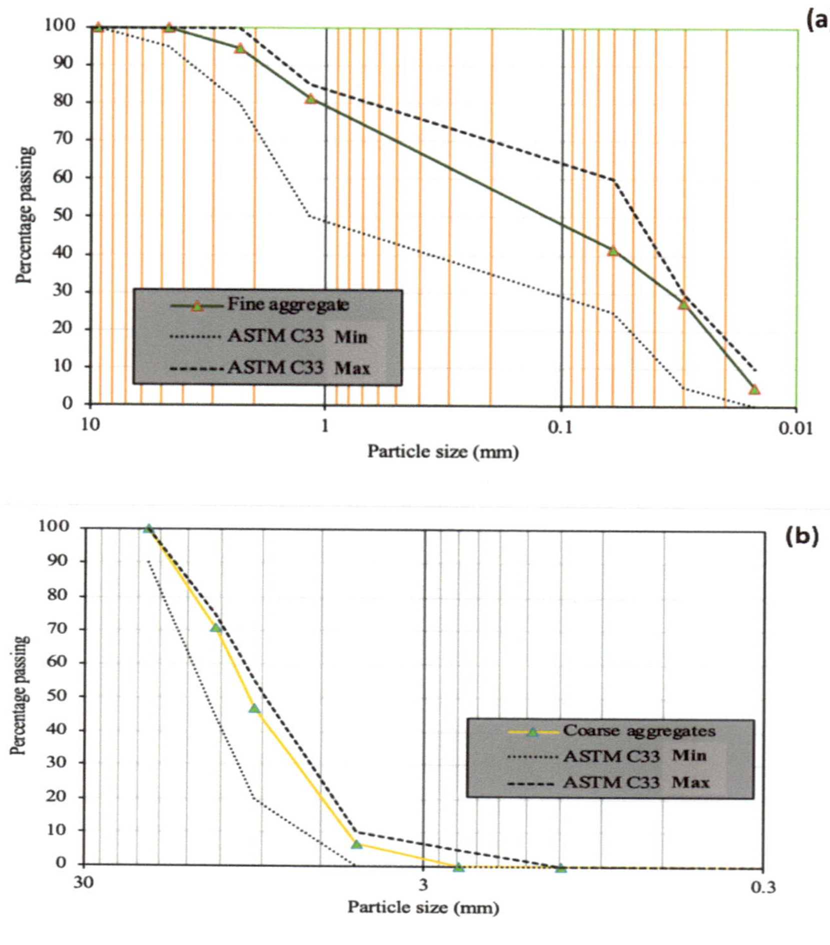

Figure 2. The gradation curves of (**a**) fine aggregates (**b**) RCA and NCA.

Table 3. Properties of fine and coarse aggregates.

Characteristic	Fine Aggregates	Coarse Aggregates	
		RCA	NCA
Maximum aggregate size (mm)	4.75	19.50	19.50
24-h water absorption (%)	0.59	6.25	0.63
10% Fine value (kN)	-	129.5	172.5
Dry rodded density (kg/m^3)	1615	1377	1575

2.2. Mix Proportions

In this study, the mixes were divided into two groups, GNAC and GRAC, based o the aggregates used, NCA and RCA, respectively, as shown in Table 4. The HCFA wa used as the primary source material to produce the geopolymer binder. For studying th influence of bentonite clay, HCFA was replaced by bentonite clay by 0%, 5%, 10%, 15% and 20% (by weight) in both the GNAC and GRAC. The GNAC and GRAC specimen with 0% replacement level of bentonite are considered as primary (PC) and secondary (SC) control mixes, respectively. The quantity of the binder (425 kg/m^3) was kept the same for all the mixes, and calculations for the replacement of HCFA by the bentonite were performed as per their mass ratio in the mix. The proportions of each mix are shown in Table 5. All the mixes are proportioned for a target slump of 75 mm to maintain the workability. An additional 20 kg/m^3 of water was used in the GRAC mixes to overcome the water absorption of RCA.

Table 4. Nomenclature for GNAC and GRAC mixes.

Type of Concrete	Specimen ID	Replacement of HCFA with Bentonite
Geopolymer natural aggregate concrete (GNAC)	GNAC/B0(PC)	0
	GNAC/B5	5%
	GNAC/B10	10%
	GNAC/B15	15%
	GNAC/B20	20%
Geopolymer recycled aggregate concrete (GRAC)	GRAC/B0(SC)	0
	GRAC/B5	5%
	GRAC/B10	10%
	GRAC/B15	15%
	GRAC/B20	20%

Table 5. Composition of GNAC and GRAC mixes.

Mix ID	Sand (kg/m^3)	NCA (kg/m^3)	RCA (kg/m^3)	Fly Ash (kg/m^3)	Bentonite (kg/m^3)	NaOH (kg/m^3)	Na_2SiO_3 (kg/m^3)	Superplasticizer (SP) (kg/m^3)
GNAC/B0 (PC)	640	1200		425	-	53	107	4
GNAC/B5	640	1200		404	21	53	107	4
GNAC/B10	640	1200		383	43	53	107	6
GNAC/B15	640	1200		361	64	53	107	8
GNAC/B20	640	1200		340	85	53	107	10
GRAC/B0 (SC)	640	-	1080	425	-	53	107	8
GRAC/B5	640	-	1080	404	21	53	107	10
GRAC/B10	640	-	1080	383	43	53	107	12
GRAC/B15	640	-	1080	361	64	53	107	14
GRAC/B20	640	-	1080	340	85	53	107	16

To maintain the workability of the mixes due to the negative effect of the increasing level of bentonite on the slump, superplasticizer was used. The reduction in the slump value is due to a relative increase in the angular and flaky shapes (having more surface area) of the bentonite particles as compared to the spherical shaped particles (having more flow-ability) of FA. Due to the larger surface area of the bentonite particles, more water is required in wetting the bentonite particles.

2.3. Mixing and Curing Process

All the mixes were prepared in a 400 L drum mixer following the conventional mixing procedure. The SH and SS were mixed in predetermined ratios to make an alkaline activator solution. This activity was performed about one hour before the mixing operation due to heat evolution resulting from the exothermic reaction. The mixing procedure and the time of mixing were kept the same for all the mixes. The sand was added at the start, followed by the binder materials (fly ash and bentonite) and coarse aggregates. After 3 min of dry mixing, the alkaline solution was added to the premixed dry ingredients. Then, water and superplasticizer were added, if required, and the mixing operation was continued for another 3 min to ensure homogeneity. The freshly mixed geopolymer concrete was poured into the molds of specimens. The compaction of specimens was carried out by th needle-type vibrator during casting. The specimens were kept at ambient temperatu for one hour to enable setting and hardening prior to the heat curing [45]. Before curi the samples at 60 °C for two days, they were covered in plastic sheets to prevent loss moisture due to evaporation during the process of curing at high temperature [30,31,4 The specimens were then demolded, wrapped with plastic film, and placed in the curir room under the ambient environment with a temperature of 25 ± 2°C and relative humidi of 70 ± 5% until the testing age.

2.4. Testing Procedures

The fresh properties of the concrete mixes such as workability, fresh density, and a content were measured soon after completion of the mixing process. The workability of th fresh mixes was measured by the slump cone test as per the ASTM C143/C143M-12 [5 The fresh density of the GNAC and GRAC mixes was measured as per ASTM C138/C138M 17a wherein the mass of the concrete used to completely fill the container was divide by the volume of the cubical container. For this test, a cubical container of 150 mm wa filled in three layers with fresh concrete. Each layer was compacted by 25 strokes of a ste rod to remove air voids [54]. The air content of freshly mixed geopolymer concrete mix was evaluated by using the pressure method following the ASTM C231/C231M-10 [5 using the Forney apparatus shown in Figure 3. As per the mentioned standard, the bas of the air content test apparatus was filled in three equal layers with a proper compactic via 25 strokes of a steel rod for each layer. After rodding, the apparatus was struck fro outside to remove any remaining air voids. We fastened the top portion of the apparatu having air meter to its base containing concrete. The device was then pressurized via hand pump until the air meter was stabilized. The stabilized reading was noted down an subtracted from aggregate correction factor to calculate the air content in the prepared mi

Figure 3. Air content apparatus.

The mechanical properties of the mixes were evaluated by compressive and spli tensile strength tests. The compressive strength test was performed on 150 × 150 mm cubes following BS EN 12390-3:2009 [56] at the age of 3, 28, and 90 days. The cubes teste for compression were placed on the face perpendicular to the cast face and the constantly increasing force at the rate of 0.6 MPa/s was applied until the cubes failed. The reading at which the cube failed was noted at the compressive strength of respective cube. The split tensile strength tests were performed on 150 × 300 mm^2 cylinders according to ASTM C496/C496M-11 [57] at the age of 28 and 90 days. The cast cylinder was placed in in a jig for alignment and adjust bearing strips, and finally placed in compressive testing machine to apply load until the specimen failed. The loading rate was kept between 0.7 and 1.4 MPa/s and the reading gave us the load at which the specimen failed. This reading was incorporated into Equation (1) to derive the splitting tensile strength of the respective specimen.

$$T = \frac{2P}{\pi LD} \quad (1)$$

where T is splitting tensile strength, P is load at which specimen failed, L is longitudinal length of specimen, and D is diameter of specimen

The durability of the GNAC and GRAC mixes was evaluated by measuring water absorption, chloride migration coefficient, and sulfuric acid resistance of specimens. The water absorption and chloride migration coefficient of the mixes were determined by 100 mm diameter discs of 50 mm thickness at the age of 90 days according to ASTM C948-81 [58] and NT Build 492 [59] (using SAMYON RCM apparatus with the test setup as shown in Figure 4), respectively. The water absorption was calculated by recording the weight of freshly cast specimen and weight of specimen after 90 days. Finally, the percentage increase in the weight of specimen was found and reported as percentage water absorbed by the respective specimen. The chloride migration test was performed on 90 days cured cylindrical specimens. The specimens are the pre-conditioned by placing them in vacuum chamber and then sealed to allow the penetration of chloride at the specimen ends. The sealed specimens were placed in cells with reservoirs having sodium hydroxide and sodium chloride solution at each end. To initiate chloride migration, a specific voltage was applied across each specimen. Finally, the specimens were taken out of cells and split using compression machine and sprayed split face with silver nitrate, which reacts with chloride to form a distinctive boundary identifying chloride penetration through the respective specimen. The depth of this distinctive boundary gave chloride migration. The cubical specimens of 100 mm were used to measure the sulfuric acid attack resistance. After demolding, the cubical specimens were placed in potable water for days [37]. These specimens were then submerged in 4% H2SO4 solution [36]. The specimens were taken out of the solution after 28, 56, and 90 days of immersion, and loss in mass of the sample was determined. It should be noted that the solution was changed after 14 days to keep the concentration of the acid stable. The number of specimens prepared to evaluate the different rheological, mechanical, and durability properties of GNAC and GRAC mixes are listed in Table 6.

Table 6. Number of specimens conducted for each test.

Test	**Testing Day**				**Total Specimens**
	Day 3	**Day 28**	**Day 56**	**Day 90**	
	Number of Specimens				
Compressive strength	30	30	-	30	90
Splitting tensile strength	-	30	-	30	60
Water absorption	-	-	-	30	30
Chloride migration coefficient	-	-	-	30	30
Percentage mass loss	-	30	30	30	90

Figure 4. Experimental setup for chloride migration.

3. Results and Discussions

3.1. Fresh Properties

3.1.1. Workability

The slump test was used to determine the workability of mixes. The test results of bot GNAC and GRAC mixes are shown in Figure 5. It can be observed that the incorporatio of bentonite has a negative effect on the workability of both GNAC and GRAC mixe This reduction in a slump is due to the high specific surface area of the bentonite cla ash as compared to the fly ash particles [38]. The larger specific surface area of bentoni clay particles increases the water/solution demand in the mix to wet the particle surfac completely, reducing the workability of the mix. Moreover, the particle size of bentonite angular (see Figure 1) rather than circular in other supplementary cementitious materia (SCMs) (e.g., cenosphere) to provide a ball-bearing effect to increase workability [60].

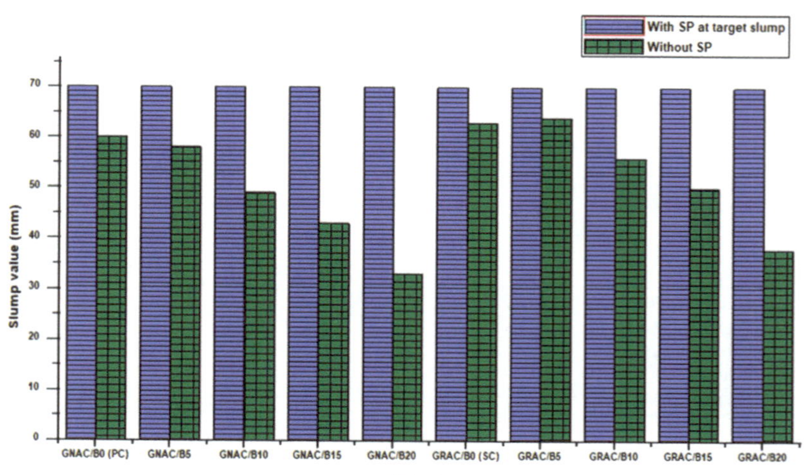

Figure 5. The slump values of GNAC and GRAC mixes.

Although the workability of the mix decreases by increasing the bentonite content, the mix remains cohesive. Superplasticizer (SP) was used to improve the lower slump values and to achieve the target slump. The target slump was selected as 70 mm for all the mixes. As depicted in Table 5, the SP dosage increases with a higher percentage replacement value of bentonite. The decreasing trend of slump values against increasing bentonite content is consistent with the previous studies [38,61]. It is important to notice that the reduction in workability is more pronounced in GNAC as compared to GRAC due to the presence of more free water in GRAC. A similar trend was also observed by Bilal et al. [41] and Kou et al. [6] in conventional concrete.

3.1.2. Air Content

Air content is a property of fresh concrete that can play an important role in increasing or decreasing the concrete strength. The presence of air content in the concrete also determines its resistance to freeze and thaw cycles and other weathering agents. Generally, the strength of concrete decreases with an increase in the amount of air content.

The result of air content tests of the mixes is shown in Figure 6. It can be observed from Figure 6 that the GNAC mixes have lower air content than the corresponding GRAC mixes. The SC mix of GRAC has 5.5% more air content than the PC mix of GNAC. The amount of water used in the GRAC mixes to meet the water absorption demand of RCA is higher than their counterparts in the GNAC mixes, resulting in slightly higher workability than the GNAC mixes. However, the air contents of GRAC mixes are greater than the corresponding mixes of GNAC. This phenomenon can be attributed to the presence of unsaturated fractured zones in RCA, which are not connected to the permeable pores of aggregates [36].

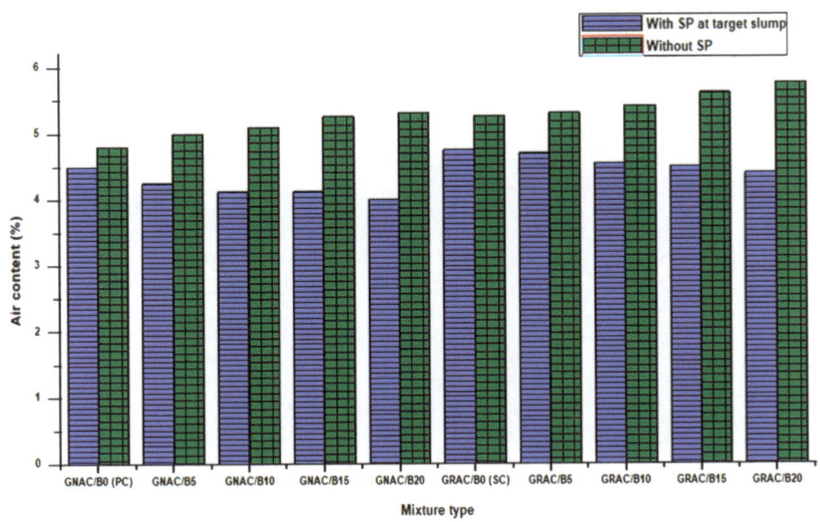

Figure 6. The amount of percent air content in the GNAC and GRAC mixes with and without SP.

It can be noticed from Figure 6 that the air content of the GNAC and GRAC mixes increased with an increase in the percentage replacement of bentonite when SP is not used to attain the target slump. This behavior can be due to the reduction in workability in the presence of bentonite. The less workable concrete easily allows the air to penetrate the fresh concrete, creating air voids.

On the other hand, when SP is used to achieve the target slump of mixes, reduction in the air content was observed with an increase in the bentonite replacement level for both the GRAC and GNAC mixes. The reduction in the air content can be ascribed to the fineness of bentonite particles that fill the spaces of air voids, resulting in a decrease in tl volume of air voids in the mix [41]. A slight increase in the air content was also noted f the mixes containing RCA than the mixes containing natural aggregates, due to the poro structure of the RCA that can easily entrap air [41]. Due to the porous structure of RC the maximum increase in air content was found to be 6.4% in the GRAC mix with 20 bentonite content when compared to the PC without the SP addition. The GRAC mix wi 20% bentonite content exhibits 4% higher air content than the PC mix at the target slum as can be observed from Figure 6. It can be inferred that the incorporation of bentonite the GRAC mixes somehow overcome the deficits due to the RCA.

3.1.3. Fresh Density

The fresh density of the GRAC and GNAC mixes (with and without the addition SP) is shown in Figure 7. The fresh density of mixes (without SP) reduces with an increase percentage of bentonite for both the GRAC and GNAC mixes.

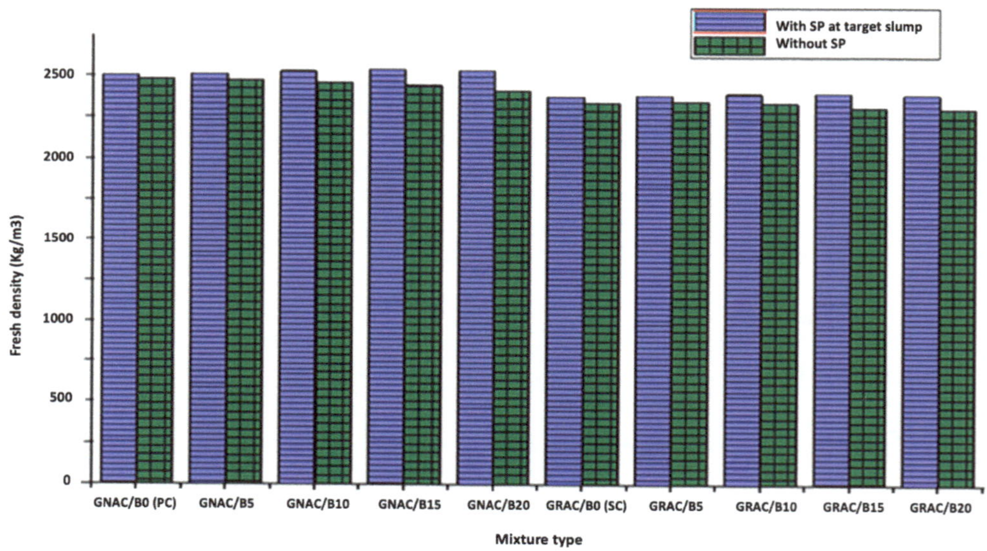

Figure 7. The fresh density of the GNAC and GRAC mixes with and without SP.

At the highest replacement of bentonite (20%), a reduction in fresh density by 3 and 2% was observed in the GNAC and GRAC mixes, respectively, compared to the corresponding control mixes. The workability is an important factor that determines th density of concrete with a given compaction effort. Consequently, less workable concret will result in a lower density of concrete. Therefore, it can be stated that the decrease i the fresh density of concrete mix is due to the lower workability of the mix resulting fro an increase in the bentonite content in the mix. On the other hand, when SP is used t attain the target slump of mixes, the fresh density of both the GRAC and GNAC mixes i improved with an increase in the bentonite content when compared to their correspondin control mixes. This increase in the fresh density can be ascribed to the fineness of lov calcium bentonite particles that fill the pores of RCA, resulting in a reduction in air conten Moreover, the particles of bentonite are finer than the HCFA and can occupy the por spaces of RCA and enhance the distribution of particle sizes in the geopolymer matrix c the mix [41,42].

3.2. Mechanical Properties

3.2.1. Compressive Strength

The compressive strength is considered to be a vital property of concrete for ascertaining the load-carrying ability of structural elements. The compressive strength of the GRAC mix generally depends on replacement level of RCA, the addition of mineral/chemical admixture, w/c ratio, concrete age, properties of parent materials, etc. The GRAC mix generally gives less compressive strength than the GNAC due to the higher porosity of RCA. However, the compressive strength of the GRAC can be improved by incorporating supplementary materials such as metakaolin, bentonite, silica fume, etc.

In the present study, the compressive strength of the GNAC and GRAC mixes with varying replacement levels of bentonite is evaluated at the age of 3, 28, and 90 days. The compressive strength tests were conducted on cubes according to BS EN 12390-3:2009 [56]. The strength of each mix was measured by conducting tests on three identical specimens of 150 × 150 × 150 mm size and the average strength of the three specimens is reported in Figure 8. It can be observed that an increase in the compressive strength was observed for 10% bentonite replacement than the corresponding control mixes. For example, 90-day compressive strength of GRAC/B10 and GNAC/B10 increased by 5.71% and 2.32%, respectively, as compared to the respective specimens containing 5% bentonite. When bentonite replacement is further increased from 10%, a decrease in the compressive strength was observed for both the GNAC and GRAC. For instance, 90-day compressive strength of GRAC/B15 and GNAC/B15 decreased by 8.11% and 4.52%, respectively, as compared to the specimens containing 10% bentonite. The probable explanation for this decreased compressive strength at the age of 3 and 28 days is slower pozzolanic reaction in the early ages, attributed to the formation of calcium silicate hydrate (C-S-H) and calcium aluminate silicate hydrate (CASH) gel at a slow speed. However, the rate of development of strength for bentonite mixes was found to be higher for older ages, i.e., 90 days for both the GNAC and GRAC mixes when compared to their corresponding control mixes. The control mixes SC and PC had lower strength than their corresponding group mixes. Similar findings were observed by Lima-Guerra et. al. [40] as well for conventional concrete. Moreover, it is interesting to note that the addition of bentonite has a more pronounced effect (increase or decrease) on the compressive strength of GRAC as compared to GNAC.

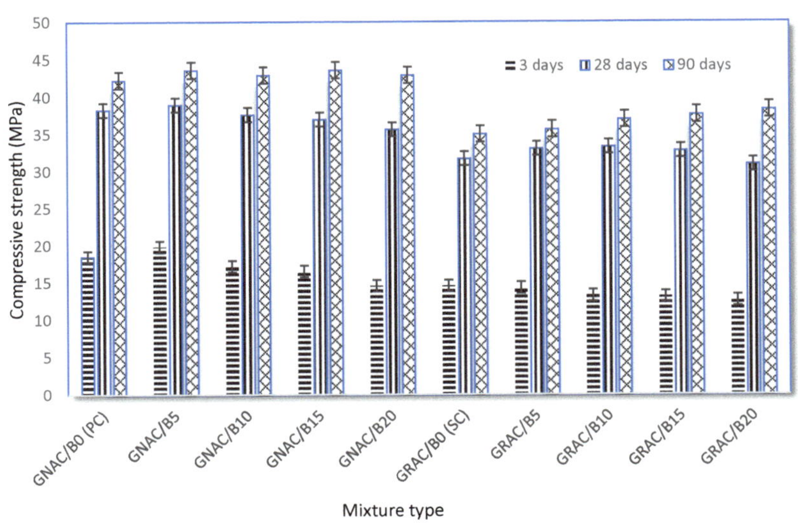

Figure 8. Compressive strength of the GNAC and GRAC mixes at 7, 28, and 90 days.

The RCA usually have old mortar attached to their surface, which makes them weak and increases their water absorption capability. Due to the higher porosity of RCA, th GRAC requires additional water to compensate for RCA absorption. Although extra wat helps the GRAC to maintain workability, it turns out to lower its hardened properti Therefore, higher porosity of RCA can be one of the main reasons for the reduction in th compressive strength of the GRAC mixes [8,62]. Furthermore, due to high porosity an water absorption of RCA, the GRAC might have weak interfacial transition zone (ITZ) tha the GNAC mixes [63].

The strength activity index (SAI) of each mix has been calculated according to AST C618-17a [51], as shown in Figure 9. SAI for each bentonite added mix, i.e., the GRAC an GNAC, should be higher than 75% (shown by red line as per ASTM C618-17a) of the relevant control mixes, i.e., SC and PC, respectively, at 28-day strength. It can be observe from Figure 9 that all the mixes of GNAC and GRAC have SAI well above the limit set b ASTM C618-17a.

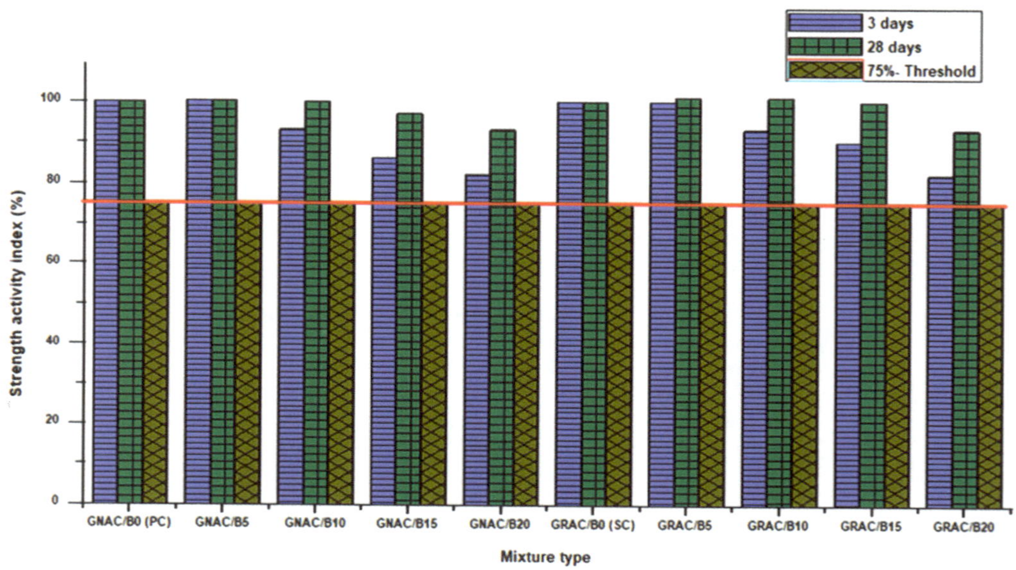

Figure 9. SAI of the GNAC and GRAC specimens.

3.2.2. Splitting Tensile Strength

Tensile strength is a mechanical property that is utilized in design aspects of concret structures such as those associated with the initiation and propagation of cracks, shea and anchorage of steel reinforcement in concrete [64]. The splitting tensile strength give a specimen's failure pattern under splitting action of loads rather than the true tensil strength of concrete.

The splitting tensile strength tests were performed on 150 × 300 mm cylinders, after 2 and 90 days of casting according to ASTM C496/C496M-11 [57]. The tensile strength value of both the GNAC and GRAC mixes are shown in Figure 10. It can be seen that the effect c varying levels of bentonite on splitting tensile strength of the GRAC and GNAC mixes i similar to that of the compressive strength results. The difference in tensile strength value of both the mixes is evident, though not huge. The GRAC mixes have lower strength tha their GNAC counterparts. As compared to the compressive strength, the tensile strengt is relatively less affected by the replacement of natural coarse aggregates with recycle coarse aggregates.

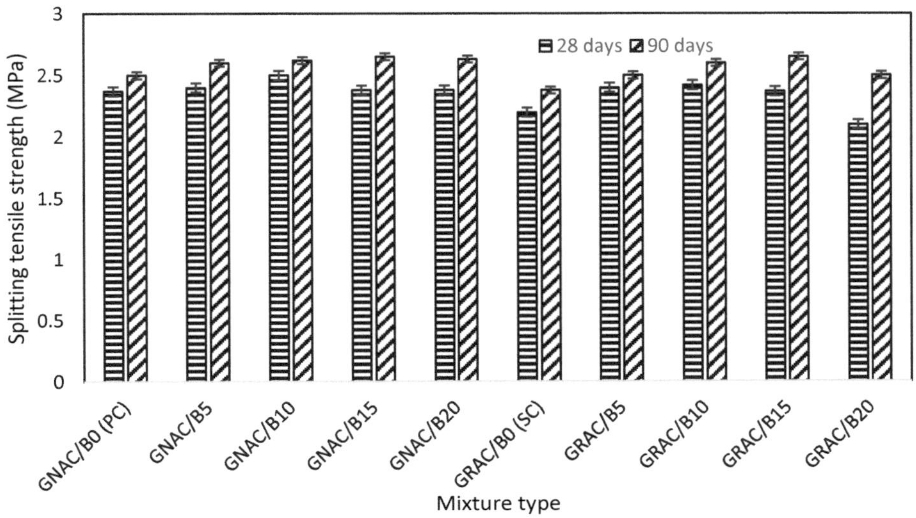

Figure 10. Splitting tensile strength of the GNAC and GRAC specimens at 28 and 90 days.

The GRAC mixes contain porous and irregular-shaped aggregates that can be responsible for developing better bonding of RCA with the geopolymer matrix. This strong bonding is effective in resisting tensile stresses. The tensile strength of the GRAC and GNAC mixes was observed to be improving for bentonite replacement up to 10%, after which a decrease was observed. For instance, 90-day splitting tensile strength of GRAC/B10 and GNAC/B10 increased by 8.69% and 6.79%, respectively, as compared to the specimens containing 5% bentonite. The decreasing trend of tensile strength beyond 10% bentonite replacement is similar to the compressive strength results. For example, 90-day compressive strength of GRAC/B15 and GNAC/B15 decreased by 12.72% and 9.01%, respectively, as compared to the respective specimens containing 5% bentonite. However, all the bentonite-incorporated GRAC and GNAC mixes showed higher tensile strength than their corresponding control mixes at 90 days. Like compressive strength, bentonite has more pronounced effect (increase or decrease) on the splitting tensile strength of GRAC as compared to GNAC. Considering the performance criteria of mechanical properties, i.e., compressive strength and splitting tensile strength, 10 wt % of bentonite can be considered optimum content for both GNAC and GRAC mixes. The bentonite with 10 wt % can be incorporated to improve the mechanical properties of geopolymer concrete with RCA.

3.3. Durability Properties

3.3.1. Water Absorption

Water absorption is a critical property in determining the durability of concrete since it measures water-accessible porosity of concrete. The water absorption of concrete controls the movement of outside moisture that could carry an aggressive chemical into the concrete.

There are various chemicals and salt solutions that can easily penetrate through pores of the concrete and can react with its ingredients to affect the material properties. The water absorption test was carried out on all mixes of GNAC and GRAC at the age of 90 days. The results obtained from the tests are shown in Figure 11. The incorporation of RCA in geopolymer concrete resulted in increased water absorption. The water absorption of GRAC mixes was observed to be increased by 29% than the GNAC mixes. This increase in water absorption can be attributed to the porous and permeable characteristics of old mortar attached to the surface of RCA that may have enhanced the connectivity between micro-channels in the microstructure of the material. Moreover, the old cement paste ma have established routes for water to stream into the material [36,46].

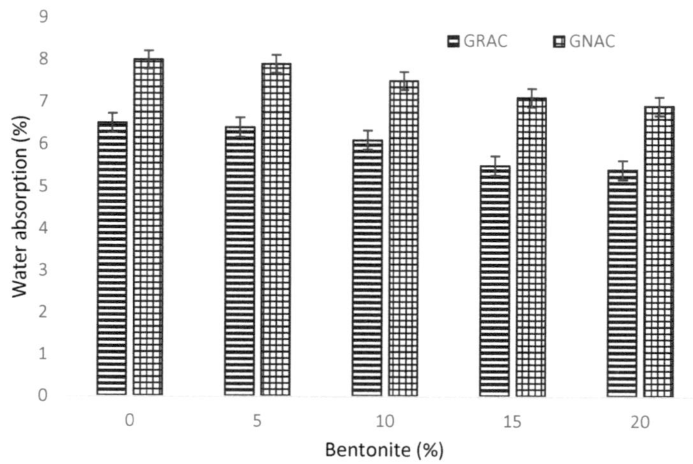

Figure 11. Percent water absorption of the GNAC and GRAC specimens with varyir bentonite percentages.

The bentonite addition in both the GRAC and GNAC mixes has a positive effect c water absorption. The maximum reduction in the water absorption for both the GNA and GRAC mixes was observed at 20% bentonite content. The water absorption of th GNAC mixes with 5%, 10%, 15%, and 20% bentonite decreased by 2.5%, 8%, 14%, and 18 respectively, compared to the PC mix; the water absorption of the GRAC mixes with simil bentonite replacements decreased by 3.5%, 8%, 17%, and 20.5%, respectively, compared the SC mix.

Due to the finer particle size of bentonite than the fly ash, it can have a unifor distribution in the mix matrix of both the GNAC and GRAC, thus providing a mo compact and denser microstructure. In addition, the pozzolanic reactivity of bentonite ma have resulted in the development of C-S-H gel in combination with geopolymeric gel th can provide a further denser microstructure of geopolymer matrix [65,66]. Therefore, can be inferred that the incorporation of bentonite in both the GNAC and GRAC mixes beneficial in terms of reducing the water absorption capabilities of the geopolymer matri

3.3.2. Chloride Migration Coefficient

The reinforced concrete's durability and serviceability are greatly affected by chlorid penetration from seawater or deicing salts. The reinforcing bars in the concrete corrod when chloride migration progresses on the steel bar surface. The formation of oxides a the steel surface results in increasing the volume of surrounding concrete, spalling an deteriorating the concrete cover. The determination of the chloride migration coefficier is an important technique to ascertain the ability of concrete to allow the penetratio of chlorides.

The results of the chloride migration test on all the mixes in terms of a coefficient ar shown in Figure 12. The GRAC mixes have an appreciably higher coefficient of chlorid migration due to their high porosity and poor interfacial transition zone (ITZ). The GRA mixes showed 27% more penetration to chlorides than their counterparts of GNAC mixe However, incorporation of bentonite in both the GRAC and GNAC mixes helped t improve the resistance against chloride migration. The better resistance with bentonit can be attributed to an improvement in impermeability and water tightness of concret and enhancement of bonding capacity of bentonite with chlorides [41]. It is reported b

Detwiler et al. [67] that supplementary materials, i.e., silica fume and GGBS, can be used to improve the resistance of concrete against chloride penetration. Nuaklong et al. [33] suggested the use of metakaolin as a supplementary material to improve the chloride penetration resistance of geopolymer concrete containing NA and RCA.

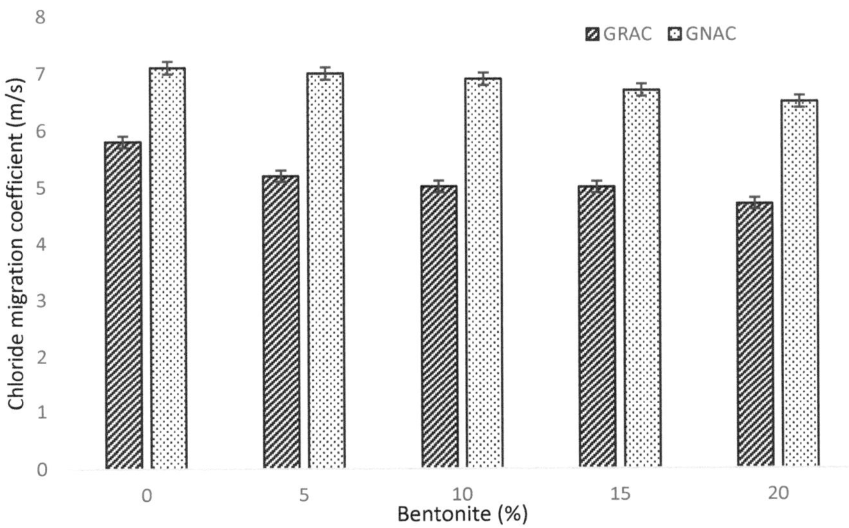

Figure 12. Chloride migration coefficient of the GNAC and GRAC specimens with varying bentonite percentages.

The inclusion of bentonite in the GNAC and GRAC mixes resulted in increasing the total alumina content, which in turn produced more calcium–alumina–silicate hydrates (C-A-S-H) in the concrete. The alumina content and C-A-S-H gel produce a protecting film in the concrete, which resisted the chloride penetration. While the chloride-carrying compounds enter into the pore structure of concrete, C-A-S-H reacts with these chlorides, producing Friedel's salt by chemical reaction. Friedel's salt increases the alkalinity of concrete by producing hydroxyl ions in the pore structure and acts as passivation for the steel reinforcement.

3.3.3. Sulfuric Acid Attack Resistance

The nature of conventional concrete is alkaline, as indicated by its pH value. When the cement paste comes in contact with acids, its components break down. Therefore, in case of an acid attack on concrete, both hydrated and unhydrated cement compounds are dissolved along with the calcareous aggregates [68].

When the pH of the concrete is reduced, the ability of the binder to bind all the ingredients of concrete is affected and concrete will be more vulnerable to the acid attack. For an ideal concrete, the pH value should lie between 9 and 11. When the pH value is below 7, the acids may attack the concrete and make it unfit. The concrete structure starts deteriorating when water containing sulfuric acid penetrates the concrete. It produces sulfate ions which react with the portlandite to form gypsum. The gypsum produced reacts with calcium–aluminate hydrate to produce ettringite. The formation of the ettringite leads to the expansion of concrete, which creates internal pressure, resulting in deterioration. In this study, the resistance of concrete mixes against sulfuric acid is measured by determining mass loss in water containing 5% sulfuric acid and the results are shown in Figure 13. The specimens were submerged in 4% H_2SO_4 solution [41]. The specimens were taken out of the solution after 28, 56, and 90 days of immersion, and loss in mass of the cubes was noted. To keep the concentration of the sulfuric acid stable in the solution, the solution was changed every 14 days. It can be observed from the Figure 13 that loss in mass of cubic specimens increased with age.

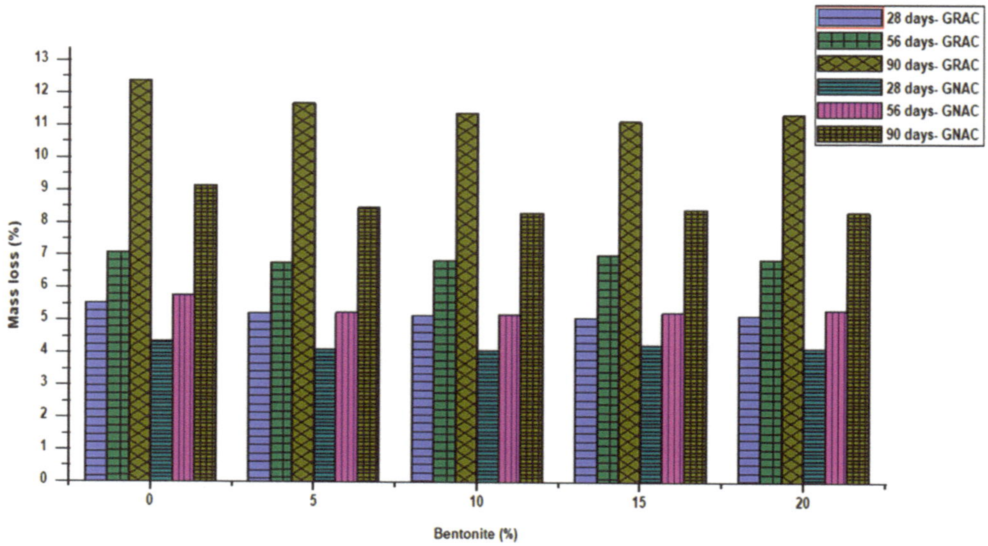

Figure 13. Percent mass loss due to sulfuric acid attack in the GNAC and GRAC specimens at varying bentonite percentages.

The GRAC mixes experienced a higher loss of mass in acid solution as compared t the GNAC, due to the mortar attached to the surface of the RCA. The inclusion of bentoni improved the resistance of both the GNAC and GRAC mixes. All the bentonite mixe exhibited a lower mass loss in the acid solution than their corresponding control mixes.

The higher resistance to acid solution of both the GNAC and GRAC mixes is due to th reduction in calcium oxide in concrete, owing to the replacement of bentonite. The contr mixes PC and SC form more portlandite than the mixes with bentonite. The consumptio of free portlandite by the bentonite particles in the pozzolanic reaction forms more CSH This mechanism would have caused an improvement in acid attack resistance of both th GNAC and GRAC mixes. It can also be noticed that the inclusion of bentonite in the GRA mixes resulted in an appreciable improvement in acid resistance. It can be noticed fror Figure 13 that bentonite has a pronounced effect on GRAC mixes as compared to GNA mixes because 15% bentonite incorporation reduces percentage mass loss (at 90 day) to 9 and 12% for GNAC and GRAC mixes, respectively.

4. Conclusions

In this paper, we presented the results of an experimental program conducted t evaluate the influence of low-calcium bentonite content on the rheological, mechanica and durability properties of HCFA-based geopolymer concrete containing natural an recycled coarse aggregates. The mixes were divided into two groups, GNAC and GRAC based on the aggregates used, i.e., NCA and RCA, respectively. For studying the influenc of bentonite clay, HCFA was replaced by bentonite clay by 0%, 5%, 10%, 15%, and 20 (by weight) in both the GNAC and GRAC mixes. A series of experiments (workabilit fresh density, air content, compressive strength, split tensile strength, water absorptio chloride migration coefficient, and acid resistance tests) were conducted to study the effe of bentonite content on rheological, mechanical, and durability properties of both GNA and GRAC mixes. The following key conclusions have been drawn from this study:

1. The use of RCA in GRAC mixes results in increasing the workability as compared to corresponding GNAC mixes. The incorporation of bentonite has a detrimental effect on the workability of both the GRAC and GNAC mixes. The inclusion of bentonite reduced the workability of all concrete mixes; increased content of bentonite further intensified the reduction in workability. Therefore, to keep the same slump of all mixes, a small dosage of super plasticizer was added.
2. The air content of GNAC mixes was relatively lower as compared to the corresponding GRAC mixes. Air content of both GNAC and GRAC mixes increased with an increase in the percentage replacement of bentonite when SP is not used to attain the target slump. A significant reduction in the air content was observed with an increase in the bentonite content for both GRAC and GNAC mixes (with SP). The fresh density of GRAC mixes with RCA was lower when compared to the corresponding GNAC mixes with NCA. The fresh density of mixes (without SP) decreased with increased content of bentonite for both GRAC and GNAC mixes.
3. The use of RCA in GRAC mixes resulted in reduced mechanical properties compared to the counterparts of GNAC mixes. The replacement ratio of bentonite with HCFA had a significant effect on the strength properties of both GNAC and GRAC mixes. Replacement level of bentonite with up to 10% improved the strength properties of both GNAC and GRAC mixes. Beyond the 10% bentonite replacement level, the strength properties of both GNAC and GRAC mixes were effected.
4. The incorporation of RCA in GRAC mixes resulted in increasing the water absorption by 20–29% compared to the corresponding GNAC mixes. An appreciable reduction in the water absorption capacity of the GRAC mixes was observed at 90 days of specimen age for all replacement levels of bentonite. The maximum reduction in the water absorption for both the GNAC and GRAC mixes was observed at 20% bentonite content.
5. The GRAC mixes have appreciably higher coefficient of chloride migration than their counterparts of GNAC mixes. The mixes with bentonite addition showed greater resistance to chloride penetration than the control mixes (without bentonite). Furthermore, the GRAC mixes experienced higher loss of mass in acid solution as compared to the corresponding GNAC mixes due to the mortar attached to the surface of the RCA. The inclusion of bentonite resulted in improving the resistance of both the GNAC and GRAC mixes. All the bentonite mixes exhibited lower mass loss in the acid solution than their corresponding control mixes.

Author Contributions: Conceptualization, R.M.W., F.B. and A.D.; Methodology, R.M.W., F.B., A.D. and B.M.; Formal analysis, R.M.W., A.D. and B.M.; Funding acquisition, M.A., E.E.H. and F.B.; Investigation, R.M.W., F.B., A.D., M.A., M.A.M., B.M. and E.E.H.; Supervision, F.B. and R.M.W.; Validation, R.M.W., F.B. and A.D.; Writing—original draft, R.M.W., F.B. and A.D.; Writing—review and editing, F.B., M.A., M.A.M. and E.E.H. All authors have read and agreed to the published version of the manuscript.

Funding: This research was funded by Taif University Researchers Supporting Project (grant number TURSP-2020/324).

Institutional Review Board Statement: Not applicable.

Informed Consent Statement: Not applicable.

Data Availability Statement: All the data are available within this manuscript.

Acknowledgments: The authors would like to acknowledge the finical support provided from Taif University Researchers Supporting Project (grant number TURSP-2020/324).

Conflicts of Interest: All authors declare no conflict of interest.

Abbreviations

C&DW: Construction and demolition waste, C-A-S-H: calcium–alumina–silicate hydrate, CO_2: C bon dioxide, C-S-H: Calcium–silicate gel, FA: Fly ash, GGBS: Ground granulated blast furnace sla GHG: Greenhouse gas, GNAC: Geopolymeric natural aggregate concrete, GPC: Geopolymer concre GRAC: Geopolymeric recycled aggregate concrete, HCFA: High-calcium fly ash, ITZ: Interfacial tra sition zone, NCA: Natural coarse aggregates, OPC: Ordinary Portland cement, PC: Primary contr mix, RCA: Recycled coarse aggregates, RHA: Rice husk ash, SC: Secondary control, SAI: Streng activity index, SCMs: Supplementary cementitious materials, SEM: Scanning electron microscop SH: Sodium hydroxide, SP: Superplasticizer, SS: Sodium silicate, SSD: Surface dry condition.

References

1. Al Bakri, M.M.; Mohammed, H.; Kamarudin, H.; Niza, I.K.; Zarina, Y. Review on fly ash-based geopolymer concrete witho Portland Cement. *J. Eng. Technol. Res.* **2011**, *3*, 1–4.
2. Mehta, P.K. High-performance, high-volume fly ash concrete for sustainable development. In Proceedings of the Internation Workshop on Sustainable Development and Concrete Technology, University of California, Berkeley, CA, USA, 14 May 20C pp. 3–14.
3. Schneider, M.; Romer, M.; Tschudin, M.; Bolio, H. Sustainable cement production—Present and future. *Cem. Concr. Res.* **2011**, *4* 642–650. [CrossRef]
4. Lei, Y.; Zhang, Q.; Nielsen, C.; He, K. An inventory of primary air pollutants and CO_2 emissions from cement production China, 1990–2020. *Atmos. Environ.* **2011**, *45*, 147–154. [CrossRef]
5. Shaikh, F.U.A. Mechanical and durability properties of fly ash geopolymer concrete containing recycled coarse aggregates. *Int. Sustain. Built Environ.* **2016**, *5*, 277–287. [CrossRef]
6. Kou, S.-C.; Poon, C.S.; Wan, H.-W. Properties of concrete prepared with low-grade recycled aggregates. *Constr. Build. Mater.* **201** *36*, 881–889. [CrossRef]
7. Zaharieva, R.; Buyle-Bodin, F.; Skoczylas, F.; Wirquin, E. Assessment of the surface permeation properties of recycled aggrega concrete. *Cem. Concr. Res.* **2003**, *25*, 223–232. [CrossRef]
8. Tabsh, S.W.; Abdelfatah, A.S. Influence of recycled concrete aggregates on strength properties of concrete. *Constr. Build. Mat* **2009**, *23*, 1163–1167. [CrossRef]
9. Sagoe-Crentsil, K.; Brown, T.; Taylor, A. Performance of concrete made with commercially produced coarse recycled concre aggregate. *Cem. Concr. Res.* **2001**, *31*, 707–712. [CrossRef]
10. Corinaldesi, V.; Moriconi, G. Influence of mineral additions on the performance of 100% recycled aggregate concrete. *Cons Build. Mater.* **2009**, *23*, 2869–2876. [CrossRef]
11. Etxeberria, M.; Vázquez, E.; Mari, A.; Barra, M. Influence of amount of recycled coarse aggregates and production process o properties of recycled aggregate concrete. *Cem. Concr. Res.* **2007**, *37*, 735–742. [CrossRef]
12. Marinković, S.; Radonjanin, V.; Malešev, M.; Ignjatović, I. Comparative environmental assessment of natural and recycle aggregate concrete. *Waste Manag.* **2010**, *30*, 2255–2264. [CrossRef]
13. de Juan, M.S.; Gutiérrez, P.A. Study on the influence of attached mortar content on the properties of recycled concrete aggrega *Constr. Build. Mater.* **2009**, *23*, 872–877. [CrossRef]
14. Debieb, F.; Courard, L.; Kenai, S.; Degeimbre, R. Mechanical and durability properties of concrete using contaminated recycle aggregates. *Cem. Concr. Compos.* **2010**, *32*, 421–426. [CrossRef]
15. Tam, V.W.; Gao, X.; Tam, C.M. Microstructural analysis of recycled aggregate concrete produced from two-stage mixing approac *Cem. Concr. Res.* **2005**, *35*, 1195–1203. [CrossRef]
16. Ahmed, S.F.U.; Shaikh, F. Existence of Dividing Strength in Concrete Containing Recycled Coarse Aggregate. *J. Mater. Civ. En* **2014**, *26*, 784–788. [CrossRef]
17. Shaikh, F.U.A.; Nguyen, H.L. Properties of concrete containing recycled construction and demolition wastes as coarse aggregate *J. Sustain. Cem. Mater.* **2013**, *2*, 204–217. [CrossRef]
18. De Domenico, D.; Faleschini, F.; Pellegrino, C.; Ricciardi, G. Structural behavior of RC beams containing EAF slag as recycle aggregate: Numerical versus experimental results. *Constr. Build. Mater.* **2018**, *171*, 321–337. [CrossRef]
19. Majhi, R.K.; Nayak, A.N. Production of sustainable concrete utilising high-volume blast furnace slag and recycled aggregate wit lime activator. *J. Clean. Prod.* **2020**, *255*, 120188. [CrossRef]
20. Duxson, P.; Provis, J.L.; Lukey, G.C.; Van Deventer, J.S. The role of inorganic polymer technology in the development of gree concrete. *Cem. Concr. Res.* **2007**, *37*, 1590–1597. [CrossRef]
21. Wallah, S.; Rangan, B.V. *Low-Calcium Fly Ash-Based Geopolymer Concrete: Long-Term Properties*; Curtin University of Technolog Perth, Australia, 2006.

2. Sumajouw, M.; Rangan, B.V. *Low-Calcium Fly Ash-Based Geopolymer Concrete: Reinforced Beams and Columns*; Curtin University of Technology: Perth, Australia, 2006.
3. Kong, D.L.; Sanjayan, J.G. Effect of elevated temperatures on geopolymer paste, mortar and concrete. *Cem. Concr. Res.* **2010**, *40*, 334–339. [CrossRef]
4. Pavithra, P.E.; Reddy, M.S.; Dinakar, P.; Rao, B.H.; Satpathy, B.; Mohanty, A. A mix design procedure for geopolymer concrete with fly ash. *J. Clean. Prod.* **2016**, *133*, 117–125. [CrossRef]
5. Shi, X.; Collins, F.G.; Zhao, X.L.; Wang, Q. Mechanical properties and microstructure analysis of fly ash geopolymeric recycled concrete. *J. Hazard. Mater.* **2012**, *237*, 20–29. [CrossRef] [PubMed]
6. Nuaklong, P.; Sata, V.; Chindaprasirt, P. Influence of recycled aggregate on fly ash geopolymer concrete properties. *J. Clean. Prod.* **2016**, *112*, 2300–2307. [CrossRef]
7. Nath, P.; Sarker, P.K. Use of OPC to improve setting and early strength properties of low calcium fly ash geopolymer concrete cured at room temperature. *Cem. Concr. Compos.* **2015**, *55*, 205–214. [CrossRef]
8. Nikolić, V.C.; Komljenović, M.; Baščarević, Z.; Marjanović, N.; Miladinović, Z.; Petrović, R. The influence of fly ash characteristics and reaction conditions on strength and structure of geopolymers. *Constr. Build. Mater.* **2015**, *94*, 361–370. [CrossRef]
9. Phoo-Ngernkham, T.; Sata, V.; Hanjitsuwan, S.; Ridtirud, C.; Hatanaka, S.; Chindaprasirt, P. High calcium fly ash geopolymer mortar containing Portland cement for use as repair material. *Constr. Build. Mater.* **2015**, *98*, 482–488. [CrossRef]
0. Nath, P.; Sarker, P. Effect of GGBFS on setting, workability and early strength properties of fly ash geopolymer concrete cured in ambient condition. *Constr. Build. Mater.* **2014**, *66*, 163–171. [CrossRef]
1. Nuaklong, P.; Jongvivatsakul, P.; Pothisiri, T.; Sata, V.; Chindaprasirt, P. Influence of rice husk ash on mechanical properties and fire resistance of recycled aggregate high-calcium fly ash geopolymer concrete. *J. Clean. Prod.* **2019**, *252*, 119797. [CrossRef]
2. Nuaklong, P.; Sata, V.; Wongsa, A.; Srinavin, K.; Chindaprasirt, P. Recycled aggregate high calcium fly ash geopolymer concrete with inclusion of OPC and nano-SiO_2. *Constr. Build. Mater.* **2018**, *174*, 244–252. [CrossRef]
3. Nuaklong, P.; Sata, V.; Chindaprasirt, P. Properties of metakaolin-high calcium fly ash geopolymer concrete containing recycled aggregate from crushed concrete specimens. *Constr. Build. Mater.* **2018**, *161*, 365–373. [CrossRef]
4. Ahmad, S.; Barbhuiya, S.A.; Elahi, A.; Iqbal, J. Effect of Pakistani bentonite on properties of mortar and concrete. *Clay Miner.* **2011**, *46*, 85–92. [CrossRef]
5. Rehman, S.U.; Yaqub, M.; Noman, M.; Ali, B.; Khan, M.N.A.; Fahad, M.; Abid, M.M.; Gul, A. The Influence of Thermo-Mechanical Activation of Bentonite on the Mechanical and Durability Performance of Concrete. *Appl. Sci.* **2019**, *9*, 5549. [CrossRef]
6. Murray, H. Industrial clays case study. *Min. Miner. Sustain. Develop.* **2002**, *64*, 1–9.
7. Karthikeyan, M.; Ramachandran, P.R.; Nandhini, A.; Vinodha, R. Application on partial substitute of cement by bentonite in concrete. *Int. J. ChemTech Res.* **2015**, *8*, 384–388.
8. Grishin, V.A.; Deryugin, L.M. Experience in the use of bentonite-cement concrete for repairing the core of the earthfill dam of Kureiskaya HPP. *Power Technol. Eng.* **2006**, *40*, 90–95. [CrossRef]
9. Shabab, M.E.; Shahzada, K.; Gencturk, B.; Ashraf, M.; Fahad, M. Synergistic effect of fly ash and bentonite as partial replacement of cement in mass concrete. *KSCE J. Civ. Eng.* **2015**, *20*, 1987–1995. [CrossRef]
0. Lima-Guerra, D.J.; Mello, I.; Resende, R.; Silva, R. Use of Bentonite and Organobentonite as Alternatives of Partial Substitution of Cement in Concrete Manufacturing. *Int. J. Concr. Struct. Mater.* **2014**, *8*, 15–26. [CrossRef]
1. Masood, B.; Elahi, A.; Barbhuiya, S.; Ali, B. Mechanical and durability performance of recycled aggregate concrete incorporating low calcium bentonite. *Constr. Build. Mater.* **2019**, *237*, 117760. [CrossRef]
2. Memon, S.; Arsalan, R.; Khan, S.; Lo, T.Y. Utilization of Pakistani bentonite as partial replacement of cement in concrete. *Constr. Build. Mater.* **2012**, *30*, 237–242. [CrossRef]
3. Mirza, J.; Riaz, M.; Naseer, A.; Rehman, F.; Khan, A.; Ali, Q. Pakistani bentonite in mortars and concrete as low cost construction material. *Appl. Clay Sci.* **2009**, *45*, 220–226. [CrossRef]
4. Petrus, H.T.B.M.; Hulu, J.; Dalton, G.S.; Malinda, E.; Prakosa, R.A. Effect of Bentonite Addition on Geopolymer Concrete from Geothermal Silica. *Mater. Sci. Forum* **2016**, *841*, 7–15. [CrossRef]
5. Yang, Y.; Jiang, J.; Hou, L.; Lu, Z.; Li, J.; Wang, J. Pore structure and properties of porous geopolymer based on pre-swelled bentonite. *Constr. Build. Mater.* **2020**, *254*, 119226. [CrossRef]
6. Huang, Y. Influence of Calcium Bentonite Addition on the Compressive Strength, Efflorescence Extent and Drying Shrinkage of Fly-Ash Based Geopolymer Mortar. *Trans. Indian Ceram. Soc.* **2020**, *79*, 77–82. [CrossRef]
7. Chamundeeswari, J. Experimental Study on Partial Replacement of Cement by Bentonite in Paverblock. *Int. J. Eng. Trends Technol.* **2012**, *3*, 41–47.
8. Akram, T.; Memon, S.A.; Khan, M.N. Utilization of Jehangira bentonite as partial replacement of cement. In Proceedings of the International Conference on Advances in Cement Based Materials and Applications in Civil Infrastructure ACBM-ACI, Lahore, Pakistan, 12–14 December 2007; pp. 301–311.
9. Swarup, O.K.; Reddy, P.R.; Reddy, M.A.K.; Rao, V.R. A Study on Durability of Concrete by Partial Replacement of Cement with Bentonite and Fly Ash. *Int. J. ChemTech Res.* **2017**, *10*, 855–861.
0. Reddy, G.V.K.; Rao, V.R.; Reddy, M.A.K. Experimental investigation of strength parameters of cement and concrete by partial replacement of cement with Indian calcium bentonite. *Technology* **2017**, *8*, 512–518.

51. ASTM. *Standard Specification for Coal Fly Ash and Raw or Calcined Natural Pozzolan for Use in Concrete*; ASTM International: We Conshohocken, PA, USA, 2013.
52. ASTM. *Standard Test. Method for Relative Density (Specific Gravity) and Absorption of Coarse Aggregate*; ASTM International: We Conshohocken, PA, USA, 2015.
53. ASTM. *C143/C143M-12-Standard Test. Method for Slump of Hydraulic-Cement Concrete*; ASTM International: West Conshohocke PA, USA, 2012.
54. ASTM. *Standard Test. Method for Density (Unit Weight), Yield, and Air Content (Gravimetric) of Concrete*; ASTM International: We Conshohocken, PA, USA, 2014.
55. ASTM. *C231/C231M-10, Standard Test. Method for Air Content of Freshly Mixed Concrete by the Pressure Method*; ASTM Internation West Conshohocken, PA, USA, 2010.
56. BSI. *BS EN 12390-3: 2009: Testing Hardened Concrete. Compressive Strength of Test. Specimens*; BSI: London, UK, 2009.
57. ASTM. *C496/C496M-11 Standard Test. Method for Splitting Tensile Strength of Cylindrical Concrete Specimens*; ASTM Internation West Conshohocken, PA, USA, 2004.
58. ASTM. *Standard Test. Method for Dry and Wet Bulk Density, Water Absorption, and Apparent Porosity of Thin Sections of Glass-Fib Reinforced Concrete*; ASTM C948-81; ASTM International: West Conshohocken, PA, USA, 2016.
59. NORDTEST Build. Concrete, mortar and cement-based repair materials: Chloride migration coefficient from non-steady-sta migration experiments. *Nordtest Method* **1999**, *492*, 1–8.
60. Jing, R.; Liu, Y.; Yan, P. Uncovering the effect of fly ash cenospheres on the macroscopic properties and microstructure of ult high-performance concrete (UHPC). *Constr. Build. Mater.* **2021**, *286*, 122977. [CrossRef]
61. Huang, W.-H. Properties of cement-fly ash grout admixed with bentonite, silica fume, or organic fiber. *Cem. Concr. Res.* **1998**, 2 395–406. [CrossRef]
62. Gómez-Soberón, J.M. Porosity of recycled concrete with substitution of recycled concrete aggregate: An experimental study. *Ce Concr. Res.* **2002**, *32*, 1301–1311. [CrossRef]
63. Li, W.; Xiao, J.; Sun, Z.; Kawashima, S.; Shah, S.P. Interfacial transition zones in recycled aggregate concrete with different mixir approaches. *Constr. Build. Mater.* **2012**, *35*, 1045–1055. [CrossRef]
64. Singhal, D.; Junaid, M.T.; Jindal, B.B.; Mehta, A. Mechanical and microstructural properties of fly ash based geopolymer concre incorporating alccofine at ambient curing. *Constr. Build. Mater.* **2018**, *180*, 298–307.
65. Danek, T.; Thomas, J.; Jelinek, J. Study of Interaction between Fly Ash-cement and Bentonite Matrices. *MATEC Web Conf.* **201** *26*, 1003. [CrossRef]
66. Fernández, R.; Ruiz, A.I.; Cuevas, J. Formation of C-A-S-H phases from the interaction between concrete or cement and bentoni *Clay Miner.* **2016**, *51*, 223–235. [CrossRef]
67. Detwiler, R.J.; Fapohunda, C.A.; Natale, J. Use of supplementary cementing materials to increase the resistance to chloride io penetration of concretes cured at elevated temperatures. *Mater. J.* **1994**, *91*, 63–66.
68. Devi, V.S. Durability properties of multiple blended concrete. *Constr. Build. Mater.* **2018**, *179*, 649–660. [CrossRef]

materials

rticle

nfluence of Asphalt Emulsion Inclusion on Fly Ash/Hydrated .ime Alkali-Activated Material

hanon Bualuang [1], Peerapong Jitsangiam [2,*], Teewara Suwan [2], Ubolluk Rattanasak [3], /eerachart Tangchirapat [4] and Suriyah Thongmunee [2]

1 Graduate Program in Civil Engineering, Department of Civil Engineering, Faculty of Engineering, Chiang Mai University, Huay Kaew Road, Muang, Chiang Mai 50200, Thailand; thanon_bualuang@cmu.ac.th
2 Center of Excellence in Natural Disaster Management, Department of Civil Engineering, Faculty of Engineering, Chiang Mai University, Huai Kaew Road, Mueang, Chiang Mai 50200, Thailand; teewara.s@cmu.ac.th (T.S.); suriyah@eng.cmu.ac.th (S.T.)
3 Department of Chemistry, Faculty of Science, Burapha University, Chonburi 20131, Thailand; ubolluk@go.buu.ac.th
4 Construction Innovations and Future Infrastructure Research Center (CIFIR), Department of Civil Engineering, Faculty of Engineering, King Mongkut's University of Technology Thonburi, Bangkok 10140, Thailand; weerachart.tan@kmutt.ac.th
* Correspondence: peerapong@eng.cmu.ac.th; Tel.: +66-053-944-157

itation: Bualuang, T.; Jitsangiam, P.; ıwan, T.; Rattanasak, U.; ıngchirapat, W.; Thongmunee, S. fluence of Asphalt Emulsion clusion on Fly Ash/Hydrated Lime lkali-Activated Material. *Materials* 021, *14*, 7017. https://doi.org/).3390/ma14227017

cademic Editors: Rui Vasco Silva, ligue Bravo and António P. . Duarte

eceived: 3 October 2021
ccepted: 17 November 2021
ublished: 19 November 2021

Abstract: Supplementary cementitious materials have been widely used to reduce the greenhouse gas emissions caused by ordinary Portland cement (OPC), including in the construction of road bases. In addition, the use of OPC in road base stabilization is inefficient due to its moisture sensitivity and lack of flexibility. Therefore, this study investigates the effect of hybrid alkali-activated materials (H-AAM) on flexibility and water prevention when used as binders while proposing a new and sustainable material. A cationic asphalt emulsion (CAE) was applied to increase this cementless material's resistance to moisture damage and flexibility. The physical properties and structural formation of this H-AAM, consisting of fly ash, hydrated lime, and sodium hydroxide, were examined. The results revealed that the addition of CAE decreased the material's mechanical strength due to its hindrance of pozzolanic reactions and alkali activations. This study revealed decreases in the cementitious product's peak in the x-ray diffraction analysis (XRD) tests and the number of tetrahedrons detected in the Fourier transform infrared spectroscopy analysis (FTIR) tests. The scanning electron microscope (SEM) images showed some signs of asphalt films surrounding hybrid alkali-activated particles and even some unreacted FA particles, indicating incomplete chemical reactions in the study material's matrix. However, the H-AAM was still able to meet the minimum road base strength requirement of 1.72 MPa. Furthermore, the toughness and flexibility of the H-AAM were enhanced by CAE. Notably, adding 10% and 20% CAE by weight to the hybrid alkali-activated binder produced a significant advantage in terms of water absorption, which can be explained by its influence on the material's consolidation of its matrices, resulting in significant void reductions. Hence, the outcomes of this study might reveal an opportunity for developing a new stabilizing agent for road bases with water-prevention properties and flexibility that remains faithful to the green construction material concept.

Keywords: asphalt emulsion; pozzolanic material; alkali-activated binder; cementless materials; fly ash; hydrated lime

1. Introduction

At present, ordinary Portland cement (OPC) replacements are being developed to reduce greenhouse gas emissions in the construction industry. Many studies have revealed the benefits of supplementary cementitious materials (SCMs), such as industrial fly ash and hydrated lime, when replacing some OPC in concrete, including cost reductions and fewer greenhouse gas emissions [1–5]. Many SCMs incorporating green materials, i. geopolymers and alkali-activated materials (AAMs), have been introduced. These silic and alumina-rich materials are activated using alkali solutions, such as sodium hydroxic (NaOH) and sodium silicate (Na_2SiO_3), resulting in new and green construction materia with mechanical properties equal to the general concrete for construction [6–9]. Theor ically, pozzolanic materials with appropriate alkali activators can form the cementitio products calcium aluminosilicate hydrate gel (C-A-S-H), calcium silicate hydrate gel (S-H), and sodium aluminosilicate hydrated product (N-A-S-H) [10–14]. These structur formations are similar to the hydration product resulting from mixing OPC and water [1 Some formations with lower mechanical strengths than cement-based materials have bee used in road bases and subbase structural layers. Consequently, the AAMs and pozzolan materials have been used as replacements for OPC in this specific work. However, it is st necessary to increase their long-term durability for actual road conditions.

AAMs and pozzolanic materials are suitable under compressive loading condition However, they are probably prone to more brittleness and less waterproofness than nece sary for roadway construction materials. During the rainy season, the relatively high wat can establish premature cracks in road pavement structures, allowing liquids to acce and penetrate capillary and air voids, decreasing the durability of cement-based materia and AAMs [16–18]. Stabilizing materials with high moisture sensitivity decrease stiffne caused by excess moisture from an external environment [19,20], negatively impactin crucial pavement properties [21,22]. This problem results in redundant maintenance cos and cannot be resolved in many countries. Hardiness, flexibility, and waterproofne are necessary for long-term service. For this reason, a comprehensive solution for th brittleness and lack of waterproofness in cement and SCMs is necessary for attainin long-term durability.

Asphalt emulsion (AE), a stabilized asphalt by a chemical emulsifier, is an excitin additive due to the efficient elasticity and waterproofness it brings to cement-based mate als and AAMs [23–26]. Researchers have investigated and applied AE as an additive i concrete or mortar and correlated its addition to dynamic track slabs and ballasts, amon other structures [27,28]. Some studies have found that it can improve pavement enginee ing properties in asphalt-treated base materials by increasing flexibility, resilience, an durability [29–32]. Moreover, the workability of AE in cement-based material influence the setting process [33,34], which may be appropriate for pavement constructions requirin a reasonable period of field-mixing preparation. Due to its inherent good flexibility an moisture resistance, AE could be used to enhance the flexibility and waterproofness c chemically stabilized road pavement materials.

The interaction mechanism of a cement-AE-based material causes asphalt drople with electrostatic charges to form an emulsifier that can react with the hydration produc such as Portlandite and C-A-S-H gel, by exchanging the polar from these different chemica components [35]. The asphalt molecules converge around the cement paste matrices an form a stiff film. The cement particles absorb the water in AE, leading to the reduce spaces in the pastes' matrices and reduced voids [26,36,37]. However, the hydration c cement-based materials is delayed due to the asphalt droplets covering the cement particle resulting in incomplete primary hydration with water. This retarding effect can increas the initial and the final setting times, allowing AE to significantly retard cement hydratio at a high AE to cement ratio [38]. In addition, hydration heat is blocked by thin films of A on the cement particles, resulting in decreases in mechanical strength and stiffness [39–41 No new crystalline substances or chemical functional groups have been found during th interaction process between cement and the AE [42]. However, based on the previou studies, there was almost no study of a hybrid alkali-activated material (H-AAM) prepare with fly ash (FA), hydrated lime (HL), and solid sodium hydroxide (NaOH); and the effect of adding AE into such material is still questionable. Even FA itself has toxicity manner from leaching [43,44]; the FA used in this study is from the common source of FA used i the concrete industry in Thailand. Therefore, it is categorized as a standard constituent of the cementitious mixing material.

Therefore, this study aimed to investigate the influence of AE on the mechanical properties and structural formations of such the H-AAM. The AE was added to the mixture in varying amounts. The mechanical characteristics of workability, compressive and flexural strength, water absorption, and shrinkage were examined. Meanwhile, the resulting mixtures' morphologies and microstructures were analyzed via x-ray diffraction (XRD), Fourier transform infrared spectroscopy (FTIR), scanning electron microscope (SEM), and pore size distribution tests.

2. Preliminary Study of the H-AAM for Road Bases

Preliminary research conducted at Chiang Mai University, Thailand, established that the H-AAM could be used as a stabilizing agent in road bases. The H-AAM and Portland cement (as a benchmark) were utilized as stabilizing agents, while the crushed rock was used as the parent material [45]. From Figure 1, it can be seen that the H-AAM achieved the target unconfined compressive strength (UCS) required by the Department of Highways, Thailand, on day 7 [46,47]. These UCS values showed that the H-AAM could be used to achieve the minimum strength required, even though its strength was less than that of the OPC-based material. Therefore, this H-AAM could be utilized as an OPC replacement in road stabilization. This study also explored the influence of AE on the flexibility and waterproofness of this material.

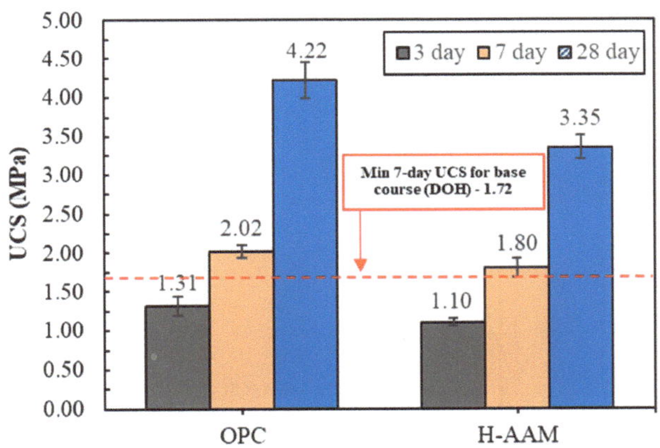

Figure 1. UCS results of road base materials stabilized by Portland cement and the H-AAM obtained during the preliminary study.

3. Materials and Methods

3.1. Materials and the Mix Design

FA and HL were used as raw materials in the production of the alkali-activated material. The FA was obtained from the Mae Moh power plant, Lampang Province, Thailand. Meanwhile, the HL came from the limestone quarry in Saraburi Province, Thailand. Figure 2 illustrates the morphology of the FA observed by using SEM (JEOL JSM-5910LV, gold coated, WD: 11 mm, EHT: 15 kV, JEOL Inc., Pleasanton, CA, USA). Table 1 shows the oxide compositions of the raw materials obtained with X-ray fluorescence (XRF, JEOL JSX3400R, JEOL Inc., Pleasanton, CA, USA). The FA was classified as a high-calcium or class C fly ash with significant CaO content (24.83%). Apart from the CaO, the FA had high contents of SiO_2 (32.47%), Al_2O_3 (15.84%), and Fe_2O_3 (14.64%). The specific gravity and Blaine fineness of the FA were 2.11 and 2400 cm^2/g, respectively.

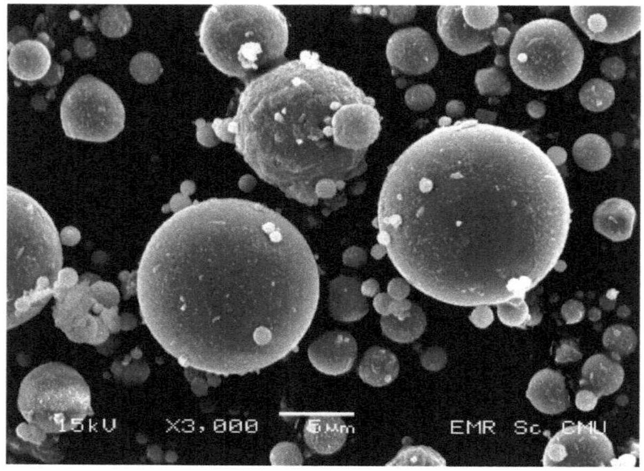

Figure 2. The SEM images of the FA in this study.

Table 1. Chemical compositions of the raw materials.

Chemical Composition (% by Total Weight)	FA	HL
SiO_2	32.47	0.6
Al_2O_3	15.84	0.5
Fe_2O_3	14.64	0.1
CaO	24.83	74.5
SO_3	4.3	-
P_2O_5	0.31	-
K_2O	1.87	-
MnO_2	0.2	-
TiO_2	0.46	-
MgO	3	1.1
Na_2O	1.99	-
LOI *	0.09	23.0

* LOI—loss of ignition.

Meanwhile, the HL contained CaO compounds with high loss on ignition (LOI High LOI can imply the evaporation of water and decomposition of $Ca(OH)_2$ at 512 °C The specific gravity and blain fineness of HL were 2.16 and 1600 cm^2/g, respectivel Commercial-grade NaOH at 98% purity was used as an alkali activator. The NaOH wa ground finely (approximately 2.00 mm particles) before being added to the mixtures.

In this study, the slow-setting CAE (CSS-1h) used as an additive was provided b the Asian Asphalt Co. of Thailand. The CAE with a positive electrostatic charge wa suspended in hydrochloric acid solution, containing 67.3% solid residue asphalt by tota modified asphalt weight. The properties of the CAE are shown in Table 2.

Based on Table 3, the hybrid alkali-activated mixtures with different CAE content were prepared using the binder-sand ratio (b/s) of 1.00:2.75, in which the sand with specifi gravity was 2.65 and the gradation shown in Figure 3. The control mixture as the H-AAM sample consisted of FA, HL, and NaOH, following Table 3. According to the previous stud the constituents of the H-AAM binder was validated the potential on cement replacemen in road base construction [45]. Proportions of CAE to the hybrid binder of 0, 0.05, 0.10, an 0.20 were used in conjunction with the added water was water-to-binder ratio (w/b) c 0.45:1.00. For the microstructure and structural formation analyses, hybrid alkali-activate pastes with different CAE contents were prepared.

Table 2. Properties of the CAE.

Property	Unit	Result
Density	(g/cm^3)	
Storage (24 h)	%	1.2
Sieve (1.18 mm)	%	0.00
Particle charge		Positive
Solid content	%	67.3
Softening point, °C	°C	61.7
Elastic recovery at 25 °C	%	64
Penetration at 25 °C	mm	65
Ductility at 25 °C	cm	95
Solubility	%	99.34

Table 3. Mortar mix design used in this study.

Mix	Material Proportions (% of Binder's Weight)				w/b	b/s	a/b Ratio
	FA	HL	SH	CSS-1h			
Control	77.3	19.3	3.4	-	0.45:1.00	1.00:2.75	-
CAE 05	77.3	19.3	3.4	5.0			0.05
CAE 10	77.3	19.3	3.4	10.0			0.10
CAE 20	77.3	19.3	3.4	20.0			0.20

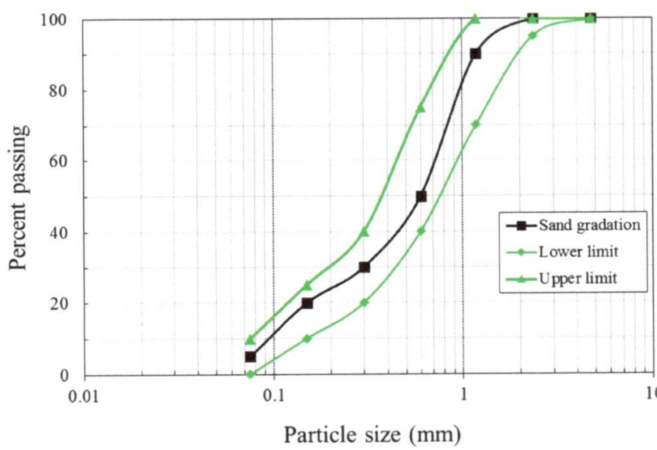

Figure 3. Sand gradation used for mortar samples.

During the mixing procedure, the dry binders were added to the saturated sand and mixed homogeneously. Subsequently, the water was added, followed by the CAE. The resulting materials were then mixed for 180 s with a Hobart mixer.

3.2. Methodology and Experiments

Figure 4 presents the perspective diagram for the research to clarify the preparation of the materials and the experiments. Mortar samples were examined to obtain the physical properties of the target specimens. Meanwhile, microstructural and structural formation analyses were performed on the hydraulic pastes.

In this study, the binder combined with varying CAE contents was examined. The strength, water absorption, elasticity, and toughness of the resulting materials were determined. All test specimens were prepared in fresh mortar form. Furthermore, microstructural analyses were conducted to support all experimental test results.

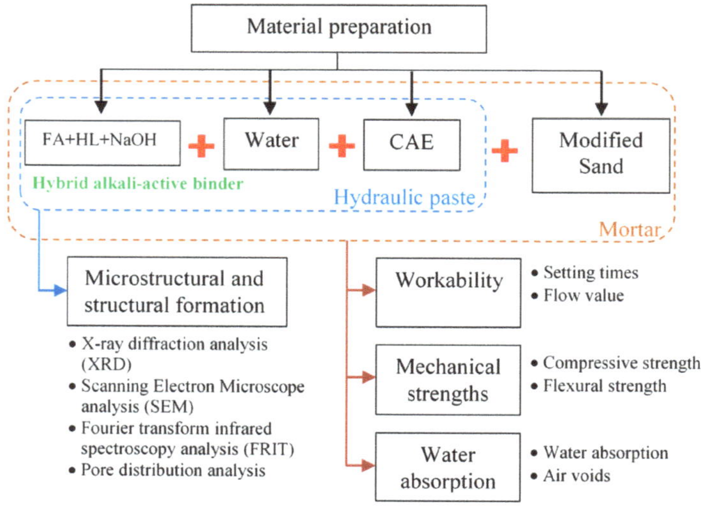

Figure 4. Perspective diagram of the research.

3.2.1. Flow and Setting Time Tests

Mortar samples were investigated for their workability within 90 s after mixin According to ASTM C1437 [48], the flow was tested, while setting time was measured via modified Vicat needle according to ASTM C807 [49]. Accordingly, the effect of free wat in the CAE on the hardening process of the H-AAM was investigated through the settin time tests.

3.2.2. Mechanical Strength Tests

The compressive and flexural strengths were investigated based on BSEN 196-1 [5 The mortar specimens were shaped into prisms measuring 40 mm × 40 mm × 160 m They were cured at ambient temperature and wrapped to prevent moisture evaporatio then tested on days 7 and 28 with the 250 kN–universal compression testing machin (UTM) of Controls-65-L12L2 Model (CONTROLS S.p.A., Milan, Italy). The force rates f the compressive and flexural strength tests were 2.4 kN/s and 0.5 kN/s, respectively. Th two specimens were examined for flexural strength. The four halves of these failed prism were tested continuously for compressive strength. The average strengths were recorde for each mixture.

3.2.3. Water Absorption and Air Voids Tests

The water absorption and air voids were evaluated based on ASTM C642 [51]. Befor the test, the prism-shaped specimens should be air-cured for 28 days. Each test sampl was kept in the oven at 105 °C for 24 h and cooled in dry air over the next 24 h. Then th oven-dried masses were measured. They were next immersed in water for not less tha 48 h, then the surface-dry mass in the air was recorded after they were removed from th water. At this point, the water absorption was calculated according to Equation (1).

$$\text{Absorption}(\%) = \frac{(W_{im} - W_{oven})}{W_{im}} \times 100 \quad (1$$

where Absorption (%) = water absorption of the specimen (%), W_{oven} = mass of oven-drie sample in the air (g), and W_{im} = mass of surface-dry sample in the air after immersion (g

3.2.4. Toughness Characteristics Definitions

The toughness, i.e., the ratio of the flexural to compressive strength, was used to describe the energy absorption of the material. It was calculated by using the data from the mechanical strength tests conducted on days 7 and 28.

3.2.5. X-ray Diffraction Analysis (XRD)

The mineral compositions of the tested pastes were analyzed using XRD (Rigaku SmartLab, 10–50 2θ, Applied Rigaku Technologies, Inc., Austin, TX, USA). A single crystal X-ray diffractometer was used to capture the XRD patterns of 28-day-old pastes in this study.

3.2.6. Scanning Electron Microscope Analysis (SEM)

SEM (JEOL JSM-5910LV, gold coated, WD: 11 mm, EHT: 15 kV, JEOL Inc., Pleasanton, CA, USA) was used to observe the pastes' microstructures. Harden specimens were coated with gold before the examination.

3.2.7. Fourier Transform Infrared Spectroscopy Analysis (FTIR)

The functional groups of the chemical compound present in the samples were examined with FTIR (Thermo Nicolet 6700, 4000–400 cm^{-1}, Thermo Fisher Scientific Inc., Waltham, MA, USA). The 28-day-old pastes were ground into powder smaller than 75 microns and mixed with KBr powder. FTIR spectrometer was used to record their spectra information as percentages of transmittance and wavenumbers.

3.2.8. Pore Distribution Analysis

The surface area analysis was conducted using the Brunauer, Emmett, and Teller (BET, Quantachrome Autosorb 1 MP, N_2 gas, 99 points Adsorption–desorption, Anton Paar GmbH, Graz, Austria) method [52] on the adsorption isotherm. The pore size distribution was determined from the desorption isotherm via the Barrett, Joyner, and Halenda (BJH) method [53]. The Nitrogen adsorption and desorption isotherms were measured at 77.35 K while using a Quatachrome Nova station to test the 28-day-old pastes with volumes of 1 cm^3. Prior to this test, each sample was degassed at 105 °C for 5 h.

4. Results and Discussion

4.1. Flow and Setting Time Tests

Figure 5 shows the test flow values and setting times of mortars with different CAE contents. The results indicate that the hybrid alkali-activated mortar's initial and final setting time values were significantly delayed with an increase in CAE content—as was true for the flow values. The mortar with 20% CAE provided the highest flow value, which was much larger than the value for the control mortar, and the setting time of this mortar was approximately twice that of the control mortar. The increase in flow was due to the extra water, which was 35% by the total amount of the CAE. From Figure 5a, the mortar with CAE contents of 5 and 10% had 187 and 275 min initial setting times, respectively. The final setting values were 458 and 468 min, respectively, both higher than 326 min recorded for the control mortar.

From Figure 5b, these results indicated that adding CAE could increase the flow values of the mixture. The CAE enhances the fluidity of the paste by providing more water for the mixing matrix through its demulsification. CAE can provide this extra water due to its free water content of approximately 35%. The emulsifier and excess water from the CAE accumulate in the free water according to the water-to-binder ratio, resulting in the excess water for the hydration of the pozzolanic materials, thus a prolonged hardening process during mixing. The H-AAM and CAE composite behaved like that of the cement-AE material and revealed the retardation effect of CAE, mainly caused by the high liquidity of the mixtures [24,25]. Hence, the addition of CAE can enhance the properties of this H-AAM remarkably.

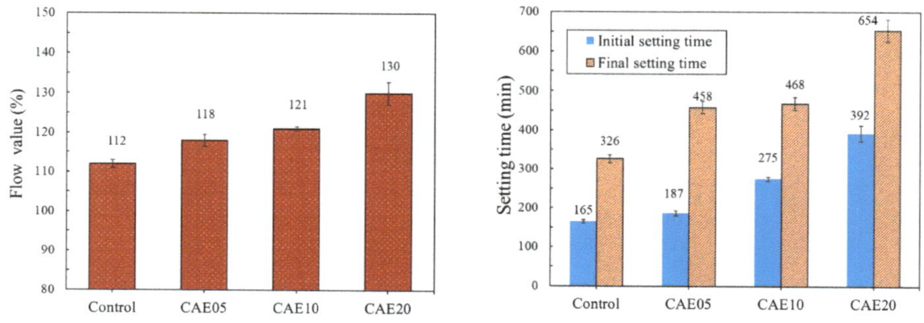

(a) (b)

Figure 5. Results for the hybrid alkali-activated mortars with different CAE contents in terms of (**a**) setting times and (**b**) flow values.

4.2. Mechanical Properties

Figure 6a illustrates the compressive strength of the test samples with different CA contents. A higher CAE content resulted in a lower compressive strength. With mor free water available with a higher CAE content, the water-to-cement ratio of the mixtu increased, leading to a decrease in the overall compressive strength of the material. Furthe more, asphalt droplets in the mixture can cover pozzolan particles, delaying the reactio and forming weak spots [54,55]. In terms of flexural strength, Figure 6b shows the relativel small decline in flexural strength with higher CAE content for both seven- and 28-da curing periods. Note the relatively small gaps in flexural strength between the seven- an 28-day curing periods, regardless of CAE content. It seems that the CAE content did nc play a leading role in developing flexural strength in the materials.

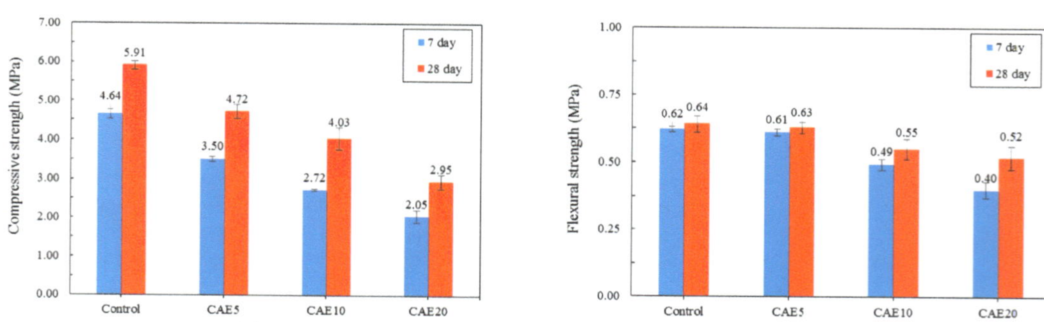

(a) (b)

Figure 6. Results for the hybrid alkali-activated mortars with different CAE contents in terms of (**a**) compressive strength and (**b**) flexural strength.

4.3. Water Absorption

Figure 7 displays the results for the water absorption of the H-AAM mortars wit different CAE contents. More CAE in the hybrid- alkali-activated matrix could reduc water absorption since the asphalt droplets in the CAE can cover and partly fill in the past matrix, resulting in a barrier to water immigration along the mortar's entire surface [56]. can be explained by the presence of CAE-induced asphalt droplets in the matrix pluggin in the free space between the binder and sand, resulting in impermeable space [37]. Fu thermore, fewer air voids in the material matrix can lead to a considerable reduction in ai porosity, resulting in improved surface barriers, the coated surface by the asphalt binder to protect the cementitious subtracts' water consumption, to contact with water caused by the hydrophobicity of the emulsified asphalt coating [34,57].

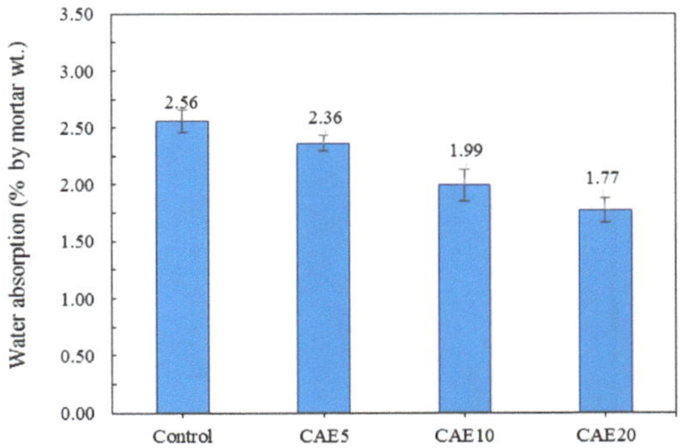

Figure 7. Water absorption values of the hybrid alkali-activated mortars with different CAE contents.

4.4. Toughness Characteristics

Figure 8 reveals the toughness characteristics of the H-AAMs with varying CAE contents. Note that adding CAE can increase the toughness of the hybrid alkali-activated mortars, with the highest CAE contents resulting in the most outstanding toughness. A material's toughness generally indicates the brittleness and ductility of the material. Relatively higher material toughness implies flexibility, and the lower one indicates brittleness. Based on the toughness results exhibited in Figure 8, the presence of CAE in the H-AAMs provides ductility. These results agree with previous research, which indicated that more flexibility is present in polymer-modified cement-based materials [58,59]. Notably, more excellent material toughness indicates greater energy absorption, which is required for road pavement materials [60,61]. Therefore, CAE can generate better energy absorption behavior in the H-AAM, making it an ideal flexible road pavement material.

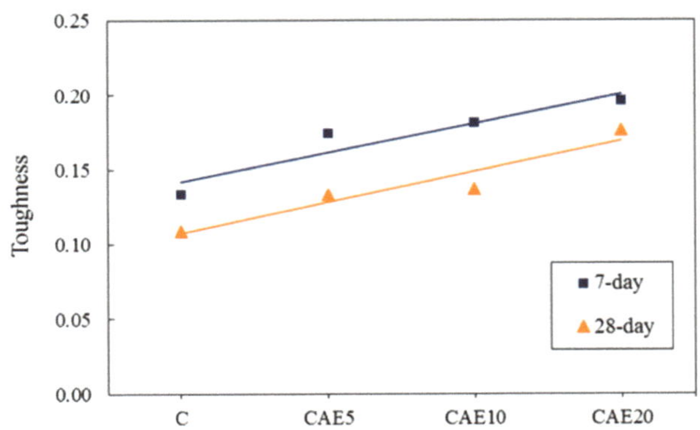

Figure 8. The linear lines of flexural-to-compressive strength ratio.

4.5. SEM Analysis

Figure 9 presents the morphologies of the hybrid alkali-activated pastes with different CAE contents obtained through the SEM analyses. Figure 9a shows the relatively smooth and continuous texture of the hybrid alkali-activated paste as a benchmark, indicating the large pieces of the solid cementitious product of the C-S-H gels. Adding CAE to the hybrid alkali-activated matrix did affect the texture. In Figure 9b–d, it is seen that a CAE layer surrounds the cluster of relatively smaller FA and HL particles. At a magnification of 5000×, agglomeration of material was found in control with 20 microns. Small sizes of particles were seen 5–10 microns with the increase in CAE content. It has been reporte that different reaction products, including C-S-H gel, AFt, and CH were formed in pas without CAE, and pores were filled by the hydration products [35]. With CAE, asphalt fil generated on the surface of fly ash particles, retardation of reactions occurred, and sma sizes of fly ash particles were observed.

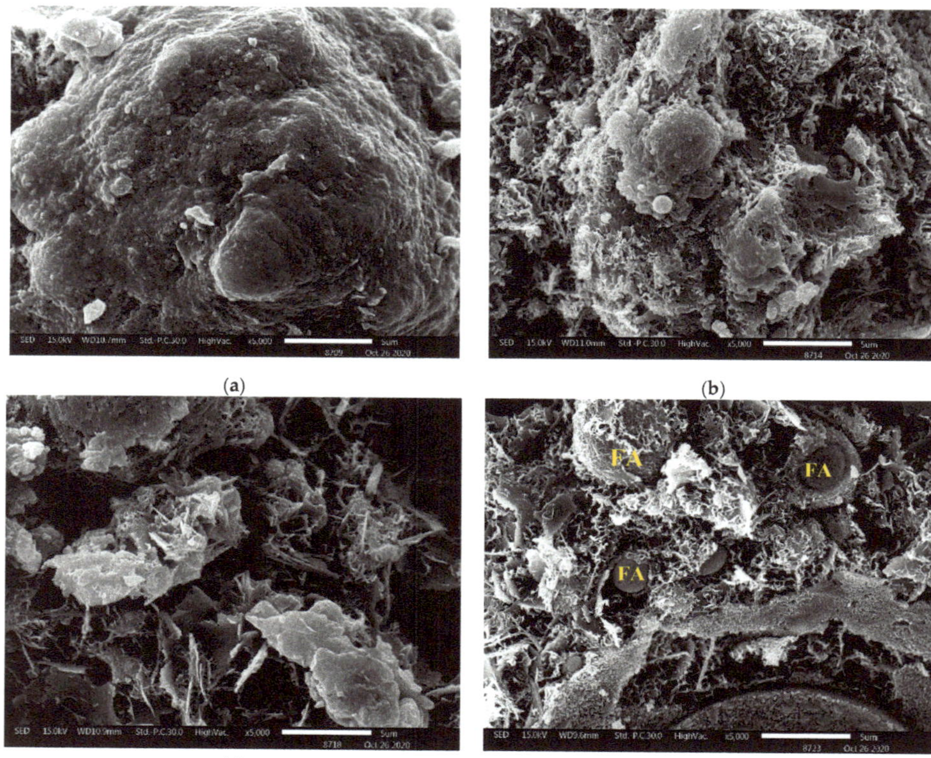

(a) (b) (c) (d)

Figure 9. The SEM images of the pastes with different CAE contents in terms of (**a**) control, (**b**) CAE05, (**c**) CAE10, and (**d**) CAE20.

This clustering pattern of the CAE-coated particles indicates that the CAE addition di interrupt and retard the continuous cementitious product formation due to the pozzolani reactions and alkali activations. Some unreacted FA particles are surrounded by CAE filr layers connected to the hybrid alkali-activated paste. The microstructures indicate tha more CAE contents had retarding effects on the chemical reaction needed for the hybri alkali-activated formation. Furthermore, the 20% CAE addition provided a smoothe consolidation in the matrices than lesser amounts of CAE due to CAE's emulsifyin property. In addition, CAE film, which was interlocked with and surrounded the hybri alkali-activated particles, can plug and reduce some void spaces in the material matri resulting in the mortar's lowest water absorption rate.

CAE layers covered the hybrid alkali-activated binder. These CAE layers governe the overall flexural resistance of the material. However, these CAE layers also played vital role in the compressive strength reduction observed as the CAE content increase (see Figure 6). The thicker CAE layers produced by more CAE contents provide lowe compressive load resistance.

4.6. XRD Analysis

Figure 10 illustrates the XRD patterns of the H-AAMs after 28 days of curing. The broad hump in the region of 26–38° 2θ was found in all samples and shows the amorphous phase of the material. The phases consist mainly of calcite, Portlandite, C-S-H, and N-(A)-S-H [9,12,14,62,63]. The control paste showed the peaks of C-S-H and N-(A)-S-H gels more clearly than the others with the CAE mixed in, indicating that the CAE retarded the alkali-activated reactions of the mixtures. Consequently, reductions in mechanical strength ensued. This result is similar to that obtained for the cement-AE system in previous research [42].

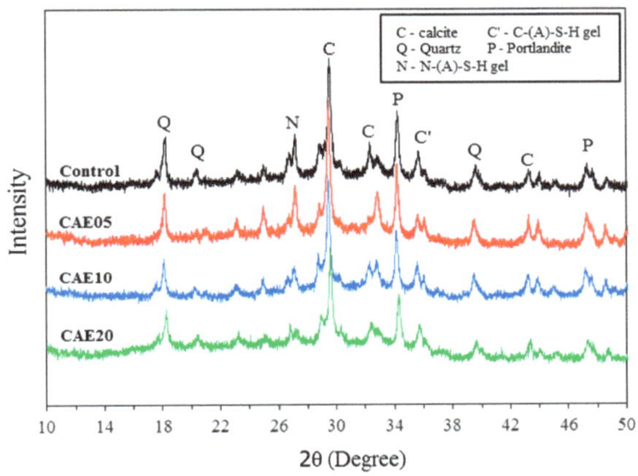

Figure 10. The XRD patterns of the hybrid alkali-activated pastes with different CAE contents.

4.7. FTIR Analysis

The FTIR results for the hybrid alkali-activated pastes with CAE contents are presented in Figure 11. Note that the intensities of the bands at about 835 and 960 cm^{-1} decreased as the CAE content increased. These two bands correspond to the asymmetric and symmetric stretching vibrations of the Si-O bonds, which can be considered part of the tetrahedrons in Q_1 and Q_2 from the structural formation of C-S-H [64,65]. The addition of CAE diminished gel formation. In contrast, the intensities of the bands at about 1375 and 2870 cm^{-1} increased as the CAE increased from 5 to 20% in the pastes. These wavenumbers refer to the stretching vibrations of the N-H bonds in ammonium chloride (NH_4Cl), i.e., the emulsifier used in the CAE [66].

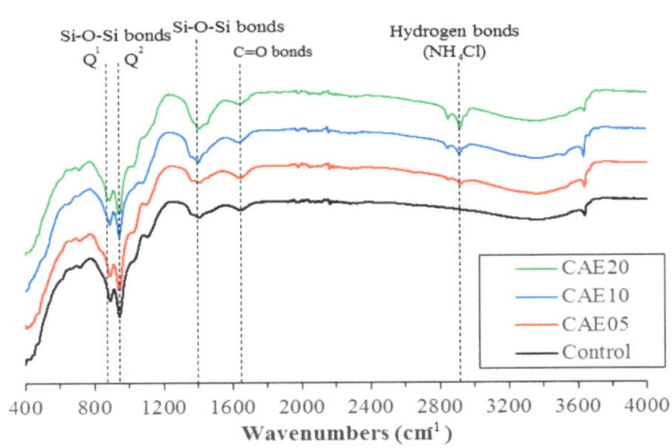

Figure 11. The FTIR spectra of the hybrid alkali-activated pastes with different CAE contents.

Furthermore, the C=O bonds indicated by the 1630 cm^{-1} band are from the hydr carbon chains in the asphalt binder. Note that the band's intensity gradually increases the CAE addition increases from 5 to 20%. Based on these FTIR results, it is confirme that the CAE did not react chemically with the pozzolanic material and AAMs. Hence, th FTIR results can validate CAE's effect on this H-AAM in terms of obstructing the stru tural formations harmful to the material's strength characteristics, as in the cement-base materials [38,67].

4.8. Pore Structure Analysis

Table 4 shows the results of BET surface area and total pore volumes following BJ pore size distribution desorption. The surface area of the hybrid alkali-activated mixtur decreased with the addition of CAE. The H-AAM with the 20% CAE addition provided th lowest surface area and had a high average pore diameter due to the asphalt droplets i CAE occupying pore spaces in the hybrid alkali-activated matrix. CAE's asphalt drople and emulsifying molecules refine the pores and are inserted into the hybrid alkali-activate matrix's free space. The newly tiny pores are perhaps aggregated, leading to an increase i the average pore diameter. Consequently, CAE prevents extraneous water from penetratin this H-AAM by reducing the total number of pores.

Table 4. BET surface and total pore volume results.

Mixture	Average Diameter (nm)	Surface Area (m^2/g)	Total Pore Volume (cm^3/g)
Control	35.752	12.705	0.114
CAE05	39.481	10.107	0.112
CAE10	44.603	9.520	0.094
CAE20	48.812	6.955	0.085

Overall, these BJH pore analyses support the mortar test and validate that the additio of CAE plays a vital role in improving the water resistance of the H-AAM. It refined th pores and reduced the total pore volume of the material. The mixture with the highe CAE content would be excellent in terms of water and moisture resistance since the CA reduced the average size of the capillary pores.

5. Conclusions

This study investigated the influence of CAE on the physical properties and structur formation of a H-AAM. According to the results, the following conclusions can be made

Augmentation with CAE content gradually prolonged both the initial and final settin times of the hybrid alkali-activated mortar as it increased from 0 to 20% of the weight c the binder. Moreover, the study material's compressive and flexural strength decrease markedly as the CAE content increased, resulting in more free water and reaction barrier to the hardening process, as confirmed by the microstructural analyses conducted i this study.

The flexural to compressive strength ratio increased with CAE content up to 20%, an the toughness of the hybrid alkali-activated mortar improved. Enhancements in materi flexibility are possible through the addition of CAE.

CAE hinders cementitious formation resulting from the pozzolanic reactions an alkali activations in the hybrid alkali-activated pastes. This study revealed decreases in th cementitious product's peak in the XRD tests and the number of tetrahedrons detected i the FTIR tests. The SEM images showed some signs of asphalt films surrounding hybri alkali-activated particles and even some unreacted FA particles, indicating incomplet chemical reactions in the study material's matrix.

The water absorption of the hybrid alkali-activated mortar decreased with the CA content from 0 to 20%. The CAE increases the average pore size of the hybrid alkal activated paste while decreasing the surface area, hindering the cementitious formation of the study material.

Based on this study's results, the H-AAM with a CAE addition could represent a sustainable construction material with water repulsion and flexibility, ideal for pavement base stabilization. This material should be subjected to further investigations concerning its environmental and economic benefits. The performance-based parameters such as resilient modulus, shrinkage, and cyclic tensile modulus are recommended for further studies of this new material to understand specific pavement properties based on more realistic conditions.

Author Contributions: Conceptualization, P.J.; methodology, T.B.; validation, P.J., T.S., U.R., W.T., and S.T.; formal analysis, T.B. and S.T.; investigation, T.B.; writing—original draft preparation, T.B. and P.J.; writing—review and editing, T.S., U.R., W.T., and S.T.; visualization, T.B.; supervision, P.J. and T.S.; project administration, P.J. All authors have read and agreed to the published version of the manuscript.

Funding: This research was funded by the Thailand Research Fund (TRF) under the "Research and Researchers for Industries (RRi)" program through grant number PHD62I0027.

Institutional Review Board Statement: Not applicable.

Informed Consent Statement: Not applicable.

Data Availability Statement: The data presented in this study are available on request from the corresponding author. The data are not publicly available due to a restriction in the funding agreement.

Acknowledgments: The authors would like to extend their gratitude and acknowledgments to Chiang Mai University for the support received for this study. The first author is honored by and thankful for the financial support of the Thailand Research Fund (TRF) under the "Research and Researchers for Industries (RRi) [grant number PHD62I0027]," collaborating with Siam Cement Group Public Company Limited. The academic members of Burapha University and King Mongkut's University of Technology Thonburi (KMUTT) are also recognized for their academic support.

Conflicts of Interest: The authors declare no conflict of interest.

References

1. Aydin, E.; Arel, H.Ş. Characterization of high-volume fly-ash cement pastes for sustainable construction applications. *Constr. Build. Mater.* **2017**, *157*, 96–107. [CrossRef]
2. Tang, S.W.; Cai, X.H.; He, Z.; Shao, H.Y.; Li, Z.J.; Chen, E. Hydration process of fly ash blended cement pastes by impedance measurement. *Constr. Build. Mater.* **2016**, *113*, 939–950. [CrossRef]
3. O'Looney, D.; Pavía, S. A study of the functionality of hydrated lime as an admixture. *J. Mater. Sci. Res.* **2014**, *4*, 1–11. [CrossRef]
4. Nguyen, T.B.T.; Chatchawan, R.; Saengsoy, W.; Tangtermsirikul, S.; Sugiyama, T. Influences of different types of fly ash and confinement on performances of expansive mortars and concretes. *Constr. Build. Mater.* **2019**, *209*, 176–186. [CrossRef]
5. Chummuneerat, S.; Jitsangiam, P.; Nikraz, H. Performances of hydrated cement treated crushed rock base for Western Australian roads. *J. Traffic Transp. Eng.* **2014**, *1*, 432–438. [CrossRef]
6. Phoo-ngernkham, T.; Hanjitsuwan, S.; Li, L.-Y.; Damrongwiriyanupap, N.; Chindaprasirt, P. Adhesion characterisation of Portland cement concrete and alkali-activated binders. *Adv. Cem. Res.* **2019**, *31*, 69–79. [CrossRef]
7. Singh, B.; Ishwarya, G.; Gupta, M.; Bhattacharyya, S.K. Geopolymer concrete: A review of some recent developments. *Constr. Build. Mater.* **2015**, *85*, 78–90. [CrossRef]
8. Imbabi, M.S.; Carrigan, C.; McKenna, S. Trends and developments in green cement and concrete technology. *Int. J. Sustain. Built Environ.* **2012**, *1*, 194–216. [CrossRef]
9. Eiamwijit, M.; Pachana, K.; Kaewpirom, S.; Rattanasak, U.; Chindaprasirt, P. Comparative study on morphology of ground sub-bituminus FBC fly ash geopolymeric material. *Adv. Powder Technol.* **2015**, *26*, 1053–1057. [CrossRef]
10. Garcia-Lodeiro, I.; Palomo, A.; Fernández-Jiménez, A.; Macphee, D.E. Compatibility studies between N-A-S-H and C-A-S-H gels. Study in the ternary diagram $Na_2O–CaO–Al_2O_3–SiO_2–H_2O$. *Cem. Concr. Res.* **2011**, *41*, 923–931. [CrossRef]
11. Somna, R.; Jaturapitakkul, C.; Mada, A.M. Effect of ground fly ash and ground bagasse ash on the durability of recycled aggregate concrete. *Cem. Concr. Compos.* **2012**, *34*, 848–854. [CrossRef]
12. Chindaprasirt, P.; Jenjirapanya, S.; Rattanasak, U. Characterizations of FBC/PCC fly ash geopolymeric composites. *Constr. Build. Mater.* **2014**, *66*, 72–78. [CrossRef]
13. Nusit, K.; Jitsangiam, P. Damage Behavior of Cement-Treated Base Material. *Procedia Eng.* **2016**, *143*, 161–169. [CrossRef]

14. Chindaprasirt, P.; Rattanasak, U.; Vongvoradit, P.; Jenjirapanya, S. Thermal treatment and utilization of Al-rich waste in hig calcium fly ash geopolymeric materials. *Int. J. Miner. Metall. Mater.* **2012**, *19*, 872–878. [CrossRef]
15. Jitsangiam, P.; Suwan, T.; Pimraksa, K.; Sukontasukkul, P.; Chindaprasirt, P. Challenge of adopting a relatively low-strength ar self-cured geopolymer for road construction application: A review and primary laboratory study. *Int. J. Pavement Eng.* **2019**, 2 1454–1468. [CrossRef]
16. Yang, K.-H.; Mun, J.-S.; Cho, M.-S. Effect of curing temperature histories on the compressive strength development of hig strength concrete. *Adv. Mater. Sci. Eng.* **2015**, *2015*, 965471. [CrossRef]
17. Jitsangiam, P.; Boonserm, K.; Phenrat, T.; Chummuneerat, S.; Chindaprasirt, P.; Nikraz, H. Recycled concrete aggregates roadways: Laboratory examination of self-cementing characteristics. *J. Mater. Civ. Eng.* **2015**, *27*, 04014270. [CrossRef]
18. Uthaichotirat, P.; Sukontasukkul, P.; Jitsangiam, P.; Suksiripattanapong, C.; Sata, V.; Chindaprasirt, P. Thermal and soun properties of concrete mixed with high porous aggregates from manufacturing waste impregnated with phase change material. *Build. Eng.* **2020**, *29*, 101111. [CrossRef]
19. Xu, Y.; Yang, G.; Zhao, H. Compressive strength gain behavior and prediction of cement-stabilized macadam at low temperatu curing. *J. Adv. Transp.* **2020**, *2020*, 2469436. [CrossRef]
20. Jitsanigam, P.; Biswas, W.K.; Compton, M. Sustainable utilization of lime kiln dust as active filler in hot mix asphalt with moistu damage resistance. *Sustain. Mater. Technol.* **2018**, *17*, e00071. [CrossRef]
21. Sebesta, S. Use of microcracking to reduce shrinkage cracking in cement-treated bases. *Transp. Res. Rec.* **2005**, *1936*, 3–1 [CrossRef]
22. Nusit, K.; Jitsangiam, P.; Kodikara, J.; Bui, H.H.; Leung, G.L.M. Advanced characteristics of cement-treated materials with respe to strength performance and damage evolution. *J. Mater. Civ. Eng.* **2017**, *29*, 04016255. [CrossRef]
23. Wang, F.; Liu, Z.; Wang, T.; Hu, S. A novel method to evaluate the setting process of cement and asphalt emulsion in CA mort *Mater. Struct.* **2007**, *41*, 643–647. [CrossRef]
24. Peng, J.; Deng, D.; Yuan, Q.; Liu, Z.; Fang, L. Study of the rheological behavior of fresh cement emulsified asphalt paste. *Cons Build. Mater.* **2014**, *66*, 348–355. [CrossRef]
25. Wang, Z.; Shu, X.; Rutherford, T.; Huang, B.; Clarke, D. Effects of asphalt emulsion on properties of fresh cement emulsifie asphalt mortar. *Constr. Build. Mater.* **2015**, *75*, 25–30. [CrossRef]
26. Bołtryk, M.; Małaszkiewicz, D. Application of anionic asphalt emulsion as an admixture for concrete. *Constr. Build. Mater.* **201** *40*, 556–565. [CrossRef]
27. Liu, Y.; Wang, F.; Hu, S.; Liu, M. Compatibility of repair materials with substrate low-modulus cement and asphalt mortar (C mortar). *Constr. Build. Mater.* **2016**, *126*, 304–312. [CrossRef]
28. Leiben, Z.; Wang, X.; Wang, Z.; Yang, B.; Tian, Y.; He, R. Damping characteristics of cement asphalt emulsion mortars. *Cons Build. Mater.* **2018**, *173*, 201–208. [CrossRef]
29. D'Angelo, G.; Thom, N.; Lo Presti, D. Bitumen stabilized ballast: A potential solution for railway track-bed. *Constr. Build. Mat* **2016**, *124*, 118–126. [CrossRef]
30. Baghini, M.S.; Ismail, A.; Bin Karim, M.R. Evaluation of cement-treated mixtures with slow setting bitumen emulsion as bas course material for road pavements. *Constr. Build. Mater.* **2015**, *94*, 323–336. [CrossRef]
31. Jenkins, K.J.; Collings, D.C. Mix design of bitumen-stabilised materials—South Africa and abroad. *Road Mater. Pavement* **2016**, 1 331–349. [CrossRef]
32. Huan, Y.; Siripun, K.; Jitsangiam, P.; Nikraz, H. A preliminary study on foamed bitumen stabilisation for Western Australia pavements. *Sci. Res. Essays* **2010**, *5*, 3687–3700.
33. Ouyang, J.; Li, H.; Han, B. The rheological properties and mechanisms of cement asphalt emulsion paste with different charg types of emulsion. *Constr. Build. Mater.* **2017**, *147*, 566–575. [CrossRef]
34. Ouyang, J.; Hu, L.; Li, H.; Han, B. Effect of cement on the demulsifying behavior of over-stabilized asphalt emulsion durin mixing. *Constr. Build. Mater.* **2018**, *177*, 252–260. [CrossRef]
35. Li, W.; Mao, Z.; Xu, G.; Chang, H.; Hong, J.; Zhao, H.; Xu, J.; Liu, Z. Study on the early cement hydration process in the presenc of cationic asphalt emulsion. *Constr. Build. Mater.* **2020**, *261*, 120025. [CrossRef]
36. Lu, C.-T.; Kuo, M.-F.; Shen, D.-H. Composition and reaction mechanism of cement–asphalt mastic. *Constr. Build. Mater.* **2009**, 2 2580–2585. [CrossRef]
37. Dareyni, M.; Mohammadzadeh Moghaddam, A.; Delarami, A. Effect of cationic asphalt emulsion as an admixture on transpo properties of roller-compacted concrete. *Constr. Build. Mater.* **2018**, *163*, 724–733. [CrossRef]
38. Li, W.; Hong, J.; Zhu, X.; Yang, D.; Bai, Y.; Liu, J.; Miao, C. Retardation mechanism of anionic asphalt emulsion on the hydratio of Portland cement. *Constr. Build. Mater.* **2018**, *163*, 714–723. [CrossRef]
39. Nanthavisit, P. Influence of cement and asphalt emulsion ratios on cement-asphalt emulsion mortar. *Int. J. Geomate* **2019**, 1 77–84. [CrossRef]
40. Yan, X.; Wang, Z.; Rao, M.; Li, M. Investigation of cement-emulsified asphalt in plastic concrete. *Adv. Mater. Sci. Eng.* **2018**, *201* 3929682. [CrossRef]
41. Rutherford, T.; Wang, Z.; Shu, X.; Huang, B.; Clarke, D. Laboratory investigation into mechanical properties of cement emulsifie asphalt mortar. *Constr. Build. Mater.* **2014**, *65*, 76–83. [CrossRef]

2. Du, S. Interaction mechanism of cement and asphalt emulsion in asphalt emulsion mixtures. *Mater. Struct.* **2014**, *47*, 1149–1159. [CrossRef]
3. Rozhina, E.; Ishmukhametov, I.; Nigamatzyanova, L.; Akhatova, F.; Batasheva, S.; Taskaev, S.; Montes, C.; Lvov, Y.; Fakhrullin, R. Comparative Toxicity of Fly Ash: An In Vitro Study. *Molecules* **2021**, *26*, 1926. [CrossRef]
4. Tian, Q.; Binglin, G.; Nakama, S.; Sasaki, K. Distributions and Leaching Behaviors of Toxic Elements in Fly Ash. *ACS Omega* **2018**, *3*, 13055–13064. [CrossRef]
5. Bualuang, T.; Jitsangiam, P.; Suwan, T.; Rattanasak, U.; Jakrawatana, N.; Kalapat, N.; Nikraz, H. Non-OPC binder based on a hybrid material concept for sustainable road base construction towards a low-carbon society. *J. Mater. Res. Technol.* **2021**, *14*, 374–391. [CrossRef]
6. DOH (Department of Highways Thailand). *DH-S204/2000 Standard of Soil Cement Base*; Department of Highways: Bangkok, Thailand, 2000. (In Thai)
7. DRR (Department of Rural Roads). *DRR244-2013 Standard of Soil Cement Subbase*; Department of Rural Roads: Bangkok, Thailand, 2013. (In Thai)
8. ASTM International. ASTM C1437-15. In *Standard Test Method for Flow of Hydraulic Cement Mortar*; ASTM International: West Conshohocken, PA, USA, 2015.
9. ASTM International. ASTM C807-20. In *Standard Test Method for Time of Setting of Hydraulic Cement Mortar by Modified Vicat Needle*; ASTM International: West Conshohocken, PA, USA, 2020.
0. British Standards Institution. BS EN 196-1. In *Methods of Testing Cement. Determination of Strength*; British Standards Institution: London, UK, 2016.
1. ASTM International. ASTM C642-13. In *Standard Test Method for Density, Absorption, and Voids in Hardened Concrete*; ASTM International: West Conshohocken, PA, USA, 2013.
2. Brunauer, S.; Emmett, P.H.; Teller, E. Adsorption of gases in multimolecular layers. *J. Am. Chem. Soc.* **1938**, *60*, 309–319. [CrossRef]
3. Barrett, E.P.; Joyner, L.G.; Halenda, P.P. The determination of pore volume and area distributions in porous substances. Computations from Nitrogen isotherms. *J. Am. Chem. Soc.* **1951**, *73*, 373–380. [CrossRef]
4. Liu, B.; Liang, D. Effect of mass ratio of asphalt to cement on the properties of cement modified asphalt emulsion mortar. *Constr. Build. Mater.* **2017**, *134*, 39–43. [CrossRef]
5. Le, T.H.M.; Park, D.-W.; Seo, J.-W. Evaluation of the mechanical properties of cement asphalt mortar with quick hardening admixture for railway maintenance. *Constr. Build. Mater.* **2019**, *206*, 375–384. [CrossRef]
6. Wang, Q.; Pan, S.; Zhan, D.; Qi, G.; Song, Y.; Wang, D. A study of the transport properties of emulsified asphalt powder-modified cement mortar. *Constr. Build. Mater.* **2020**, *248*, 118650. [CrossRef]
7. Bhatt, A.; Priyadarshini, S.; Acharath Mohanakrishnan, A.; Abri, A.; Sattler, M.; Techapaphawit, S. Physical, chemical, and geotechnical properties of coal fly ash: A global review. *Case Stud. Constr. Mater.* **2019**, *11*, e00486. [CrossRef]
8. Parghi, A.; Shahria Alam, M. Effects of curing regimes on the mechanical properties and durability of polymer-modified mortars—An experimental investigation. *J. Sustain. Cem.-Based Mater.* **2016**, *5*, 324–347. [CrossRef]
9. Wang, R.; Wang, P.-M.; Li, X.-G. Physical and mechanical properties of styrene–butadiene rubber emulsion modified cement mortars. *Cem. Concr. Res.* **2005**, *35*, 900–906. [CrossRef]
0. Purwana, Y.; Nikraz, H.; Jitsangiam, P. Experimental Study of Suction-Monitored CBR Test on Sand-Kaolin Clay Mixture. *Int. J. Geomate* **2012**, *3*, 419–422.
1. Sounthararajah, A.; Bui, H.; Nguyen, N.; Jitsangiam, P.; Kodikara, J. Early-age fatigue damage assessment of cement-treated bases under repetitive heavy traffic loading. *J. Mater. Civ. Eng.* **2018**, *30*, 04018079. [CrossRef]
2. Chindaprasirt, P.; Rattanasak, U. Fire-resistant geopolymer bricks synthesized from high-calcium fly ash with outdoor heat exposure. *Clean Technol. Environ. Policy* **2018**, *20*, 1097–1103. [CrossRef]
3. Chindaprasirt, P.; Rattanasak, U. Synthesis of polypropylene fiber/high-calcium fly ash geopolymer with outdoor heat exposure. *Clean Technol. Environ. Policy* **2017**, *19*, 1985–1992. [CrossRef]
4. Fernández-Jiménez, A.; Palomo, A. Mid-infrared spectroscopic studies of alkali-activated fly ash structure. *Microporous Mesoporous Mater.* **2005**, *86*, 207–214. [CrossRef]
5. Yu, P.; Kirkpatrick, R.; Poe, B.; McMillan, P.; Cong, X. Structure of calcium silicate hydrate (C-S-H): Near-, mid-, and far-infrared spectroscopy. *J. Am. Ceram. Soc.* **2004**, *82*, 742–748. [CrossRef]
6. Nasrazadani, S.; Mielke, D.; Springfield, T.; Ramasamy, N. *Practical Applications of FTIR to Characterize Paving Materials*; University of North Texas: Denton, TX, USA, 2010.
7. Jadidi, K.; Esmaeili, M.; Kalantari, M.; Khalili, M.; Karakouzian, M. A review of different aspects of applying asphalt and bituminous mixes under a railway track. *Materials* **2020**, *14*, 169. [CrossRef]

materials

MDPI

Article

Characteristics of Waste Iron Powder as a Fine Filler in a High-Calcium Fly Ash Geopolymer

Toon Nongnuang [1], Peerapong Jitsangiam [2,*], Ubolluk Rattanasak [3], Weerachart Tangchirapat [4], Teewara Suwan [2] and Suriyah Thongmunee [2]

1 Graduate Program in Civil Engineering, Department of Civil Engineering, Faculty of Engineering, Chiang Mai University, Huay Kaew Road, Muang, Chiang Mai 50200, Thailand; toon_n@cmu.ac.th
2 Center of Excellence in Natural Disaster Management, Department of Civil Engineering, Faculty of Engineering, Chiang Mai University, Chiang Mai 50200, Thailand; teewara.s@cmu.ac.th (T.S.); suriyah.t@cmu.ac.th (S.T.)
3 Department of Chemistry, Faculty of Science, Burapha University, Chonburi 20131, Thailand; ubolluk@go.buu.ac.th
4 Construction Innovations and Future Infrastructure Research Center (CIFIR), Department of Civil Engineering, Faculty of Engineering, King Mongkut's University of Technology Thonburi, Bangkok 10140, Thailand; weerachart.tan@kmutt.ac.
* Correspondence: peerapong@eng.cmu.ac.th

Abstract: Geopolymer (GP) has been applied as an environmentally-friendly construction materi in recent years. Many pozzolanic wastes, such as fly ash (FA) and bottom ash, are commonly used a source materials for synthesizing geopolymer. Nonetheless, many non-pozzolanic wastes are ofte applied in the field of civil engineering, including waste iron powder (WIP). WIPs are massive produced as by-products from iron and steel industries, and the production rate increases ever year. As an iron-based material, WIP has properties of heat induction and restoration, which ca enhance the heat curing process of GP. Therefore, this study aimed to utilize WIP in high-calciu FA geopolymer to develop a new type of geopolymer and examine its properties compared to th conventional geopolymer. Scanning electron microscopy and X-ray diffraction were performed o the geopolymers. Mechanical properties, including compressive strength and flexural strength, we also determined. In addition, setting time and temperature monitoring during the heat curing proce were carried out. The results indicated that the addition of WIP in FA geopolymer decreased th compressive strength, owing to the formation of tetrahydroxoferrate (II) sodium or $Na_2[Fe(OH)_4$ However, a significant increase in the flexural strength of GP with WIP addition was detected. flexural strength of 8.5 MPa was achieved by a 28-day sample with 20% of WIP addition, nearly thre times higher than that of control.

Keywords: geopolymer; geopolymer paste; waste iron powder; composite material

Citation: Nongnuang, T.; Jitsangiam, P.; Rattanasak, U.; Tangchirapat, W.; Suwan, T.; Thongmunee, S. Characteristics of Waste Iron Powder as a Fine Filler in a High-Calcium Fly Ash Geopolymer. *Materials* **2021**, *14*, 2515. https://doi.org/10.3390/ma14102515

Academic Editors: Miguel Bravo, Rui Vasco Silva and António P.C. Duarte

Received: 11 April 2021
Accepted: 10 May 2021
Published: 12 May 2021

1. Introduction

Due to global warming, conventional concrete in construction results in tons of carbo dioxide (CO_2) emissions to the atmosphere, leading to the greenhouse effect. For decade eco-friendly materials have been widely used in civil engineering work worldwide. By products or waste materials are being investigated as alternatives in civil constructio particularly as replacements for cement [1–3], because they can provide properties of rav materials and effectively reduce construction costs [4–6]. These attempts aim to achiev three of the UN Sustainable Development Goals (SDGs)—sustainable cities and commu nities (11), responsible consumption and production (12), and climate action (13). On environmentally-friendly cementitious material is geopolymer, which can partially or full replace Portland cement, emitting significantly less CO2 [7]. It can utilize any aluminosi icate by-products by incorporating alkaline solutions. Coal fly ash (FA) is a pozzolani material that is a by-product of a coal-fired power plant and widely used as a sourc material for geopolymer production [8–11]. Geopolymerization is a result of an inorganic polycondensation reaction from the aluminosilicate binders with an alkaline solution [12], which requires heat curing from an external energy source to activate the reaction and accelerate the strength development [13]. Typically, an oven is applied at a temperature ranging from 40 to 90°C for 4 to 48 h in the heat curing process [14,15]. Geopolymer could be made as geopolymer paste, mortar, or composites. Several researchers tried using additional materials, such as steel fiber [16], polyvinyl alcohol fiber (PVA) [17], and even natural fiber like hemp [18], to reinforce geopolymer properties.

Researchers have report the use of iron-rich fly ash in geopolymer, which this fly ash was classified by air classifier [19]. Results showed that iron compound in fly ash resulted in the positive effect on geopolymer properties. In addition, the high iron content in the staring material for the geopolymer synthesis can lead to the formation of ferro–silico-aluminate geopolymer in which some Al atom is substitute by Fe atom [20,21]. However, production temperature of this geopolymer type is around 600 °C–800 °C [20]. The behavior of iron (Fe^{3+}) during geopolymerization reaction relies on its chemical and mineralogical state in the starting materials [21]. Low-iron ash has been suggested to use in geopolymer since the harmful action of source ferrous (Fe^{2+}) compound can block the development of geopolymerization reaction [20]. Therefore, using of iron-rich ash in geopolymer is considered. Since there are many waste materials produced from civil engineering applications, such as waste iron powder (WIP). In Thailand, iron and steel industries produce more than 2M tons of waste iron and steel powders annually [22], and this rate is growing, creating a long-term environmental problem for the future. Some recent works applied the WIP for mixing in concrete [23] or asphalt concrete [24,25]. However, the application of WIP in geopolymer has not been reported.

Based on the outstanding properties of iron powder (i.e., good heat conduction), the WIP utilization in geopolymer can benefit from the mechanical properties and the heat transfer in oven curing. The heat energy penetrates the material's surface and transfers to the inner core by thermal convection; nevertheless, geopolymerization may take place at a slow rate, owing to the gentle thermal gradients inside the material [26]. WIP has better thermal properties than FA, such as heat conduction and heat retention. Consequently, proper WIP content in geopolymer can possibly contribute to the heat transfer.

This research is novelty which is the unique application of an iron powder as a filler in geopolymer, particularly WIP. Therefore, this study aims to discover the potential of WIP utilization in high-calcium FA geopolymer in comparison to conventional FA geopolymer through the physical and mechanical properties and microstructure. The use of the waste material is always a challenge in the civil engineering work field.

2. Materials and Methodology

2.1. Materials

2.1.1. Fly Ash

Fly ash (FA) from the Mae Moh lignite-coal–fired power plant in Lampang, Thailand, was used as source material for geopolymer production. Energy-dispersive X-ray fluorescence spectrometer (EDXRF, JSX3400R, JEOL Ltd., Tokyo, Japan) analysis and scanning electron microscopy (SEM, JSM-5910LV, JEOL Ltd., Tokyo, Japan) were performed to investigate the chemical compositions and the microstructure. The results of EDXRF and SEM are shown in Table 1 and Figure 1, respectively. FA comprises spherical particles due to the high combustion temperature of coal, leading to the sintering of ash particles and resulting in the spherical shape [10]. The median diameter (D_{50}) of particles was 19 μm. According to ASTM C618 [27], this FA was classified as type C FA because of the excessive amount of sulfur trioxide (SO_3, more than 5%) and more than 20% lime (CaO) and, therefore, could be defined as a high-calcium FA. The CaO content of 31.41% in the FA was due to the low-grade lignite coal. The calcium content of this lignite coal is rather high and increased with the depth of coal mining [28]. Iron oxide (Fe_2O_3) was also high, resulting reddish-brown color. The FA in this study had a specific gravity of 2.3.

Table 1. Oxide compositions of Mae Moh fly ash and WIP.

Component	Fly Ash	WIP
Al_2O_3	8.52	-
SiO_2	18.17	1.96
Fe_2O_3	29.85	95.95
CaO	31.41	-
SO_3	8.43	-
K_2O	2.47	-
TiO_2	0.58	-
SrO	0.29	-
MnO	0.28	0.53
Cr_2O_3	-	1.56

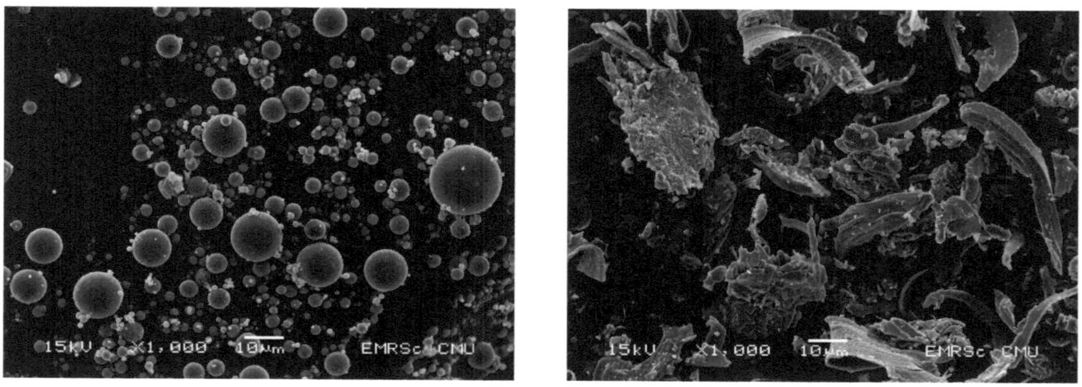

(**a**) (**b**)

Figure 1. Microstructure of materials by SEM: (**a**) FA at 1000× magnification, (**b**) Waste iron powder at 1000× magnification.

2.1.2. Waste Iron Powder

The WIP in this study was a waste material from a bearing manufacturer in Thailan Because of the industrial process, WIP was contaminated with oil, which needed to b removed with detergent and water. The particle size of WIP was analyzed by particle siz analyzer showing the median size (D_{50}) of 22 µm. The chemical compositions by EDXR analysis and the microstructure observation by SEM are also presented in Table 1 an Figure 1, respectively. The specific gravity of this WIP was 7.2.

2.1.3. Alkaline Activators

The alkaline activators used in this research were sodium hydroxide (NaOH, SH) an sodium silicate (Na_2SiO_3, SS) solutions. A laboratory-grade NaOH with 99% purity wa obtained from ACI Labscan Ltd., Thailand. For the preliminary study, using of high NaOH concentration (10 M) with high-calcium fly ash resulted in the fast setting of paste due t the reaction between NaOH and CaO. Therefore, lower NaOH concentration of 8 M wa selected to overcome this problem and provide the more workability of paste. Solid NaOH was dissolved in the distilled water to obtain the NaOH concentrations of 4, 6, and 8 M. A commercial-grade Na_2SiO_3 solution was purchased from a local distributor from Kusawa Chemical Group Ltd., Chiang Mai, Thailand with concentrations of 27 wt % SiO_2 an 8 wt % Na_2O. The alkaline activators were the blended solutions of $Na_2SiO_3$and NaOH yielding weight ratios (SS/SH) of 0.67 and 1.00.

2.2. Experimental Procedures

2.2.1. Manufacturing Process

First, FA and WIP were dried in the oven for 24 h before mixing. For the alkaline activator, Na_2SiO_3 and the desired concentration of NaOH were blended to obtain SS/SH values of 0.67 and 1.00. In the mixing process, FA was combined with the blended alkaline solution using a blender. The mass ratio of liquid and binder (L/B) used in the test was 0.45. Next, the fresh geopolymer paste was poured in $4 \times 4 \times 16$ cm^3 prisms and was left at room temperature for 1 h. All samples were then placed in the oven for heat curing at the desired temperature between 30 °C–60 °C for 24 h as shown in Table 2. The specimens were cooled down to room temperature and wrapped with a cling film after de-molding. All specimens were then cured at ambient temperature (25 °C) until they reached their testing age. The details of mixing proportion are summarized as follows.

Table 2. Mixing details of all samples.

Mixture	L/B	%FA	%WIP	FA (g)	WIP (g)	NaOH (M)	SS/SH Mass Ratio	Heat Curing Temp. (°C)	Temp. Monitoring	Testing Age (days)
CGP	0.45	100	0	1500	0	4, 6, 8	0.67, 1.00	30, 40, 50, 60	✓	3, 28
W5GP	0.45	100	5	1500	75	4, 6, 8	0.67, 1.00	60	-	3, 28
W10GP	0.45	100	10	1500	150	4, 6, 8	0.67, 1.00	60	✓	3, 28
W15GP	0.45	100	15	1500	225	4, 6, 8	0.67, 1.00	60	-	3, 28
W20GP	0.45	100	20	1500	300	4, 6, 8	0.67, 1.00	30, 40, 50, 60	-	3, 28

2.2.2. Mixing Proportion

- Controlled geopolymer paste (CGP).

The controlled geopolymer consisted of FA, NaOH, and Na_2SiO_3 solution. The oven-dried FA was mixed with the alkaline activators, which are the blended solution of 4, 6, and 8 M NaOH and Na_2SiO_3 solution with the mass SS/SH values of 0.67 and 1.00 in order to produce the high strength geopolymer [14].

- WIP mixed geopolymer mortar (WGm).

Amounts of 5%, 10%, 15%, and 20% of WIP were added to the geopolymer paste as a fine filler by %weight of binder (FA). The details of the mixing process are given in Table 2. 'WxGP' indicates WIP mixed geopolymer paste, where x is the percent mass of WIP.

2.2.3. Analytical Methods

- Physical properties.

The setting times were tested using Standard Vicat apparatus (L0028/A, Controls Group, Milan, Italy) in accordance with ASTM C191-19 [29]. The moment the raw material made contact with the alkaline activators was defined as the start of the setting time test. The bulk density of hardened geopolymers was measured at 28 days prior to the mechanical properties test by measuring a prismatic sample's dimensions and weight.

- Mechanical properties.

The mechanical properties of hardened CGPs and WGPs, including compressive strength and flexural strength, were carried out from the three and 28 day–aged samples using the CST (Compression machine, CST Instruments (Thailand) Part. Ltd., Thailand) compression machine in accordance with the BS EN 196-1 [30]. The flexural strength test was conducted on $4 \times 4 \times 16$ cm^3 prismatic samples using a three-point loading method in the 100-mm span length of two supporting points. The parts obtained from the flexural strength test were then tested for compressive strength at a loading rate of 1.0 mm/min and 4×4 cm^2 loading area.

2.2.4. Microstructure Analysis

The X-ray Diffraction (XRD, Rigaku SmartLab, Rigaku Corporation., Tokyo, Japa and SEM techniques (JSM-5910LV, JEOL Ltd., Tokyo, Japan) were performed to observ the crystalline phases and microstructure. The energy-dispersive X-ray spectroscop (EDX, JSM-5910LV, JEOL Ltd., Tokyo, Japan) technique was used to identify the chemic compositions of products in the spectrum.

2.2.5. Temperature Monitoring

To consider the thermal properties of W10GP sample, the high-calcium FA geopolym with 10% WIP sample was set up for temperature monitoring using type K thermocoupl during the 24-h heat curing process (60 °C in the oven) and for three hours after the ove curing. The thermocouples were placed to measure temperature at three locations: surfac core, and oven (ambient).

3. Results and Discussion

3.1. Physical Properties

3.1.1. Bulk Density of Hardened Geopolymer Paste

Figure 2 shows the bulk densities of the hardened geopolymers pastes with varyin WIP contents. The WIP content clearly affected the density of geopolymer, owing to WIP significantly greater specific gravity (7.2) compared to FA (2.3). The bulk density increase with WIP content.

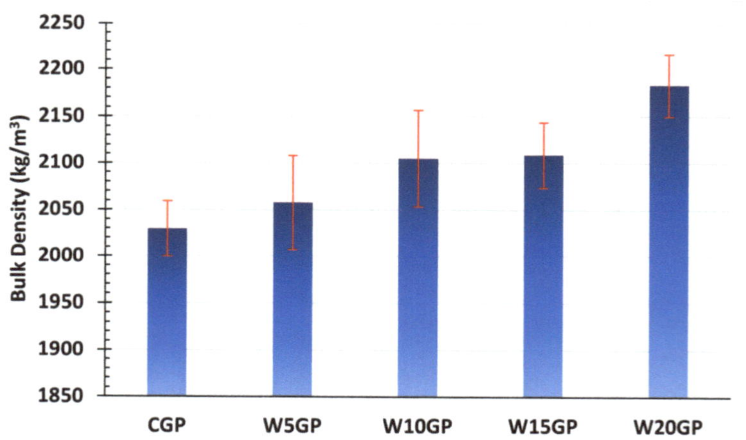

Figure 2. Bulk density of 8M NaOH, SS/SH of 1.00 (curing temperature of 60 °C).

3.1.2. Setting Time Test

Setting time is a defining characteristic of geopolymer. The hardening process i related to the formation of calcium aluminosilicate hydrate gel (C–A–S–H) and $Ca(OH)$ depending on the raw materials and alkaline activators [28,31]. The high-calcium F geopolymer had a short setting time because of the high CaO content in FA. In accordanc with Figure 3, setting time tests were performed using Vicat apparatus. The results showe that WGP samples had both lower initial and final setting times than the CGP samples, i which hematite (iron(III) oxide, Fe_2O_3) in WIP does not contribute to the geopolymer' reaction or hydration. The initial and final setting times for CGP were 576 and 1020 s respectively; meanwhile, the initial setting time of the WGP samples was 338 to 368 s, an the final setting time was 604 to 746 s. The addition of WIP reduced the setting time c the WGP samples, owing to the reaction between WIP and alkali solution, and increase i solid content with same alkali volume. The new product was formed in the system and accelerated the hardening of geopolymer. The reaction between WIP and alkali solutions could form the iron(II) oxide or ferrous oxide ($Fe(OH)_2$), as shown in Equation (1).

$$Fe(s) + 2NaOH(aq) \rightarrow Fe(OH)_2(s) + 2Na^+(aq) \quad (1)$$

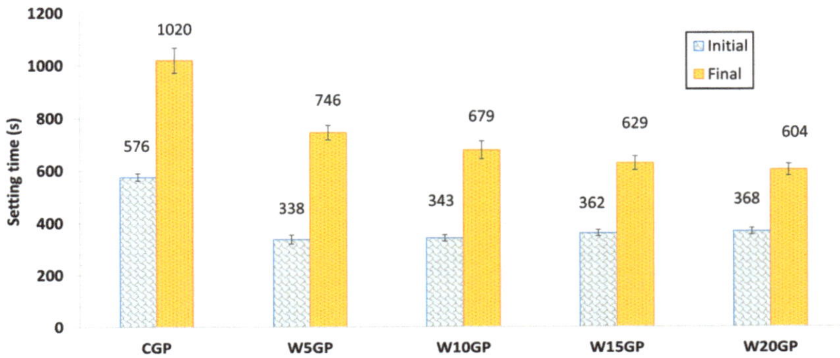

Figure 3. Setting times of CGP and WGP samples.

Comparison between samples, the higher WIP contents in the mixtures resulted in the longer initial setting times, which can be due to a hindrance of FA-alkali reaction by WIP particles. In contrast, the final setting time of WGP was reduced by adding more WIP, likely owing to the formation of $Fe(OH)_2$. In summary, the addition of WIP promoted a quicker hardening process of geopolymer.

3.2. Mechanical Properties

Figure 4a illustrates the compressive strength of samples mixed with 8 M NaOH alkaline activator with SS/SH of 1.00 at a curing temperature of 60 °C. CGP samples yielded a high early strength of 46.6 MPa at three-day age, and the strength developed with time because of the geopolymerization and pozzolanic reactions. The strength reached 66.4 MPa at 28-day curing time. The early strength of some WGP samples resulted in greater compressive strength than that of CGP. For the three-day age, W5GP obtained a compressive strength of 44.6 MPa, while W20GP achieved 59.9 MPa. This difference could be because the reaction between $Fe(OH)_2$ and NaOH solutions led to the formation of tetrahydroxoferrate(II) sodium, $Na_2[Fe(OH)_4]$, which can occur under heat curing, as shown in Equation (2). In addition, $Ca(OH)_2$ can be formed, owing to the reaction between CaO in FA with alkali solution.

$$Fe(OH)_2(s) + 2NaOH(aq) \rightarrow Na_2[Fe(OH)_4](s) \quad (2)$$

However, the compressive strength of the WGP group noticeably decreased in comparison to CGP at 28-day curing time. The compressive strength wildly dropped down to around 48.5 MPa, 76.0% of CGP's strength. The formation of both $Na_2[Fe(OH)_4]$ and $Ca(OH)_2$ may have had an adverse effect on the strength of a sample at longer age.

The results of the flexural strength test are displayed in Figure 4b. The flexural strength of CGP was 3.8 and 3.0 MPa at 3- and 28-day curing times, respectively. The development of flexural strength of the WGP group was similar to that of compressive strength. The 28-day samples provided lower flexural strength than the three-day samples. Nevertheless, the flexural strengths of WGP samples were considerably higher than those of the CGP, especially for the 28-day W20GP sample with a flexural strength of 8.5 MPa, almost thr times that of the conventional geopolymer.

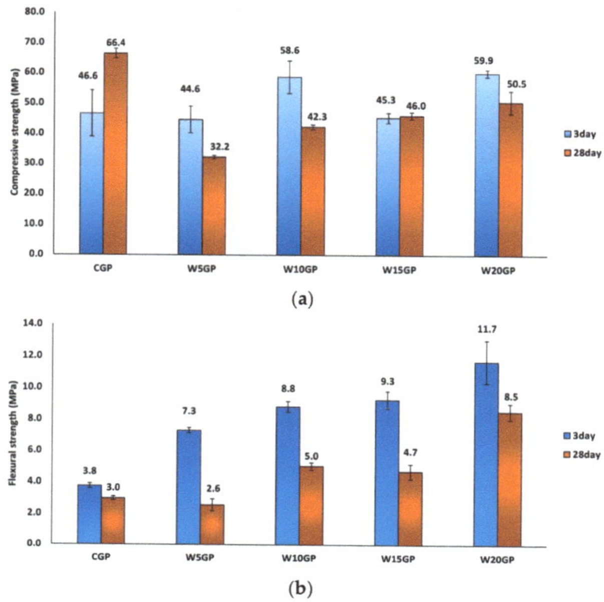

Figure 4. (**a**) Compressive and (**b**) flexural strengths of a sample at 8 M NaOH, SS/SH = 1.00 (he curing temperature of 60 °C).

The addition of WIP as a filler is a new approach to making geopolymer paste. How ever, an optimum content can effectively contribute to the best mechanical performanc for geopolymer. The results of these tests revealed that the WGP had lower compressiv strength but significantly higher flexural strength in comparison to the controlled geopol mer paste. The WIP had an adverse effect on the compressive strength since the reactio between micro-WIP particles and alkali solution occurred. WIP was the waste and had a impurity which could affect the geopolymerization reaction as the compressive strengt of WIP geopolymer was lower than that of control. However, flexural strength had muc improved due to the grain-shape of WIP.

For only WGP samples, higher amounts of WIP mixed in the geopolymer improve both compressive strength and flexural strength. The compressive strength of W5GP wa 32.2 MPa and rose to 50.5 MPa in the W20GP (20% content of WIP) sample, 76% of the CG strength. In a similar way, the flexural strength of W5GP was 2.6 MPa and significantl increased to 5.0, 4.7, and 8.5 MPa in the W10GP, W15GP, and W20GP samples, respectivel Hence, addition of WIP can reinforce the geopolymer matrix by long grain shape an rough surface of WIP, resulting in the higher flexural strength.

3.3. NaOH Concentration and SS/SH

The previous research of Rattanasak [32] showed that a higher concentration of NaO and larger amount of Na_2SiO_3 led to higher compressive strength in geopolymer becaus the high concentration of NaOH could dissolve more Si^{3+} and Al^{4+} ions out of the surfac of FA particles to form the geopolymerization gel, connecting each FA particle. Moreover, large amount of Na_2SiO_3solution added more silica to form the gel in geopolymer. To prov this theory, this research also studied the behavior of WGP on the NaOH concentratio and SS/SH, and the results are illustrated in Figures 5 and 6. Figure 5 shows that bot compressive strength and flexural strength of W20GP increased along with the higher concentration of NaOH. Furthermore, Figure 6 shows the results between two mixtures of geopolymer paste with the different SS/SH values and NaOH concentrations. The higher compressive strength was obtained from the samples with an SS/SH value of 1.00 since Na_2SiO_3 provided the dissolved SiO_2 to the system, resulting in more formations of C–S–H and aluminosilicate compound in geopolymer. However, both SS/SH values of 0.67 and 1.00 yielded the same level of flexural strength. In summary, WIP mixed with FA-based geopolymer paste behaved similarly to a conventional geopolymer.

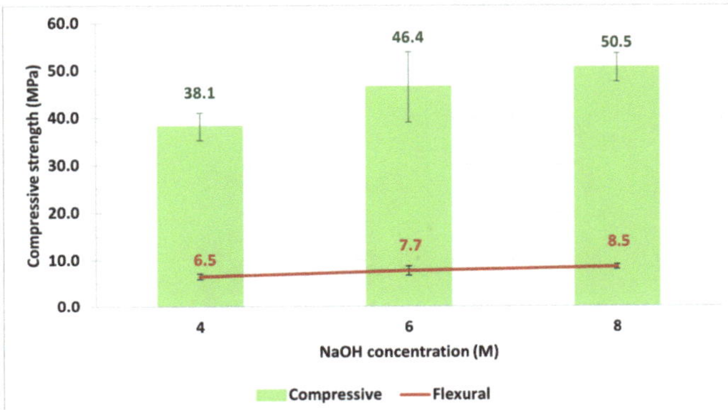

Figure 5. Effect of NaOH concentration on the W20GP sample at 28-day curing time (heat curing temperature of 60 °C) with SS/SH = 1.00.

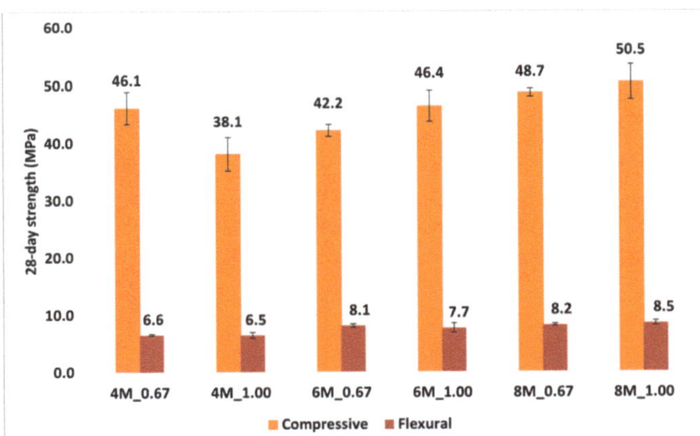

Figure 6. SS/SH ratio effect on 20%WIP at 28-day curing time.

3.4. Heat Curing Temperature

Figure 7 shows the heat curing temperature affected the W20GP samples, varying the temperature from 30 °C–60 °C. It was seen that oven curing resulted in higher strength of geopolymer. This result corresponded to the typical heat curing temperature of FA-based geopolymer [14]. A heat curing temperature range of 40 °C–75 °C could provide the heat for the formation of geopolymerization adequately. However, extremely high heat curing (over 80 °C) could negatively affect the geopolymer properties, owing to a quick loss of moisture content and the addition of micro-cavities in the structure [33]. However, when WIP was applied in geopolymer mixture, an internal temperature of sample could be higher than oven temperature due to the heat conduction of WIP. Therefore, a curing temperatu of 60 °C reduced the strength of material.

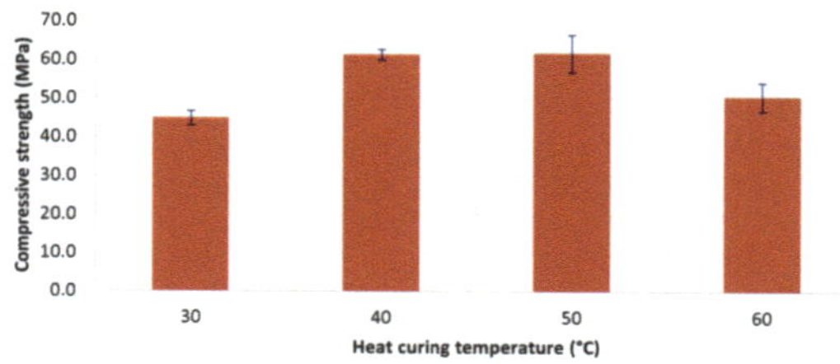

(**a**)

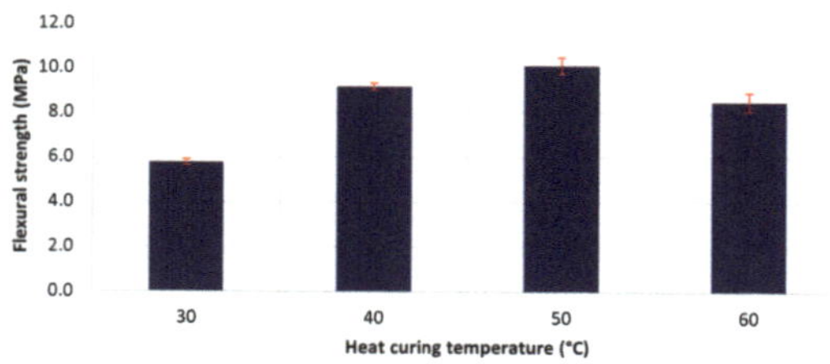

(**b**)

Figure 7. Effect of heat curing temperature on (**a**) compressive and (**b**) flexural strengths of W20G at 28-day age with the solution of NaOH 8 M, SS/SH = 1.00, and L/B = 0.45.

In summary, no remarkable difference in the appropriate temperature for heat curin between WGP and conventional FA-based geopolymer was found.

3.5. Temperature Monitoring

Figure 8a illustrates the heat monitoring result for the 24-h heat curing process. Typ K thermocouples were installed to measure the temperature at three locations: sampl core, sample surface, and oven. The temperature measurement presented no apparer difference between the core and surface temperatures in both groups of CGP and W10G therefore, the result is only displayed in terms of core temperature. According to the result the temperature increased to 60 °C after four to five hours in the heat curing proces hence, the heat transfer theoretically occurred by thermal convection [34]. Neverthele the addition of 10% WIP into the FA-based geopolymer had no impact on temperatur during the heat curing process because of the discontinuous iron matrix in geopolyme mass. However, Figure 8b shows the temperature in the three hours after oven curin The W10GP's temperature clearly lasted longer than the CGP's. Therefore, WIP played significant role in heat retention after the heat curing process and provided the longer hea curing time under this heat holding situation.

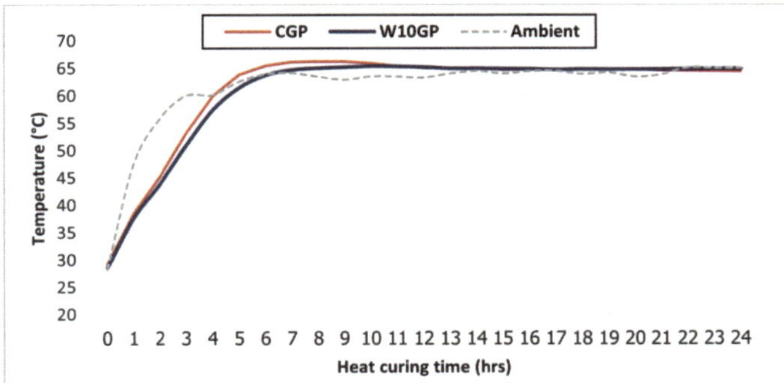

(a)

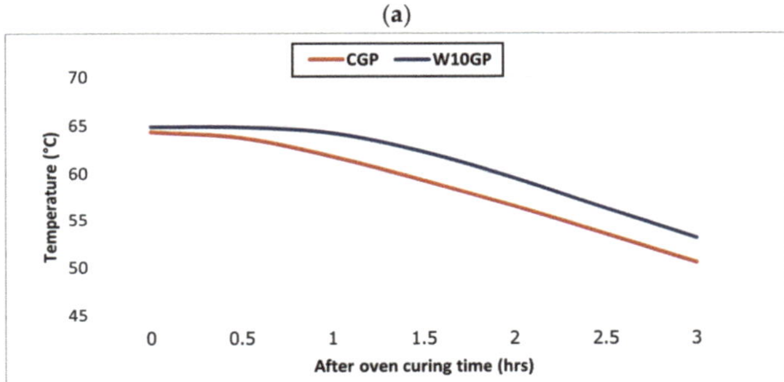

(b)

Figure 8. Temperature monitoring of (**a**) 24 h during the heat curing process and (**b**) 3 h after the heat curing process.

3.6. Morphological Analysis

The results of XRD patterns of CGP and WGP mixtures are given in Figure 9. At the testing age of 28 days, the major phase of each group was amorphous, indicating a broad hump at the region of 20°–36° 2θ. The phases mainly contained kyanite (Al_2O_5Si), tobermorite ($Ca_3HO_9Si_3$), portlandite ($Ca(OH)_2$), and magnetite (Fe_3O_4) in all groups. The peaks at 43°–45° 2θ between CGP and WGP, which were of iron (Fe) and nepheline zeolite, were notably different. Sodium iron(III) oxide was also detected in the specimen, owing to the reaction between WIP and alkali solution under heat curing. The addition of WIP in samples mainly increased the peaks of Fe. Particularly, the peak area from 20° to 40° 2θ seemed to reduce, which implied that WIP acted as a filler and no geopolymeric product was found in this system. Origin program was used to calculate the hump area under the peak between 20 and 37° 2θ; the results are shown in Table 3.

Table 3. Hump area under the peak between 20° and 37° 2θ.

Samples	Area (a.u.)
CGP	4110
W5GP	4032
W10GP	3644
W15GP	3346
W20GP	3005

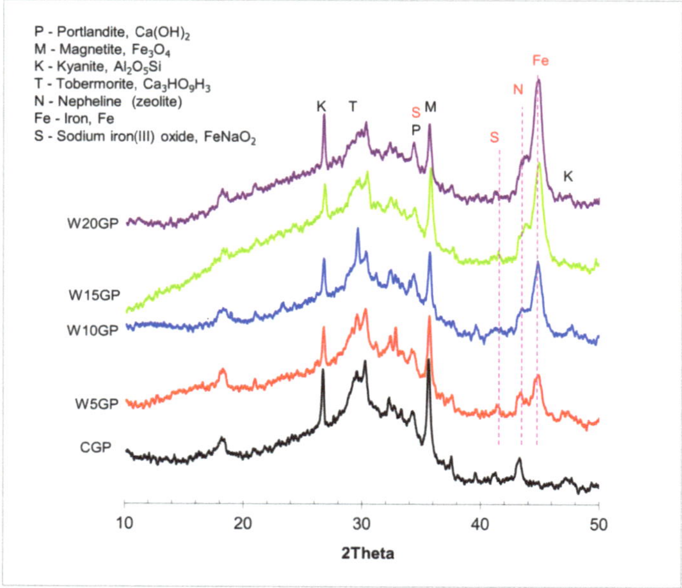

Figure 9. XRD patterns analysis of 28 day–aged CGP and WGP samples.

SEM and EDX techniques were used to investigate the microstructure of the 28 day aged WGP mixed with a solution of 8 M NaOH and SS/SH value of 1.00, as illustrate in Figures 10 and 11. Figure 10 displays the appearances of an amorphous structure wit some unreacted FA particles surrounded by geopolymeric gel. The residual shells of FA after the reaction are clearly seen in the geopolymer matrix. Large amounts of amorphou network gel product and CSH gel were generated, covering the residual particles to forn a continuous gel mass and resulting in a relatively dense microstructure [35,36]. Th EDX spectrum revealed high contents of Si, Al, and Ca, which were the products o hydration and geopolymeric reaction. There were small clusters of Fe scattering around th geopolymeric gel. The EDX spectrum also showed the high peak of Fe, which corresponde to the addition of WIP in the geopolymer paste. Moreover, in the higher WIP conten samples, many large clusters of Fe were settled.

The SEM results showed a major group of Fe coated with C–A–S–H gel and geopoly meric gel, as illustrated in Figure 11. The flat and twisted iron particles were evident. A ge formation of geopolymerization and hydration connected these uniquely shaped particle together, leading to hermetical iron clusters that had a strong interlocking and bonding between the iron particles. These clusters contributed to the resistance of flexural loac in the samples. Therefore, adding more WIP content significantly promoted the flexura strength in the geopolymer.

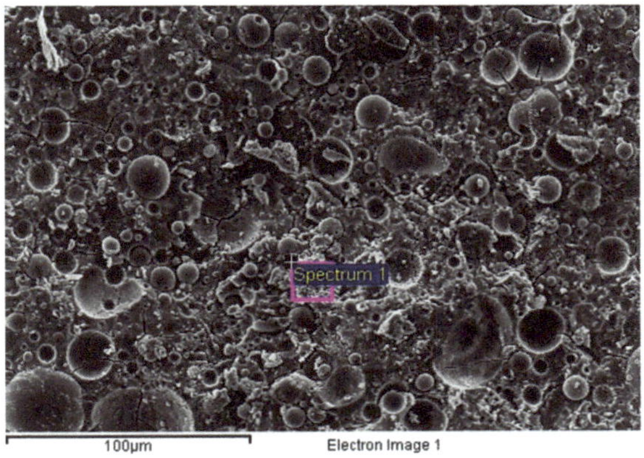

(a)

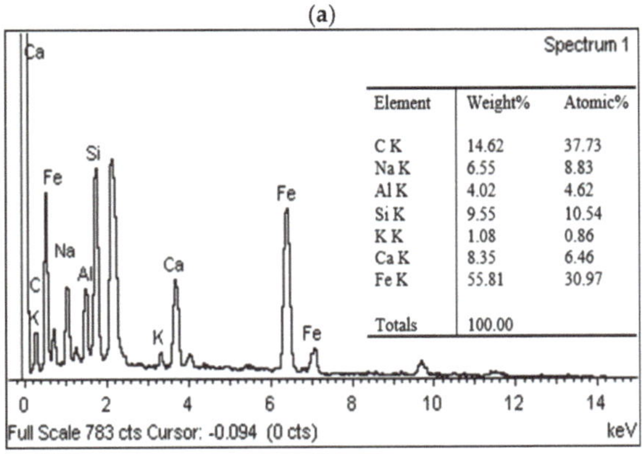

(b)

Figure 10. (**a**) SEM and (**b**) EDX results of W20GP at 28-day curing time.

Nevertheless, the reduction in compressive strength must be explained. The main component of Fe in WIP reacted with the high alkaline solution. The product from this reaction was $Na_2[Fe(OH)_4]$ under heat curing. The occurrence of $Na_2[Fe(OH)_4]$ increased in the volume of iron compounds, causing stresses in the geopolymer matrix [37–39]. Zeolite could form when an SS/SH of 1.0 was used in the mixture [40]. After that, the microcracks occurred in the structure, resulting in the reduction of compressive strength and flexural strength in the passing time. The presence of iron compound in the microstructure. These crystalline expansions weakened the completion of the geopolymer structure, according to XRD analysis. The crystalline phases of kyanite, portlandite, and C–S–H gel of tobermorite, which were the main products of geopolymerization and pozzolanic reaction, were reduced when the high amount of WIP was added. The presence of Fe and $Na_2[Fe(OH)_4]$ chemically reduced the strength. These products hindered the dissolution of the Al^{4+} and Si^{3+} ions at the FA particle's surface, which could lead to an insufficient geopolymeric reaction. However, only in WGP samples, the higher amount of WIP contributed more compressive strength because of the increased density in the samples.

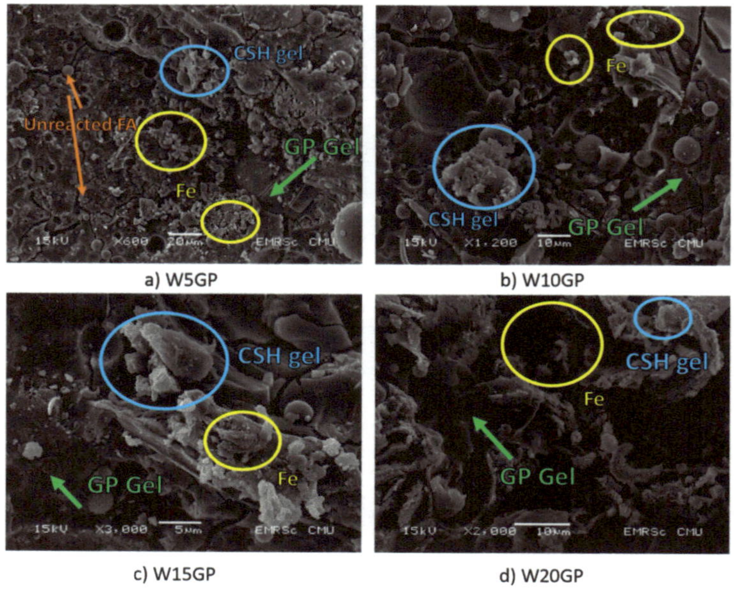

Figure 11. SEM figures of $Na_2[Fe(OH)_4]$ existence. (**a**) W5GP at 600× magnification (**b**) W10GP a 1200× magnification (**c**) W15GP at 3000× magnification (**d**) W20GP at 2000× magnification.

4. Conclusions

WIP has the potential as a filler in high-calcium FA geopolymer. This waste enhanc the geopolymer properties. WIP played a significant role in heat retention after the hea curing process and provided the longer heat curing time under the heat holding situatio The addition of WIP in geopolymer increased the bulk density of the samples; howeve compressive strength was lower due to the grain-shape of WIP than the control sinc the impurity in WIP could affect the geopolymerization reaction; however, the flexura strength significantly increased because of the interlocking of Fe clusters and its grai shape and rough surface. The occurrence of $Na_2[Fe(OH)_4]$ and zeolite phases was found i the geopolymer's matrix. The optimum content of WIP in high-calcium FA geopolyme was suggested as 20%, obtaining the 28-day compressive strength of 50.5 MPa, which wa approximately 76% of the strength of the control. This sample also yielded the high flexura strength of 8.5 MPa, which was almost three times that of the conventional geopolymer.

Author Contributions: Conceptualization, P.J.; methodology, T.N., P.J., and T.S.; validation, P.J T.S., U.R., W.T., and S.T.; formal analysis, P.J., T.N.; investigation, T.N. and P.J.; writing—origina draft preparation, T.N., P.J., and U.R.; writing—review and editing, P.J., T.S., U.R., W.T., and S.T visualization, T.N. and U.R.; supervision, P.J., T.S., U.R., W.T., and S.T.; project administration, P.J. A authors have read and agreed to the published version of the manuscript.

Funding: This research received no external funding.

Institutional Review Board Statement: Not applicable.

Informed Consent Statement: Not applicable.

Data Availability Statement: The data presented in this study are available on request from th corresponding author.

Acknowledgments: The author would like to express thanks to the research units for the hard wor and for the research supporters:(i) Tanapon Phenrat for the WIP supply; (ii) Teaching Assistant an

Research Assistant (TA&RA) Scholarship from Graduate School, Chiang Mai University; and (iii) 50th Year Civil Engineering Chiang Mai University Scholarship for funding.

Conflicts of Interest: The authors declare no conflict of interest.

References

1. Rattanashotinunt, C.; Tangchirapat, W.; Jaturapitakkul, C.; Cheewaket, T.; Chindaprasirt, P. Investigation on the Strength, Chloride Migration, and Water Permeability of Eco-Friendly Concretes from Industrial by-Product Materials. *J. Clean. Prod.* **2018**, *172*, 1691–1698. [CrossRef]
2. Abdulmatin, A.; Khongpermgoson, P.; Jaturapitakkul, C.; Tangchirapat, W. Use of Eco-Friendly Cementing Material in Concrete Made from Bottom Ash and Calcium Carbide Residue. *Arab. J. Sci. Eng.* **2018**, *43*, 1617–1626. [CrossRef]
3. Rattanachu, P.; Toolkasikorn, P.; Tangchirapat, W.; Chindaprasirt, P.; Jaturapitakkul, C. Performance of Recycled Aggregate Concrete with Rice Husk Ash as Cement Binder. *Cem. Concr. Res.* **2020**, *108*, 103533. [CrossRef]
4. Peerapong, J.; Kornkanok, B.; Tanapon, P.; Suphat, C.; Prinya, C.; Hamid, N. Recycled Concrete Aggregates in Roadways: Laboratory Examination of Self-Cementing Characteristics. *J. Mater. Civ. Eng.* **2015**, *27*, 04014270. [CrossRef]
5. Huan, Y.; Siripun, K.; Jitsangiam, P.; Nikraz, H. A Preliminary Study on Foamed Bitumen Stabilisation for Western Australian Pavements. *Sci. Res. Essays* **2010**, *5*, 23.
6. Sounthararajah, A.; Bui, H.; Nguyen, N.; Jitsangiam, P.; Kodikara, J. Early-Age Fatigue Damage Assessment of Cement-Treated Bases Under Repetitive Heavy Traffic Loading. *J. Mater. Civ. Eng.* **2018**, *30*, 04018079. [CrossRef]
7. Rattanasak, U. *Geopolymer*, 1st ed.; Thailand Concrete Association: Bangkok, Thailand, 2017; Volume 167.
8. Palomo, A.; Grutzeck, M.W.; Blanco, M.T. Alkali-Activated Fly Ashes: A Cement for the Future. *Cem. Concr. Res.* **1999**, *29*, 1323–1329. [CrossRef]
9. Chindaprasirt, P.; Jaturapitakkul, C.; Chalee, W.; Rattanasak, U. Comparative Study on the Characteristics of FA and Bottom Ash Geopolymers. *Waste Manag.* **2009**, *29*, 539–543. [CrossRef]
10. Chindaprasirt, P.; Jenjirapanya, S.; Rattanasak, U. Characterizations of FBC/PCC Fly Ash Geopolymeric Composites. *Constr. Build. Mater.* **2014**, *66*, 72–78. [CrossRef]
11. Chindaprasirt, P.; Rattanasak, U.; Vongvoradit, P.; Jenjirapanya, S. Thermal Treatment and Utilization of Al-Rich Waste in High Calcium FA Geopolymeric Materials. *Int. J. Min. Met. Mater.* **2012**, *19*, 872–878. [CrossRef]
12. Davidovits, J. Geopolymer Chemistry and Sustainable Development. In Proceedings of the World Congress Geopolymer, Saint-Quentin, France, 1 July 2005; pp. 9–15.
13. Jitsangiam, P.; Suwan, T.; Pimraksa, K.; Sukontasukkul, P.; Chindaprasirt, P. Challenge of Adopting Relatively Low Strength and Self-Cured Geopolymer for Road Construction Application: A Review and Primary Laboratory Study. *Int. J. Pavement Eng.* **2019**. [CrossRef]
14. Chindaprasirt, P.; Chareerat, T.; Sirivivatnanon, V. Workability and Strength of Coarse High Calcium FA Geopolymer. *Cem. Concr. Res.* **2007**, *29*, 224–229. [CrossRef]
15. Suwan, T.; Fan, M.; Braimah, N. Internal Heat Liberation and Strength Development of Self-Cured Geopolymers in Ambient Curing Conditions. *Constr. Build. Mater.* **2016**, *114*, 297–306. [CrossRef]
16. Carabba, L.; Santandrea, M.; Carloni, C.; Manzi, S.; Bignozzi, M.C. Steel Fiber Reinforced Geopolymer Matrix (S-FRGM) Composites Applied to Reinforced Concrete Structures for Strengthening Applications: A Preliminary Study. *Compos. Part B-Eng.* **2017**, *128*, 83–90. [CrossRef]
17. Tanyildizi, H.; Yonar, Y. Mechanical Properties of Geopolymer Concrete Containing Polyvinyl Alcohol Fiber Exposed to High Temperature. Constr. *Build. Mater.* **2016**, *126*, 381–387. [CrossRef]
18. Maichin, P.; Suwan, T.; Jitsangiam, P.; Chindaprasirt, P.; Fan, M. Effect of Self-Treatment Process on Properties of Natural Fiber-Reinforced Geopolymer Composites. *Mater. Manuf. Process.* **2020**, *35*, 1120–1128. [CrossRef]
19. Kumar, S.; Yankwa Djobo, J.N.; Kumar, A.; Kumar, S. Geopolymerization Behavior of Fine Iron-Rich Fraction of Brown Fly Ash. *J. Build. Eng.* **2016**, *8*, 172–178. [CrossRef]
20. Davidovits, J.; Davidovits, R. Ferro-Sialate Geopolymers (-Fe-O-Si-O-Al-O-). *Geopolymer Inst. Libr.* **2020**. [CrossRef]
21. Lemougna, P.N.; MacKenzie, K.J.D.; Jameson, G.N.L.; Rahier, H.; Chinje Melo, U.F. The Role of Iron in the Formation of Inorganic Polymers (Geopolymers) from Volcanic Ash: A 57Fe Mössbauer Spectroscopy Study. *J. Mater. Sci.* **2013**, *48*, 5280–5286. [CrossRef]
22. Center of Excellence on Hazardous Substance Management. 100 Types of Industrial Waste Quantities (2008–2011) in Thailand. 2012. Available online: http://recycle.dpim.go.th/wastelist/download_files/G/waste_quantity.pdf (accessed on 11 April 2021).
23. Ghannam, S.; Najm, H.; Vasconez, R. Experimental Study of Concrete Made with Granite and Iron Powders as Partial Replacement of Sand. *Sustain. Mater. Technol.* **2016**, *9*, 1–9. [CrossRef]
24. Arabani, M.; Mirabdolazimi, S.M. Experimental Investigation of the Fatigue Behaviour of Asphalt Concrete Mixtures Containing Waste Iron Powder. *Mater. Sci. Eng. A* **2011**, *528*, 3866–3870. [CrossRef]
25. Liu, K.; Fu, C.; Dai, D.; Jin, C.; Li, W.; Li, S.; Xu, X. Induction Heating Performance of Asphalt Pavements Incorporating Electrically Conductive and Magnetically Absorbing Layers. *Constr. Build. Mater.* **2019**, *229*, 116805. [CrossRef]
26. Suwan, T.; Fan, M.; Braimah, N. Micro-Mechanisms and Compressive Strength of Geopolymer-Portland Cementitious System under Various Curing Temperatures. *Mater. Chem. Phys.* **2016**, *180*, 219–225. [CrossRef]

27. *ASTM C618-19 Standard Specification for Coal FA and Raw or Calcined Natural Pozzolan for Use in Concrete*; ASTM International: We Conshohocken, PA, USA, 2019.
28. Chindaprasirt, P.; Rattanasak, U. Characterization of the High-Calcium FA Geopolymer Mortar with Hot-Weather Curing System for Sustainable Application. *Adv. Powder Technol.* **2017**, *28*, 2317–2324. [CrossRef]
29. *ASTM C191-19 Standard Test Methods for Time of Setting of Hydraulic Cement by Vicat Needle*; ASTM International: West Co shohocken, PA, USA, 2019.
30. BS EN 196-1 Methods of Testing Cement. *Determination of Strength*; British Standards Institution: London, UK, 2016.
31. Suwan, T.; Fan, M. Influence of OPC Replacement and Manufacturing Procedures on the Properties of Self-Cured Geopolyme *Constr. Build. Mater.* **2014**, *73*, 551–561. [CrossRef]
32. Rattanasak, U.; Chindaprasirt, P. Influence of NaOH Solution on the Synthesis of FA Geopolymer. *Miner. Eng.* **2009**, *22*, 1073–107 [CrossRef]
33. Mustafa Al Bakria, A.M.; Kamarudin, H.; BinHussain, M.; Khairul Nizar, I.; Zarina, Y.; Rafiza, A.R. The Effect of Curin Temperature on Physical and Chemical Properties of Geopolymers. In *Proceedings of the Physics Procedia*; Elsevier B.V.: Amsterdar The Netherlands, 2011; Volume 22, pp. 286–291.
34. Satpute Manesh, B.; Wakchaure Madhukar, R.; Patankar Subhash, V. Effect of Duration and Temperature of Curing on Compressiv Strength of Geopolymer Concrete. *Int. J. Eng. Innov. Technol.* **2012**, *1*, 152–155.
35. Song, W.; Zhu, Z.; Peng, Y.; Wan, Y.; Xu, X.; Pu, S.; Song, S.; Wei, Y. Effect of Steel Slag on Fresh, Hardened and Microstructur Properties of High-Calcium FA Based Geopolymers at Standard Curing Condition. Constr. *Build. Mater.* **2019**, *229*, 11693 [CrossRef]
36. Yip, C.K.; van Deventer, J.S.J. Microanalysis of Calcium Silicate Hydrate Gel Formed within a Geopolymeric Binder. *J. Mater. Sc* **2003**, *38*, 3851–3860. [CrossRef]
37. Jaśniok, T.; Słomka-Słupik, B.; Zybura, A. The Concrete Reinforcement Chloride Corrosion, Immediately after Its Initiation. *Cen Wapno Beton* **2014**, *3*, 158–165.
38. Tang, S.W.; Yao, Y.; Andrade, C.; Li, Z.J. Recent Durability Studies on Concrete Structure. *Cem. Concr. Res.* **2015**, *78*, 143–15 [CrossRef]
39. Zhao, Y.; Yu, J.; Hu, B.; Jin, W. Crack Shape and Rust Distribution in Corrosion-Induced Cracking Concrete. *Corros. Sci.* **2012**, *5* 385–393. [CrossRef]
40. Eiamwijit, M.; Pachana, K.; Kaewpirom, S.; Rattanasak, U.; Chindaprasirt, P. Comparative Study on Morphology of Groun Sub-Bituminus FBC FA Geopolymeric Material. *Adv. Powder Technol.* **2015**, *26*, 1053–1057. [CrossRef]

materials

rticle

'erformance of Mortars Made Using Ternary Binders with \ddition of Slag, Fly Ash and Limestone Exposed to a Real :nvironmental Condition Compatible with Exposure Class XC3

ıvier Ibáñez-Gosálvez [1], Teresa Real-Herraiz [2] and José Marcos Ortega [1,*]

[1] Departamento de Ingeniería Civil, Universidad de Alicante, Ap. Correos 99, 03080 Alicante, Spain; javier.ibanez@ua.es
[2] Instituto de Matemática Multidisciplinar, Universidad Politécnica de Valencia, Camino de Vera s/n, 46022 Valencia, Spain; tereaher@upv.es
* Correspondence: jm.ortega@ua.es; Tel.: +34-96-5903-400 (ext. 2470)

Abstract: The use of eco-friendly cements prepared with ternary binders could contribute to improving the sustainability of cement production. However, their use for manufacturing commercial cements is very low, at least in Spain. The purpose of this research is to study the behavior in the long term of mortars made with ternary binders which incorporated ground granulated blast furnace slag, fly ash, and limestone, exposed to environmental conditions compatible with the specifications of exposure class XC3 of the Eurocode 2, compared to mortars without additions and mortars with binary binders. The exposure station was placed in an underground floor of a building used as a garage with circulation of vehicles and moderately high CO_2 concentration. The ternary and binary binders verified the prescriptions of cement type CEM II/B. The microstructure was characterized using mercury intrusion porosimetry and electrical resistivity. Water absorption, diffusion coefficient, carbonation depth, mechanical strengths, and ultrasonic pulse velocity were determined. A loss of microstructure refinement with time was noted for all the analyzed binders, probably related to the development of carbonation and drying shrinkage. The binary mortars with slag and fly ash and the ternary binder which combined them showed the best mechanical performance at 250 days.

Keywords: ternary binders; real condition exposure; class XC3; ground granulated blast furnace slag; carbonation; fly ash; limestone; microstructure; mechanical properties; durability

itation: Ibáñez-Gosálvez, J.; eal-Herraiz, T.; Ortega, J.M. erformance of Mortars Made Using ernary Binders with Addition of ag, Fly Ash and Limestone Exposed a Real Environmental Condition ompatible with Exposure Class XC3. *laterials* **2021**, *14*, 5937. https:// oi.org/10.3390/ma14205937

cademic Editors: Miguel Bravo, Rui asco Silva and António P.C. Duarte

eceived: 24 August 2021
ccepted: 1 October 2021
ublished: 10 October 2021

1. Introduction

Reaching a more environmentally sustainable cement production is currently one of the main aims of the cement industry to increase its contribution to global sustainable development goals [1]. Between the different strategies for achieving this objective, the use of additions for replacing clinker has become increasingly popular [2–4]. The incorporation of these additions into blended cements provides environmental advantages, such as the reduction of CO_2 emissions and the reduction of energy consumption throughout the cement manufacturing process [5], because a lower quantity of clinker is required. Additionally, the majority of these additions are wastes originating in other industrial sectors [6–8], so their reutilization also helps to solve the problem of their storage in landfills or specific installations, avoiding other environmental impacts. The use of additions may also have an influence in the development of the pore network and properties of cement-based materials [7,9–11]. For that reason, the effects of additions for developing eco-friendly cements and their applicability in real construction are currently a wide field of investigation.

In relation to the different standardized additions which are available, among the most commonly used are ground granulated blast-furnace slag and fly ash [5,12]. Many studies [7,13–15] have shown that these additions provide a better performance in cementitious materials in comparison with those made using ordinary Portland cement without additions. On one hand, for slag, this adequate behavior is linked to the development o hydration reactions of this addition, forming new CSH phases, which would produce more refined pore network [7,13,16]. On the other hand, the fly ash has pozzolanic activit which means that it is able to react with portlandite produced along the clinker hydratio resulting in the formation of additional hydrated products, which entails an increase o the pore network refinement of mortars and concretes [9,14,16–18]. The microstructu refinement produced by both additions improves the performance of cement-based m terials, such as their resistance to chloride ingress [18–23] and sulphate attack [18,24], a well as their permeability [25]. Slag and fly ash additions belong to the group of activ additions, due to their hydraulic and pozzolanic activity previously explained. Limeston is another popular addition which is widely used in cement production for substitutin clinker. This addition lacks hydraulic or pozzolanic activity, so it mainly has a filler effect i cement-based materials, which also influences their microstructure and properties [26,27

Currently, the use of additions in commercial cements is generally focused on binar binders [28], in which clinker is partially replaced by one addition. Despite that, th standards which regulate the manufacture of commercial cements [12] permit the use o ternary binders. In these binders, clinker is substituted in part by two additions. Howeve traditionally, the use of ternary binders in the production of commercial cements is sti very low, at least in Spain [28,29]. The presence of two additions in these binders coul improve the performance of cement-based materials, caused by the probable synergisti effects of blending them [30–33]. Subsequently, to analyze the performance of cemen based materials which incorporate ternary binders [26,34–36], focusing on their use fo commercial cement manufacture, could constitute an interesting field of research in order t develop additional options to lessen the harmful pollutant effects of cement manufactur

Within the field related to the real applications of commercial cements with addition their behavior in construction elements could be influenced by the different environmenta conditions to which they are exposed depending on their geographical location [20,37]. I this regard, the active additions seemed to be more sensitive to variations in the harden ing conditions, in particular the temperature and relative humidity, because the change in these parameters could influence the development of hydration and pozzolanic re actions [38–40]. With respect to the characterization of the performance of eco-friendl cementitious materials exposed to real in-situ environments [19,20,41–44], in the majorit of them, binary binders were studied and the results obtained in general showed certai changeability based on the climate conditions, as well as with the presence of harmfu agents in the exposure site. Therefore, to analyze the behavior of cementitious material made using ternary binders subject to real in-situ conditions may be interesting in order t evaluate if they are adequate for being used in commercial cement production.

Therefore, the purpose of this research is to study the behavior in the long term o mortars made with ternary binders, which incorporated the additions of ground granulate blast furnace slag, fly ash, and limestone, exposed to environmental conditions compatibl with the specifications of exposure class XC3, according to the Eurocode 2 [45], regardin their microstructure, durability, and mechanical properties. The mortars were expose to the real environment of an underground floor of a building, used as a garage wit moderate circulation of vehicles, with a moderately high concentration of CO_2. Thi condition would be a frequent service environment of cement-based materials used i buildings. On the other hand, the additions of ground granulated blast furnace slag fly ash, and limestone were selected for preparing the ternary binders because they ar mostly used in the manufacture of blended commercial cements in Spain. In addition, th analyzed ternary binders accomplished the prescriptions for a standardized commercia cement type CEM II/B-M, according to Spanish and European standard UNE-EN 197 1 [12], in order to facilitate the possible real practical application of the obtained result by cement manufacturers. The performance of mortars prepared with ternary binder has been compared to that noted for reference mortars made using ordinary Portlan cement without additions, as well as with other mortars prepared with binary binders, incorporating only one of the analyzed additions as clinker substitution.

2. Materials and Methods

2.1. Materials and Sample Preparation

The materials tested were mortars made with different binders. The first one was the reference mortar, prepared with an ordinary Portland cement without additions, type CEM I 42.5 R according to Spanish and European standard UNE-EN 197-1 [12]. This reference mortar was designated as REF in the results section.

With respect to mortars which incorporated additions, firstly, three different binary binders were studied, in which 30% of CEM I 42.5 R cement was substituted by ground granulated blast furnace slag, fly ash, and limestone. S, F, and L, respectively, were the designations of these binders in the results section.

Furthermore, mortars made with three ternary binders were prepared. One of these binders, named SL, included 15% ground granulated blast furnace slag and 15% limestone as cement CEM I 42.5 R substitution. In the second ternary binder, designated as SF, the previously mentioned cement CEM I 42.5 R was in part replaced by 15% ground granulated blast furnace slag and 15% fly ash. The third ternary binder consisted of incorporating 15% fly ash and 15% limestone as cement CEM I 42.5 R replacement, named FL. In Table 1, the designations of the analyzed binders and the percentage of their components have been compiled.

Table 1. Designation of the mortars prepared and percentage (in weight) of the binder components.

Designation	CEM I 42.5 R	Blast Furnace Slag	Limestone	Fly Ash
REF	100%	-	-	-
F	70%	-	-	30%
S	70%	30%	-	-
L	70%	-	30%	-
FL	70%	-	15%	15%
SF	70%	15%	-	15%
SL	70%	15%	15%	-

Limestone, ground granulated blast furnace slag and fly ash verified the specifications of the UNE-EN 197-1 standard [12] for their use in commercial cement manufacture as additions. The three additions analyzed in this work were facilitated by the company Cementos Portland Valderrivas (Madrid, Spain) and they are currently used for manufacturing the commercial blended cements produced by this company. The chemical compositions of blast furnace slag, fly ash, and limestone can be observed in Table 2.

Table 2. Chemical composition of the additions used in this work.

Components	Fly Ash	Blast Furnace Slag	Limestone
SiO_2	54.40%	31.50%	2.85%
MgO	1.40%	6.98%	0.47%
K_2O	3.12%	0.52%	0.18%
Al_2O_3	27.70%	10.10%	1.22%
SO_3	0.53%	1.94%	0.10%
TiO_2	1.05%	0.94%	0.11%
Fe_2O_3	8.06%	0.37%	0.54%
CaO	2.55%	46.80%	94.40%
P_2O_5	0.46%	0.02%	0.02%
ZnO	0.11%	-	-
MnO	0.06%	0.17%	-
Na_2O	-	0.30%	-

The studied binders were selected for verifying the prescriptions for a standardize commercial cement type CEM II/B, according to the standard UNE-EN 197-1 [12]. The pu pose of choosing this type of cement is related to the fact that today, type II cements [12] ar mostly produced in Spain. Therefore, this could allow a more widespread real applicatio of the results derived from this work.

The mortars were made with water to binder ratio 0.5. In addition, for all of them, th aggregate to binder ratio was 3:1. The fine aggregate used verified the prescriptions of th standard UNE-EN 196-1 [46].

Three kinds of specimens were made. The first kind of sample consisted of cylinder with dimensions 5 cm diameter and 6 cm height. The second kind was also cylinders, bu now with 10 cm diameter and 22 cm height. Finally, prismatic samples with dimension 4 cm × 4 cm × 16 cm were cast.

During the initial 24 h, the samples were stored in a chamber under an optimum lab ratory condition (20 °C temperature and 95% relative humidity). Once finished that tim they were de-molded and kept in that optimum condition until 7 days, when they wer moved to the real in-situ exposure site. It has been reported by several authors [20,40,47,48 that an adequate curing is important for satisfactorily developing the properties of ceme titious and pozzolanic materials exposed to non-optimum and harmful conditions. In view of that, here, a 7-days curing period was selected. Lastly, the tests were performed at 2 and 250 days.

2.2. Environmental Exposure Condition

The real in-situ condition consisted of placing the samples in an underground floo of a building, used as a garage with moderate circulation of vehicles. This site would b compatible with the specifications of exposure class XC3 (corrosion induced by carbonatio moderate humidity) defined by the Eurocode 2 [45].

The exposure period started at the age of 7 hardening days, after ending the curin time of the samples, as explained in the previous subsection, and finished at 250 harde ing days.

The average CO_2 concentration measured in the exposure site along the studied tim period was around 2000 ppm, reaching an absolute maximum value slightly higher tha 5000 ppm. On the other hand, the temperatures registered were overall in the rang between 18 °C and 23 °C, whereas the average relative humidity was in the interval fro 65% to 70%.

2.3. Mercury Intrusion Porosimetry

This technique has been widely used for characterizing the microstructure of differen types of materials [49–51]. A Poremaster-60 GT porosimeter commercialized by Quan tachrome Instruments (Boynton Beach, FL, USA) was used. Prior to performing the tes the samples were dried in an oven at 50 °C for 48 h. Here, the total porosity and the por size distributions obtained with this technique were analyzed. Regarding the pore siz distributions, the next ranges of pore diameters were considered: <10 nm, 10–100 nm 100 nm to 1 μm, 1–10 μm, 10 μm to 0.1 mm, and >0.1 mm [52,53]. Two measurements wer performed on each series at both testing ages. Pieces taken from cylindrical specimens wit 5 cm diameter and 6 cm height were tested.

2.4. Electrical Resistivity

Electrical resistivity is a useful parameter for obtaining additional data of the por network of cementitious materials [54,55]. The resistivity of the mortars was obtaine with the non-destructive Wenner four-point test, described in the Spanish standard UN 83988-2 [56]. A Proceq analyzer was used to directly obtain the electrical resistivity o cylindrical specimens with 22 cm height and10 cm diameter. The measurements wer performed at several ages up to 250 days. The surface of the cylinders was prepare according to the technical recommendation number 5 of Alconpat International [57], prio to each measurement. For each binder, three cylinders were tested and four measurements were performed per sample at each age.

2.5. Water Absorption

The water absorption after immersion was obtained according to the procedure explained in the ASTM Standard C642-06 [58]. Six pieces taken from cylinders with dimensions 5 cm diameter and 6 cm height were tested for each binder at 28 and 250 days.

2.6. Steady-State Chloride Diffusion Coefficient

The steady-state chloride diffusion coefficient was obtained from the electrical resistivity of the water-saturated samples. The electrical resistivity was measured according to the procedure explained in Section 2.4. Before the measurements, the specimens were saturated in water along 24 h following the standard ASTM C1202-97 [59]. For each series, three cylindrical specimens with 22 cm height with 10 cm diameter were tested at 28 and 250 days. Four measurements were performed per sample at both testing ages. Lastly, the steady-state diffusion coefficient was calculated with the following equation [60]:

$$D_S = \frac{2 \times 10^{-10}}{\rho} \quad (1)$$

where D_s is the chloride steady-state diffusion coefficient through the sample (m^2/s) and ρ is the electrical resistivity of the specimen ($\Omega \cdot m$).

2.7. Carbonation Depth

The carbonation front depths in the mortars were obtained following the RILEM recommendation CPC-18 [61]. Pieces extracted from the cylinders with 5 cm diameter and 6 cm height were sprayed with a 1% phenolphthalein solution. The depth of the colorless carbonated part from the external surface of the sample was measured. For each series, six pieces taken from the abovementioned cylinders were tested at 28 and 250 days.

2.8. Mechanical Strengths

The compressive and flexural strengths were determined following the procedure included in the Spanish and European standard UNE-EN 1015-11 [62]. For each series, three different prismatic specimens with dimensions 4 cm × 4 cm × 16 cm were tested at 28 and 250 days.

2.9. Ultrasonic Pulse Velocity

The ultrasonic pulse velocity (UPV) constitutes a helpful parameter for getting additional data about the mechanical behavior of the material [63]. This parameter was obtained according to the standard UNE-EN 12504-4 [64]. In this work, the propagation time of the ultrasonic waves was determined in the largest dimension of the sample (160 mm) with direct transmission, using a Pundit Lab model commercialized by Proceq company (Schwerzenbach, Switzerland). Contact transducers which emitted ultrasonic pulses at 54 kHz were attached to the top and bottom base sides of the samples with a coupling gel. The UPV was calculated from the propagation time and the length of the sample. This parameter was obtained at several hardening times until 250 days. At each age, for the same mortar series, three prismatic specimens with dimensions 4 cm × 4 cm × 16 cm were tested and three determinations were performed per specimen.

3. Results

3.1. Mercury Intrusion Porosimetry

Regarding the mercury intrusion porosimetry results, the total porosities noted for the binders analyzed at 28 and 250 days are shown in Figure 1. At 28 days, this parameter was relatively similar for all the mortars. Between 28 and 250 days, a reduction in total porosity was observed for REF, S, F, and SL mortars, whereas it increased for L, SF, and

FL mortars. At the last age studied, the lowest total porosities were noted for REF and binders, followed by F and SL ones. Finally, the highest values of this parameter at 250 da were noted for L, SF, and FL mortars.

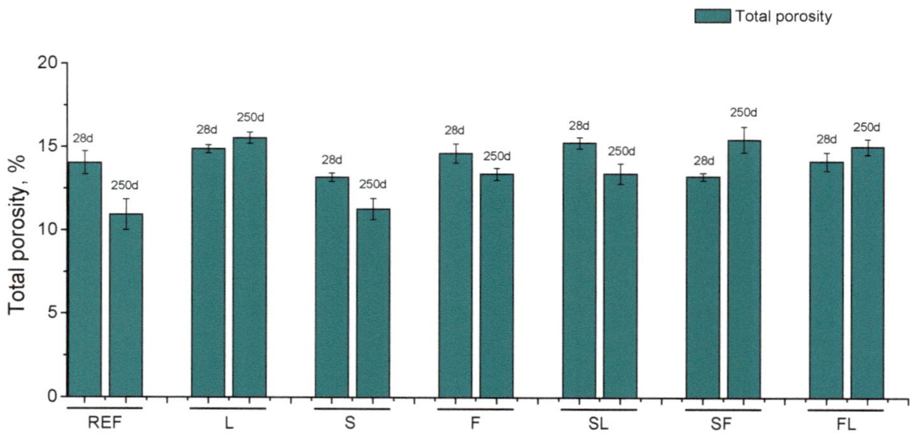

Figure 1. Results of total porosity for the studied mortars. In this figure and in the next figures, the mean value of the parameter obtained for each type of mortar is represented. The error bars represent the standard deviation.

The pore size distributions are depicted in Figure 2. At 28 days, the relative volum of finer pores (diameters in the ranges <10 nm and 10–100 nm) was higher for S, F, an SF binders compared to REF ones, being lower for the rest of series analyzed. In genera the proportion of pores with diameters lower than 100 nm decreased for all the mortar studied from 28 to 250 days. Despite that, the percentage of pores with sizes lower tha 1 μm at 250 days was higher for S, F, SF, and FL mortars in comparison with referenc specimens. This percentage was similar at that age for REF, L and SL mortars, although th REF and SL series showed higher relative volume of pores lower than 100 nm compared t L mortars.

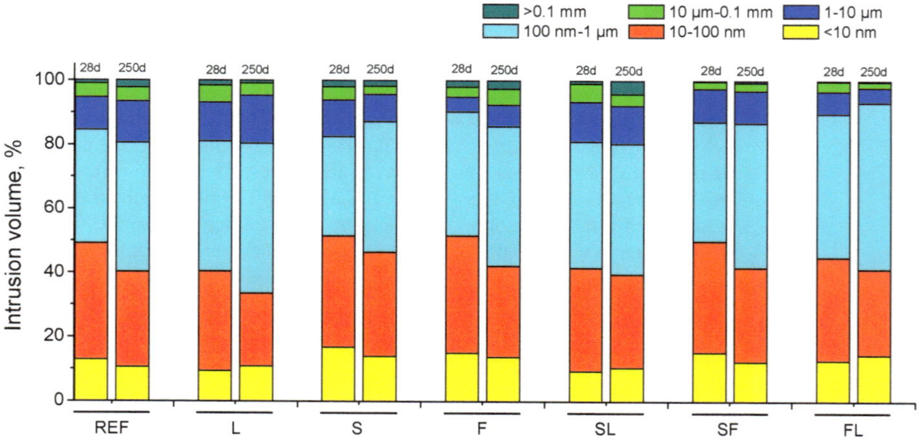

Figure 2. Pore size distributions noted for the analyzed binders.

3.2. Electrical Resistivity

The results of electrical resistivity are represented in Figure 3. In general, this parameter rose with age for the different types of mortars analyzed. Around 28 days, the highest values of electrical resistivity were noted for S mortars, followed by SF ones. From then, mortars with fly ash (F, FL, and SF) developed a noticeable growth of the electrical resistivity, with F and SF mortars showing the highest values of this parameter at 250 days, followed by FL binder. The other binary and ternary binders with slag (S and SL) presented lower resistivity values in the long term, compared to those specimens with fly ash. At 250 days, this parameter was very similar for both S and SL series. However, the increase with time of the electrical resistivity was slower for SL mortars in comparison with S ones. On the other hand, from 28 to 250 hardening days, REF and L mortars showed the lowest resistivity values, and the increasing rate with time of this parameter was also lower compared to the other binders studied.

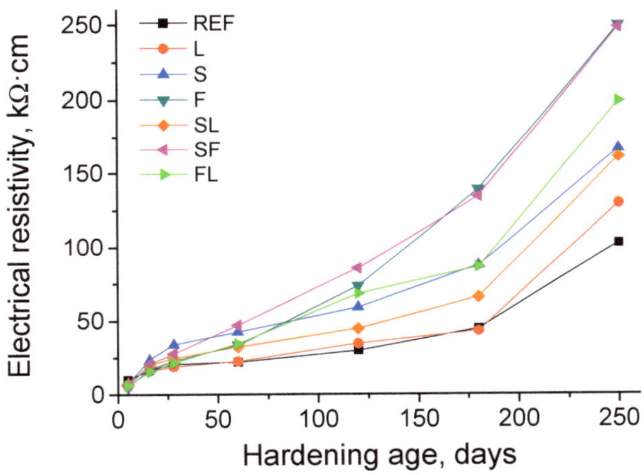

Figure 3. Electrical resistivity results for the different types of mortars tested.

3.3. Water Absorption

The results of water absorption after immersion are depicted in Figure 4. A slight decrease with time of this parameter was noted for all the mortars studied. In general, few differences between the analyzed binders were observed at 28 and 250 days regarding the water absorption after immersion.

3.4. Steady-State Chloride Diffusion Coefficient

The results of steady-state chloride diffusion coefficient obtained from sample's resistivity for the analyzed mortars can be observed in Figure 5. All the binary and ternary binders studied showed higher values of this parameter at 28 days compared to the reference specimens. The highest diffusion coefficients at that age were noted for F and L series, followed by the three ternary binders analyzed (SL, SF, and FL series). On the other hand, the lowest 28-days value of this parameter for mortars with additions corresponded to the S series. Between 28 and 250 days, the diffusion coefficient decreased for all the mortars studied, and this reduction was more noticeable for those which incorporate at least one active addition in the binder (S, F, SF, SL, and FL series). The lowest coefficient at 250 days was noted for F and SF mortars, closely followed by S and FL ones, while it was slightly higher for SL series. In addition, for all the binary and ternary binders already mentioned, the diffusion coefficient was lower at 250 days in comparison with reference mortars. Lastly, at that last age, L mortars presented the highest value of this coefficient.

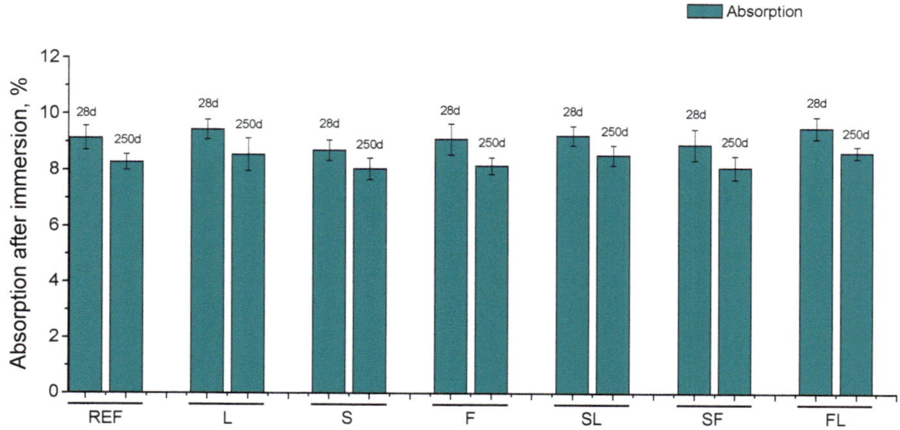

Figure 4. Results of absorption after immersion noted for the studied series.

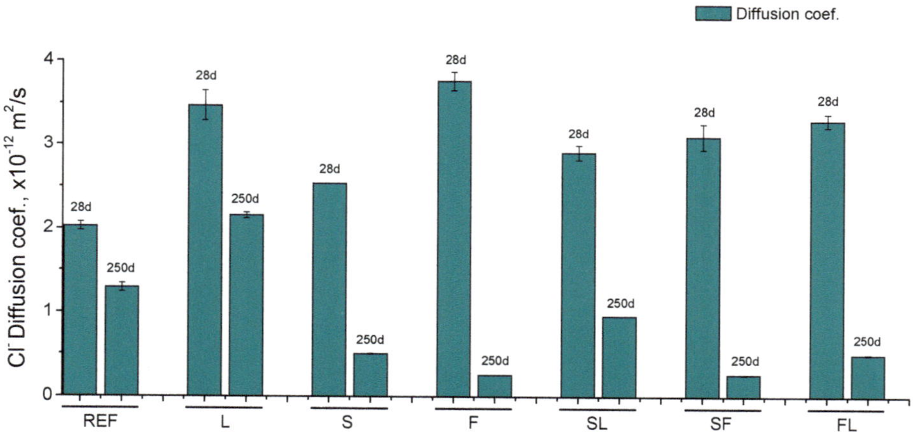

Figure 5. Steady-state chloride diffusion coefficient results obtained for the analyzed mortars.

3.5. *Carbonation Front Depth*

The depths of carbonation front obtained for the studied binders are represented in Figure 6. Reference mortars showed the lowest carbonation front depths at 28 days followed by S ones, whereas these depths were higher for the other binders studied but with slight differences between them. From 28 to 250 days, the carbonation depth increased for all the analyzed mortars. At 250 days, this parameter was again lower for REF series, compared to the binders with additions. The highest carbonation depths at that age were noted for the binary binders, particularly for L mortars. Regarding the ternary binders (SL, SF, and FL series), they presented slight lower depths than binary ones.

3.6. *Mechanical Strengths*

The compressive strength results are depicted in Figure 7. This parameter was lower for the mortars with additions at 28 days in comparison with REF specimens. With respect to the binders with active additions, the highest compressive strength in the short term was observed for SF mortars, whereas this parameter was barely lower for S, F, and SL series, followed by FL ones. At 28 days, the smallest compressive strength corresponded to

L mortars. Between 28 and 250 days, the compressive strength decreased for the REF and L series, while for binders with at least one active addition, it remainedpractically constant (S, SL, SF, and FL series) or even increased (F series). The S, F, and SF mortars showed very similar compressive strength at 250 days compared to REF mortars. Nevertheless, at the abovementioned age, the ternary binders with limestone (SL and FL series) had lower values than reference mortars, although their compressive strength was higher than that observed for the binary binder with only limestone as addition (L series).

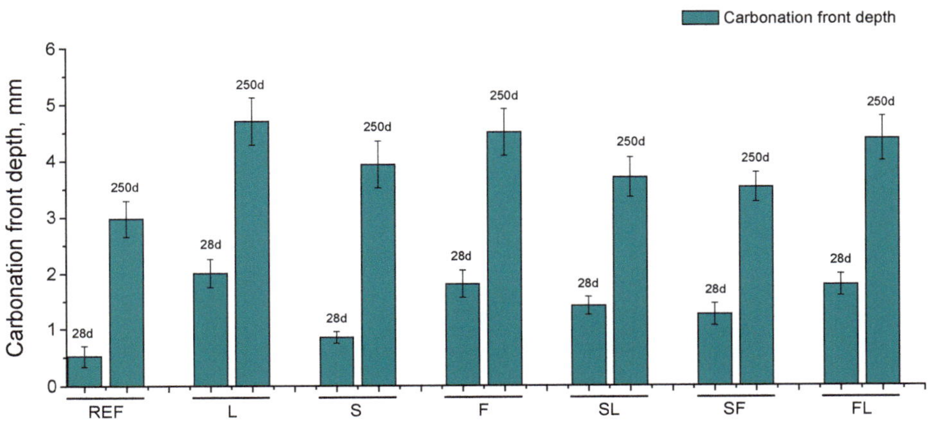

Figure 6. Carbonation front depths observed at 28 and 250 days for the characterized binders.

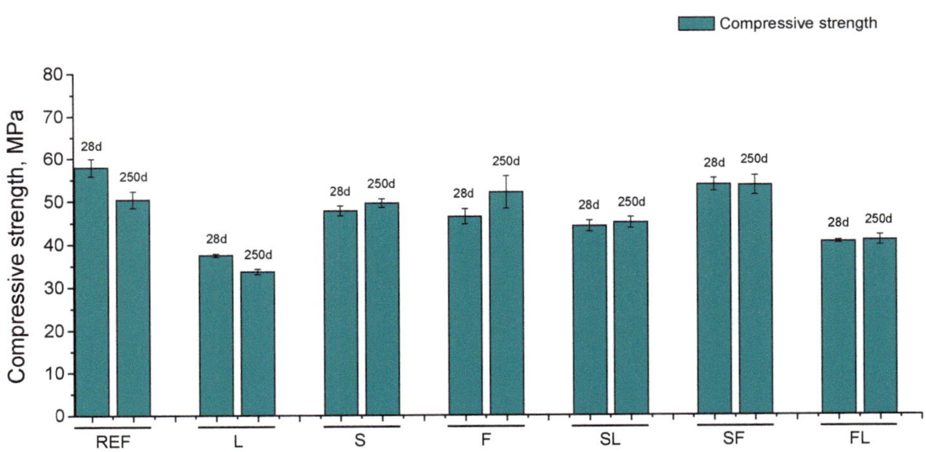

Figure 7. Compressive strength results noted for the analyzed mortars.

In relation to the flexural strength, its results can be observed in Figure 8. Scarce differences have been noted in this parameter between the mortars tested, being in the range from 7.5 to 8.5 MPa for most of them at the studied hardening ages. This strength hardly changed with time for REF, SL, SF, and FL series, and it decreased slightly for S and F ones. The most noticeable fall of flexural strength from 28 to 250 days was observed for L mortars, showing the lowest value of this parameter at 250 days of all the studied series. At that last testing age, the highest flexural strength was noted for REF and FL mortars, closely followed by the S, F, SL, and SF series.

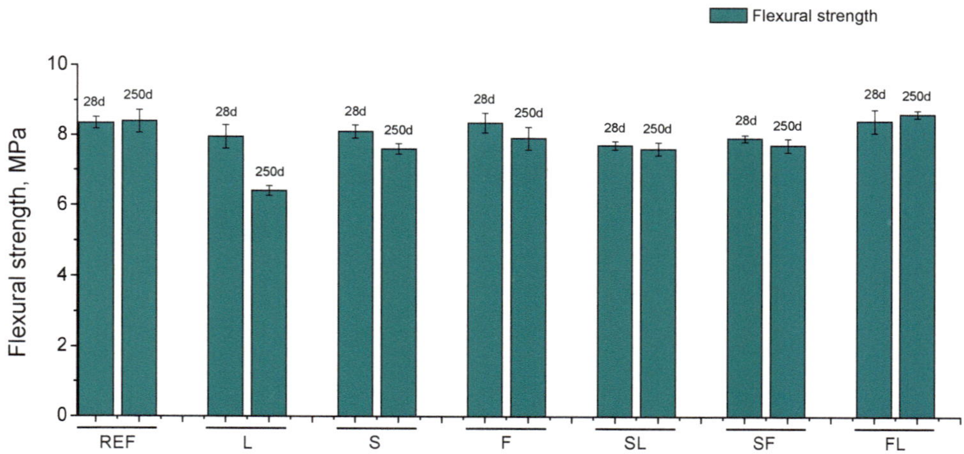

Figure 8. Flexural strength results for the different mortars tested.

3.7. Ultrasonic Pulse Velocity

The evolution of the ultrasonic pulse velocity (UPV) is represented in Figure 9. In general, the main increase of this parameter for most of the analyzed binders was observed in the very short term. At initial hardening times, REF mortars showed higher UPV than the other series studied. With respect to the mortars with active additions, at those early ages, the binary binders S and F presented slight greater values of this parameter than the ternary binders SF, SL, and FL. At later ages, the UPV for the mortars which incorporate active additions overall rose with time, although with a relatively slow rate. In contrast, this parameter slightly decreased in the long term for reference mortars. At 250 days, the UPV was higher for S and F series compared to REF ones. Moreover, for the tested ternary binders (SF, SL, and FL series), it was similar to reference mortars. During the studied hardening time, the lowest values of UPV parameter were noted for L mortars.

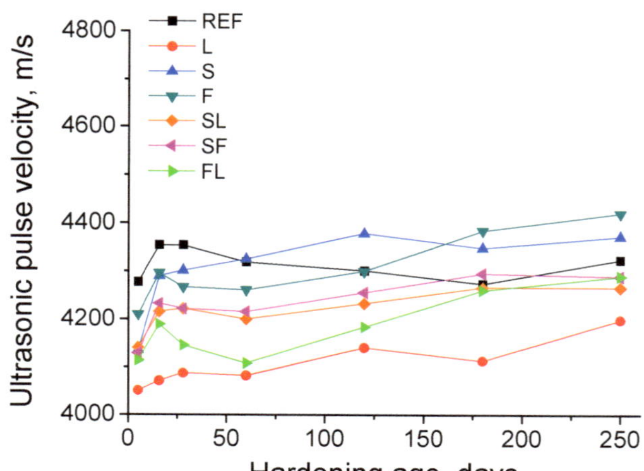

Figure 9. Ultrasonic pulse velocity results for the analyzed binders.

4. Discussion

4.1. Microstructure Characterization

Firstly, in relation to total porosity results (see Figure 1), the scarce differences at 28 days in this parameter between the different mortars analyzed could be related to the environmental relative humidity (RH) in the exposure location of the samples. As was explained in Section 2.2, the average RH registered in the exposure station was in the range 65–70%. This RH was relatively high, although it was lower than the RH of an optimum standardized laboratory condition (95–100%). Therefore, this lower RH in the environment could affect the slag and clinker hydration reactions [39,65], because they need the presence of water to be developed [66,67]. Then, with a lower availability of water in the environment, these hydration reactions would slow down [19,39,65], giving as a result that the total porosity differences were less noticeable in the short term comparing the binders analyzed. Furthermore, this could affect the binders with fly ash too, because the development of pozzolanic reactions of fly ash could also be slowed down by a lower environmental RH [68,69], in a similar way to that explained for slag and clinker hydration [19,39,65]. In addition, the slower development of those hydration reactions could produce a delay in the starting of fly ash pozzolanic reactions [69], because they need the presence of enough portlandite to be developed [9,14].

The differences in pore size distributions at 28 days between the binders analyzed (see Figure 2) were not high, which could be related to the abovementioned effects of the lower environmental RH, although several effects of the studied additions in the microstructure of mortars could be observed. On one hand, the pore network was more refined at 28 days for S, F, and SF binders, as suggests their higher percentage of pores with sizes lower than 100 nm, and particularly those pores in the range <10 nm, compared to reference specimens. This could be related to the short-term influence in the microstructure of slag hydration [39,65,70] and fly ash pozzolanic reactions [9], which produced additional solid phases [7], and their effects were noticeable despite the lower RH provided by the environment. Moreover, when both slag and fly ash additions were combined in the same binder, their combined effects also gave an improvement of the microstructure refinement, at least in the short term and under the environmental conditions analyzed, as would indicate the pore size distribution of SF series at 28 days, compared to the reference one. On the other hand, the slightly less refined pore structure noted for ternary binders with limestone (SL and FL series) at short times, in comparison with the other binders with slag and fly ash, would be related to the inert character of the limestone, because it is not an active addition, without hydraulic or pozzolanic activity [71]. Therefore, the limestone addition only has a filler effect [71] and it does not produce additional solid phases, quite the opposite of what happened with slag and fly ash [7], so its influence in the microstructure is more limited. This would also explain the less refined microstructure of all the mortars studied for binary binder with limestone (L series).

Regarding the evolution with time of the pore size distributions, a loss of the microstructure refinement was observed for all the mortars studied from 28 to 250 days, as suggested by the reduction of the relative volume of finer pores size intervals (see Figure 2). On one hand, this could be due to development of the carbonation phenomenon in the mortars, produced by the CO_2 present in the environment (see Section 2.2), as revealed by the carbonation front depth measurements (see Figure 6). Several authors [66,67,72] have reported that this coarser pore network due to the carbonation development could be related to the additional formation of silica during the decomposition of C-S-H gel caused by the exposure to CO_2. On the other hand, the lower RH in the environment could also have an influence in the reduction of microstructure refinement with the hardening time, producing the formation of shrinkage microcracks by drying [65,67,73]. According to several works [67,74], the RH of the exposure medium greatly affects the magnitude of shrinkage. Therefore, the development of shrinkage microcracking could contribute to produce a coarser pore structure in the long term for the analyzed binders. Additionally, as has been already explained, the lower RH would make the development of hydration and pozzolanic reactions slower and more difficult once the setting water was consumed [19,39,65,68,69], so the additional solid phases formation as products of those reactions could not totally counteract the effects of carbonation and shrinkage in the microstructure. Despite that, the overall higher pore refinement observed at 250 days for ternary and binary binders with at least one active addition (slag and/or fly ash) incorporated, compared to reference specimens, would reveal their beneficial long-term effects in the microstructure [7], even under the exposure to a non-optimum real condition.

With respect to the evolution of total porosity between 28 and 250 days (see Figure 1), it would also be influenced by both shrinkage and carbonation phenomena. For several binders studied (REF, S, F, and SL series), this parameter decreased with time and this reduction was more noticeable for REF mortars. In this regard, several authors [72,75,76] have reported similar results in cement-based materials exposed to carbonation. This could be related to the formation of $CaCO_3$ inside the pores, as products of carbonation development, because the $CaCO_3$ occupies a higher volume than the initial hydrated phases from which it is formed, giving as a result a reduction of porosity [66]. The higher reduction of total porosity noted for REF mortars would also be in keeping with other researchers [72]. These reference mortars had greater content of clinker than those with additions. Therefore, the reference mortars would have a higher availability of portlandite (formed by clinker hydration) for reacting with the environmental CO_2, so a greater amount of $CaCO_3$ as a consequence of the carbonation process would be produced [72], giving as a result the more notable reduction of total porosity noted for REF series. Furthermore, the loss of microstructure refinement due to carbonation, already discussed in terms of the pore size distribution results, while a reduction of total porosity was simultaneously produced, would agree with other works [72,76].

Nevertheless, the possible development of shrinkage microcracking in the long term due to the lower RH in the environment would produce an increase of total porosity [65,67,73,74], as has been previously explained for the results of pore size distribution. Then, the effect of the shrinkage would counteract the decrease of porosity produced by carbonation. The more noticeable influence of shrinkage could explain the increase of total porosity with age noted for some of the analyzed binders (L, SF, and FL series).

Electrical resistivity is a useful parameter for getting information related to the pore connectivity and the changes in the pore structure. For all the studied mortars, an increase of the electrical resistivity with hardening time was observed (see Figure 3), which would suggest a progressive microstructure refinement [77], with a rise of the relative volume of finer pores. This result would not be in agreement with those obtained with mercury intrusion porosimetry, particularly with the pore size distributions previously discussed, which revealed an overall coarsening of the microstructure with age. Firstly, the saturation degree of the material is one of the most important factors which can affect the electrical resistivity values [57,77]. This saturation degree changes when a drying of the material is produced, for example under non-optimum RH environments. Several works have concluded that the electrical resistivity considerably increases as the saturation degree of the material decreases [78,79]. In the analyzed environment, the average RH was around 65% (see Section 2.2), so it would produce a progressive drying of the samples, which could explain the gradual rise with time of electrical resistivity noted for the mortars.

Another possible factor that could have contributed to the differences between electrical resistivity and porosimetry results would be related to the characteristics of each technique and the geometry of the samples used for each one. On one hand, small pieces taken from cylindrical specimens with 5 cm diameter and 6 cm height were tested with mercury intrusion porosimetry. On the other hand, the electrical resistivity was measured with the Wenner four-point test on cylinders with 22 cm height and 10 cm diameter. In the case of cylindrical specimens with 5 cm diameter and 6 cm height, due to their relatively small volume, after 250 days of exposure, the development of their microstructure would be more globally affected by the environmental conditions, and as a consequence by the abovementioned harmful processes developed in this environment, previously described [illegible]

For the cylinders with 22 cm height and 10 cm diameter, their higher size and volume would entail that the external part of the sample would be more influenced by the environment, whereas its influence would be lower and delayed in the core of the sample. Then, the development of the microstructure towards finer pores could be higher in the core part of the sample, because the effects of environmental drying would affect this part with a lesser degree, allowing a greater development of hydration and pozzolanic reactions [39,65–67]. The electrical resistivity measurements provide more global information about the microstructure of the sample [77] compared to mercury intrusion porosimetry, so these possible differences in the pore network between different parts of the specimen could explain the lack of coincidences between the resistivity measurements and pore size distributions.

Despite the abovementioned arguments, all the samples used for measuring the electrical resistivity were subject to the same environment during the studied time period, so the results of this parameter for the analyzed mortars can be compared. At early ages, the greater resistivity values showed by S and SF binders, could be related to the hydration of slag [7], already explained. The low resistivity noted in the short term for binary and ternary binders with fly ash (F and FL series) and their progressive growth with time (F, SF, and FL series), even overtaking the values of the other binders with slag (S and SL series) at the latest testing age, would show the effects of fly ash pozzolanic reactions [9,80], reducing the pore size, as well as the delay of their starting compared to slag and clinker hydration, previously mentioned in the discussion of total porosity results at 28 days. Finally, the greater electrical resistivity noted for binders which incorporate at least one active addition (fly ash and/or slag) would be in keeping with other authors [77]. This would indicate a higher proportion of pores with small sizes in their porous network, agreeing with results of porosimetry, which generally showed a higher pore refinement in those mortars in contrast with REF mortars, and particularly with L ones.

4.2. *Durability and Mechanical Parameters*

Regarding the durability-related parameters, the highest values of the steady-state chloride diffusion coefficient noted at 28 days (see Figure 5) for the binary binder with fly ash (F series) would be related to the abovementioned delay of the initiation of fly ash pozzolanic reactions [9,14], compared to slag and clinker hydration, in which the lower RH in the environment could also have an effect. This delay was also noticeable in the ternary binders with fly ash (FL and SF series), although their diffusion coefficient at 28 days was lower than that noted for F series, probably due to the influence of the other addition present in these binders, such as the filler effect of limestone [26,71] and the slag hydration [7,66]. The lowest diffusion coefficients noted for REF and S mortars at 28 days could be explained in terms of clinker and slag hydration, particularly their sooner starting [39,65], despite the lower environmental RH. In the case of binary binder with limestone (L series), the high diffusion coefficient in the short term could be related to the lack of hydraulic or pozzolanic activity of this addition [26,71], already discussed for pore size distributions results. The non-active character of this addition would also explain the higher coefficient at 28 days noted for SL binder in comparison with S one.

With respect to the evolution of the diffusion coefficient, a decrease of this parameter from 28 and 250 days was observed. This tendency would be overall in agreement with the rise with time of electrical resistivity, although it would not coincide with the evolution of pore size distributions. As was described in Section 2.6, the steady-state chloride diffusion coefficient was determined from the electrical resistivity of water-saturated samples. These samples were cylinders with 22 cm height and 10 cm diameter, similar to those used for following the changes in the electrical resistivity in non-saturated samples. Therefore, part of the arguments previously given to justify the differences between the porosimetry and resistivity results would also be valid for explaining the evolution of diffusion coefficient. On one hand, the effect of saturation degree of the material would not be considered in the results of this parameter because water-saturated specimens were used for measuring the electrical resistivity, from which the diffusion coefficient was calculated. On the oth hand, the different development of the microstructure in the external and core part of th sample, more notable in those specimens used for studying the electrical resistivity and th diffusion coefficient than in those used for taking the pieces tested with porosimetry, cou be compatible with the diffusion coefficient results obtained. Then, the lower influence the environment, particularly the drying process produced by the lower RH, would allo the pore network to become more refined in the inside part of the samples, giving as result a lower global chloride diffusion coefficient, despite being superficial parts of th sample with a coarser microstructure and more affected by the exposure condition and i harmful processes.

It is interesting to highlight that all the binary and ternary binders with slag and/ fly ash tested showed lower diffusion coefficients at 250 days than reference mortars und the studied environment. The appreciable fall of this parameter for samples with fly ash (SF, and FL series) would show the effect of the pozzolanic activity of fly ash [9,14] in th pore size reduction in the long term, previously explained. The slag hydration produce a similar effect [7], also entailing an important reduction of the diffusion coefficient wi time, as suggested by the results of the S, SF, and SL series. In REF and L mortar the clinker hydration would contribute to reducing this parameter, although their les refined microstructure would entail the higher diffusion coefficients noted for them in th long term.

In relation to the water absorption after immersion (see Figure 4), scarce difference between the studied binders have been noted at the testing ages, and a slight decrease this parameter was observed for all of them from 28 to 250 days. This result would no coincide with those previously discussed, particularly the similar values for all the binde This lack of differences could be related to the experimental procedure described in th ASTM Standard C642-06 [58], which was followed to determine the water absorption. Th first step of this procedure consisted of drying the samples in an oven at a temperatu of 100 to 110 °C for not less than 24 h. This conditioning of the samples could induc the formation of shrinkage cracks in the mortars [25,67,73], which would homogenize th behavior of the mortars regarding the water absorption by immersion (the samples wer water-saturated once finished the drying process in a second step of the experiment procedure [58]). This would mask and notably reduce the effects produced by the exposu environment in each binder, giving the uniform result obtained in this parameter.

The higher carbonation depths observed at 28 and 250 days for the binary and ternar binders studied in comparison with reference specimens (see Figure 6) would agree wit other authors [66,67,72,76]. This could be explained in relation to the lower content portlandite in those mortars prepared with binary and ternary binders [67], which woul imply a faster carbonation of the material for the same concentration CO_2 in the enviro ment, reaching higher depths in the samples, as has been noted. Comparing the morta with additions, at both ages studied, the carbonation depth was slight higher for the L, and FL series. In the case of fly ash, this could be related to its pozzolanic activity, whic would entail a lower presence of portlandite, because it is consumed in the pozzolani reactions [9,14,66], giving as result the deeper carbonation observed, being in keeping wit the results pointed out by other authors [76]. Regarding the limestone, it has no pozzolani activity, so the greater carbonation depths, particularly for the L binder, could also b explained in terms of less pore refinement produced by this addition [67]. The increase carbonation front depths between 28 and 250 days for all the binders would be expecte due to the exposure to the analyzed environment (see Section 2.2), in an undergroun garage with moderate circulation of vehicles. Finally, as has been explained in the discu sion of mercury intrusion porosimetry results, the progressive development of carbonatio produced a loss of microstructure refinement with time, probably due to the reaction o C-S-H compounds with the environmental CO_2, producing silica formation, as has bee reported by several authors [66,67,72].

The effects of the exposure environment were also noticeable in the compressive strength (see Figure 7). In general, lower values of this parameter were obtained for mortars which incorporated limestone (L, SL, and FL series). As was explained in the microstructure characterization, this would be due to the fact that this addition does not have hydraulic or pozzolanic activity [71], having a negative effect on the compressive strength of the materials, even for the ternary binders in which it was combined with slag or fly ash. It is also interesting to highlight the good performance of SF binders in relation to this parameter, which would be due the synergetic effects combining two active additions in the binder [15]. The flexural strength showed a similar order of magnitude for all the mortars (see Figure 8), which would suggest a similar influence of the environment, independently of the binder, with only the lower values of this parameter noted for L mortars being remarkable, attributable to the abovementioned non-active character of this addition.

The scarce change or the decrease of both compressive and flexural strengths with exposure time would be overall in agreement with the pore size distributions, which showed a progressive reduction of pore refinement. This would be related to the harmful processes developed in the environment. On one hand, the lower RH in the environment would make the development of hydration and pozzolanic reactions more difficult [19,39,65,68,69], limiting the strength improvement with time. This lower RH would also produce the formation of drying shrinkage, which could lead to the formation of microcracks, which would contribute to the scarce rise or worsening of the mechanical performance of the materials [40,73,81]. On the other hand, the carbonation process, already described, would also have an influence in the recorded strength results, because their development also leads to a loss of the mechanical behavior of cement-based materials [82,83]. Lastly, it is interesting to highlight the adequate compressive strength of S, F, and SF at 250 days, with very similar values to reference ones.

The ultrasonic pulse velocity (UPV) results showed similarities with the mechanical strength ones. In the short term, the UPV was higher for reference mortars (see Figure 9), whereas it decreased with time after 28 days, in keeping with the results of compressive strength. This would indicate the initial effects of clinker hydration, closing the microstructure and improving the mechanical behavior of the material, and thus raising the UPV, as well as the damaging influence of the environment in the long term, which was revealed by the reduction with time of this parameter. The rest of the binders studied showed a similar trend, with a UPV increase in the very short term, linked to the early development of hydration and pozzolanic reactions, which reduced the voids in the mortars [7], and keeping practically constant or barely increasing this parameter from 28 days on, due to the harmful effects of the environment, as already discussed. In the particular case of the S and F series, the UPV rose with time, but with a slower rate for F, coinciding with their compressive strength results and showing the different behaviors of slag and fly ash additions. Ternary mortars with limestone overall showed slight lower UPV values in the long term compared to reference ones, which would also agree with their lower compressive strength results. The lowest UPV was noted for L mortars, coinciding with their lowest compressive strength of all the studied binders, due to the explained influence of limestone addition [71]. Finally, S and F mortars presented greater UPV values than REF specimens at 250 days, and they were very similar for SF series compared to REF ones, so this would also be in accordance with their compressive strength results at that age.

5. Conclusions

The main conclusions that can be drawn from the results previously discussed can be summarized as follows:

- In the short term, the differences regarding the total porosity and the pore size distributions were not high between the studied mortars. Despite that, the pore network at 28 days was more refined for binary binders with slag and fly ash (S and F series) and for ternary binder which incorporate both additions (SF series). The presence c limestone in the binder overall reduced this pore refinement.

- A loss of microstructure refinement with time was noted for all the analyzed binder This could be related to the development of carbonation process, due to the CO present in the environment, as well as by the possible formation of drying shrinkag microcracks in the long term caused by the lower environmental relative humidity.
- The slight differences between the studied binders regarding the water absorptio after immersion would be influenced by the experimental procedure used for i determination, which would homogenize the effects of exposure environment i each binder.
- The carbonation front depths were higher for the analyzed binary and ternary binder in comparison with reference mortars. This was particularly noticeable for mortar with fly ash and limestone.
- The mechanical strengths hardly changed or even decreased with time depending o the binder, which would also be related to the harmful processes developed in th environment, such as carbonation and drying shrinkage. In addition, the ultrasoni pulse velocity results were compatible with the mechanical strength ones.
- The main novelty of this research was its analysis of the behavior of several mo tars prepared with binary and ternary blended cements, which accomplished th prescriptions required for a standardized commercial cement type CEM II/B whe they were exposed to environmental conditions compatible with the specification of exposure class XC3, defined by Eurocode 2. In view of the results obtained, it i interesting to highlight that the binary and ternary binders with at least one activ addition overall showed a higher pore refinement and lower steady-state chlorid diffusion coefficient in the long term compared to reference mortars. In relation to th mechanical properties at later exposure times, the binary mortars with slag and fl ash (S and F series) and the ternary binder with both additions (SF series) showed th best mechanical performance, similar to reference mortars.

Author Contributions: Conceptualization, J.M.O., J.I.-G. and T.R.-H.; methodology, J.I.-G., J.M.C and T.R.-H.; investigation, J.I.-G. and J.M.O.; data curation, J.I.-G. and J.M.O.; writing—original dra preparation, J.I.-G.; writing—review and editing, J.M.O.; supervision, J.M.O. and T.R.-H.; fundin acquisition, J.M.O. The results included in this paper were obtained in the PhD thesis carried out b J.I.-G. at University of Alicante (Spain), under the supervision of J.M.O. and T.R.-H. All authors hav read and agreed to the published version of the manuscript.

Funding: This work was supported by the Conselleria de Educación, Investigación, Cultura Deporte (presently re-named as Conselleria de Innovación, Universidades, Ciencia y Socieda Digital) de la GeneralitatValenciana, Spain (grant code GV/2019/070).

Institutional Review Board Statement: Not applicable.

Informed Consent Statement: Not applicable.

Data Availability Statement: The data that support the findings of this study are available from th corresponding author, J.M.O., upon reasonable request.

Acknowledgments: Authors wish to thank Cementos Portland Valderrivas S.A. for providing th ordinary Portland cement CEM I 42.5 R and the limestone, fly ash, and ground granulated blas furnace slag additions used in this study.

Conflicts of Interest: The authors declare no conflict of interest.

References

1. Benhelal, E.; Shamsaei, E.; Rashid, M.I. Challenges against CO_2 abatement strategies in cement industry: A review. *J. Environ. Sc* **2021**, *104*, 84–101. [CrossRef]
2. Ponikiewski, T.; Gołaszewski, J. The effect of high-calcium fly ash on selected properties of self-compacting concrete. *Arch. Ci Mech. Eng.* **2014**, *14*, 455–465. [CrossRef]

Hadji, T.; Guettala, S.; Quéneudec, M. Mix design of high performance concrete with different mineral additions. *World J. Eng.* **2021**, *18*, 767–779. [CrossRef]

Mounanga, P.; Khokhar, M.I.A.; El Hachem, R.; Loukili, A. Improvement of the early-age reactivity of fly ash and blast furnace slag cementitious systems using limestone filler. *Mater. Struct. Constr.* **2011**, *44*, 437–453. [CrossRef]

Yang, K.-H.; Jung, Y.-B.; Cho, M.-S.; Tae, S.-H. Effect of supplementary cementitious materials on reduction of CO_2 emissions from concrete. *J. Clean. Prod.* **2015**, *103*, 774–783. [CrossRef]

Puertas, F.; Barba, A.; Gazulla, M.F.; Gómez, M.P.; Palacios, M.; Martínez-Ramírez, S. Ceramic wastes as raw materials in portland cement clinker fabrication: Characterization and alkaline activation. *Mater. Constr.* **2006**, *56*, 73–84. [CrossRef]

Bijen, J. Benefits of slag and fly ash. *Constr. Build. Mater.* **1996**, *10*, 309–314. [CrossRef]

Faleschini, F.; Zanini, M.A.; Brunelli, K.; Pellegrino, C. Valorization of co-combustion fly ash in concrete production. *Mater. Des.* **2015**, *85*, 687–694. [CrossRef]

Wang, A.; Zhang, C.; Sun, W. Fly ash effects. *Cem. Concr. Res.* **2004**, *34*, 2057–2060. [CrossRef]

10. Schackow, A.; Stringari, D.; Senff, L.; Correia, S.L.; Segadães, A.M. Influence of fired clay brick waste additions on the durability of mortars. *Cem. Concr. Compos.* **2015**, *62*, 82–89. [CrossRef]
11. Tang, W.C.; Wang, Z.; Liu, Y.; Cui, H.Z. Influence of red mud on fresh and hardened properties of self-compacting concrete. *Constr. Build. Mater.* **2018**, *178*, 288–300. [CrossRef]
12. AENOR UNE-EN 197-1:2011. *Composición, Especificaciones y Criterios de Conformidad de los Cementos Comunes*; Asociación Española de Normalización y Certificación: Madrid, Spain, 2011.
13. Wedding, P.; Manmohan, D.; Mehta, P. Influence of Pozzolanic, Slag, and Chemical Admixtures on Pore Size Distribution and Permeability of Hardened Cement Pastes. *Cem. Concr. Aggreg.* **1981**, *3*, 63. [CrossRef]
14. Papadakis, V.G. Effect of fly ash on Portland cement systems. *Cem. Concr. Res.* **1999**, *29*, 1727–1736. [CrossRef]
15. Hale, W.M.; Freyne, S.F.; Bush, T.D., Jr.; Russell, B.W. Properties of concrete mixtures containing slag cement and fly ash for use in transportation structures. *Constr. Build. Mater.* **2008**, *22*, 1990–2000. [CrossRef]
16. Leng, F.; Feng, N.; Lu, X. An experimental study on the properties of resistance to diffusion of chloride ions of fly ash and blast furnace slag concrete. *Cem. Concr. Res.* **2000**, *30*, 989–992. [CrossRef]
17. Nochaiya, T.; Wongkeo, W.; Chaipanich, A. Utilization of fly ash with silica fume and properties of Portland cement–fly ash–silica fume concrete. *Fuel* **2010**, *89*, 768–774. [CrossRef]
18. Falaciński, P.; Machowska, A.; Szarek, Ł. The impact of chloride and sulphate aggressiveness on the microstructure and phase composition of fly ash-slag mortar. *Materials* **2021**, *14*, 4430. [CrossRef]
19. Geiseler, J.; Kollo, H.; Lang, E. Influence of blast furnace cements on durability of concrete structures. *ACI Mater. J.* **1995**, *92*, 252–257.
20. Thomas, M.D.A.; Scott, A.; Bremner, T.; Bilodeau, A.; Day, D. Performance of slag concrete in marine environment. *ACI Mater. J.* **2008**, *105*, 628–634.
21. Pastor, J.L.; Ortega, J.M.; Flor, M.; López, M.P.; Sánchez, I.; Climent, M.A. Microstructure and durability of fly ash cement grouts for micropiles. *Constr. Build. Mater.* **2016**, *117*, 47–57. [CrossRef]
22. Jain, J.A.; Neithalath, N. Chloride transport in fly ash and glass powder modified concretes—Influence of test methods on microstructure. *Cem. Concr. Compos.* **2010**, *32*, 148–156. [CrossRef]
23. Kamali, M.; Ghahremaninezhad, A. An investigation into the hydration and microstructure of cement pastes modified with glass powders. *Constr. Build. Mater.* **2016**, *112*, 915–924. [CrossRef]
24. Ortega, J.M.; Esteban, M.D.; Rodríguez, R.R.; Pastor, J.L.; Ibanco, F.J.; Sánchez, I.; Climent, M.A. Long-Term Behaviour of Fly Ash and Slag Cement Grouts for Micropiles Exposed to a Sulphate Aggressive Medium. *Materials* **2017**, *10*, 598. [CrossRef] [PubMed]
25. Ortega, J.M.; Pastor, J.L.; Albaladejo, A.; Sánchez, I.; Climent, M.A. Durability and compressive strength of blast furnace slag-based cement grout for special geotechnical applications. *Mater. Constr.* **2014**, *64*, e003. [CrossRef]
26. Dhandapani, Y.; Santhanam, M.; Kaladharan, G.; Ramanathan, S. Towards ternary binders involving limestone additions—A review. *Cem. Concr. Res.* **2021**, *143*, 106396. [CrossRef]
27. Marchetti, G.; Irassar, E.F.; Rahhal, V.F. Effects of packing density and water film thickness on fresh and hardened properties of ternary cement pastes. *Adv. Cem. Res.* **2020**, *32*, 444–455. [CrossRef]
28. Agrupación de Fabricantes de Cemento de España OFICEMEN. Available online: https://www.oficemen.com/ (accessed on 20 September 2021).
29. Sanjuán, M.Á. Los cementos de adición en España del año 2000 al 2005. *Cem. Hormigón* **2007**, *909*, 4–55.
30. Dadsetan, S.; Bai, J. Mechanical and microstructural properties of self-compacting concrete blended with metakaolin, ground granulated blast-furnace slag and fly ash. *Constr. Build. Mater.* **2017**, *146*, 658–667. [CrossRef]
31. Ban, C.C.; Sern, L.J.; Jasme, N. The mechanical strength and drying shrinkage behavior of high performance concrete with blended mineral admixture. *J. Teknol.* **2019**, *81*, 59–67. [CrossRef]
32. Radlinski, M.; Olek, J. Investigation into the synergistic effects in ternary cementitious systems containing portland cement, fly ash and silica fume. *Cem. Concr. Compos.* **2012**, *34*, 451–459. [CrossRef]
33. De Weerdt, K.; Kjellsen, K.O.; Sellevold, E.; Justnes, H. Synergy between fly ash and limestone powder in ternary cements. *Cem. Concr. Compos.* **2011**, *33*, 30–38. [CrossRef]

34. Cordeiro, G.C.; Toledo Filho, R.D.; Tavares, L.M.; Fairbairn, E.M.R. Experimental characterization of binary and ternary blende cement concretes containing ultrafine residual rice husk and sugar cane bagasse ashes. *Constr. Build. Mater.* **2012**, *29*, 641–64 [CrossRef]
35. Radwan, M.K.H.; Onn, C.C.; Mo, K.H.; Yap, S.P.; Ng, C.G.; Yusoff, S. Eco-mechanical performance of binary and ternary ceme blends containing fly ash and slag. *Proc. Inst. Civ. Eng. Eng. Sustain.* **2020**, *174*, 23–36. [CrossRef]
36. Bumanis, G.; Zorica, J.; Bajare, D. Properties of foamed lightweight high-performance phosphogypsum-based ternary syste binder. *Appl. Sci.* **2020**, *10*, 6222. [CrossRef]
37. Moffatt, E.G.; Thomas, M.D.A.; Fahim, A. Performance of high-volume fly ash concrete in marine environment. *Cem. Concr. R* **2017**, *102*, 127–135. [CrossRef]
38. Detwiler, R.J.; Kjellsen, K.O.; Gjorv, O.E. Resistance to chloride intrusion of concrete cured at different temperatures. *ACI Mater.* **1991**, *88*, 19–24.
39. Çakır, Ö.; Aköz, F. Effect of curing conditions on the mortars with and without GGBFS. *Constr. Build. Mater.* **2008**, *22*, 308–31 [CrossRef]
40. Ramezanianpour, A.A.; Malhotra, V.M. Effect of curing on the compressive strength, resistance to chloride-ion penetration ar porosity of concretes incorporating slag, fly ash or silica fume. *Cem. Concr. Compos.* **1995**, *17*, 125–133. [CrossRef]
41. Shattaf, N.R.; Alshamsi, A.M.; Swamy, R.N. Curing/environment effect on pore structure of blended cement concrete. *J. Mat Civ. Eng.* **2001**, *13*, 380–388. [CrossRef]
42. Polder, R.B.; De Rooij, M.R. Durability of marine concrete structures—Field investigations and modelling. *Heron* **2005**, *50*, 133–15
43. Thomas, M.D.; Matthews, J. Performance of pfa concrete in a marine environment—-10-year results. *Cem. Concr. Compos.* **200** *26*, 5–20. [CrossRef]
44. Chalee, W.; Jaturapitakkul, C.; Chindaprasirt, P. Predicting the chloride penetration of fly ash concrete in seawater. *Mar. Stru* **2009**, *22*, 341–353. [CrossRef]
45. European Committee for Standardization. *EN 1992-1-1 Eurocode 2: Design of Concrete Structures—Part 1-1: General Rules and Rul for Buildings*; Committee European Normalization (CEN) Ed.: Brussels, Belgium, 2004.
46. AENOR UNE-EN 196-1:2005. *Métodos de Ensayo de Cementos. Parte 1: Determinación de Resistencias Mecánicas*; Asociación Españo de Normalización y Certificación: Madrid, Spain, 2005.
47. Ortega, J.M.; Sánchez, I.; Cabeza, M.; Climent, M.A. Short-Term Behavior of Slag Concretes Exposed to a Real In Situ Medite ranean Climate Environment. *Materials* **2017**, *10*, 915. [CrossRef]
48. Monaco, M.; Aurilio, M.; Tafuro, A.; Guadagnuolo, M. Sustainable mortars for application in the cultural heritage field. *Materia* **2021**, *14*, 598. [CrossRef]
49. Ouellet, S.; Bussière, B.; Aubertin, M.; Benzaazoua, M. Microstructural evolution of cemented paste backfill: Mercury intrusic porosimetry test results. *Cem. Concr. Res.* **2007**, *37*, 1654–1665. [CrossRef]
50. Horpibulsuk, S.; Rachan, R.; Chinkulkijniwat, A.; Raksachon, Y.; Suddeepong, A. Analysis of strength development in cemen stabilized silty clay from microstructural considerations. *Constr. Build. Mater.* **2010**, *24*, 2011–2021. [CrossRef]
51. Olson, R.A.; Neubauer, C.M.; Jennings, H.M. Damage to the pore structure of hardened portland cement paste by mercur intrusion. *J. Am. Ceram. Soc.* **1997**, *80*, 2454–2458. [CrossRef]
52. Pedeferri, P.; Bertolini, L. *La Durabilità del Calcestruzzo Armato*; Mc Graw-Hill Education: Milano, Italy, 2000; ISBN 978883860845
53. Ho, L.S.; Nakarai, K.; Duc, M.; Kouby, A.L.; Maachi, A.; Sasaki, T. Analysis of strength development in cement-treated soi under different curing conditions through microstructural and chemical investigations. *Constr. Build. Mater.* **2018**, *166*, 634–64 [CrossRef]
54. Polder, R.; Andrade, C.; Elsener, B.; Vennesland, Ø.; Gulikers, J.; Weidert, R.; Raupach, M. Test methods for on site measureme of resistivity of concrete. *Mater. Struct.* **2000**, *33*, 603–611. [CrossRef]
55. Lübeck, A.; Gastaldini, A.L.G.; Barin, D.S.; Siqueira, H.C. Compressive strength and electrical properties of concrete with whi Portland cement and blast-furnace slag. *Cem. Concr. Compos.* **2012**, *34*, 392–399. [CrossRef]
56. AENOR UNE 83988-2:2014. *Durabilidad del Hormigón. Métodos de Ensayo. Determinación de la Resistividad Eléctrica. Parte 2: Méto de las Cuatro Puntas o de Wenner*; Asociación Española de Normalización y Certificación: Madrid, Spain, 2014.
57. Andrade, C.; Climent, M.A. *Recomendaciones sobre determinación de la resistividad eléctrica del hormigón in situ mediante el método las cuatro puntas y del disco*; ALCONPAT Internacional: Mérida, Yucatán, Mexico, 2020; p. 11. [CrossRef]
58. ASTM C642-06. *Standard Test Method for Density, Absorption, and Voids in Hardened Concrete*; ASTM International: West Co shohocken, PA, USA, 2006; p. 3.
59. ASTM C1202-12. *Standard Test Method for Electrical Indication of Concretes Ability to Resist Chloride Ion Penetration*; ASTM Intern tional: West Conshohocken, PA, USA, 2012.
60. Andrade, C.; Alonso, C.; Arteaga, A.; Tanner, P. Methodology based on the electrical resistivity for the calculation of reinforceme service life. In *Proceedings of the 5th CANMET/ACI International Conference on Durability of Concrete, Supplementary Papers*; Malhotr V.M., Ed.; American Concrete Institute: Barcelona, Spain, 2000; pp. 899–915.
61. DE LA RILEM RE. CPC-18 Measurement of hardened concrete carbonation depth. *Mater. Struct.* **1988**, *21*, 453–455. [CrossRef]
62. AENOR UNE-EN 1015-11:1999. *Métodos de Ensayo de los Morteros Para Albañilería. Parte 11: Determinación de la Resistencia a Flexió y a Compresión del Mortero Endurecido*; Asociación Española de Normalización y Certificación: Madrid, Spain, 1999.

3. Hong, G.; Oh, S.; Choi, S.; Chin, W.-J.; Kim, Y.-J.; Song, C. Correlation between the compressive strength and ultrasonic pulse velocity of cement mortars blended with silica fume: An analysis of microstructure and hydration kinetics. *Materials* **2021**, *14*, 2476. [CrossRef]
4. AENOR UNE-EN 12504-4:2006. *Ensayos de Hormigón en Estructuras. Parte 4: Determinación de la Velocidad de los Impulsos Ultrasónicos*; Asociación Española de Normalización y Certificación: Madrid, Spain, 2006.
5. Ortega, J.M.; Sánchez, I.; Climent, M.A. Impedance spectroscopy study of the effect of environmental conditions in the microstructure development of OPC and slag cement mortars. *Arch. Civ. Mech. Eng.* **2015**, *15*, 569–583. [CrossRef]
6. Taylor, H.F.W. *Cement Chemistry*, 2nd ed.; Thomas Telford Publishing: London, UK, 1997; ISBN 978-0727725929.
7. Neville, A.M. *Properties of Concrete*, 4th ed.; Pearson Education Limited: Harlow, UK, 1995.
8. Shafiq, N.; Cabrera, J.G. Effects of initial curing condition on the fluid transport properties in OPC and fly ash blended cement concrete. *Cem. Concr. Compos.* **2004**, *26*, 381–387. [CrossRef]
9. Ortega, J.M.; Sánchez, I.; Climent, M.A. Impedance spectroscopy study of the effect of environmental conditions on the microstructure development of sustainable fly ash cement mortars. *Materials* **2017**, *10*, 1130. [CrossRef] [PubMed]
0. Bouikni, A.; Swamy, R.N.; Bali, A. Durability properties of concrete containing 50% and 65% slag. *Constr. Build. Mater.* **2009**, *23*, 2836–2845. [CrossRef]
1. Meddah, M.S.; Lmbachiya, M.C.; Dhir, R.K. Potential use of binary and composite limestone cements in concrete production. *Constr. Build. Mater.* **2014**, *58*, 193–205. [CrossRef]
2. Díaz, B.; Freire, L.; Nóvoa, X.R.; Pérez, M.C. Chloride and CO_2 transport in cement paste containing red mud. *Cem. Concr. Compos.* **2015**, *62*, 178–186. [CrossRef]
3. Kanna, V.; Olson, R.A.; Jennings, H.M. Effect of shrinkage and moisture content on the physical characteristics of blended cement mortars. *Cem. Concr. Res.* **1998**, *28*, 1467–1477. [CrossRef]
4. Ma, B.-G.; Wen, X.-D.; Wang, M.-Y.; Yan, J.-J.; Gao, X.-J. Drying Shrinkage of Cement-Based Materials Under Conditions of Constant Temperature and Varying Humidity. *J. China Univ. Min. Technol.* **2007**, *17*, 428–431. [CrossRef]
5. Anstice, D.J.; Page, C.L.; Page, M.M. The pore solution phase of carbonated cement pastes. *Cem. Concr. Res.* **2005**, *35*, 377–383. [CrossRef]
6. Ngala, V.T.; Page, C.L. Effects of carbonation on pore structure and diffusional properties of hydrated cement pastes. *Cem. Concr. Res.* **1997**, *27*, 995–1007. [CrossRef]
7. De Grazia, M.T.; Deda, H.; Sanchez, L.F.M. The influence of the binder type & aggregate nature on the electrical resistivity of conventional concrete. *J. Build. Eng.* **2021**, *43*, 102540. [CrossRef]
8. Andrade, M.C.; Bolzoni, F.; Fullea, J. Analysis of the relation between water and resistivity isotherms in concrete. *Mater. Corros.* **2011**, *62*, 130–138. [CrossRef]
9. Martínez-Molina, W.; Torres-Acosta, A.A.; Jáuregui, J.C.; Chávez-García, H.L.; Alonso-Guzmán, E.M.; Graff, M.; Arteaga-Arcos, J.C. Predicting concrete compressive strength and modulus of rupture using different NDT techniques. *Adv. Mater. Sci. Eng.* **2014**, *2014*, 742129. [CrossRef]
0. Chindaprasirt, P.; Jaturapitakkul, C.; Sinsiri, T. Effect of fly ash fineness on compressive strength and pore size of blended cement paste. *Cem. Concr. Compos.* **2005**, *27*, 425–428. [CrossRef]
1. Ortega, J.M.; Tremiño, R.M.; Sánchez, I.; Climent, M.Á. Effects of environment in the microstructure and properties of sustainable mortars with fly ash and slag after a 5-year exposure period. *Sustainability* **2018**, *10*, 663. [CrossRef]
2. Atiş, C.D. Accelerated carbonation and testing of concrete made with fly ash. *Constr. Build. Mater.* **2003**, *17*, 147–152. [CrossRef]
3. Khan, M.I.; Lynsdale, C.J. Strength, permeability, and carbonation of high-performance concrete. *Cem. Concr. Res.* **2002**, *32*, 123–131. [CrossRef]

materials

Article

Utilization of Waste Glass in Autoclaved Silica–Lime Materials

Katarzyna Borek * and Przemysław Czapik

Faculty of Civil Engineering and Architecture, Kielce University of Technology, Al. Tysiąclecia Państwa Polskiego
25-314 Kielce, Poland; p.czapik@tu.kielce.pl
* Correspondence: k.komisarczyk@tu.kielce.pl

Abstract: This paper aims to investigate the possibility of using waste glass of different colours as complete substitute for quartz sand in autoclaved silica–lime samples. On the one hand, this increase the possibility of recycling waste glass; on the other hand, it allows obtaining autoclaved material with better properties. In this research, reference samples with quartz sand (R) and white (WG brown (BG), and green (GG) waste container glass were made. Parameters such as compressiv strength, bulk density, and water absorption were examined on all samples. The samples wer examined using a scanning electron microscope with an energy dispersive spectroscopy detectc (SEM/EDS) and subjected to X-ray diffraction (XRD) analysis. The WG samples showed 187% highe compressive strength, BG by 159%, and GG by 134% compared to sample R. In comparison to th reference sample, volumetric density was 16.8% lower for sample WG, 13.2% lower for BG, and 7.1 lower for GG. Water absorption increased as bulk density decreased. The WG sample achieved th highest water absorption value, 15.84%. An X-ray diffraction analysis confirmed the presence c calcite, portlandite, and tobermorite phases. Depending on the silica aggregate used, there wer differences in phase composition linked to compressive strength. Hydrated calcium silicates wit varying crystallisation degrees were visible in the microstructure image.

Keywords: waste glass aggregate; container glass; stained glass; autoclaved bricks; autoclavin SEM; XRD

Citation: Borek, K.; Czapik, P. Utilization of Waste Glass in Autoclaved Silica–Lime Materials. *Materials* **2022**, *15*, 549. https://doi.org/10.3390/ma15020549

Academic Editors: Daniel Georgiev and Andrea Petrella

Received: 22 November 2021
Accepted: 9 January 2022
Published: 12 January 2022

1. Introduction

The overall amount of waste generated within the EU-27 (27 Member States of th European Union) amounted to 2317 million tonnes in 2018. This represents 5.2 tonnes o waste per one EU27 inhabitant [1]. Different economic and household activities contribut to waste generation in the following sectors: construction (36% of the total amount in 2018 mining and quarrying (26.2%), processing (10.6%), wastewater services (9.9%), household (8.2%), and other (9.1%). Waste packaging has a significant share in the waste quantit generated by households. Between 2008 and 2018, the main packaging waste component in the EU included paper and cardboard (a total of 31.8 million tonnes in 2018), followed b plastic and glass (14.8 million tonnes of plastic and 14.5 million tonnes of glass in 2018) [2 In 2018, the total amount of packaging waste generated in the EU27 was estimated to b 77.7 million tonnes, representing a slight increase of 0.3% compared to the previous yea That was mainly due to an increase in the quantity of packaging made of glass (+3.1% compared to 2017), but also plastic (+1.7%) and paper and cardboard (+1.0%). In 200 the total per capita waste originating from packaging materials within the EU average 161.6 kg. In the decade following on from 2008, that figure increased by 12.4 kg to record of 174.0 kg per capita waste (from 67.8 kg in Croatia to 227.5 kg per capita waste i Germany) [2].

Container glass, such as bottles and jars, can be recycled almost completely and wit no time restrictions, without compromising its quality or purity [3]. The Glass Packagin Institute, representing the US glass packaging industry, has determined that a tonne o recycled glass saves over a tonne of natural resources. Furthermore, recycling six tonnes o container glass results in one tonne less carbon emissions, which has significant environmental benefits [4].

There are two approaches adopted in the market for glass cullet recycling, i.e., open and closed-loop recycling. The closed-loop approach focuses on material reuse. However, due to high temperature and relatively long silica melting time, glass production consumes considerable amounts of energy [5]. Whilst, theoretically, glass waste can be fully recycled into a new glass product, minor impurities with a distinct composition or colour are not accepted in the glassmaking process. Furthermore, the complex sorting of fine glass particles makes recycling more expensive [6]. More efficient options for waste glass reuse are therefore being sought. The answer to that is the open-loop approach. This recycling mode usually involves the one-time use of waste glass as an ingredient in materials with long service life, e.g., wall construction materials. Up to now, waste glass has been utilised for the production of concrete [7–10], mortars [11–14], ceramic components [15,16], autoclaved concrete [17], or sand–lime products [18–20], to name a few. Crushed waste glass can be used as a partial substitute for cement/lime, sand, and/or coarse aggregate. Utilised in that manner, the glass requires no heat treatment, which is beneficial in reducing energy consumption. As a result, the processing of glass waste is greatly simplified, and its utilisation share is increasing along with the demand for construction materials. In the era of construction industry development with the growing demand for residential buildings, the increase in production, e.g., masonry wall materials, is a normal effect. According to GUS (Statistics Poland) of 2018 [21], bricks made of cellular concrete, ceramics, and silicate are the most popular materials. Two of the above also have specific production methods in common. These include the autoclaved materials: autoclaved cellular concrete and sand–lime products.

The sand–lime products, also known as silicates or "white bricks", are made from three basic materials: sand (92%), lime (8%), and water. Quartz sand as a crystalline silica carrier acts as the reactive aggregate, which reacts with hydrated quicklime, a binder [22]. In the autoclaving process, the lime, as a carrier of Ca^{2+} cations and OH^- anions, after hydration, reacts with silicate ions which originate from the dissolution of SiO_2, forming products included in the group of hydrated calcium silicates [23]. A solid bond between the unreacted aggregate and the binder matrix is thus created. As the production process uses only natural raw materials with no chemical additives and low energy consumption, autoclaved materials are fully sustainable.

In the literature are known modifications of silica–lime products based on various types of waste [20,24–30]. Waste materials have been often utilised as substitutes for aggregate in conventional products. Studies have demonstrated that ternary blends of lime, sand, and recycled waste glass had higher compressive strength and lower density than the reference sample. The increase in those parameters was commensurate with the amount of substitute in the blends [18].

The use of amorphous glass sand as a substitute for crystalline quartz sand (90% by weight) was investigated by A. Stępień et al. [19]. As for the physical and chemical properties, the compressive strength of silica–lime samples with glass addition was improved (20 MPa with glass addition vs 6.5 MPa in the reference samples). With the increase in the percentage of the above additive, the bulk density decreased slightly (1.7 kg/dm^3–>1.65 kg/dm^3), while the moisture content increased (from 0.5% to 2.1%). The share of glass sand in silicate products is therefore considered beneficial.

Since the glass industry has a diversified and complicated production structure, it is estimated that the number of developed glass types (compositions) currently amounts to 70–80,000. The range of this industry includes 20–30,000 varieties of products. The variety of manufacturing methods and applications and chemical and granulometric compositions of glass products have resulted in the need to divide the glass industry into four basic sectors: construction glass, household and lighting glass, technical glass and container glass [31]. Soda-lime glass is used for the majority of glass packaging products. The chemical composition of the glass used for glass packaging production also depends on the

colour. Stained glass can be achieved by adding appropriate metal oxides to the glass mas
Table 1 presents the chemical compositions of glass by colours.

Table 1. Chemical compositions of glass by colours (content in % by weight) [32,33].

	White Glass			Brown Glass			Green Glass		
	Jin, W., Meyer, C., Baxter, S.	**Karamberi A., Moutsatsou A.**	**Poland/HS "Pollena Czechy"**	**Jin, W., Meyer, C., Baxter, S.**	**Karamberi A., Moutsatsou A.**	**Poland/HS "Ujście"**	**Jin, W., Meyer, C., Baxter, S.**	**Karamberi A., Moutsatsou A.**	**Poland/HS "Ujście"**
SiO_2	73.20–73.50	70.65	71.20	71.90–72.40	71.20	71.30	71.30	70.5	71.00
$Na_2O + K_2O$	13.60–14.10	13.80	13.00	13.80–14.40	13.75	14.00	13.10	13.40	14.00
$CaO + MgO$	10.70–10.80	13.15	14.50	11.60	12.95	11.40	12.20	12.90	11.20
Al_2O_3	1.70–1.90	1.75	1.55	1.70–1.80	1.90	2.50	2.20	1.80	2.50
SO_3	0.20–0.24	0.45	-	0.12–0.14	0.30	-	0.05	0.25	-
Fe_2O_3	0.04–0.05	0.45	0.03	0.30	0.35	0.30	0.56	0.45	0.60
Cr_2O_3	-	-	-	0.01	0.06	-	0.43	0.25	0.20

There are many studies on the influence of colour and glass type on the physicochem cal properties of construction materials. While testing cement blends, Dyer and Dhir [34 found that mortar containing 10% white and green cullet had approximately 27 MPa an 20 MPa higher compressive strength after 28 days than the reference mortar. The morta containing 10% powdered amber glass (a different name for brown glass) achieved onl approximately 2 MPa higher compressive strength than the reference mortar. Karambe and Moutsatsou [33] used standard soda-lime glass from recycled packaging in their stud The authors observed that the compressive strength of mortar containing 25% powdere white glass was about 34.5% higher than that containing 25% brown glass. They attribute this discrepancy to the basic elements colouring the glass particles and claimed that the presence would cause a chemical reaction responsible for compressive strength develo ment. Mirzahosseini and Riding [35] reported that mortar containing green glass cull below 25 μm in size had approximately 3% higher compressive strength after 91 days tha white glass. The authors concluded that the solubility of silicon and aluminium in gree glass is high, which results in an increased pozzolanic reaction at the high pH of the cemer matrix. Furthermore, Bignozzi et al. [36] explained that the glass dissolution rate increase with the increase in the $PbO + Na_2O$ content and with the decrease of both glassy form $SiO_2 + Al_2O_3$ and $CaO + MgO$ glass stabilisers. Glass with a dusty fraction containing higher quantity of modifiers and a lower amount of glassy substances and glass stabilise does not favour the pozzolanic reaction. As a result, the authors stated that the mortar containing ground soda-lime glass showed approximately 4% higher compressive strengt after 90 days than mortars made of crystal glass. Al-Zubaid et al. [37] reported that concret with neon glass revealed approx. 9.5% higher compressive strength than concrete usin green waste glass for 13% cement replacement. They attributed that effect to a higher sha of SiO_2 (68.2%) and CaO (22.6%) in the neon glass compared to the green glass. On th other hand, Zimmer and Braganca [38] stated that the chemical compositions of clear an stained glass powders are very similar and assumed that the finished products woul not demonstrate varying properties due to the addition of glass with different colours. I should also be noted that when glass is used in cement products curing under norma conditions, apart from the favourable pozzolanic reaction, the so-called alkali-silica reactio may occur, which is a destructive process [11–13]. The research conducted so far show that the colour of the glass may also affect its course. Whether the pozzolanic or alkali-silic reaction occurs is primarily determined by the fragmentation of the material.

Until now, most of the studies show that glass colour, due to the variation in chemica composition, can significantly impact the development of compressive strength in morta and concrete. So far, most studies have been based on the use of waste glass as a partia replacement for sand while increasing the percentage of binder in the blend. Howeve there are no studies in which the only aggregate is waste glass, and the mass amount c binder is the same as in the standard blend. The primary objective of this study is to verif the possibility of utilizing waste container glass as a total substitute for quartz sand i sand–lime products. Given the discrepancies resulting from the literature review, anothe objective is to identify the effect of the used glass colour on the physical and mechanical properties of autoclaved bricks.

2. Materials and Methods

2.1. Materials Characterization

2.1.1. Lime

Ground quicklime is a product obtained by crushing and grinding lumps of quicklime. The lime used in this study comes from Sitkówka Trzuskawica A CRH Company (Poland). The chemical composition presented in Table 2 meets the requirements for quicklime used to produce lime and silica products. The bulk density of the lime is 0.79 kg/dm^3. Reactivity 600 °C $\leq$ 2.0 min. A positive soundness test proves that the lime is not overburned and does not contain excessive magnesium oxide (MgO) with dolomite admixture. Lime sieve analysis with 2 mm mesh = 100% passing, 0.2 mm mesh $\geq$ 97% passing by weight and 0.09 mm mesh $\geq$ 90% passing by weight complies with PN-EN 459-2:2010 [39].

Table 2. Chemical composition of lime.

Chemical Composition of Lime	Requirements for Lime	Lime Used in the Study
CaO + MgO	≥90%	≥91%
MgO	≤5%	≤2.0%
CO_2	≤4%	≤3.0%
SO_3	≤2%	≤0.50%

2.1.2. Quartz Sand

Quartz sand with a continuous grain size curve without a clear dominance of any grain fraction is considered suitable for producing sand–lime products. The grain size composition of quartz sand was determined following PN-EN 933-1:2012 [40]. The percentages of individual fractions with specifications for quartz sand and the sand used in this study are listed in Table 3.

Table 3. The percentages of individual fractions with specifications for quartz sand and of the sand used in this study.

Individual Fractions	Requirements for Sand	Sand Used in the Study
2.5–0.5 mm	<30%	2.5–0.5 mm 19%
0.5–0.05 mm	≥65%	0.5–0.05 mm 81%

The quality of quartz sand also depends on its chemical composition. Sand types with a silica content of at least 80% are considered helpful in producing sand–lime products. The remainder is composed of elements listed in Table 4, where their mass percentage in the tested sand sample is given. The quartz sand used in the study meets the quantitative specifications for the chemical composition. The grains are irregular in shape with a medium roundness degree. The density of the sand is 2650 kg/m^3.

Table 4. Weight % of elements in quartz sand.

Element	Weight (%)	Atomic (%)	Net Int.
C	3.73	6.43	53.28
O	41.79	54.14	1976.71
Mg	0.44	0.38	34.33
Al	4.50	3.46	363.93
Si	46.60	34.39	3655.03
K	0.71	0.38	32.18
Fe	2.23	0.83	30.50

2.1.3. Water

According to PN-EN 1008:2004 [41], potable water, which is not subject to additiona quality tests, is readily suitable for the production of silicate products. Such water wa therefore used as an ingredient in the silica–lime mix.

2.1.4. Container Glass

Two types of container glass from two sources were used in the study. Those we narrow-bore glass from green and brown bottles and wide-bore glass from clear (whit jars. Glass cullet was obtained from the solid input material by mechanical crushing. Th grain sizing was adjusted to the distribution of the quartz sand fraction as the aggregat was replaced at a 1:1 weight ratio.

Irregular and sharp-edged shapes of the waste glass particles are found in the m crostructural image (Figure 1). The EDS (Energy Dispersive Spectroscopy) elementa analysis (Figure 2) reveals the content of Si, Na, Mg, Al, S, K, Ca in the material. The specifi density of the glass types with respect to colour is as follows: white glass 2479 kg/m green glass 2494 kg/m^3, and brown glass 2495 kg/m^3.

Figure 1. Microstructure of packaging glass.

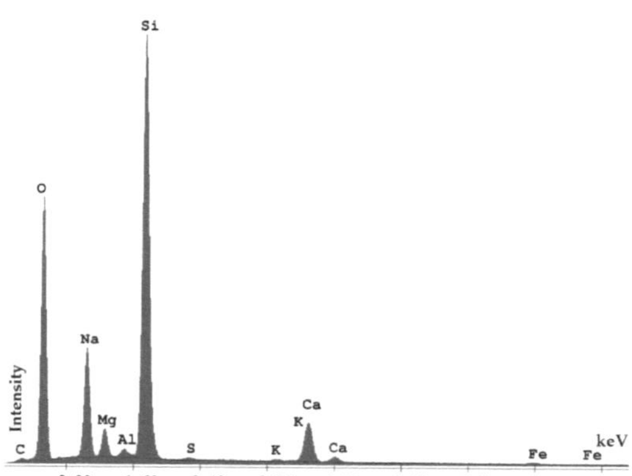

Figure 2. EDS of packaging glass.

2.2. Preparation of Silica–Lime Samples with the Addition of Coloured Container Glass

The quartz sand, constituting 92% of the sand–lime mixture, was replaced by three types of waste glass (clear/white glass—WG, green glass—GG, and brown glass—BG) (Table 5). The container glass cullet replaced the sand fractions in a 1:1 ratio. Three series of samples were prepared. Silicate, in which quartz sand was the aggregate, was used as a reference sample—R. The same preparation scheme was followed each time (Figure 3).

Table 5. Quantitative summary of the silica–lime mixture in each series [% mass].

	Reference (R)	White Glass (WG)	Brown Glass (BG)	Green Glass (GG)
Glass	0	92	92	92
Quartz sand	92	0	0	0
Lime	8	8	8	8

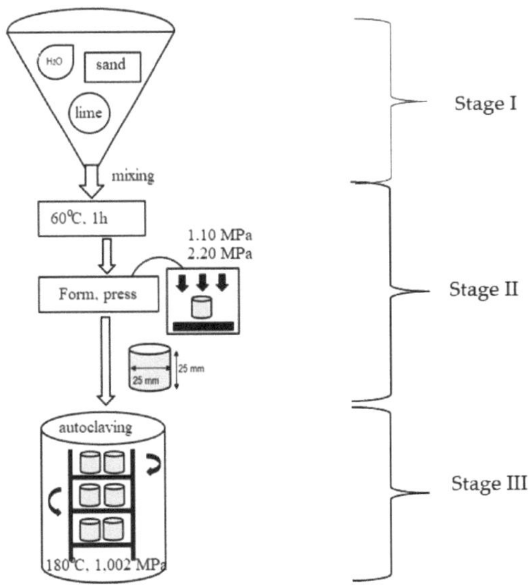

Figure 3. Scheme for the preparation of silica–lime samples involving packaging glass.

First, the individual dry ingredients were weighed. After mixing, water was added to obtain mass with a 6–8% moisture content. The mass was then placed in a sealed glass vessel and heated at 65 °C. After 1 h, the mass was cooled to ambient temperature, and the ingredients were mixed again. The next step was to form cylindrical samples with a height and diameter of 25 mm. The mass was placed in a steel mould and subjected to two-stage pressing at 10 MPa and 20 MPa with venting between the stages. The last stage, autoclaving, was divided into three phases (Figure 4). In the first stage, the samples were heated in an autoclave to a temperature of 180 °C for 2.5 h. When the assumed temperature and steam pressure were obtained, the second stage began, in which the samples were autoclaved at 180 °C under a saturated steam pressure of 1.002 MPa for 8 h. The third stage lasted 12 h and involved cooling the samples to ambient temperature.

2.3. Testing Methods

The experimental samples were tested for physical and mechanical parameters. For each recipe, 12 cylindrical 2.5 × 2.5 cm samples were made. Six of these samples were used in bulk density and water absorption test. Compressive strength tests were carried out on the remaining six samples. Single pieces formed after the compressive strength test were used to investigate the phase composition and microstructure of the samples. The tests were performed on six samples each time. Mean values of the obtained results, including the standard deviation, are shown in individual charts. The samples prepared for testing compressive strength, volumetric density, and water absorption were stored in laboratory conditions at room temperature. The SEM analysis of the microstructure was carried out on crumb specimens prepared as a result of compressive strength tests. The methodology of conducted research is presented in Table 6.

Figure 4. Broken silica–lime sample with GG after compressive strength test.

Table 6. The methodology of conducted research.

Compressive strength	• Laboratory conditions at room temperature • Equipment: Controls 50-C9030 hydraulic press, Manchester, Barcelona • PN-EN 772-1+A1:2015-10 [42]
Volumetric density	• Hydrostatic method
Water absorption	• PN-EN 772-21:2011 [43]
Microstructure	• Scanning electron microscopy (SEM) • Equipment: SEM-type, Quanta 250 FEG, Brno, Czech Republic • Signals collected by secondary electron (SE) detectors • The microstructures of samples were examined under low vacuum conditions without sputtering pre-treatment, using 5-kV voltage. • EDS (Energy Dispersive Spectroscope—EDAX)
Phase composition	• X-ray diffractometry (XRD) method • X-ray analysis was performed using the DSH (Deby-Scherrer-Hull) powder method. The samples were previously ground by hand in a mortar. The tests were carried out in the range of 5–70° (2θ) at the registration speed of 0.0005° (2θ)/s. • Equipment: Empyrean, PANALYTICAL, Almelo, Netherlands with a Cu lamp.

3. Results and Discussion

Figure 4 shows the broken silica–lime GG sample after compressive strength test. The compressive strength results of the silica–lime products modified with different colour glass are shown in Figure 5. Replacing quartz sand with waste glass, regardless of the colour, results in a compressive strength increase of the samples prepared, compared to the samples of standard composition. The highest increase of 10.95 MPa was recorded for the WG sample, which was 187% above R, followed by the BG sample with 159% increase (9.86 MPa) above R. The lowest increase, of 8.91 MPa (134% above R) was recorded for the GG sample. The compressive strength of the WG samples was 11.1% higher than that of the GG samples. The use of waste glass as a substitute for quartz sand was also found to have improved the compressive strength in the study of Stępień and all [44]. The hydration temperature of lime with the increasing share of WG was lowered, hence it can be deducted that hydration of lime during the production of autoclaved samples provided adequate chemical bonds, and as a result strength of samples increased. It should be also noted that calcium hydroxide shows retrograde solubility with respect to the temperature, so it is possible that at this lower temperature more calcium ions were released upon dissolution and available for the reaction to form hydrates.

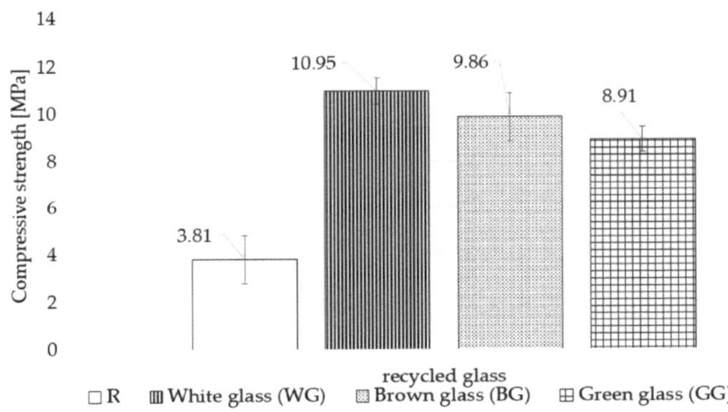

Figure 5. Compressive strength test results.

Figure 6 shows the volumetric density and Figure 7 the water absorption. Volumetric density examination proves that the replacement of quartz sand by waste container glass of different colours positively affects the volumetric density results of the autoclaved samples. The glass colour is also important in this case since the lowest volumetric density was achieved by the WG sample (16.8% decrease). The second result of 1.70 g/cm^3 had the BG sample, representing a decrease of 13.3% relative to the R sample. The lowest decrease was recorded for the GG sample—1.82 g/cm^3 (7.1% decrease). The WG and GG samples demonstrate the most significant differences in volumetric density. The volumetric density of the GG sample is 10.4% above that of the WG sample. These differences can be attributed to the density of the aggregate used in the sample, as the density of the green glass cullet is 0.6% higher than that of the white glass cullet. The differences in the densities of the studied samples may be due to their different chemical properties. They may form other phases and microstructures, the determination of which is the subject of the XRD and SEM/EDS studies presented in Figures 8 and 9, respectively. The phases formed can affect the porosity of the autoclaved samples.

In each case studied, replacing quartz sand with waste glass increases water absorption. The brown and green glass samples show comparable values of 12.41% (BG) and 12.48% (GG). The WG samples recorded the most significant increase in water absorption (15.84%) compared to the standard composition. The increased water absorption in the studied case should be attributed not as much to the material substituting the quartz sand but to the increased open porosity. In the performed study, there is an evident relationship between the compressive strength and bulk density and water absorption due to the colour of used waste glass. The WG samples show the highest compressive strength and absorption, yet the lowest density relative to the BG and GG samples. The GG samples, in turn, have the lowest compressive strength and water absorption and the highest density compared t the WG and BG ones.

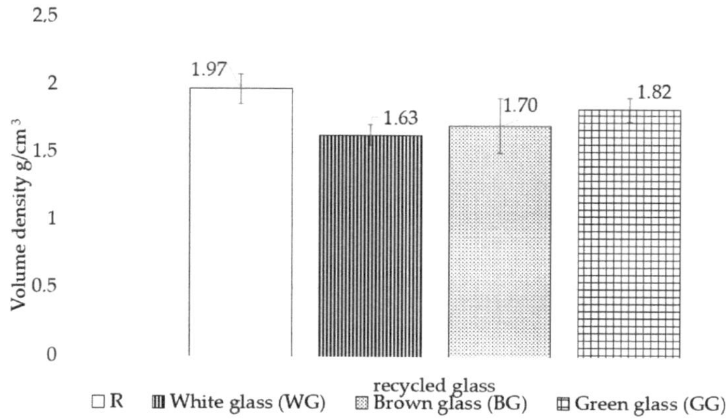

Figure 6. Volume density results.

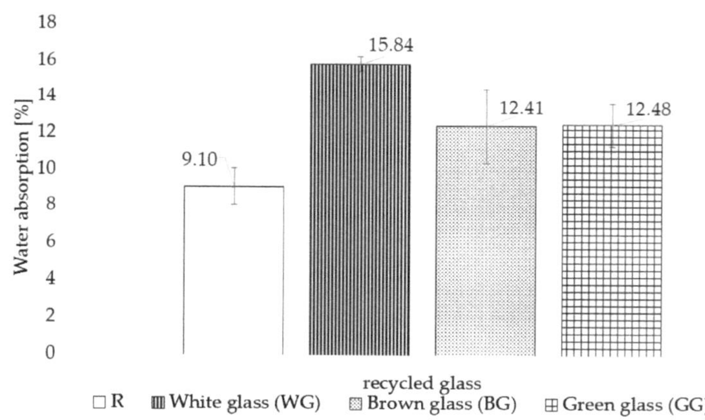

Figure 7. Results of water absorption.

The phase composition of the silica–lime products shown in Figure 8 indicates tha calcium hydroxide (portlandite) formed in all samples with glass aggregate. Peaks typica of calcite are also found, confirming portlandite carbonation. Carbonation may caus the lack of cohesion of the sample subjected to compression, Figure 4. In the analyse samples, hydrated calcium silicates appear as hydration products. A peak characteristic o tobermorite (5–10° 2θ) reaches the highest intensity in GG sample. In the other samples this peak does not occur. In all samples containing glass, however, a raised backgroun is visible, which may be caused by the presence of unreacted glass and the low degree o structural order of formed hydrated calcium silicates. Their structure may be similar to th C-S-H phase structure formed in natural conditions [45]. This confirms the observation tha in the autoclaving process, the presence of reactive silica in the amorphous state increase the amount of the C-S-H phase, but at the same time, the transformation of this phase int a more stable form is hindered [46].

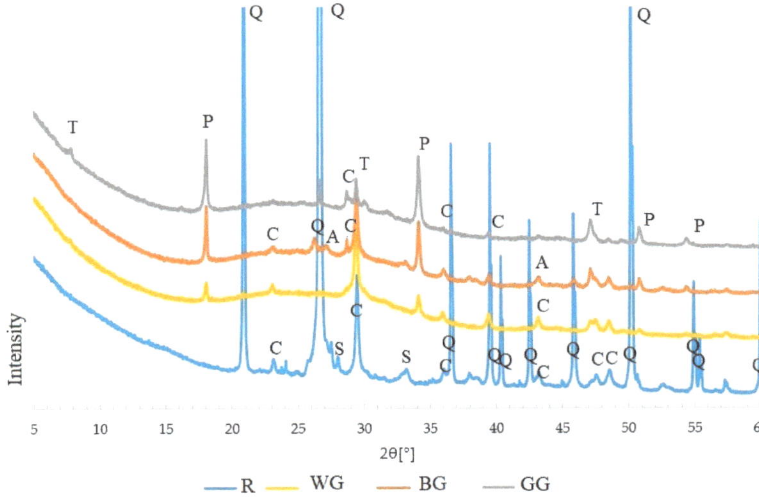

Figure 8. X-ray-pattern of WG, BG, and GG samples (symbols: A—aragonite, C—calcite, P—portlandite, S—spurrite, T—tobermorite, Q—quartz).

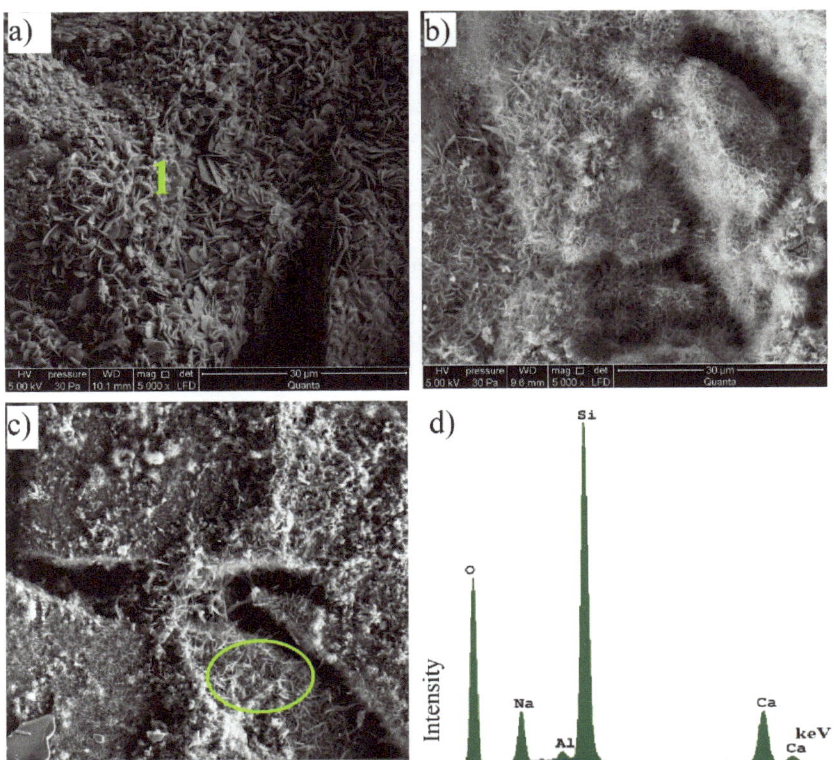

Figure 9. Microstructure of samples (**a**) WG, (**b**) BG, (**c**) GG, and (**d**) EDS analysis at point 1 in (**a**).

When samples with different glass colours were tested, the main differences in th diffractograms were the intensities of the portlandite peaks. Significantly, for sample for which higher compressive strengths were achieved (Figure 5), correspondingly low intensities of peaks of portlandite were found. The decrease in intensity of these peak can be related to a higher degree of portlandite reactivity, which indicates a more efficie autoclaving process. However, portlandite also reacted in the previously mentione carbonation process, which is also associated with a decrease in its content.

Peaks of unreacted quartz were also detected in the GG sample, which, togeth with high portlandite peaks, indicate the lowest reactivity degree of the green glass cull Despite the similarities in the chemical composition of white, brown, and green was glass, the results obtained do not confirm the assumptions of Zimmer and Braganca [3 Silica–lime samples with the above-mentioned glass colours show variable properties.

No evident changes that could indicate the presence of hydrated calcium silicat were detected on the reference sample X-ray pattern. Instead, the most intense peaks fro calcite were observed, which may indicate a significant degree of carbonation. This ma be the reason for this sample's low compressive strength performance, where the calciu binder has carbonated instead of reacting with the silica [47,48]. In addition, due to quar sand, very intense quartz peaks are observed in this sample.

Figure 9 shows SEM images of silica–lime samples containing the WG, BG, and GG a 5000× magnification. Hydrothermal synthesis products with a lamellar pattern forme between glass-derived silica and lime were found in each analysed sample. These produc cover the entire surface of the samples. The WG and BG samples show densely packe formations surrounding the glass aggregate. In the GG sample, the synthesis produc were formed in the spaces between and on the surface of the glass aggregate; howeve they are much less dense/packed and reveal a lower crystallisation degree. The produc formed in the studied samples resemble hydrated calcium silicates typical of the standar sand–lime products [44]. Based on the microstructure observation, it can be conclude that their development degree varies according to the waste glass colour used. In the W sample (Figure 9a), the hydrated lime silicates took the form of small plates with rounde edges, resembling clover leaves in shape, which are closely spaced and arranged at differe angles. These silicates can be identified as tobermorite (Figure 9d), whose high content ma be related to the increased compressive strength of this sample [46,49–51]. In this sampl however, a lot of pores occurring between the produced plates are observed. There are mor of these pores than in other samples tested and they may be the reason for increased wate absorption results (Figure 7). According to Černý et al. [49], the high strength achieved b autoclaved materials may be due to the presence of tobermorite, which can be increase by using suitable raw materials. However, this effect may be diminished when the ra materials used increase the porosity of the material, as may be the case with fluidised be ash in the zone around the ash grains. In the case of white glass (WG), the pores visibl in the SEM images did not cause such an effect. As in the study by Černý et al. [49], th use of container glass contributed to a significant increase in compressive strength. In th present study, the increase in strength of the material is observed, which is also linked t the formation of tobermorite during the autoclaving process [50,51].

In the BG sample (Figure 9b), the hydrated lime silicates resemble narrow plate with an irregular arrangement (grass shape). The analysed GG sample (Figure 9c) i mostly covered with a spongy formation typical of the C-S-H phase. Its higher level o crystallization was found in a small area visible in the marked circle.

4. Conclusions

Based on the findings of the study, the following conclusions can be drawn:

- a complete substitution of quartz sand with glass waste aggregate in silica–lim products is possible and has a beneficial effect on selected physical properties of th finished products;

- the use of different colours of waste glass has a variable impact on the physical properties of the autoclaved silica–lime products;
- regardless of the glass colour used, the tested samples demonstrate a significant increase in the compressive strength of finished products; the use of white waste glass (WG) results in a 184% increase in strength with a 159% increase for brown glass (BG) and 134% increase for green glass, compared to the samples with the standard composition;
- regardless of the glass colour used, the samples demonstrate a decrease in volumetric density and an increase in water absorption;
- in samples containing waste glass, the study reveals the formation of portlandite and calcite is the main product, regardless of the colour of the glass, additionally, based on the SEM/EDS analysis, a tobermorite phase was found;
- the microstructure image shows the formation of phases typical of autoclaved sand–lime products, the colour of the glass affects to a varying extent the crystallisation degree of hydrated calcium silicates;
- white glass aggregate is the most reactive and has the strongest effect on the properties of silica–lime products. Green glass proved to be the least reactive.

Author Contributions: Conceptualization, K.B. and P.C.; methodology, K.B.; software, K.B.; formal analysis, K.B. and P.C.; investigation, K.B. and P.C. (XRD).; resources, K.B., data curation, K.B. and P.C.; writing—original draft preparation, K.B.; writing—review and editing, K.B. and P.C.; visualization, K.B. All authors have read and agreed to the published version of the manuscript.

Funding: The work is supported by the program of the Minister of Science and Higher Education under the name: Regional Initiative of Excellence in 2019–2022 project number 025/ RID/2018/19 financing amount PLN 12.000.000.

Institutional Review Board Statement: Not applicable.

Informed Consent Statement: Not applicable.

Data Availability Statement: No new data were created or analyzed in this study. Data sharing is not applicable to this article.

Conflicts of Interest: The authors declare no conflict of interest.

References

1. Eurostat. Waste Statistics, Statistics Explained. Available online: https://ec.europa.eu/eurostat/statistics-explained/index.php?title=Waste_statistics (accessed on 29 April 2021).
2. Eurostat. Packaging Waste Statistics, Statistics Explained. Available online: https://ec.europa.eu/eurostat/statistics-explained/index.php?title=Packaging_waste_statistics (accessed on 18 October 2021).
3. Glass Packing Institute. Glass Packaging European Production Continues to Steadily Grow. Available online: https://feve.org/about-glass/statistics/ (accessed on 2 July 2019).
4. Kim, K.; Kim, K.; Hwang, J. Thin Film Transistor-Liquid Crystal Display Cullet: A Raw Material for Production of Commercial Soda Lime Silicate Glasses. *J. Clean. Prod.* **2014**, *79*, 276–282. [CrossRef]
5. Olofinnade, O.M.; Ndambuki, J.M.; Ede, A.N.; Booth, C. Application of Waste Glass Powder as a Partial Cement Substitute towards more Sustainable Concrete Production. *Int. J. Eng. Res.* **2017**, *31*, 77–93. [CrossRef]
6. Flood, M.; Fennessy, L.; Lockrey, S.; Avendano, A.; Glover, J.; Kandare, E.; Bhat, T. Glass Fines: A review of cleaning and up-cycling possibilities. *J. Clean. Prod.* **2020**, *267*, 121875. [CrossRef]
7. Ali, B.; Qureshi, L.A. Influence of glass fibers on mechanical and durability performance of concrete with recycled aggregates. *Constr. Build. Mater.* **2019**, *228*, 116783. [CrossRef]
8. Shayan, A.; Xu, A. Value-Added utilisation of waste glass in concrete. *Cem. Concr. Res.* **2004**, *34*, 81–89. [CrossRef]
9. Topçu, I.B.; Canbaz, M. Properties of concrete containing waste glass. *Cem. Concr. Res.* **2004**, *34*, 267–274. [CrossRef]
10. Ismail, Z.Z.; Al-Hashmi, E.A. Recycling of waste glass as a partial replacement for fine aggregate in concrete. *Waste Manag.* **2009**, *29*, 655–659. [CrossRef] [PubMed]
11. Czapik, P. Microstructure and Degradation of Mortar Containing Waste Glass Aggregate as Evaluated by Various Microscopic Techniques. *Materials* **2020**, *13*, 2186. [CrossRef]
12. Alducin-Ochoa, J.; Martín-Del-Río, J.; Torres-González, M.; Flores-Alés, V.; Hernández-Cruz, D. Performance of mortars based on recycled glass as aggregate by accelerated decay tests (ADT). *Constr. Build. Mater.* **2021**, *300*, 124057. [CrossRef]

13. Hamzah, H.K.; Huseien, G.F.; Asaad, M.A.; Georgescu, D.P.; Ghoshal, S.; Alrshoudi, F. Effect of waste glass bottles-derive nanopowder as slag replacement on mortars with alkali activation: Durability characteristics. *Case Stud. Constr. Mater.* **202** *15*, e00775. [CrossRef]
14. Guo, M.-Z.; Tu, Z.; Poon, C.S.; Shi, C. Improvement of properties of architectural mortars prepared with 100% recycled glass b CO_2 curing. *Constr. Build. Mater.* **2018**, *179*, 138–150. [CrossRef]
15. Bohn, B.P.; Von Mühlen, C.; Pedrotti, M.F.; Zimmer, A. A novel method to produce a ceramic paver recycling waste glass. *Clea Eng. Technol.* **2021**, *2*, 100043. [CrossRef]
16. Matteucci, F.; Dondi, M.; Guarini, G. Effect of soda-lime glass on sintering and technological properties of porcelain stonewa tiles. *Ceram. Int.* **2002**, *28*, 873–880. [CrossRef]
17. Walczak, P.; Małolepszy, J.; Reben, M.; Szymański, P.; Rzepa, K. Utilization of Waste Glass in Autoclaved Aerated Concret *Procedia Eng.* **2015**, *122*, 302–309. [CrossRef]
18. Borek, K.; Czapik, P.; Dachowski, R. Recycled Glass as a Substitute for Quartz Sand in Silicate Products. *Materials* **2020**, *13*, 103 [CrossRef]
19. Dachowski, R.; Stepien, A. Impact of modification of sand-lime mass with organic compounds on the microstructure an mechanical features of silicate bricks. In Proceedings of the 9th International Conference Environmental Engineering, Vilniu Lithuania, 22–23 May 2014. [CrossRef]
20. Jasińska, I. Effect of foam glass granules fillers modification of lime-sand products on their microstructure. *Open Eng.* **201** *9*, 299–306. [CrossRef]
21. Główny Urząd Statystyczny. *Rynek Producentów Wyrobów Budowlanych w Latach 2015–2018*; Główny Urząd Statystyczny: Warsa Poland, 2020.
22. Stepien, A.; Kostrzewa, P.; Dachowski, R. Influence of barium and lithium compounds on silica autoclaved materials propertie and on the microstructure. *J. Clean. Prod.* **2019**, *236*, 117507. [CrossRef]
23. Pytel, Z. *Modyfikowanie Składu Fazowego i Mikrostruktury Autoklawizowanych Tworzyw Wapienno-Piaskowych*; Ceramika: Krako Poland, 2014.
24. Komisarczyk, K.; Dachowski, R.; Czapik, P. Silicate composites modified with the use of diabase and barite aggregate as a alternative for low-energy construction. *IOP Conf. Ser. Earth Environ. Sci.* **2019**, *214*, 12072. [CrossRef]
25. Komisarczyk, K.; Czapik, P.; Komisarczyk, K. Quartz bentonite sandmix in sand-lime products. *Open Eng.* **2019**, *9*, 367–37 [CrossRef]
26. Li, J.; Lv, Y.; Jiao, X.; Sun, P.; Li, J.; Wuri, L.; Zhang, T.C. Electrolytic manganese residue based autoclaved bricks with Ca(OH) and thermal-mechanical activated K-feldspar additions. *Constr. Build. Mater.* **2020**, *230*, 116848. [CrossRef]
27. Zhao, Y.; Zhang, Y.; Chen, T.; Chen, Y.; Bao, S. Preparation of high strength autoclaved bricks from hematite tailings. *Constr. Buil Mater.* **2012**, *28*, 450–455. [CrossRef]
28. Fang, Y.; Gu, Y.; Kang, Q.; Wen, Q.; Dai, P. Utilization of copper tailing for autoclaved sand–lime brick. *Constr. Build. Mater.* **201** *25*, 867–872. [CrossRef]
29. Li, J.; Li, X.; Liang, S.; Zhang, Y.; Ye, Q.; Zhang, L.; Pan, C.; Lv, R.; Li, Q.; Xiao, K.; et al. Preparation of water-permeable brick derived from fly ash by autoclaving. *Constr. Build. Mater.* **2021**, *271*, 121556. [CrossRef]
30. Yatsenko, E.; Smoliy, V.; Klimova, L.; Yatsenko, L.A. Use of TPP Ash and Slag Waste in Silicate Technologies. In Proceedings o the IEEE International Conference Management of Municipal Waste as an Important Factor of Sustainable Urban Developmen St. Petersburg, Russia, 4–6 October 2018. [CrossRef]
31. Wójcicki, J.; Ciągło, A.; Czerwiński, Z.; Galewicz, M.; Hasslinger, K.; Hilgertner, A.; Skirgajłło, J.; Sztandera, W.; Świketlik, W Tomaszewski, J.; et al. *Technologia Szkła*, 3rd ed.; Arkady: Warsaw, Poland, 1987.
32. Jin, W.; Meyer, C.; Baxter, S. Glascrete—Concrete with glass aggregate. *ACI Struct. J.* **2000**, *97*, 208–213.
33. Karamberi, A.; Moutsatsou, A. Participation of coloured glass cullet in cementitious materials. *Cem. Concr. Compos.* **200** *27*, 319–327. [CrossRef]
34. Dyer, T.D.; Dhir, R.K. Chemical Reactions of Glass Cullet Used as Cement Component. *J. Mater. Civ. Eng.* **2001**, *13*, 412–41 [CrossRef]
35. Mirzahosseini, M.; Riding, K.A. Effect of curing temperature and glass type on the pozzolanic reactivity of glass powder. *Cem Concr. Res.* **2014**, *58*, 103–111. [CrossRef]
36. Bignozzi, M.; Saccani, A.; Barbieri, L.; Lancellotti, I. Glass waste as supplementary cementing materials: The effects of glas chemical composition. *Cem. Concr. Compos.* **2014**, *55*, 45–52. [CrossRef]
37. Al-Zubaid, A.B.; Shabeeb, K.M.; Ali, A.I. Study The Effect of Recycled Glass on The Mechanical Properties of Green Concret *Energy Procedia* **2017**, *119*, 680–692. [CrossRef]
38. Zimmer, A.; Bragança, S.R. A review of waste glass as a raw material for whitewares. *J. Environ. Manag.* **2019**, *244*, 161–17 [CrossRef]
39. *PN-EN 459-2:2010. Wapno Budowlane—Część 2: Metody Badań*; Polish Committee for Standarization: Warsaw, Poland, 2010.
40. *PN-EN 933-1:2012. Badania Geometrycznych Właściwości Kruszyw—Część 1: Oznaczanie Składu Ziarnowego—Metoda Przesiewani* Polish Committee for Standarization: Warsaw, Poland, 2012.
41. *PN-EN 1008:2004. Woda Zarobowa do Betonu—Specyfikacja Pobierania Próbek, Badanie i Ocena Przydatności Wody Zarobowej do Beton w tym Wody Odzyskanej z Procesów Produkcji Betonu*; Polish Committee for Standarization: Warsaw, Poland, 2004.

2. *PN-EN 772-1+A1:2015-10. Metody Badań Elementów Murowych- Część 1: Określenie Wytrzymałości na Ściskanie*; Polish Committee for Standarization: Warsaw, Poland, 2015.
3. *PN-EN 772-21:2011. Metody Badań Elementów Murowych- Część 21: Określenie Absorpcji Wody Ceramicznych i Silikaotwych Elementów Murowych Przez Absorpcję Zimnej Wody*; Polish Committee for Standarization: Warsaw, Poland, 2011.
4. Stepien, A.; Potrzeszcz-Sut, B.; Prentice, D.P.; Oey, T.J.; Balonis, M. The Role of Glass Compounds in Autoclaved Bricks. *Buildings* **2020**, *10*, 41. [CrossRef]
5. Sasui, S.; Kim, G.; Nam, J.; van Riessen, A.; Eu, H.; Chansomsak, S.; Alam, S.F.; Cho, C.H. Incorporation of Waste Glass as an Activator in Class-C Fly Ash/GGBS Based Alkali Activated Material. *Materials* **2020**, *13*, 3906. [CrossRef] [PubMed]
6. Drochytka, R.; Černý, V. Influence of fluidized bed combustion fly ash admixture on hydrothermal synthesis of tobermorite in the mixture with quartz sand, high temperature fly ash and lime. *Constr. Build. Mater.* **2020**, *230*, 117033. [CrossRef]
7. Chen, Y.; Liu, P.; Yu, Z. Effects of Environmental Factors on Concrete Carbonation Depth and Compressive Strength. *Materials* **2018**, *11*, 2167. [CrossRef]
8. Tan, K.H.; Du, H. Use of waste glass as sand in mortar: Part I—Fresh, mechanical and durability properties. *Cem. Concr. Compos.* **2013**, *35*, 109–117. [CrossRef]
9. Černý, V.; Drochytka, R.; Kulísek, K.; Húšt'Avová, J. Influence of secondary raw materials on synthesis of tobermorite in lime-silicate composite. In Proceedings of the ICAAC—6th International Conference on Autoclaved Aerated Concrete, Podstam, Germany, 4–6 September 2018. [CrossRef]
0. Wongkeo, W.; Chaipanich, A. Compressive strength, microstructure and thermal analysis of autoclaved and air cured structural lightweight concrete made with coal bottom ash and silica fume. *Mater. Sci. Eng. A* **2010**, *527*, 3676–3684. [CrossRef]
1. Wongkeo, W.; Thongsanitgarn, P.; Pimraksa, K.; Chaipanich, A. Compressive strength, flexural strength and thermal conductivity of autoclaved concrete block made using bottom ash as cement replacement materials. *Mater. Des.* **2012**, *35*, 434–439. [CrossRef]

materials

MDPI

Article

Stabilized High Clay Content Lateritic Soil Using Cement-FGD Gypsum Mixtures for Road Subbase Applications

Phattharachai Maichin [1], Peerapong Jitsangiam [2,*], Toon Nongnuang [1], Kornkanok Boonserm [3], Korakod Nusit [4], Suriyavut Pra-ai [5], Theechalit Binaree [6] and Chuchoke Aryupong [2]

1 Department of Civil Engineering, Faculty of Engineering, Chiang Mai University, Huai Kaew Road, Mueang, Chiang Mai 50200, Thailand; phattharachai_mai@cmu.ac.th (P.M.); toon_n@cmu.ac.th (T.N.)
2 Center of Excellence in Natural Disaster Management, Department of Civil Engineering, Faculty of Engineering, Chiang Mai University, Huai Kaew Road, Mueang, Chiang Mai 50200, Thailand; cendru1@gmail.com
3 Department of Applied Chemistry, Faculty of Sciences and Liberal Arts, Rajamangala University of Technology Isan, Nakhon Ratchasima 30000, Thailand; Kornkanok.bo@rmuti.ac.th
4 Centre of Excellence on Energy Technology and Environment, Department of Civil Engineering, Faculty of Engineering, Naresuan University, Tha-Po, Mueang, Phitsanulok 65000, Thailand; korakodn@nu.ac.th
5 GWR Research Unit, School of Engineering, University of Phayao, Phayao 56000, Thailand; suriyavut.pr@up.ac.th
6 Department of Mechanical Engineering, Faculty of Engineering, Chiang Mai University, Huai Kaew Road, Mueang, Chiang Mai 50200, Thailand; theechalit_b@cmu.ac.th
* Correspondence: peerapong@eng.cmu.ac.th; Tel.: +66-053-944-157

Citation: Maichin, P.; Jitsangiam, P.; Nongnuang, T.; Boonserm, K.; Nusit, K.; Pra-ai, S.; Binaree, T.; Aryupong, C. Stabilized High Clay Content Lateritic Soil Using Cement-FGD Gypsum Mixtures for Road Subbase Applications. *Materials* **2021**, *14*, 1858. https://doi.org/10.3390/ma14081858

Academic Editor: Miguel Bravo

Received: 15 March 2021
Accepted: 5 April 2021
Published: 8 April 2021

Abstract: With a lack of standard lateritic soil for use in road construction, suitable economical and sustainable soil-stabilization techniques are in demand. This study aimed to examine flue gas desulfurization (FGD) gypsum, a by-product of coal power plants, for use in soil–cement stabilization, specifically for ability to strengthen poor high-clay, lateritic soil but with a lower cement content. A series of compaction tests and unconfined compressive strength (UCS) tests were performed in conjunction with scanning electron microscope (SEM) analyses. Therefore, the strength development and the role of FGD gypsum in the soil–cement–FGD gypsum mixtures with varying cement and FGD gypsum contents were characterized in this study. The study results showed that adding FGD gypsum can enhance the strength of the stabilized substandard lateritic soil. Extra FGD gypsum added to the cement hydration system provided more sulfate ions, leading to the formation of ettringite and monosulfate, which are the hardening cementitious products from the cement hydration reaction. Both products contributed to the strength gain of the soil–cement–FGD gypsum material. However, the strength can be reduced when too much FGD gypsum is added because the undissolved gypsum has a weak structure. Examinations of FGD gypsum in the soil–cement–FGD gypsum mixtures by SEM confirmed that adding FGD gypsum can reduce the cement content in soil–cement mix to achieve a given UCS value.

Keywords: clay–cement stabilization; lateritic soil; FGD gypsum; unconfined compressive strength; SEM

1. Introduction

Lateritic soil is iron- and aluminum-rich and formed in a special environment of relatively high temperature and humidity in a tropical area (Figure 1). It is formed from the weathering process of the parent rock, the laterization process, which begins with the removal of silica (SiO_2) and the accumulation of some oxides from the parent rock, such as iron oxide (Fe_2O_3) and aluminum oxide (Al_2O_3) [1,2]. Lateritic soil is defined as having a ratio of SiO_2/Al_2O_3 between 1.33 and 2.00 [3] and is noticeably reddish or brownish-red with a high clay content [4]. Such soil is also generally used in most tropical countries as a road construction material [5] because of its ready availability and relatively low cost

Furthermore, good lateritic soil has some satisfying engineering properties, and material specifications for the quality control of lateritic soil for road construction have been established (Table 1) [6–11]. However, over the past decade a shortage of standard natural road materials, including standard lateritic soil, has become a significant problem for the road construction industry. High-quality materials complying with the specifications established by road authorities are in short supply. Most of the available natural materials do not meet the standards for road construction because they have undesirable characteristics of high plasticity and substandard grading [12–15]. Consequently, soil improvement techniques, especially soil–cement stabilization, have become the new method for overcoming the scarcity of standard materials by improving the essential properties of poor soil to the level needed for practical use in road construction.

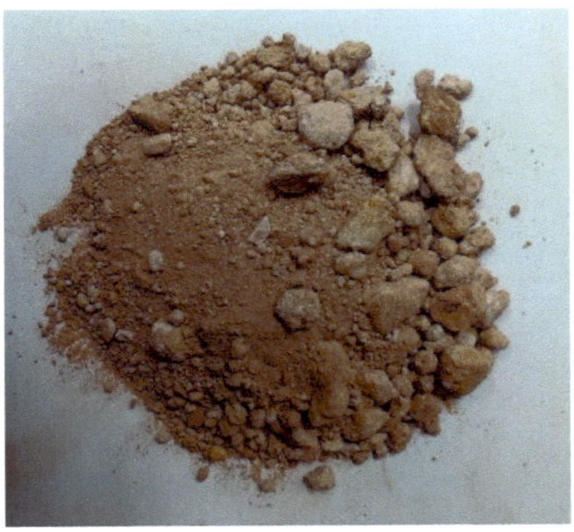

Figure 1. Lateritic soils.

Table 1. General Requirements for standard subbase materials in tropical areas.

Properties	Thailand: Department of Highways	Brazil: Brazilian National Highway Department [7]		Nigeria: Federal Ministry of Works and Housing	Malaysian: Jabatan Kerja Raya Malaysia	Recommended Criteria Material Selection for Lateritic Soil by Charman's
	[6]	Heavy Traffic	Light Traffic	[8]	[9]	[10]
California Bearing Ratio, (%)	≥25	≥80	≥60	≥20	≥30	≥25
Percent passing, (%)	-	-	-	No. 200 sieve ≤12	No. 4 sieve ≥45	-
Liquid limit, (%)	≤35	<35	<40	≤25	-	-
Plasticity index, (%)	≤11	<10	<12	≤6	≤12	≤25
Los Angeles abrasion loss, (%)	≤60	<65	<65	-	-	-
Swell, (%)	-	<0.2	<0.2	-	-	-

Soil improvement for road construction can be achieved by using techniques such as preloading, soil replacement, using recycled concrete aggregates, and using soil stabilizing agents [16–20]. The most popular technique is the use of ordinary Portland cement (OPC) as a stabilizing agent, leading to the technique of "soil–cement stabilization" [21], the goal of which is to increase soil strength. Some previous research [22–28] confirmed that using OPC to stabilize different soil types, including soft clay and lateritic soil, can improve their unconfined compressive strength (UCS). Furthermore, those UCS values increase as cement content and curing time increase. When this OPC clinker (Table 2) chemically reacts with water, the cementitious compounds of C–S–H (calcium silicate hydrate) and others are produced, resulting in a strength gain characteristic of stabilized soil.

In addition, in Table 2, the presence of calcium sulfate dihydrate (gypsum) of approximatel 5% by weight, can play a role in delaying the hydration reaction of calcium aluminate (C_3A) in the OPC clinker [29–31] when reacting with water. This retarding effect cause by the gypsum in OPC could enhance the strength and durability of the cementitiou products. However, even though OPC is one of the most popular construction material at present, its production process and application generate carbon dioxide (CO_2). Th high CO_2 footprint of OPC and OPC-oriented material production could profoundl affect global warming [32–34]. From a sustainability point of view, OPC would not be a environmentally friendly construction material. Moreover, for soil stabilization, OPC i relatively costly compared to the cost of the soil. Some recent research used only wast materials without OPC, such as rice husk ash [35], palm oil fuel ash [36], calcium carbid residue [37], or even waste beverage cans [38] that can be applied to improve soil propertie These waste materials have played a gradually more prominent role by avoiding cemen consumption, which is the main problem with CO_2 emissions. However, these material will not be widely acceptable without OPC. For reasons of sustainability and cost, th development of cement substitutions for various construction applications is an activ research field. There is potential to use industrial by-products, such as coal ash an furnace slag, or waste materials instead of OPC substitution. These materials maintai or increase the soil's material strength while reducing the amount of OPC and the CO_2 i generates [39–42].

Table 2. Chemical composition of the OPC clinker [29,30].

Compound	Formula (Notation)	Wt.(%)
Tricalcium silicate	$3CaO{\cdot}SiO_2$ (C_3S)	49–55
Dicalcium silicate	$2CaO{\cdot}SiO_2$ (C_2S)	20–25
Tricalcium aluminate	$3CaO{\cdot}Al_2O_3$ (C_3A)	10–12
Tetracalcium aluminoferrite	$4CaO{\cdot}Al_2O_3{\cdot}Fe_2O_3$ (C_4AF)	8
Calcium sulfate dihydrate	$CaSO_4{\cdot}2H_2O$ (CSH_2)	5
Other Sodium oxide Potassium oxide	Na_2O (N) K_2O (K)	1–2

Coal-fired power plants generate electricity in many countries around the worlc Based on their electricity generation (Figure 2), the flue gas desulfurization (FGD) plan plays a vital role in reducing sulfur dioxide (SO_2) emissions [43,44]. This minimizes advers environmental effects. Consequently, a by-product called FGD gypsum ($CaSO_4{\cdot}2H_2O$ is generated in abundance (Figure 3). Its properties, including physical and chemica properties and mineral composition, are nearly identical to those of natural gypsun (Table 3) [43,45–47]. FGD gypsum has been used for specific applications such as landfill and agriculture [48,49]. The American Coal Ash Association (ACAA) reported that the U.S produced around 23 million tons of FGD gypsum in 2019, with a 50% rate of reuse [50 In Europe, the European Coal Combustion Products Association reported FGD gypsun production of approximately 10 million tons in 2016 [51]. Therefore, FGD gypsum i an industrial by-product that could be used more effectively. Using it as fill or roac construction material would be a high-volume usage.

For soil–cement stabilization, FGD gypsum can be used as a cement substitute basec on the additional gypsum concept. It can decrease the cement content and increase th compressive strength of the soil–cement material [52–55]. An optimum amount of gypsun in the soil–cement system is crucial because too much additional gypsum can cause lef over gypsum, which then becomes the weak element in the soil–cement material, referrec to as "supersaturated gypsum". FGD gypsum would be a viable cement substitute for soil cement stabilization. Moreover, water, OPC, and FGD gypsum are essential parameter for the design mixture. Still, there are no clear guidelines on designing mixtures for soi stabilization when using stabilizing agents of more than one type. Therefore, in thi study the relationship of those parameters has been established for soil–cement–FGD gypsum application. A soil–cement–FGD gypsum mixture is proposed in this study and is compared to the traditional soil–cement process (Figure 4).

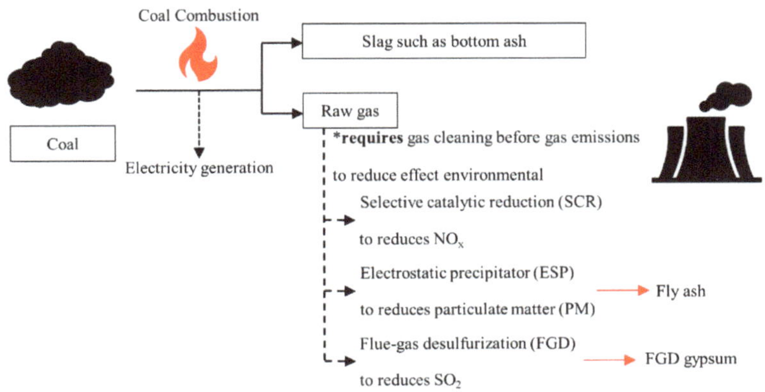

Figure 2. Pollution-control systems used to clean emissions from a coal-fired power plant.

Figure 3. Flue gas desulfurization gypsum (FGD gypsum).

Table 3. Properties of FGD gypsum and natural gypsum, adapted from [43,45–47].

Property	FGD-Gypsum	Natural Gypsum
Mineral composition	Gypsum, Quartz	Gypsum, Quartz, Dolomite
CaSO4·2H_2O, (%)	84–99.6	87.1
Specific density, (g/cm^3)	~2.2	~2.3
Main chemical properties, (%)		
CaO	38	59
SO_3	54	41

This study aimed to examine the use of a soil–cement–FGD gypsum mixture for improving substandard lateritic soil with a relatively high clay content for the purpose of road construction. This research could be a guideline for using FGD gypsum in other soil–cement applications. In this study, a series of modified compaction tests [56] and UCS tests [57] were performed in the laboratory on the soil–cement–FGD gypsum mixtures with varying cement content (1–3% by dry weight), varied FGD gypsum content (1, 3, and 5% by dry weight), and curing periods (3, 7, and 28 days). In addition, Californi Bearing Ratio (CBR) tests [58] were carried out on soaked and unsoaked forms of selecte mixes to establish the CBR index (%), the percentage between the CBR values of the soake sample and the unsoaked sample. This CBR index was used to evaluate the moistu sensitivity of mixtures and soil. Microstructure observations through the scanning electro microscope (SEM) analyses were also performed to examine chemical interactions in soil–cement–FGD gypsum matrix. Finally, a process to determine an optimal cement–FG gypsum content corresponding to a target compressive strength was proposed.

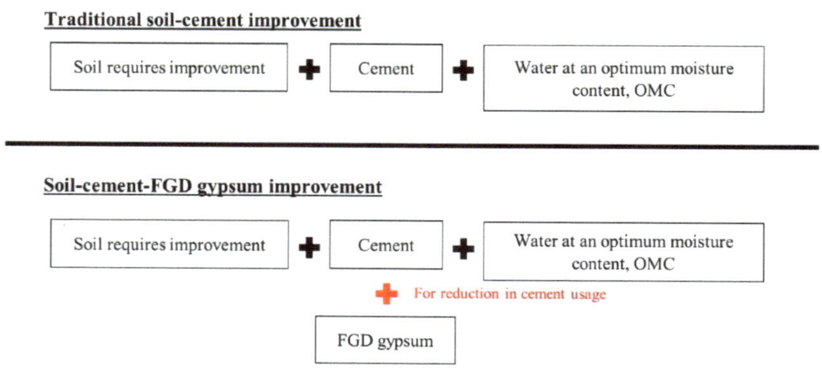

Figure 4. Traditional soil–cement process and the soil–cement–FGD gypsum process.

2. Materials and Methodology

2.1. Materials

2.1.1. Lateritic Soil

Lateritic soils A, B, and C were sourced from three local areas in Chiang Mai, Thailan The physical and engineering properties of those soils were measured in accordance wit the standards set by the Department of Highway (DOH), Thailand, for use as a subbas material for the DOH [6]. All the lateritic soil samples had relatively high clay content wit plastic index (PI) values higher than 11, the specified standard value for road constructio (Table 4). These elevated PI values confirmed that the majority of the lateritic soils wer substandard. As a result, those soils had to be modified and improved before they could b used as road construction material. In this research, substandard lateritic soil from locatio C (Figure 5a) was selected for use in laboratory tests because it had the highest PI valu which indicated the most substandard condition compared to the other two sources i this study. Hexagonal plates of clayey soil particles stacked into some spaces between th soil particles were observed in the SEM image of the lateritic soil from C (Figure 5b). Thi lateritic soil has a grain-sized distribution (Figure 6) and can be classified as low plasticit clay (CL) according to the Unified Soil Classification System (USCS) [59].

Table 4. The physical and engineering properties of lateritic soils.

Properties	A	B	C	DOH Standard (Subbase Materials)	Standard Method
Specific gravity, Gs	-	-	2.86	Not specified	ASTM D854
Passing No. 200 (%)	48	52	64	≤40	ASTM D422
Liquid Limit LL. (%)	30	33	41	≤35	
Plastic Limit PL. (%)	20	20	19	Not specified	ASTMD4318
Plastic Index PI. (%)	10	13	22	≤11	
Classification (as per USCS)	CL	CL	CL	Not specified	ASTM D2487

(a) (b)

Figure 5. Lateritic soil (**a**) in overview; (**b**) microstructure.

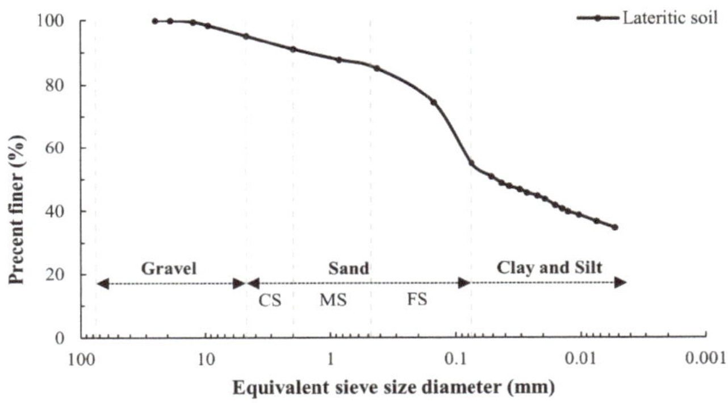

Figure 6. The grain size distribution of lateritic soil C.

2.1.2. Stabilizing Agents

A commercial type of Portland Cement (Type 1) complying with ASTM C150 [60] was used.

FGD gypsum was collected from the Mae Moh coal power plant, Lampang, Thailand, where around 2.25 million tons of FGD gypsum is produced annually [61]. Figure SEM analysis of the FGD gypsum used in this study shows rod-shaped particles with sizes between approximately 5 and 20 µm (Figure 7).

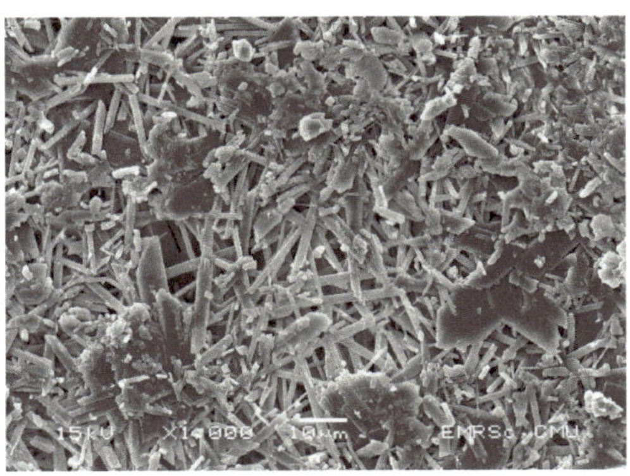

Figure 7. The microstructure of FGD gypsum.

2.2. Methodology and Experimental Works

2.2.1. Test Standard

The experiments to determine the physical properties of lateritic soil and the strength development of soil–cement–FGD gypsum mixtures are detailed as follows:

Physical properties

- Specific gravity tests in accordance with ASTM D854 [62];
- Soil particle size distribution by wet sieving tests in accordance with ASTM D422-63e2 [63];
- Soil particle size distribution by hydrometer tests in accordance with ASTM D422-63e2 [63];
- Liquid limit tests and plastic limit tests in accordance with ASTM D4318 [64].

Strength development of soil–cement–FGD gypsum mixtures

- Modified Proctor compaction tests in accordance with ASTM D1557 [56];
- Unconfined compressive strength (UCS) tests in accordance with ASTM D1633 [57];
- California Bearing Ratio (CBR) tests in accordance with ASTM D1883 [58];
- Microstructure analyses by a scanning electron microscope (SEM), JEOL JSM-5910LV (JEOL Ltd., Tokyo, Japan).

2.2.2. Methodology

The study process started with the determination of the optimum moisture content (OMC) and the maximum dry density (MDD) values of all mixtures in this study using the modified Proctor compaction tests [56]. For each mix, the densification at the OMC and MDD was used to prepare test samples for UCS tests [57]. Next, UCS tests were performed to assess the strength development of the soil–cement–FGD gypsum mixtures. CBR tests [58] were also carried out on selected mixes for both soaked and unsoaked conditions, and then the CBR index (%) was established (see Equation (1)) to measure the moisture sensitivity of selected mixtures and soil indirectly. Lastly, the test samples of the highest compressive strength in each cement group were subjected to the microstructure observation through the SEM analyses (Figure 8). Based on a series of test samples in this study, the sample symbols were specified as CxGy, where x is the percentage amount of cement by dry weight soil, and y is the percentage amount of FGD-gypsum by dry weight soil.

$$\text{CBR index (\%)} = (\text{Soaked CBR Value}/\text{Unsoaked CBR value}) \times 100 \quad (1)$$

2.2.3. Sample Preparation and Testing

The UCS tests by ASTM D1633 (Method A) [57] were followed based on cylindrical samples of size 101.6 mm in diameter and 116.4 mm in height. UCS test samples were compacted in five layers for 25 blows per layer to achieve their MDD values as target densities. After that, each sample was removed from the Proctor Mold and wrapped in a plastic bag to prevent moisture loss during the curing period and kept in at ambient temperature until the target curing periods of 3, 7, and 28 days were reached. Then, the UCS samples which had the highest UCS values were selected to observe the micro-interactions of the mixture matrix using the SEM analyses. The CBR tests by ASTM D1883 [58] were conducted on samples having the same mixture condition as those with the highest UCS values at a 28-days curing time. The substandard lateritic soils were stabilized by mixing OPC (C) in the proportions of 0, 1, 2, and 3% by dry weight of soil. In addition, each sample was mixed with FGD gypsum (G) in the proportions of 0, 1, 3, and 5% by dry weight soil respectively (Table 5). Please note that all the results in this study are the average values from the three samples.

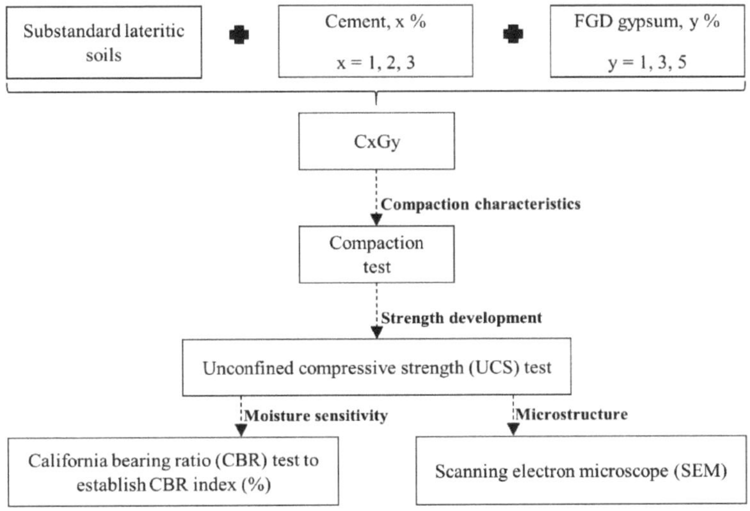

Figure 8. Methodology process of this research.

Table 5. Testing scheme.

Purpose	Tests	Moisture	Cement Contents (% by Dry Weight Soil)	FGD-Gypsum Contents (% by Dry Weight Soil)	Curing Days (Days)
Compaction characteristics	Modified Proctor	Various moisture contents to find OMC	0, 1, 2, and 3	0, 1, 3, and 5	-
Strength development	Unconfined compressive strength	OMC	0, 1, 2, and 3	0, 1, 3, and 5	3, 7, and 28
Moisture sensitivity (CBR index)	California bearing ratio		The highest UCS specimens of each cement group at 28 days		28
Microstructure of soil–cement–FGD gypsum	Scanning electron microscope				

3. Results and Discussion

3.1. Compaction

The compaction curve referred to the relationship between the dry density, and the moisture contents of a soil sample and is generally used to represent the compaction characteristics of all soil–cement–FGD gypsum mixtures in this study. In Figure 9, the OMCs of all mixes are in between 11.20 and 12.19%, all of which are lower than that of the substandard lateritic soil. Furthermore, as expected, lower OMCs and higher MDDs were observed for all soil–cement–FGD gypsum mixtures compared to the soil. This lower OMC–higher MDD characteristic of the study mixtures was in agreement with previous research [65,66]. The compaction curves of the soil–cement–FGD gypsum mixtures were examined. Significant findings were as follows:

- Adding cement and FGD gypsum to the substandard lateritic soil altered its compaction characteristics as indicated by a higher OMC and a lower OMC (Figure 9). The cause of the lower OMC–higher MDD characteristic was that some of the water in the mixture was consumed by the hydration reaction between cement, FGD gypsum, and water, which yielded solid cementitious products, the generation of which results in more solids in the mixture system, and these cementitious products occupied available void spaces in the mixture system. Consequently, a less porous matrix of the soil–cement–FGD gypsum mixture was obtained, leading to fewer void spaces for water to occupy and lower OMC characteristics of the mixture. Furthermore, those solid cementitious products have a higher specific gravity which led to a much larger overall specific gravity of the mixture compared to the substandard lateritic soil. Therefore, the soil–cement–FGD gypsum mixtures provided a higher MDD than that of the reference soil.

- The soil–cement–FGD gypsum mixtures which were compacted at the larger OMC conditions (the wet-side compaction) showed much higher dry densities compared to those of the substandard lateritic soil. At the MDD, the air void is equal to 8% (see the air void line of 8% in Figure 9) for the reference soil, but all study mixtures demonstrated noticeably fewer air voids. Apart from more solid fractions caused by the new cementitious products in the mixture system, the leftover cement, and FGD gypsum after the hydration reaction assisted in filling available air voids in the mixture system, which led to higher dry densities in the wet-side compaction. This would indicate the addition of a cement–FGD gypsum mixture to the soil provides the lower moisture sensitivity characteristics of a geomaterial.
- A very narrow range (11.20–12.19%) of OMC values was derived for the soil–cement–FGD gypsum mixtures, as seen in a series of compaction curves in Figure 9. This almost identical OMC value caused the presence of Ca^{2+} in the soil–cement–FGD gypsum matrix, which resulted in a reduction of repulsive forces between clay particles due to Ca^{2+} absorption on the surfaces of clay particles in the system. Therefore, the formation of a new edge-to-face arrangement of platy-shaped particles of clay existed [67]. For this unique clay particle arrangement, the unchangeable liquid limit phenomenon of the soil–cement–FGD gypsum mixtures would be expected. This phenomenon can govern a relatively constant OMC value of the mix, based on the findings of [68,69]. They reported that for the clayey soil, its liquid limit was a governing factor in controlling its OMC value.

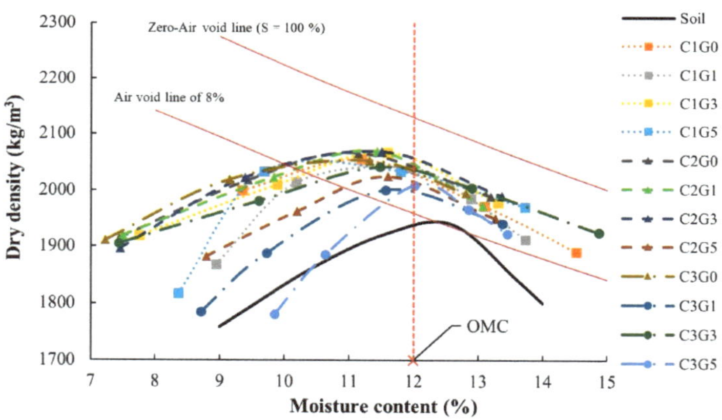

Figure 9. Compaction test results of all samples. **Note**: Soil refers to the substandard lateritic soil used as a benchmark.

As a result, an OMC of 12% was used to prepare test samples for all tests in this study.

3.2. *Strength Development*

Test results of the mixes through UCS values are summarized in Table 6 and illustrated in Figure 10. The values of CBR and the CBR index of the target mixes are also presented in Figure 10. Important findings in the strength development and the moisture sensitivity of the mixes are described as follows.

Table 6. Results of unconfined compression strength test and CBR test.

No.	Specimen Symbol	Cement (% of the Dry Weight of Soil)	FGD Gypsum (% of the Dry Weight of Soil)	Test Result						
				UCS (ksc.)				CBR (%)		CBR Index (%)
				0 Day	3 Days	7 Days	28 Days	Unsoaked	Soaked	
1	Soil	-	-	9.43		-		40	17	42.5
2	C1G0	1	0	-	9.56	14.18	17.94	44	26	59.1
3	C1G1		1		18.54	19.54	23.16	50	25	50.0
4	C1G3		3		15.54	17.46	21.27	-	-	-
5	C1G5		5		11.29	15.63	20.42	-	-	-
6	C2G0	2	0		20.16	20.74	29.92	67	44	65.7
7	C2G1		1		21.01	24.47	30.01	-	-	-
8	C2G3		3		24.21	28.30	37.67	-	-	-
9	C2G5		5		28.26	28.54	41.17	107	78	72.9
10	C3G0	3	0		24.82	26.65	30.17	80	55	68.7
11	C3G1		1		26.04	28.52	31.73	-	-	-
12	C3G3		3		30.36	31.97	37.12	78	41	52.4
13	C3G5		5		18.88	31.95	34.92	-	-	-

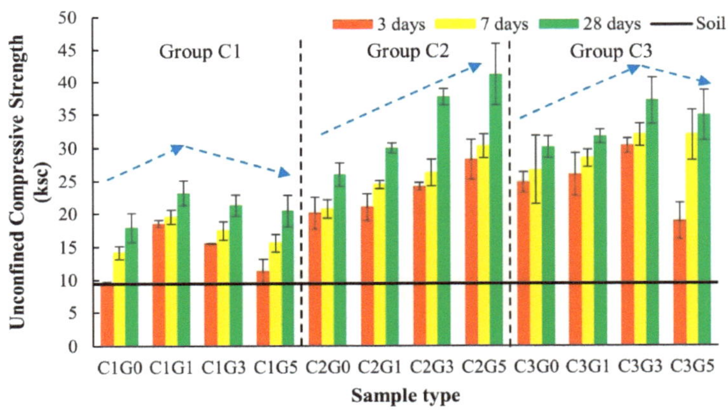

Figure 10. Unconfined compressive strength of the soil–cement–FGD gypsum mixtures.

3.2.1. Strength Development Concerning Cement Content, FGD Gypsum Content, and Curing Period

The effects of cement on the strength development of the mixture were considered by three contents of 1, 2, and 3% by dry weight of soil, namely Groups C1, C2, and C3, respectively, as shown in Figure 10. Based on the UCS test results in Figure 10, the higher cement contents yielded higher UCS levels for all mixes. Additional cement added to the mixes increased the UCS value. A series of UCS test results revealed that Group C3 had a higher UCS level than Groups C2 and C1, and Group C2 had a higher UCS level than Group C1, respectively. The strength development of the cement-stabilized soil results from the main solid cementitious compounds of C–S–H was generated from the hydration reaction between cement and water. Regarding the strength development with respect to the UCS of the soil–cement material by varying cement contents, the different zone concept was proposed following previous research work [70–73]. The soil–cement strength development can be divided into three zones; (1) soil–cement interaction zone (cement content of approximately 0–20%), (2) transitional zone (cement content of roughly 20–30%), and (3) cement–soil interaction zone (cement content of more than 30%). For the cement contents of 1–3% used in this study, the mixtures could be categorized in the soil–cement interaction zone. Within this zone, for a clay cluster in soil, cement had a role in welding water-filled spaces in a clay matrix with inherent cementitious characteristics. As a consequence, the soil–cement interaction zone was established. Also within this zone, interactions among water, clay particles, and cement particles, which have a particular manner in nature, took place by hydration reactions and the physicochemical action from cement with water and clay with water, respectively. Therefore, the continuity c a hardened cement paste in which soil particles were embedded existed, as a result c which a higher cement content caused more hardening continuity among soil clusters i the soil–cement matrix because the increase of cement led to higher UCS characteristics.

FGD gypsum content did affect strength development (Figure 10). Adding more FGD gypsum (1–5% of dry soil by weight) generated two patterns in the strength developmer trends. A rise in UCS values peaked at a certain cement content and then declined wit additional FGD gypsum content. Such trends of strength development were observed fc Groups C1 and C3. For Group C2, a gradual rise in UCS values over a more FGD gypsur rate was recorded. However, based on the results shown in Figure 10, adding FGD gypsur into a soil–cement mixture resulted in better gains in compressive strength compared t the traditional soil–cement material.

For the curing time, the more extended the period, the higher the recorded UCS value for all mixtures in this study. The strength development through UCS values against curin time indicated and confirmed the on-going chemical reactions for all mixtures within th curing times of 3, 7, and 28 days in this study (Figure 10).

3.2.2. Strength Increase Rate (SIR)

C2G5 provides the highest strength increase rate (SIR) of all the samples (Figure 11 Overall, the SIR range was 0.6–1.4 ksc per day. Group C2 demonstrated a higher SIR tha Groups C1 and C3, respectively. In addition, when the baseline mixtures of C1G0, C2G0 and C3G0 were considered, adding FGD gypsum to the soil–cement–FGD gypsum mixe tended to show higher SIRs than those of the baseline materials.

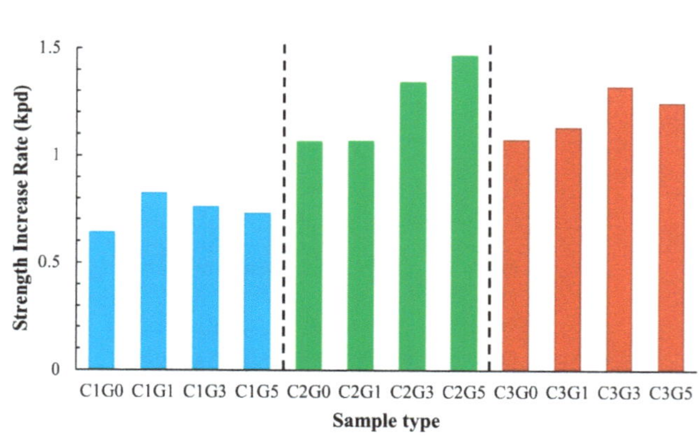

Figure 11. Strength increase rate (SIR) of all mixtures in this study.

3.2.3. Strength Development with Respect To the Water to Cement and FGD Gypsum (w/(c + g)) Ratio

The w/c ratio is a crucial factor in concrete technology since the strength of concrete crucially depends upon such the w/c ratio [74]. Therefore, a newly defined parameter o the water to cement and FGD Gypsum (w/(c + g)) ratio was established in this study. Thi w/(c + g) ratio was used to capture the strength characteristics of the soil–cement–FGD gypsum mixes relative to the influential parameters of water content, cement content, and FGD-Gypsum content. The trend of the w/c ratio and the compressive strength of concrete were inversely related and the reduction of the w/(c + g) ratio tended to increase the strength of the soil–cement–FGD gypsum mixtures (Figure 12). Furthermore, in Figure 12 UCS values of all mixes in the study were well above 6.9 ksc, the UCS values at seven days which is based on the specification of the UCS value of the soil–cement material for the road subbase [75].

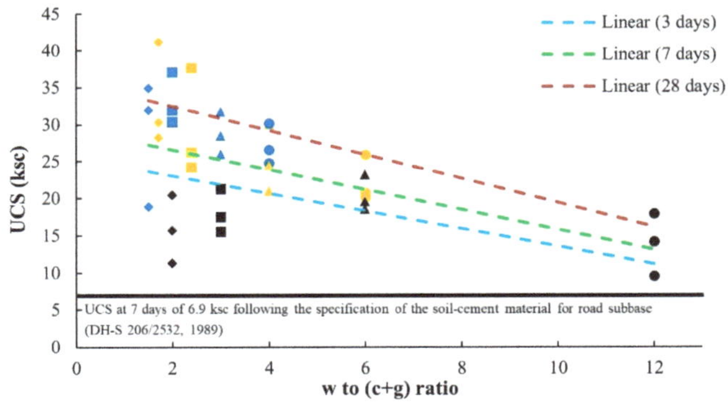

Figure 12. Unconfined compressive strength vs w/(c + g) ratio. **Note**: Line color: light blue, green, and dark red represent 3, 7, and 28 days, respectively. The color of shapes: dark, orange, and blue represents C1, C2, and C3, respectively. Shape: circle, triangle, square, and diamond represent G0, G1, G3, and G5, respectively.

3.3. Moisture Sensitivity

In this study, the moisture sensitivity of the target mixes was evaluated through the CBR index following Equation (1). The CBR index is not a general parameter indicating the moisture sensitivity characteristics of road base materials; it described the difference in CBR values as a material bearing resistance between relatively dry (unsoaked) and wet (soaked) conditions. C1G1 and C3G3 had the highest USC values in Group C1 and Group C3, respectively, and demonstrated lesser CBR index values than those of C1G0 and C3G0. These lower CBR index values would indicate higher moisture sensitivity of the mixture with the inclusion of FGD gypsum (Figure 13). However, for Group C2, C2G5 had the highest UCS, and SIR also provided a higher CRB index than that of C2G0 and all target mixes. Based on the results shown in Figure 13, that would imply that relatively high UCS values (e.g., C1G1 and C3G3) of the soil–cement–FGD gypsum mixes cannot guarantee a sound moisture sensitivity. Only the addition of a proper amount of FGD gypsum could assist the moisture resistance of the mixture, and C2G5 is the case in this study.

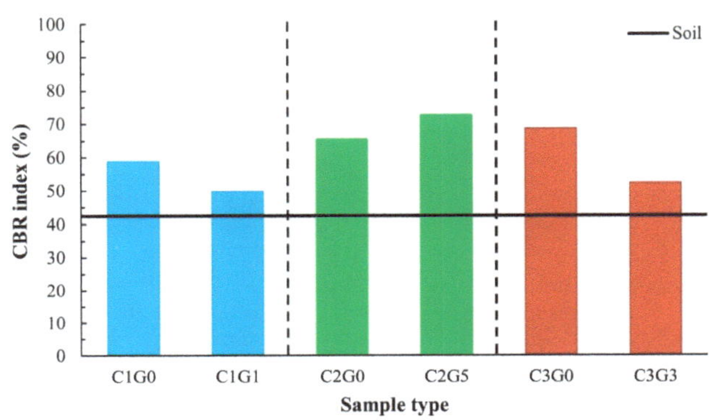

Figure 13. CBR indexes of the target mix in this study.

4. Roles of FGD-Gypsum Added in the Soil–cement–FGD Gypsum Mixture through SEM Analyses

UCS results illustrated the strength development of the soil–cement–FGD gypsu mixes corresponding to cement content, FGD gypsum content, and curing time (Figure 1 This strength development indicated the effect of the hardening generated from the chem ical reactions of cement, FGD gypsum, and water. As was recognized, the well-know theory of conventional cement hydration exists. This theory described the mechanism o how Portland cement compounds of alite or tricalcium silicate (C_3S), Belite (C_2S), Alum nate (C_3A), and Ferrite (C_4AF) react with water in association with gypsum ($CaSO_4{\cdot}2H_2O$ to generate the cementitious compounds like C–S–H, calcium hydroxide (CH), ettringit (AFt), and monosulfate (AFm). Gypsum is generally added during cement production as retarding agent to delay the setting time of the cement paste for workability purposes. I this section, the roles of FGD-gypsum added in the soil–cement–FGD gypsum matrix we explained through the results of SEM analyses.

Starting from the role of the gypsum in cement for the hydration of cement, wit the presence of such gypsum, ettringite is formed by the reaction of C_3A (the majorit compound in Portland cement), gypsum, and water, as shown in Equation (2) [76,77 The ettringite, which has a needle-shaped crystal morphology, could occupy availab pore spaces together with C–S–H in a cement-based material, which would lead t the strength gain characteristics of such a material. The ettringite can also further re act with the rest of the C_3A in a paste-formed matrix and then, a more stable stage o monosulfate ($3CaO{\cdot}Al_2O_3{\cdot}CaSO_4{\cdot}12H_2O$), a flower-shaped crystal, was formed followin Equation (3) [76,77].

$$3CaO{\cdot}Al_2O_3 + 3(CaSO_4{\cdot}2H_2O) + 26H_2O \rightarrow 3CaO{\cdot}Al_2O_3{\cdot}3CaSO_4{\cdot}32H_2O$$

$$\text{(Tricalcium aluminate)} + \text{(Gypsum)} + \text{(Water)} \rightarrow \text{(Ettringite)} \quad (2$$

$$3CaO{\cdot}Al_2O_3 + 3CaO{\cdot}Al_2O_3{\cdot}3CaSO_4{\cdot}32H_2O + 4H_2O \rightarrow 3(3CaO{\cdot}Al_2O_3{\cdot}CaSO_4{\cdot}12H_2O)$$

$$\text{(Tricalcium aluminate)} + \text{(Ettringite)} + \text{(Water)} \rightarrow \text{(Monosulfate)} \quad (3$$

At this stage, the chemical products of ettringite and monosulfate were obtained b adding gypsum to the cement and mixing it with water. Furthermore, ettringite generall demonstrated better strength enhancement in a cement-based material—-in this case, th soil–cement–FGD gypsum mixtures—-than in monosulfate [76,78]. Both ettringite an monosulfate are compounds of C_3A, $CaSO_4$, and water with formation dependent upo the ratios of available alumina to sulfate (Al_2O_3/SO_4, the A/S ratio) in the cement-base material. Ettringite is formed following the condition that a sufficient amount of gypsur exists in the hydration reaction system (high sulfate with the low A/S ratio). Moreove ettringite can further react with gypsum to convert to monosulfate, a more stable state forr than ettringite itself, based on the condition of low sulfate with a high A/S ratio [76,79 Based on the hydration reaction process, with a continuous increase manner of alumin (Al_2O_3), ettringite tends to decrease with a corresponding increase in monosulfate b the time corresponding to a fixed amount of gypsum (a higher A/S ratio condition). T maintain enough ettringite in the hardening process to enhance the strength gain of th soil–cement–FGD gypsum mixtures, a relatively low A/S condition is required. By doin this, more gypsum (a higher amount of sulfate) must be added. In this study, more FGI gypsum was added to the soil–cement–FGD gypsum system. With a sufficient amoun of water in the system, the added FGD gypsum provided extra sulfate ions to keep th condition of a low A/S ratio, suitable for ettringite formation. Based on the low A/S rati environment, the strength gain characteristic of the mixtures was observed. Howeve adding too much FGD gypsum to the hydration system can cause an adverse effect to th strength gain of the soil–cement–FGD gypsum mixtures. With a limited amount of water i the system to dissolve the sulfate ions from the FGD gypsum, in the case of too much FGI gypsum, gypsum dissolution could take place. This dissolution situation may generat a supersaturated gypsum crystallization in the soil–cement–FGD gypsum mixtures. The FGD gypsum structure itself revealed a weaker structure than that of monosulfate and ettringite. Therefore, the strength-reduction mechanism of the soil–cement–FGD gypsum mixtures can be lost in case of too much gypsum added into the mixture system.

In this study, samples having maximum UCS values of each group (C1G1, C2G5, and C3G3) were examined using SEM.

The ettringite and C–S–H gels were observed to confirm the strength gain manner of such a mixture as mentioned above (Figure 14). The further reaction products of monosulfate did not appear with a relatively low gypsum content at this mix of C1G1. However, the strength reduction trend of Group C1 illustrated a range of the FGD gypsum content beyond that of C1G1 (Figure 10). This reduction trend probably resulted from the adverse effect of the excess FGD gypsum, as explained.

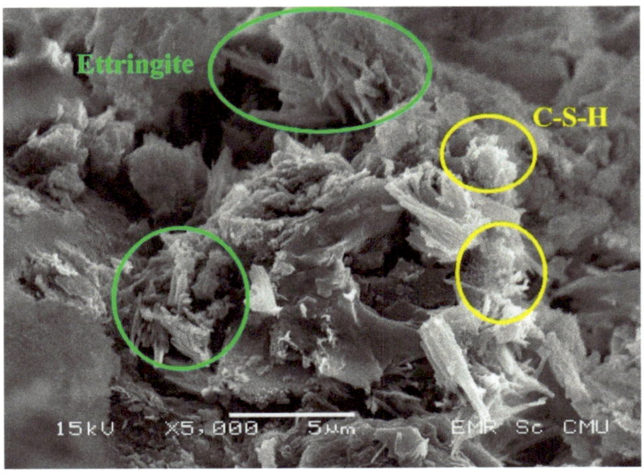

Figure 14. The SEM image of C1G1.

According to Group C2 in Figure 10, a gradual increase in strength exists with no maximum UCS value within the entire range of the mixtures of Group C2. C2G5 demonstrated the highest UCS value in Group C2 and all mixtures in this study. C–S–H, ettringite, and monosulfate are observed in C2G5 to explain its relatively high UCS value (Figure 15). Furthermore, excessive gypsum in the form of supersaturated gypsum, a weaker structure, was not seen. A proper combination among the cement content of 2%, the FGD gypsum contents of 2, 3, and 5%, and active water would play a role in generating C–S–H, ettringite, and monosulfate which are compounds of sound compounds to strengthen the compressive strength of a material. Furthermore, the presence of monosulfate would indicate the interchangeable stage from monosulfate to ettringite under the excessive gypsum environment.

For Group C3, the strength development trend revealed a peak of the UCS value (Figure 10), which was the same as that of Group C1. C3G3 provided the highest UCS value in this group; the strength reduction occurred at the 5% FGD content of C3G5. Apart from C–S–H and ettringite with no monosulfate, unreacted FGD gypsum was not observed in C3G3 (Figure 16). More unreacted FGD gypsum could cause the strength reduction of the mixtures having the FGD gypsum content greater than that of C3G3. In Group C3, with the highest cement content of 3%, an unbalanced amount of active water in the chemical reactions occurred. The water in the mixtures played multiple roles in lubricating soil particles, getting closer, and sharing for the cement hydration during mixing. In the case of C3G5 (the strength reduction), with excessive FGD gypsum, a limited amount of active water to dissolve more sulfate ions existed for some reason. For example, more activ water was required to react with more of the 3% cement content, leading, in this case, t less active water in the cement hydration to counteract more FGD gypsum. Active wate content for the hydration of the cement would be another influential factor to control th strength gain of the mixture if excessive FGD gypsum were added.

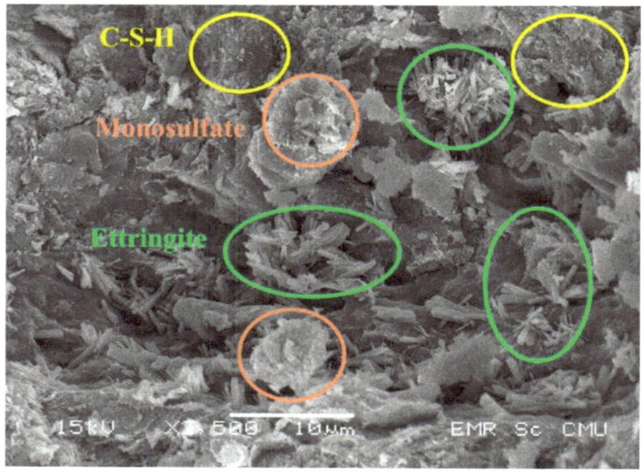

Figure 15. The SEM image of C2G5.

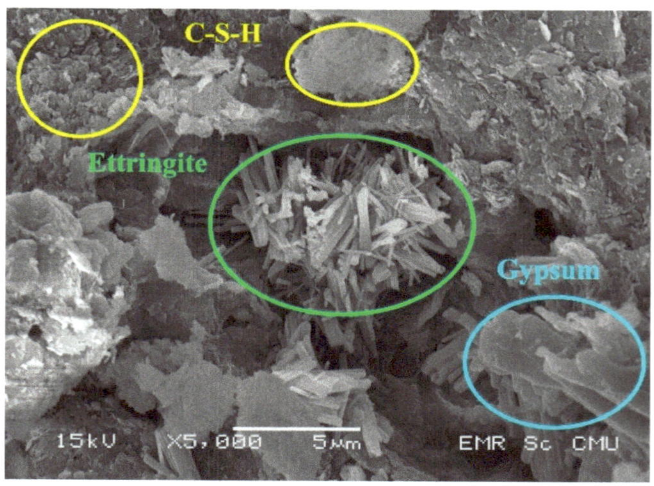

Figure 16. The SEM image of C3G3.

5. Application of the Cement–FGD Gypsum Soil Stabilization

Based on the results of this study, FGD gypsum could enhance the use of cement in soil stabilization to obtain a stronger stabilized material. For a target compressive strength of the soil–cement material, adding FGD gypsum into a soil–cement mixture could reduce the cement content of a mix while maintaining such a given compressive strength value.

Based on the w/(c + g) ratio established in this study and the test results of this study its relationship with UCS values can be expressed through Equation (4).

$$USC = k_1 \, [w/(c+g)] + k_2 \quad (4$$

The relationships between the UCS values of the soil–cement–FGD gypsum mixes in this study and the w/(c + g) ratios in Figure 17 are based on the constant w parameter of 12% (the optimum moisture content) and the target cement contents of 1% (C1), 2% (C2), and 3% (C3) as divided zones. For most applications, when a target UCS value was set, according to the linear relationships in Figure 17, a corresponding w/(c + g) to a given UCS value could be found, and then with w = 12% and the correct amount of cement, an optimum amount of FGD gypsum could be calculated for use in a mix. Table 7 illustrates the examples of different proportions of cement and FGD gypsum calculated following given UCS values. This table demonstrates that in the comparison between the traditional soil–cement (TSC) materials (the cement contents of 1, 2, and 3% used in this study) and the soil–cement–FGD gypsum (SCF) materials, adding FGD gypsum reduced the use of cement in a mix to achieve a given UCS value.

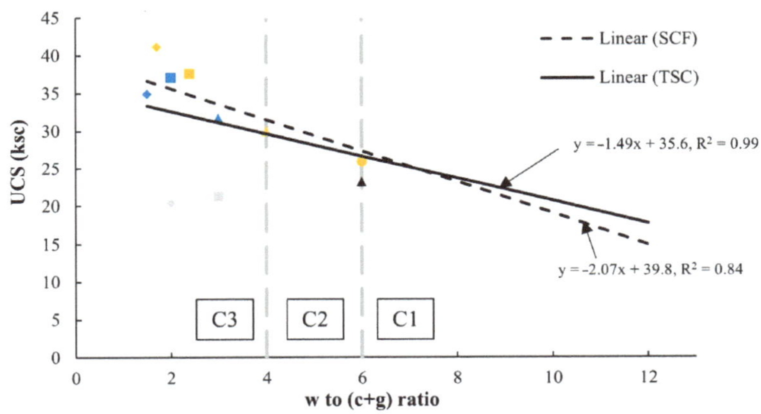

Figure 17. UCS–w/(c + g) relationships. **Note:** Color of shape: dark, orange, and blue represents C1, C2, and C3, respectively. Special color of shape: gray means the value is not used to create the USC–w/(c + g) relationships. Shape: circle, triangle, square, and diamond represent G0, G1, G3, and G5, respectively.

Table 7. The different proportions of the traditional soil–cement (TSC) materials and the soil–cement–FGD gypsum (SCF) materials corresponding to given USC values.

Process	Target USC	w/c	w/(c + g)	OMC	Cement Content	FGD Gypsum Content
	(ksc)				(% by dry weight soil)	
TSC	20 *	10.5	-	12	1.14	-
SCF		-	9.6		1	0.25
TSC	30 *	3.8	-		3.16	-
SCF		-	4.7		2	0.55
TSC	32.6	2 *	-		6	-
SCF	35.7	-	2 *		3	3

Note: the star symbol (*) is used to define (1) target USC or (2) target w/c and w/(c + g) ratio.

6. Conclusions

Based on the preliminary examination of the use of FGD gypsum to enhance the effective implementation of the soil–cement mixture in improving poor lateritic soil with a relatively high clay content for road construction purposes in this study, the essential findings of this study can be summarized as follows:

The addition of cement and FGD gypsum into substandard lateritic soil altered its compaction characteristics (see Figure 9). For strength development, higher cement content and higher UCS values were found for all mixes. Furthermore, adding more FGD gypsu (1–5% of dry soil by weight) provided two trends in the strength development: (1) a rise i UCS values peaked at a certain cement content and then declined with greater FGD gypsur content and (2) a gradual increase in UCS values over a more extensive range of FG gypsum content. For curing time, higher UCS values were observed with more extende curing periods. However, the soil–cement–FGD gypsum mixes could not guarantee soun moisture sensitivity. Only the proper amount of added FGD gypsum could assist th moisture resistance of the mixture. These test results can be well confirmed through th SEM examinations in this study.

The water to cement and FGD gypsum (w/(c + g)) ratio was newly established i this study. This w/(c + g) ratio captured the strength characteristics of the soil–cemen FGD gypsum mixes in relation to the influential parameters of water content, cemer content, and FGD gypsum content (Figure 12). The relation between UCS values and th w/(c + g) ratios of the mixtures in this study showed the same trend of the w/c rati and the compressive strength of concrete (e.g., the lower the w/c ratio, the higher th compressive strength): the reduction of the w/(c + g) ratio tended to increase the strengt of the soil–cement–FGD gypsum mixtures.

For the exercise of how to assign the right amount of FGD gypsum to the soil–cemer mixture based on the TSC materials (the cement contents of 1, 2, and 3% used in this stud and the SCF materials, we found that FGD gypsum can reduce the use of cement in a mi to achieve a given UCS value.

Author Contributions: Conceptualization, P.J.; methodology, P.M. and P.J.; validation, P.J., K.B., K. and S.P.-a.; formal analysis, P.J., P.M. and T.N.; investigation, T.N. and T.B.; writing—original dra preparation, P.M., P.J. and T.N.; writing—review and editing, P.J., K.N., S.P.-a., and C.A.; visualizatio P.M. and T.N.; supervision, K.B., T.B., and C.A.; project administration, P.J. All authors have read an agreed to the published version of the manuscript.

Funding: This research received no external funding.

Institutional Review Board Statement: Not applicable.

Informed Consent Statement: Not applicable.

Data Availability Statement: All data and information that support the outputs of this study ar available from the corresponding author upon reasonable request.

Acknowledgments: The authors would like to give sincere thanks to Chiang Mai University fc its significant support of this project and to Benjarach Boontifong, a master's student and researc assistant for his hard work. The group of Chiang Mai University students who were involve in this study should be strongly acknowledged and thanked to their main contributions to th laboratory experiments.

Conflicts of Interest: The authors declare no conflict of interest.

References

1. Mitchell, J.K.; Soga, K. *Fundamentals of Soil Behavior*; John Wiley & Sons: New York, NY, USA, 2005; Volume 3, ISBN 978-0-47 46302-3.
2. De Carvalho, J.C.; de Rezende, L.R.; Cardoso, F.B.d.F.; Lucena, L.C.d.F.L.; Guimarães, R.C.; Valencia, Y.G. Tropical soils fo highway construction: Peculiarities and considerations. *Transp. Geotech.* **2015**, *5*, 3–19. [CrossRef]
3. Schaetzl, R.J.; Thompson, M.L. *Soils: Genesis and Geomorphology*; Cambridge University Press: New York, NY, USA, 2015.
4. Ko, T.H. Nature and properties of lateritic soils derived from different parent materials in Taiwan. *Sci. World J.* **2014**, *2014*, 1– [CrossRef] [PubMed]
5. Oyelami, C.A.; van Rooy, J.L. A review of the use of lateritic soils in the construction/development of sustainable housing i Africa: A geological perspective. *J. Afr. Earth Sci.* **2016**, *119*, 226–237. [CrossRef]
6. *DH-S 205/2532 Standard Specification of Subbase Aggregate Material*; Department of Highways: Bangkok, Thailand, 1989. (In Tha
7. *DNER-ES Mechanical Stabilization of Subbase Layers Using Lateritic Soils*; Brazilian National Highway Department: Rio de Janeir Brazil, 1974; pp. 48–74.
8. *Highway Manual Part 1: Design. Volume III: Pavements and Materials Design*; Federal Ministry of Works: Abuja, Nigeria, 2013.

9. Jabatan Kerja Raya (JKR). *Standard Specification for Road Works;* Section 4: Flexible Pavement; Jabatan Kerja Raya (JKR): Kuala Lumpur, Malaysia, 2008.
10. Charman, J.H. *Laterite in Road Pavements;* Construction Industry Research and Information Association: London, UK, 1988; ISBN 0860172759.
11. Ruenkrairergsa, T. *Proposed Specifications of Soil Aggregates for Low-Volume Roads;* Department of Highways: Bangkok, Thailand, 1987.
12. Simon, A.B.; Giesecke, J.; Bidlo, G. Use of lateritic soils for road construction in North Dahomey. *Eng. Geol.* **1973**, *7*, 197–218. [CrossRef]
13. Amu, O.O.; Bamisaye, O.F.; Komolafe, I.A. The suitability and lime stabilization requirement of some lateritic soil samples as pavement. *Int. J. Pure Appl. Sci. Technol.* **2011**, *2*, 29–46.
14. Otunyo, A.W.; Chukuigwe, C.C. Investigation of the impact of palm bunch ash on the stabilization of poor lateritic soil. *Niger. J. Technol.* **2018**, *37*, 600–604. [CrossRef]
15. Onyelowe, K.C. Nanosized palm bunch ash (NPBA) stabilisation of lateritic soil for construction purposes. *Int. J. Geotech. Eng.* **2019**, *13*, 83–91. [CrossRef]
16. Eisazadeh, A.; Kassim, K.A.; Nur, H. Solid-state NMR and FTIR studies of lime stabilized montmorillonitic and lateritic clays. *Appl. Clay Sci.* **2012**, *67*, 5–10. [CrossRef]
17. Mengue, E.; Mroueh, H.; Lancelot, L.; Eko, R.M. Mechanical improvement of a fine-grained lateritic soil treated with cement for use in road construction. *J. Mater. Civ. Eng.* **2017**, *29*. [CrossRef]
18. Musec, P. Soil improvement methods for road construction. *Int. Multidiscip. Sci. GeoConference SGEM* **2018**, *18*, 637–644. [CrossRef]
19. Huan, Y.; Siripun, K.; Jitsangiam, P.; Nikraz, H. A preliminary study on foamed bitumen stabilisation for Western Australian pavements. *Sci. Res. Essays* **2010**, *5*, 3687–3700.
20. Jitsangiam, P.; Boonserm, K.; Phenrat, T.; Chummuneerat, S.; Chindaprasirt, P.; Nikraz, H. Recycled concrete aggregates in roadways: Laboratory examination of self-cementing characteristics. *J. Mater. Civ. Eng.* **2015**, *27*, 04014270. [CrossRef]
21. Das, B.M. *Principles of Foundation Engineering, Cengage Learning;* Cengage Learning: Boston, MA, USA, 2011.
22. Moh, Z.C.; Chin, Y.P.; Ng, S.C. Cement Stabilization of lateritic soils. In Proceedings of the Third Asian Regional Conference on Soil Mechanics and Foundation Engineering, Haifa, Israel, 25–28 September 1967.
23. Chen, H.; Wang, Q. The behaviour of organic matter in the process of soft soil stabilization using cement. *Bull. Eng. Geol. Environ.* **2006**, *65*, 445–448. [CrossRef]
24. Joel, M.; Agbede, I.O. Mechanical-cement stabilization of laterite for use as flexible pavement material. *J. Mater. Civ. Eng.* **2011**, *23*. [CrossRef]
25. Goodarzi, A.R.; Akbari, H.R.; Salimi, M. Enhanced stabilization of highly expansive clays by mixing cement and silica fume. *Appl. Clay Sci.* **2016**, *132–133*, 675–684. [CrossRef]
26. Apampa, O.A. Environmental benefits of corn cob ash in lateritic soil cement stabilization for road works. *African J. Sci. Technol. Innov. Dev.* **2019**, *11*, 427–431. [CrossRef]
27. Li, W.; Ni, P.; Yi, Y. Comparison of reactive magnesia, quick lime, and ordinary Portland cement for stabilization/solidification of heavy metal-contaminated soils. *Sci. Total Environ.* **2019**, *671*, 741–753. [CrossRef] [PubMed]
28. Sounthararajah, A.; Bui, H.H.; Nguyen, N.; Jitsangiam, P.; Kodikara, J. Early-age fatigue damage assessment of cement-treated bases under repetitive heavy traffic loading. *J. Mater. Civ. Eng.* **2018**, *30*. [CrossRef]
29. Soroka, I. *Portland Cement Paste and Concrete;* Macmillan International Higher Education: London, UK, 1979; ISBN 1349039942.
30. Dunuweera, S.P.; Rajapakse, R.M.G. Cement types, composition, uses and advantages of nanocement, environmental impact on cement production, and possible solutions. *Adv. Mater. Sci. Eng.* **2018**, *2018*, 1–11. [CrossRef]
31. Wansom, S.; Chintasongkro, P.; Srijampan, W. Water resistant blended cements containing flue-gas desulfurization gypsum, Portland cement and fly ash for structural applications. *Cem. Concr. Compos.* **2019**, *103*, 134–148. [CrossRef]
32. Chen, C.; Habert, G.; Bouzidi, Y.; Jullien, A. Environmental impact of cement production: Detail of the different processes and cement plant variability evaluation. *J. Clean. Prod.* **2010**, *18*, 478–485. [CrossRef]
33. Benhelal, E.; Zahedi, G.; Shamsaei, E.; Bahadori, A. Global strategies and potentials to curb CO2 emissions in cement industry. *J. Clean. Prod.* **2013**, *51*, 142–161. [CrossRef]
34. He, Z.; Zhu, X.; Wang, J.; Mu, M.; Wang, Y. Comparison of CO 2 emissions from OPC and recycled cement production. *Constr. Build. Mater.* **2019**, *211*, 965–973. [CrossRef]
35. Brooks, R.M. Soil stabilization with flyash and rice husk ash. *Int. J. Res. Rev. Appl. Sci.* **2009**, *1*, 209–217.
36. Pourakbar, S.; Asadi, A.; Huat, B.B.K.; Fasihnikoutalab, M.H. Stabilization of clayey soil using ultrafine palm oil fuel ash (POFA) and cement. *Transp. Geotech.* **2015**, *3*, 24–35. [CrossRef]
37. Horpibulsuk, S.; Phetchuay, C.; Chinkulkijniwat, A. Soil stabilization by calcium carbide residue and fly ash. *J. Mater. Civ. Eng.* **2012**, *24*. [CrossRef]
38. Canakci, H.; Celik, F.; Bizne, M.O.A.; Bizne, M.O.A. Stabilization of clay with using waste beverage can. *Procedia Eng.* **2016**, *161*, 595–599. [CrossRef]
39. Peyronnard, O.; Benzaazoua, M. Estimation of the cementitious properties of various industrial by-products for applications requiring low mechanical strength. *Resour. Conserv. Recycl.* **2011**, *56*, 22–33. [CrossRef]

40. Crossin, E. The greenhouse gas implications of using ground granulated blast furnace slag as a cement substitute. *J. Clean. Pro* **2015**, *95*, 101–108. [CrossRef]
41. Rafieizonooz, M.; Mirza, J.; Salim, M.R.; Hussin, M.W.; Khankhaje, E. Investigation of coal bottom ash and fly ash in concrete a replacement for sand and cement. *Constr. Build. Mater.* **2016**, *116*, 15–24. [CrossRef]
42. Yoobanpot, N.; Jamsawang, P.; Horpibulsuk, S. Strength behavior and microstructural characteristics of soft clay stabilized wi cement kiln dust and fly ash residue. *Appl. Clay Sci.* **2017**, *141*, 146–156. [CrossRef]
43. Srivastava, R.K.; Jozewicz, W. Flue gas desulfurization: The state of the art. *J. Air Waste Manag. Assoc.* **2001**, *51*, 1676–168 [CrossRef]
44. Huang, J.T. Sulfur dioxide (SO2) emissions and government spending on environmental protection in China2014Evidence fro spatial econometric analysis. *J. Clean. Prod.* **2018**, *175*, 431–441. [CrossRef]
45. Dontsova, K.; Lee, Y.B.; Slater, B.K.; Bigham, J.M. *Gypsum for Agricultural Use in Ohio: Sources and Quality of Available Products*; Th Ohio State University: Columbus, OH, USA, 2005.
46. Songpoon, J. *Effects of Additives on Growth Mechanism of FGD-Gypsum Crystals from Mae Moh Power Plant*; Chiang Mai Universit Chiang Mai, Thailand, 2011.
47. Lei, D.Y.; Guo, L.P.; Sun, W.; Liu, J.P.; Miao, C.W. Study on properties of untreated FGD gypsum-based high-strength buildin materials. *Constr. Build. Mater.* **2017**, *153*, 765–773. [CrossRef]
48. Watts, D.B.; Dick, W.A. Sustainable uses of FGD gypsum in agricultural systems: Introduction. *J. Environ. Qual.* **2014**, *43*, 246–25 [CrossRef]
49. Kaweewong, J. Using Flue Gas Desulfurization (FGD) Gypsum, By-Product of Coal-Fired Power Plants in Agriculture. *Kin Mongkuts Agric. J.* **2018**, *36*, 161–172.
50. American Coal Ash Association. *An American Recycling Success Story: Beneficial Use Of Coal Combustion Products*; American Coa Ash Association: Farmington Hills, MI, USA, 2019.
51. European Coal Combustion Products Association (ECOBA). *Production and Utilisation of CCPs in 2016 in Europe*; European Coa Combustion Products Association: Essen, Germany, 2016.
52. Burkart, B.; Goss, G.C.; Kern, J.P. The role of gypsum in production of sulfate-induced deformation of lime-stabilized soil *Environ. Eng. Geosci.* **1999**, 173–187. [CrossRef]
53. Degirmenci, N.; Okucu, A.; Turabi, A. Application of phosphogypsum in soil stabilization. *Build. Environ.* **2007**, *42*, 3393–339 [CrossRef]
54. Shen, Z.; Cao, Y.; Fang, L. Experimental Investigation of Rapid Stabilization of Soft Clay Soils Using Chemical Admixtures. *So Mech. Found. Eng.* **2017**, *54*, 202–210. [CrossRef]
55. Kuttah, D.; Sato, K. Review on the effect of gypsum content on soil behavior. *Transp. Geotech.* **2015**, *4*, 28–37. [CrossRef]
56. *ASTM D1557-12e1 Standard Test Methods for Laboratory Compaction Characteristics of Soil Using Modified Effort (56,000ft-lbf/ft3 (270 kN-m/m3))*; ASTM International: West Conshohocken, PA, USA, 2012. [CrossRef]
57. *ASTM D1633-17 Standard Test Methods for Compressive Strength of Molded Soil-Cement Cylinders*; ASTM International: Wes Conshohocken, PA, USA, 2017. [CrossRef]
58. *ASTM D1883-16 Standard Test Method for California Bearing Ratio (CBR) of Laboratory-Compacted Soils*; ASTM International: Wes Conshohocken, PA, USA, 2016. [CrossRef]
59. *ASTM D2487-17 Standard Practice for Classification of Soils for Engineering Purposes (Unified Soil Classification System)*; ASTM International: West Conshohocken, PA, USA, 2017. [CrossRef]
60. *ASTM C150-16 Standard Specification for Portland Cement*; ASTM International: West Conshohocken, PA, USA, 2019. [CrossRef]
61. Boonserm, K. *Effects of Organic and Inorganic Additives on Hydration Reaction*; Chiang Mai University: Chiang Mai, Thailand, 200
62. *ASTM D854 Standard Test for Specific Gravity of Soil Solids by Water Pycnometer*; ASTM International: West Conshohocken, PA USA, 2010.
63. *ASTM D422-63e2 Standard Test Method for Particle-Size Analysis of Soils*; ASTM International: West Conshohocken, PA, USA, 200
64. *ASTM D4318—17e1 Standard Test Methods for Liquid Limit, Plastic Limit, and Plasticity Index of Soils*; ASTM International: Wes Conshohocken, PA, USA, 2017. [CrossRef]
65. Osula, D.O.A. A comparative evaluation of cement and lime modification of laterite. *Eng. Geol.* **1996**, *42*, 71–81. [CrossRef]
66. Mengue, E.; Mroueh, H.; Lancelot, L.; Eko, R.M. Physicochemical and consolidation properties of compacted lateritic soil treate with cement. *Soils Found.* **2017**, *57*, 60–79. [CrossRef]
67. Chew, S.H.; Kamruzzaman, A.H.M.; Lee, F.H. Physicochemical and engineering behavior of cement treated clays. *J. Geotec Geoenviron. Eng.* **2004**, *130*, 696–706. [CrossRef]
68. Nagaraj, T.S.; Banu, Z. Generalization of Abrams' law. *Cem. Concr. Res.* **1996**, *26*, 933–942. [CrossRef]
69. Horpibulsuk, S.; Miura, N.; Nagaraj, T.S. Assessment of strength development in cement-admixed high water content clays wit Abrams' law as a basis. *Geotechnique* **2003**, *53*, 439–444. [CrossRef]
70. Uddine, K. *Strength and Deformation Behaviour of Cement Treated Bankok Clay*; Asian Institute of Technology: Bangkok, Tha land, 1994.
71. Horpibulsuk, S.; Katkan, W.; Sirilerdwattana, W.; Rachan, R. Strength development in cement stabilized low plasticity and coars grained soils: Laboratory and field study. *Soils Found.* **2006**, *46*, 351–366. [CrossRef]

2. Horpibulsuk, S.; Rachan, R.; Chinkulkijniwat, A.; Raksachon, Y.; Suddeepong, A. Analysis of strength development in cement-stabilized silty clay from microstructural considerations. *Constr. Build. Mater.* **2010**, *24*, 2011–2021. [CrossRef]
3. Kampala, A.; Jitsangiam, P.; Pimraksa, K.; Chindaprasirt, P. An investigation of sulfate effects on compaction characteristics and strength development of cement-treated sulfate bearing clay subgrade. *Road Mater. Pavement Des.* **2020**, 1–14. [CrossRef]
4. Abrams, D.A. *Design of Concrete Mixtures*; Structural Materials Research Laboratory, Lewis Institute: Chicago, IL, USA, 1919.
5. *DH-S 206/2532 Standard Specification of Soil Cement Subbase*; Department of Highways: Bangkok, Thailand, 1989. (In Thai)
6. Portland Cement Association. *Ettringite Formation and the Performance of Concrete*; Portland Cement Association: Skokie, IL, USA, 2001.
7. Peng, J.; Zhang, J.; Qu, J. Mechanism of the formation and transformation of ettringite. *J. Wuhan Univ. Technol. Mater. Sci. Ed.* **2006**, *21*, 158–161. [CrossRef]
8. Dermatas, D. Ettringite-induced swelling in soils: State-of-the-art. *Appl. Mech. Rev.* **1995**, *48*, 659–673. [CrossRef]
9. Christensen, A.N.; Jensen, T.R.; Hanson, J.C. Formation of ettringite, Ca6Al2 (SO4) 3 (OH) 12·26H2O, AFt, and monosulfate, Ca4Al2O6 (SO4)·14H2O, AFm-14, in hydrothermal hydration of Portland cement and of calcium aluminum oxide—calcium sulfate dihydrate mixtures studied by in situ synchrotron X-ray. *J. Solid State Chem.* **2004**, *177*, 1944–1951. [CrossRef]

materials

MDPI

Article

Thermal Properties of Calcium Sulphoalumínate Cement as an Alternative to Ordinary Portland Cement

Małgorzata Gołaszewska *, Barbara Klemczak and Jacek Gołaszewski

Faculty of Civil Engineering, Silesian University of Technology, Akademicka 5, 44-100 Gliwice, Poland; barbara.klemczak@polsl.pl (B.K.); jacek.golaszewski@polsl.pl (J.G.)

* Correspondence: Malgorzata.golaszewska@polsl.pl

Abstract: This paper presents the results of research into the heat of hydration and activation energ of calcium sulphoaluminate (CSA) cement in terms of the dependence on curing temperature an water/cement ratio. Cement pastes with water/cement ratios in the range of 0.3–0.6 were teste by isothermal calorimetry at 20 °C, 35 °C and 50 °C, with the evolved hydration heat and its ra monitored for 168 h from mixing water with cement. Reference pastes with ordinary Portlan cement (OPC) were also tested in the same range. The apparent activation energy of CSA and OP was determined based on the results of the measurements. CSA pastes exhibited complex therm behaviour that differed significantly from the thermal behaviour of ordinary Portland cement. Th results show that both the w/c ratio and elevated temperature have a meaningful effect on the he emission and the hydration process of CSA cement pastes. The determined apparent activati energy of CSA revealed its substantial variability and dependence, both on the w/c ratio and th curing temperature.

Keywords: hydration heat; isothermal heat flow; CSA cement; Portland cement; apparent activatio energy

Citation: Gołaszewska, M.; Klemczak, B.; Gołaszewski, J. Thermal Properties of Calcium Sulphoaluminate Cement as an Alternative to Ordinary Portland Cement. *Materials* **2021**, *14*, 7011. https://doi.org/10.3390/ma14227011

Academic Editor: Lizhi Sun

Received: 22 October 2021
Accepted: 17 November 2021
Published: 19 November 2021

1. Introduction

Environmental concerns are one of the most important challenges in current concret technology, due to the high amount of CO_2 emitted to the atmosphere during Portlan clinker production. The production of 1 ton of clinker emits from 0.6 to 1 ton of CO depending on the applied technology [1,2]. To decrease the impact of the concrete industr on carbon dioxide emissions, many different approaches are currently taken, includin decreasing the amount of Portland clinker in cement by its substitution with other material such as blast furnace slags [3], pozzolans [4], fly ash [5], or limestone [6], or by using othe binders not based on Portland clinker, such as geopolymers [7] or calcium sulphoaluminat (CSA) clinker [8,9].

CSA cement is a relatively new and uncommon type of cement, which was invented b Alexander Klein in the 1950s and introduced to industrial production in the 1960s [9]. CS clinker is a hydraulic binder obtained by burning limestone, bauxite and gypsum [10]. Du to the lower amounts of limestone involved in the process, as well as lower burning tempe ature, it has lower CO_2 emissions per 1 ton of clinker in comparison to Portland clinker [11 Its phase composition, and thus its hydration process, differs significantly from Portlan clinker. The mineral composition of CSA cement consists mostly of ye'elemite ($C_4A_3\hat{S}$ with smaller amounts of belite (C_2S) and calcium sulphate ($C\hat{S}$), while Portland clinke consists mostly of alite (C_3S), belite (C_2S), tricalcium aluminate (C_3A) and brownmillerit (C_4AF). Therefore, instead of mostly CSH phases and portlandite (CH), as in Portlan clinker, the main product of CSA clinker hydration is ettringite ($C_6A\hat{S}_3H_{32}$), stratlingit ($C_2A\hat{S}H_8$) and monosulphate, with low amounts of CSH also appearing due to the belit hydration process [12–14].

Due to ettringite being the main product of the hydration process, CSA cement is usually characterized by short initial and final setting times, high early strength, and low shrinkage [15,16]. Although all those properties are usually considered desirable in practical concrete technology, CSA cement is not widely used due to the high costs of the materials needed to produce the clinker, especially bauxite. However, with the increasing cost of Portland clinker production, this issue may prove to be less pressing. A more prevalent issue preventing the widespread use of CSA cement is the limited research on its properties, leading to an incomplete understanding of CSA cement as a material, especially in terms of the parameters used in the designing of concrete structures.

One of the relatively poorly recognized characteristics is its hydration heat and hydration heat development. Generally, the main reaction of CSA cement is connected to a rapid reaction of ye'elimite and gypsum, which form ettringite (AFt), monosulphate (AFm) and crystalline aluminium hydroxide (AH3), resulting in a rapid heat release and the high hydration heat main peak, usually in the first few hours after mixing cement with water [17–19]. Small amounts of belite present in the clinker react at a slower rate, forming limited amounts of the C-S-H phase, CH and gehlenite hydrate (C_2ASH_8) [20]. The research made by Winnefeld et al. [19] and Hargis et al. [17] has shown that the exact hydration heat development is dependent on the amount and form of the calcium sulphate added to the CSA clinker, with the addition of gypsum increasing hydration rate in comparison to anhydrite. It is therefore hard to draw comparisons between the hydration heat development between CSA cement and Portland cement.

In research by Li et al. [21], hydration heat evolution was characterized by both a higher hydration peak than in the case of Portland cement and its faster appearance. A similar effect was observed in research by Batog and Synowiec [22], Qin et al. [23] and Ramanathan [24]. However, the research of Li et al. [25] has shown that, while the main peak was higher, it could be later than in the case of Portland cement. The water–cement ratio also appears to affect the hydration heat of CSA. Research by Doval et al. [26] has shown that, with the increase in water-solids ratio (which was adopted in this research), the hydration heat can decrease. The same research found that, with higher temperatures, the hydration heat development is accelerated, and the hydration peak is greater. It must be noted, however, that those tests were performed for a very limited number of water-solids ratios (0.5, 0.7 and 1.0) that do not correspond well with water–cement ratios, which are more commonly used. The issue of choosing the right w/c ratio can prove to be more difficult in the case of CSA cement than Portland cement because, according to stoichiometric calculations, the right amount of water needed for the CSA cement to fully hydrate is 0.64, while for Portland cement it is less than 0.3 [14].

CSA hydration in a high curing temperature has been the subject of the research of Jeong et al. [27], in which hydration heat of CSA cement was tested at 30 °C, 60 °C and 90 °C. It has been found that the composition of the CSA cement plays an important role in the effect of temperature on the hydration, with the amount of ye'elemite in the cement affecting the response to the high temperature. In the case of higher ye'elemite content, the hydration heat at 90 °C was significantly lower than at 60 °C, while for the lower amount, with the increase of temperature the hydration heat also increased. Wang et al. [28] found that the hydration degree of CSA cement and clinker increases with increasing temperature. Additionally, with the increasing temperature, the type of hydration products changed in CSA clinker; however, in CSA cement no such effect was observed.

The thermal properties of cement, related to its hydration heat, are important in a variety of technological situations, such as concreting in low or high temperatures, and knowledge of these properties allows a better choice of cement for the given environmental situation. Hydration heat tests are the basis for the classification of cement as low heat (LH) or very low heat (VLH) cement according to the standard BS EN 197-1 [29]. Moreover, the time and size of the hydration peaks in the hydration flow charts can be the basics of understanding when to expect any heat phenomena, which can lead to better planning of the curing procedures.

Hydration heat development is also vital information needed to design concrete for structure, also in the context of massive structures, where the high-temperature gradier between the centre of the structure and its surface may easily occur, leading to increase tensile stress in the element [30]. If the tensile stress induced by the uneven volume change in the element, resulting from the temperature gradient exceeding the tensile strength c concrete, cracking may occur, with the simultaneous decrease in the overall load-bearin capacity of the element and loss of durability [31].

Moreover, high temperatures of the concrete can affect strength development at earl stages [32]. Therefore, if the need arises to model the thermal behaviour of concrete an to assess its actual maturity, the equivalent age approach based on the Arrhenius concep is usually used. In this case, the value of apparent activation energy is required. It value can be determined through both mechanical and calorimetric testing and subsequer calculations; however, the calorimetric method is often preferred [32–36]. In the case c CSA cement, this subject is poorly recognized.

The presented research focused on the hydration heat evolution of CSA cement paste with different water/cement ratios by means of isothermal calorimetry at three differer curing temperatures: 20 °C, 35 °C and 50 °C. Reference pastes with ordinary Portlan cement (OPC) were also tested in the same range, which highlighted the distinct therma behaviour of CSA cement. Next, the apparent activation energy of CSA and ordinar Portland cement was determined based on the results of the measurements. The effect of temperature and w/c ratio on the apparent activation energy based on the Arrheniu concept were also examined.

The addressed paper consists of five sections. After the introduction (Section 1), th experimental program with the basic characteristics of materials and testing methods ar provided in Section 2. Next, the results obtained from the conducted study are presente and discussed in Section 3. Finally, the main conclusions are specified in Section 4.

2. Experimental Program

In the conducted research, commercially available CSA cement was used. A referenc paste with ordinary Portland cement CEM I 42.5R (CEM I) was also tested. The chemica composition of the tested types of cement is presented in Table 1. The basic properties o both types of cement are depicted in Table 2.

Table 1. Composition of CSA and CEM I (OPC) cements.

Cement Type	Constituent, %									
	LOI	SiO_2	Al_2O_3	Fe_2O_3	CaO	MgO	SO_3	Na_2O	K_2O	Na_2O_{eq}
CSA	0.46	9.2	28.1	1.52	39.2	3.5	11.4	0.08	0.35	-
CEM I (OPC)	-	20.6	4.7	2.8	64.4	1.2	2.8	0.2	0.4	0.46

Table 2. Basic properties of CSA and CEM I (OPC) cements.

Cement Property	Unit	Value	
		CSA	CEM I (OPC)
Initial setting time	min	50	191
The soundness of cement, by Le Chatelier's method	mm	1	0.4
Compressive strength after 2 days	MPa	42.6	30.1
Compressive strength after 28 days	MPa	67.7	57.4
Specific surface area	cm^2/g	5600	4400

Experimental tests were performed to study the effect of two factors influencing the hydration process: the water-cement ratio and the temperature. The applied water/cemen ratio was equal to 0.3, 0.35, 0.4, 0.45, 0.5, 0.55 and 0.6, both for CSA and for CEM I cement. The chosen w/c ratios reflect the commonly used proportion between cement and water.

Three curing temperatures were used to measure the evolved heat. A temperature of 20 °C was applied as a reference value, typically used in heat of hydration studies. Additionally, the influence of increased curing temperatures (35 °C, 50 °C) on the hydration process of CSA cement was investigated. The curing temperature of 35 °C can easily occur in concrete structures cast in the summer season, while 50 °C may be applied to structures of larger cross-sectional dimensions with internal self-heating and elevated curing temperature, directly affecting the amount of heat released. Furthermore, the heat development at these three curing temperatures enabled the determination of the activation energy, which is described later in this section.

Hydration heat was measured in an isothermal calorimeter, TAM Air. The tests were performed according to the European standard PN-EN 196-11 [37] with the internal mixing procedure, allowing for the measurement of heat development from the moment of adding water to cement. As a reference, quartz sand was used, in an amount calculated based on its heat capacity corresponding to the cement paste samples.

The measurement was conducted on cement paste samples consisting of 5 g of CSA or CEM I cement and distilled water in an amount calculated from the assumed w/c ratio. The ampoules with cement and water were inserted into the calorimeter at least 24 h before the start of the measurement so that the temperature of the sample was equal to the set temperature (20 °C, 35 °C or 50 °C) at the start of the procedure. The measurement lasted 168 h from the time of adding water to the cement. Measurements for each instance were performed three times, and the mean value was calculated. For all of the instances, the standard deviation was lower than 6%.

Subsequently, based on the obtained results of the released heat at different temperatures, the activation energy of the tested types of cement was determined. The apparent activation energy is a crucial measure in the equivalent age approach used to estimate the combined effects of time and temperature on the property development of cementitious materials. This approach is a simple and common method for a reliable prediction of in-place strength during the construction of structures. Furthermore, the hydration heat development at different curing temperatures can also be accounted for, which is important in massive structures.

Hence, the sensitivity of the cement hydration process to the curing temperature can be determined using a function introduced by Hansen and Pedersen [32] which is based on the Arrhenius rate concept, originally describing the effect of temperature on the rate of a chemical reaction [34,38]. Following this function, the equivalent age, t_{eq}, is computed as follows:

$$t_{eq} = \int_{0}^{t} e^{\frac{-E_a}{R}\left(\frac{1}{T(\tau)} - \frac{1}{T_{ref}}\right)} d\tau \tag{1}$$

with:

$T(\tau)$, the actual temperature of the cementitious material, °C;
T_{ref}, the reference temperature usually equal to 20 °C;
R, the universal gas constant equals 8.314 J $mol^{-1} \cdot °C^{-1}$;
E_a, the activation energy, J mol^{-1};
τ, time, h or day.

The application of Equation (1) involves the conversion of the actual age of the cementitious material to its equivalent age at the reference temperature. The temperature dependence in Equation (1) is described by the value of the apparent activation energy, E_a, thus, its value is essential. The reported experimental tests revealed its dependence on the cement type, curing temperature and also, to a lesser extent, on the water/cement ratio [34,39–41]. Although the activation energy can be determined by strength tests of samples curing at different temperatures [32,42], isothermal calorimetry is appraised as a better method to quantify the activation energy [33,35]. Considering the computation of the activation energy based on the isothermal tests, the different procedures are di cussed in [35,42]. However, regardless of the method of its determination, it should b remembered that only an approximate description of the influence of temperature o cement hydration is obtained, which is reflected in the name "apparent" of the determine activation energy [33].

In this study, the rate method [43–45] has been applied to determine the discusse energy. Following this method, the apparent activation energy, E_a, is calculated based o heat evolution rates q_1 and q_2, consistent with the same evolved heat, Q_i, and two curin temperatures, T_1 and T_2:

$$E_a(Q_i) = -\frac{R}{\frac{1}{T_1(Q_i)} - \frac{1}{T_2(Q_i)}} ln\left[\frac{q_1(Q_i)}{q_2(Q_i)}\right] \quad (2$$

3. Results and Discussion

3.1. Evolution and the Rate of Hydration Heat

The curves of the heat evolution for CSA and CEM I (OPC) cements with differer water/cement ratios measured in isothermal tests at 20 °C, 35 °C and 50 °C are show in Figure 1. The values of the released heat are listed in Table 3 (CSA) and Table 4 (OPC To facilitate the interpretation of the obtained values of the released heat, they are als presented graphically in Figure 2.

Generally, a significant impact of the water/cement ratio on the amount of the release heat of CSA cement can be noticed. For all applied curing temperatures, the evolved hea increases with the increase of the w/c ratio (Figure 1a–c). In the initial period of 12 h, thi effect is not so meaningful, and the heat emission increases about 6% for each w/c rati increase of 0.05. In the later period, after 41, 72 and 168 h, the average value of heat growt is 10% under the curing temperature equal to 20 °C and 35 °C. Simultaneously, the greates increase in heat release can be seen between CSA pastes with a water/cement ratio of 0. and 0.35 (Figure 1a,b). This is over 20%, considering the heat released after 41, 72 and 168 For higher values of w/c, this growth successively decreases. For w/c = 0.6, it is only 4 and 1.5% for tests at 20 °C and 35 °C, respectively.

Considering ordinary Portland cement, CEM I, the water/cement effect is not s significant and clear as in CSA pastes (Figure 2). In detail, at a curing temperature of 20 °C the heat released is practically the same for different w/c. Slight differences are visible onl after 120 h, and they are less than 3% (Figure 1d). At a higher curing temperature of 35 °C the w/c effect is visible, but it is still less pronounced than for CSA cement (Figure 1e). I this case, after 168 h, the greatest growth (12%) in the released heat was observed betwee OPC pastes with a w/c of 0.3 and 0.35 (Figure 1e).

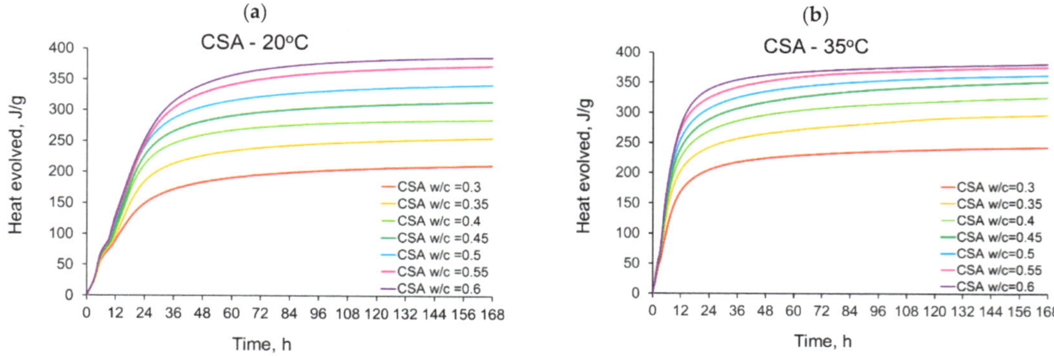

Figure 1. *Cont.*

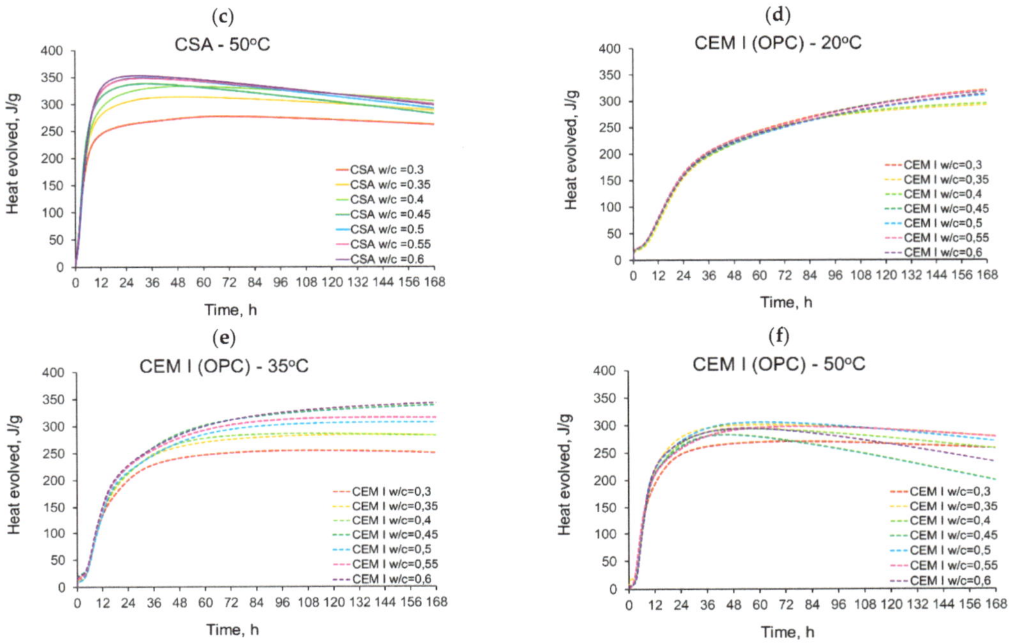

Figure 1. The heat evolved from cement pastes with different water/cement ratios in isothermal tests at different curing temperatures: (**a**) CSA at 20 °C, (**b**) CSA at 35 °C, (**c**) CSA at 50 °C, (**d**) CEM I (OPC) at 20 °C, (**e**) CEM I (OPC) at 35 °C, (**f**) CEM I (OPC) at 50 °C.

Table 3. Characteristic values of the heat evolved in CSA cement pastes.

Sample, Curing Temperature	The Heat Evolved, Jg^{-1}				The Peak Value of Heat Evolution Rate, $Jg^{-1} \cdot h^{-1}$		
	12 h	41 h	72 h	168 h	Second Peak	Third Peak	Fourth Peak
CSA w/c = 0.3							
20 °C	88.9	177.3	195.3	211.0	14.2	6.6	5.9
35 °C	168.7	220.8	231.7	243.0	21.8	19.2	-
50 °C	243.1	270.0	276.9	261.1	35.5	48.0	-
CSA w/c = 0.35							
20 °C	96.5	219.4	238.9	255.2	15.7	9.3	9.2
35 °C	203.0	260.3	275.5	296.2	25.1	26.1	-
50 °C	278.9	312.6	311.0	287.5	40.1	51.0	-
CSA w/c = 0.4							
20 °C	105.6	253.0	272.3	284.6	16.8	10.5	11.2
35 °C	225.1	290.6	307.2	325.4	26.5	30.2	-
50 °C	291.0	331.7	330.6	305.0	41.0	51.2	-
CSA w/c = 0.45							
20 °C	109.5	273.9	296.5	314.4	17.1	11.3	11.6
35 °C	240.4	311.6	330.5	351.6	26.8	32.4	-
50 °C	311.5	337.2	326.0	281.6	47.1	57.2	-
CSA w/c = 0.5							
20 °C	118.1	296.0	322.4	341.8	19.2	12.6	11.7
35 °C	255.8	330.0	346.8	362.5	28.1	35.3	-
50 °C	321.4	347.6	336.3	291.8	33.9	56.3	-
CSA w/c = 0.55							
20 °C	124.3	315.7	350.1	372.3	19.5	13.2	10.9
35 °C	271.8	346.6	362.7	375.8	29.4	38.5	-
50 °C	322.0	347.9	337.8	299.5	31.2	56.8	-
CSA w/c = 0.6							
20 °C	129.2	327.9	365.8	386.4	20.7	14.3	11.2
35 °C	278.5	358.1	369.9	388.3	29.7	38.7	-
50 °C	327.4	351.1	340.3	297.6	36.7	57.4	-

Table 4. Characteristic values of the heat evolved in CEM I (OPC) cement pastes.

Sample, Curing Temperature	The Heat Evolved, Jg^{-1}				The Peak Value of Heat Evolution Rate, $Jg^{-1} \cdot h^{-1}$		
	12 h	41 h	72 h	168 h	Second Peak	Third Peak	Fourth Peak
CEM I w/c = 0.3							
20 °C	73.8	213.6	257.0	320.8	8.2	-	-
35 °C	132.1	235.2	250.9	250.2	16.1	-	-
50 °C	191.3	263.8	270.4	258.0	32.1	-	-
CEM I w/c = 0.35							
20 °C	67.5	204.7	252.4	291.6	8.3	-	-
35 °C	136.5	254.2	275.3	282.4	16.9	-	-
50 °C	218.4	297.2	299.9	279.1	33.9	-	-
CEM I w/c = 0.4							
20 °C	73.9	206.1	251.3	295.4	8.3	-	-
35 °C	137.6	258.9	282.1	282.3	17.1	-	-
50 °C	209.6	290.4	291.2	257.7	33.5	-	-
CEM I w/c = 0.45							
20 °C	76.1	211.3	254.3	317.8	8.3	-	-
35 °C	148.5	273.3	311.4	338.6	17.2	-	-
50 °C	209.5	282.3	272.8	199.8	33.9	-	-
CEM I w/c = 0.5							
20 °C	74.8	208.2	249.7	311.5	8.1	-	-
35 °C	133.9	257.7	294.0	306.9	15.7	-	-
50 °C	218.7	299.5	304.7	271.1	35.2	-	-
CEM I w/c = 0.55							
20 °C	78.0	214.3	255.4	319.3	8.3	-	-
35 °C	146.6	266.6	302.1	315.1	17.5	-	-
50 °C	208.4	285.4	297.0	278.7	32.1	-	-
CEM I w/c = 0.6							
20 °C	75.7	209.7	251.2	313.5	8.1	-	-
35 °C	149.7	271.2	311.2	342.6	17.1	-	-
50 °C	219.2	290.9	290.4	233.4	34.6	-	-

(a)

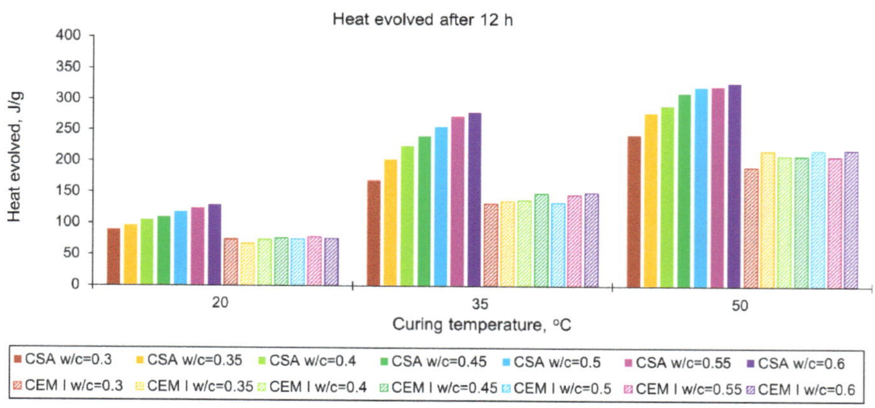

Figure 2. *Cont.*

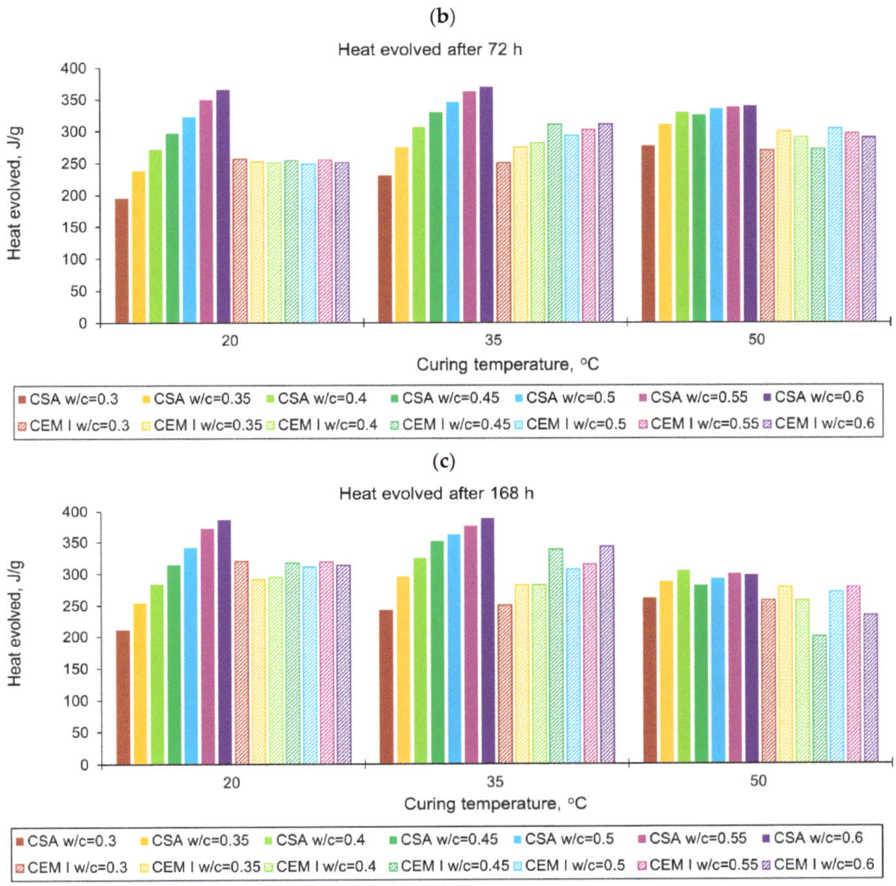

Figure 2. The heat evolved from CSA and CEM I (OPC) cement pastes with different w/c ratios in isothermal tests at different curing temperatures after: (**a**) 12 h, (**b**) 72 h, (**c**) 168 h.

The same observations can also be made for isothermal tests at 50 °C (Figure 1c,f), but in this case, the heat-releasing characteristic revealed unexpected thermal behaviour of the tested pastes. In this context, the typical progress of heat evolution is rapid growth during the initial period of hydration and final stabilization as hydration slows. Such development of the hydration heat is visible for curing temperatures of 20 °C and 35 °C (Figure 1a,b,d,e), while at 50 °C, a decrease in the amount of heat released is noticed (Figure 1c,f). The time after which the amount of released heat begins to decrease depends on the water/cement ratio of CSA paste and reaches 68 h for a w/c of 0.3, and only 30 h for a w/c of 0.6. For OP cement (CEM I) this time is delayed to 72 h, except for the paste with a w/c equal to 0.45. Various reasons for such development of the heat may be considered.

The first reason may be an incorrect measurement or a defect in the measuring device. This was eliminated by repeated measurements which showed the same results. Next, the decreased amount of the released heat may be connected to the stopped hydration process; however, isothermal calorimetry cannot precisely explain this phenomenon. Furthermore, the most probable reason for the decrease in the released heat is due to the technical issue of water evaporation in the samples, which can occur at higher temperatures. The process of evaporation requires energy, which causes the temperature of the sample to drop, and therefore to show lower heat values. The process may have been occurring from the sta of the test, but with the high amount of heat produced in the exothermic reaction of ceme with water, it was not possible to notice the lowering of the generated heat. This clai can be substantiated by the fact that the higher the w/c ratio of the tested CSA cemen pastes, the higher the difference in the heat after 168 h between samples cured at 20 ° and 50 °C. In the case of low w/c ratios, between 0.3 and 0.4, the heat loss is minima and the heat produced during the first 168 h of hydration by samples cured at 50 °C slightly higher than that of samples cured at 20 °C. With the increasing w/c ratio, th difference between the cumulative heat after 168 h of samples cured at 50 °C and 20 ° increases, with samples cured at 50 °C reaching much lower values, up to over 30% low in the case of a w/c ratio of 0.6. This effect can be connected to a significantly high amount of water available for the process of evaporation. It should be noted that the CS cement requires a high amount of water for the reaction of hydration to fully take plac According to stoichiometric calculations, CSA clinker requires a high w/c ratio of ov 0.6 to fully react [13], and therefore, for lower w/c ratios, more water may be chemicall bound into ettringite.

Considering the overall effect of temperature on the amount of heat released in CS pastes, a substantial increase in heat emission for the hydration temperature rise fro 20 °C to 35 °C or 50 °C can be observed. Moreover, hydration in the first hours is mo sensitive to temperature. After 12 h, the evolved heat is nearly two times higher at 35 ° and 50 °C, comparing the corresponding values of heat measured at 20 °C. Later, this about 6–20%, with the lowest values for 168 h. It can be remarked that the increase in th amount of heat released under the higher curing temperature slows for the higher w/ The temperature also influences the hydration of OP cement, but this effect is less than i the case of CSA cement.

Finally, comparing the amount of the heat released after 168 h and at 20 °C, OP pastes are characterized by higher values for w/c equal to 0.3–0.45 (Figure 2c). In th range of w/c = 0.5–0.6, the heat evolved by CSA pastes is higher, especially in the cas of w/c = 0.6; 73 J/g was more evolved in the CSA paste. At the same time, at the curin temperature equal to 35 °C, only in the case of w/c equal to 0.3 was more heat release in OP cement; in other w/c, CSA cement was characterized by higher evolved heat (fro 7 J/g to 60 J/g).

The rate of the hydration heat measured for all tested CSA and OPC pastes i isothermal tests at 20 °C, 35 °C and 50 °C are depicted in Figure 3 and summarize in Tables 3 and 4. All curves plotted in Figure 3 show the distinct behaviour of CSA paste compared to CEM I (OPC) paste. The main differences are in the number of exothermi peaks, their values and the occurring time. The several heat flow peaks in CSA cemen are reported in other tests [19,46,47], and the results from the presented study confir this tendency.

It should be noted that there are significant differences in the behaviour of OPC an CSA cements, which can be traced to their different hydration processes. OPC cemen hydration is based on the reaction of four main clinker phases: alite C_3S, belite C_2S tricalcium aluminate C_3A and brownmillerite C_4AF, with water, producing C-S-H gel an portlandite [48]. The reaction is usually characterized by a high heat evolution in the firs several minutes from adding water to the cement, due to dissolution and C_3A reactio with gypsum, resulting in ettringite. After that, the reaction slows down for up to severa hours (induction phase), to restart with alite hydration, followed by the belite and C_3A hydration [46,48]. In the case of CSA cement, the main constituents of the CSA clinker ar ye'elimite and gypsum, with low amounts of belite present. The process of hydration o CSA cement is swift and results mostly in the production of ettringite (AFt), monosulphat (AFm), crystalline aluminum hydroxide (AH_3) and stratlingite (C_2ASH_8), with sma amounts C-S-H phase appearing late in hydration process due to belite reaction [46]. Thi results in a significantly faster start of the reaction than in the case of OPC, a short inductio phase and high hydration peaks [18,20,25].

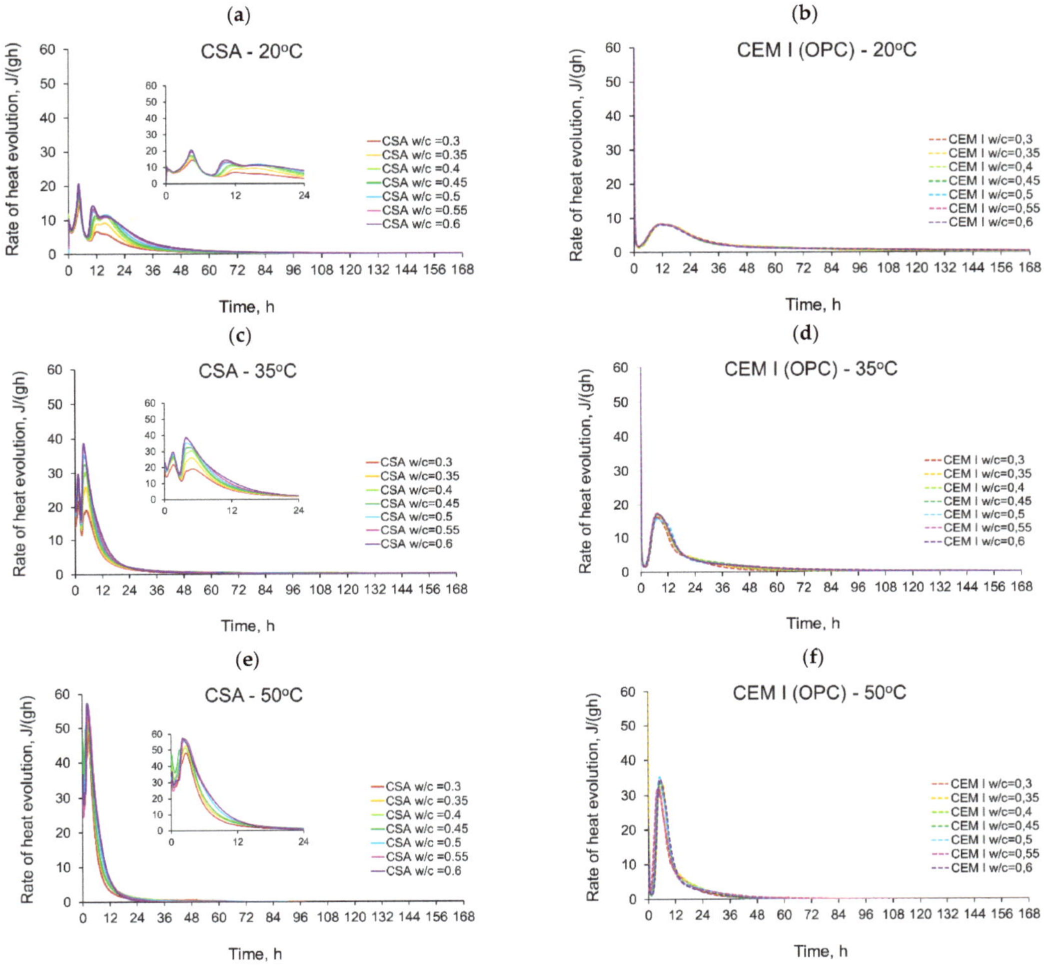

Figure 3. Rate of heat evolution of cement pastes with different water/cement ratios in isothermal tests at different curing temperatures: (**a**) CSA at 20 °C, (**b**) CEM I (OPC) at 20 °C, (**c**) CSA at 35 °C, (**d**) CEM I (OPC) at 35 °C, (**e**) CSA at 50 °C, (**f**) CEM I (OPC) at 50 °C.

Considering the isothermal tests of CSA at 20 °C (Figure 3a), four peaks with values depending on the w/c ratio can be seen. The first peak in the initial period occurs at 0.25 h and reaches 8–10 $Jg^{-1}{\cdot}h^{-1}$. Contrary to the CEM I cement paste (Figure 3b), this initial maximum represents the smallest value in all peaks detected in the curve of the heat release rate. The first peak may be attributed to the dissolution of the solid grains in water [48]. It has been noted that the CEM I dissolution peak might be higher due to the extremely fast C_3A dissolution and the production of ettringite in a reaction that differs greatly from the one taking place in the CSA cement [46]. Next, a certain induction period can be observed in the case of both the CSA cement and CEM I. It should be noted that, while the reaction of CSA cement is slow during this period, the hydration of ye'elemite is progressing, as the heat flow rate is much higher than in the case of CEM I, and in 50 °C the induction period cannot be clearly separated from the initial peak connected with dissolution. The induction phase of Portland cement is longer than that of CSA cement at all temperatures; at 20 °C it is twice as long, and at 35 °C it is around 5 times longer, which points to a faster reactio speed of the phases present in CSA cement.

The second peak in CSA cement takes the highest values (14.2–20.7 $Jg^{-1} \cdot h^{-1}$), in creasing with the increase of the water/cement ratio. The time of its occurrence is almos identical and equal to 4.5 h for all CSA pastes. The peak is connected to the initial dissolu tion of the sulphates and the consequent reaction of calcium sulphoaluminate, anhydrit and the water in which ettringite is formed [48].

The rate of hydration heat of CSA at the third local maximum is lower (6.6–14.3 $Jg^{-1} \cdot h^{-}$ and occurs at a similar time as the maximum for CEM I paste. The third local maximur appears after 10.5–12 h, which is earlier the higher the w/c. After the third peak, the rate c reaction gently falls, and the fourth local maximum with a strength similar to the third pea (5.9–11.7 $Jg^{-1} \cdot h^{-1}$) appears after the next 3–4 h. The third peak is a result of the main reactio of main ettringite production, followed by depletion of the gypsum and the formation c monosulphates [18]. In the case of 20 °C, a fourth peak can be observed, which may b attributed to the second ettringite formation or point to the possibility that the monosulphat formation in 20 °C starts later than during the main peak of ettringite reaction; however, i higher temperatures, this process is accelerated and does not show separately in the hea flow [19,20].

When the curing temperature increases from 20 °C to 35 °C, the induction period i significantly shortened for all tested CSA pastes (Figure 3c). Consequently, the time of th second exothermic peak is shortened to 1.4–1.7 h, and for the third peak to 3.7–4.0 h. Th second and third exothermic peaks show a very narrow shape, with much higher value compared to 20 °C. Contrary to the isothermal tests at 20 °C, the third peak revealed greate values (17.0–38.7 $Jg^{-1} \cdot h^{-1}$) than the second peak (21.8–29.7 $Jg^{-1} \cdot h^{-1}$), except for paste with w/c ratio equal to 0.3 and 0.35. A clear fourth maximum was not observed for tests a 35 °C.

Similar observations for CSA cement may be seen for the isothermal tests at 50 °C (Figure 3e). The strength of the second and third peaks seriously increases, and for al water/cement ratios the third peak takes the highest value. The third exothermic effec grows from 4.0 (w/c = 0.6) to 7.3 (w/c = 0.3) times, with the curing temperature increasin from 20 °C to 50 °C. Thus, the rate of reaction increases rapidly, reaching maximun values at a time of 0.08–0.17 h (second peak) and 2.1–2.7 h (third peak) after initial mixin Simultaneously, in the case of CEM I, the maximum value of the exothermic peak occurs a 6 h and increases by about 4 times at 50 °C (Figure 3f).

It should be noted that the presence of additional peaks of significant power in the cas of CSA cement is complex and of significant importance. On the one hand, it accelerates th maturation process and the development of the mechanical properties of the cementitiou material, which is generally a desirable feature. On the other hand, the maturation of th cement material at a too high initial curing temperature may damage its structure anc consequently, reduce its final strength [49], despite its initially rapid growth.

Furthermore, the large amount of rapidly released heat and the prolonged heat releas time to the third maximum may cause numerous problems such as long-term elevatec hardening temperatures in the structure, which often initiates early thermal cracks.

3.2. Apparent Activation Energy

The activation energy was calculated for all tested CSA cement pastes with differen water/cement ratios as well as for reference pastes with ordinary Portland cement, CEM I Two sets of the results from isothermal calorimetry are used in Equation (2): 20 °C/35 °C and 20 °C/50 °C. The necessary development of the heat evolution rates needed fo Equation (2), q(20 °C), q(35 °C) and q(50 °C) related to the same amount of the evolvec hydration heat, is depicted in Figures 4–10, respectively for all tested w/c. These diagram highlight once again the different thermal behaviours of CSA compared to OPC (CEM I).

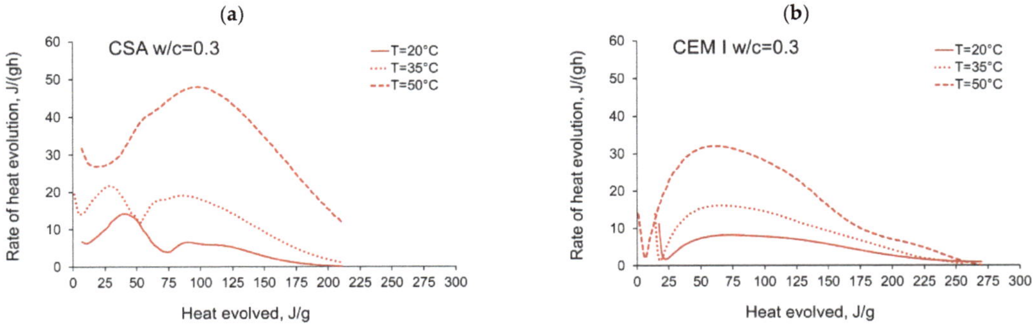

Figure 4. Rate of heat evolution in the function of heat evolved for cement pastes with w/c = 0.3: (**a**) CSA, (**b**) CEM I (OPC).

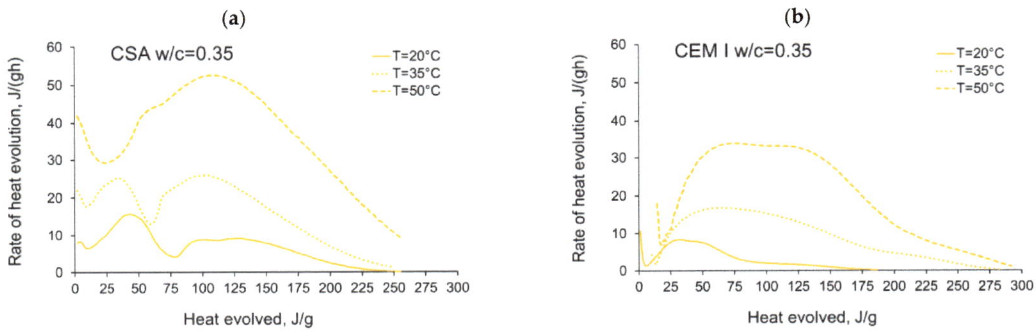

Figure 5. Rate of heat evolution in the function of heat evolved for cement pastes with w/c = 0.35: (**a**) CSA, (**b**) CEM I (OPC).

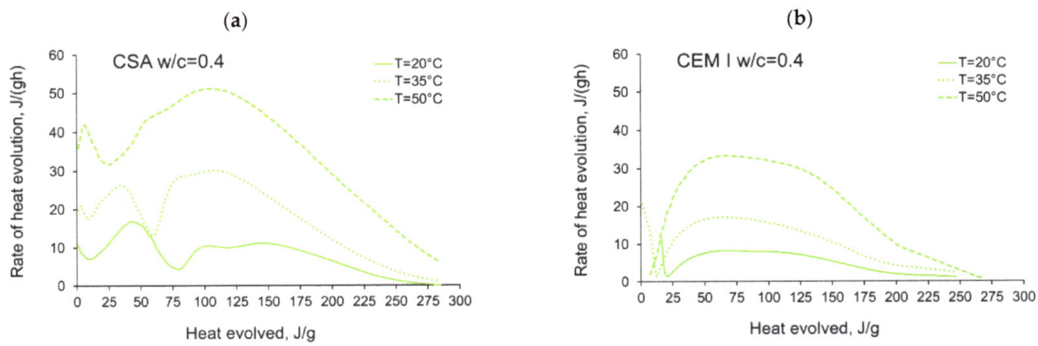

Figure 6. Rate of heat evolution in the function of heat evolved for cement pastes with w/c = 0.4: (**a**) CSA, (**b**) CEM I (OPC).

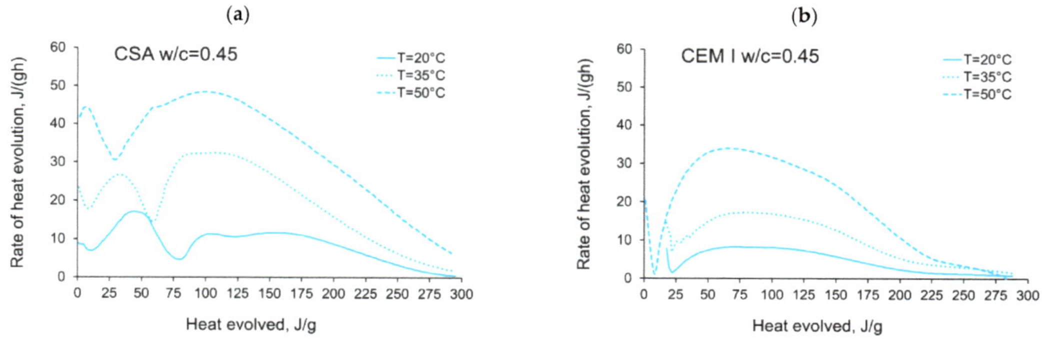

Figure 7. Rate of heat evolution in the function of heat evolved for cement pastes with w/c = 0.45: (**a**) CSA, (**b**) CEM I (OPC).

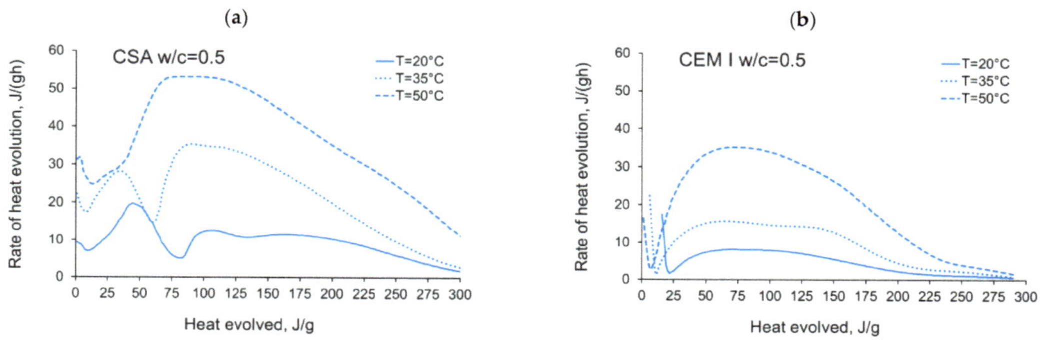

Figure 8. Rate of heat evolution in the function of heat evolved for cement pastes with w/c = 0.5: (**a**) CSA, (**b**) CEM I (OPC).

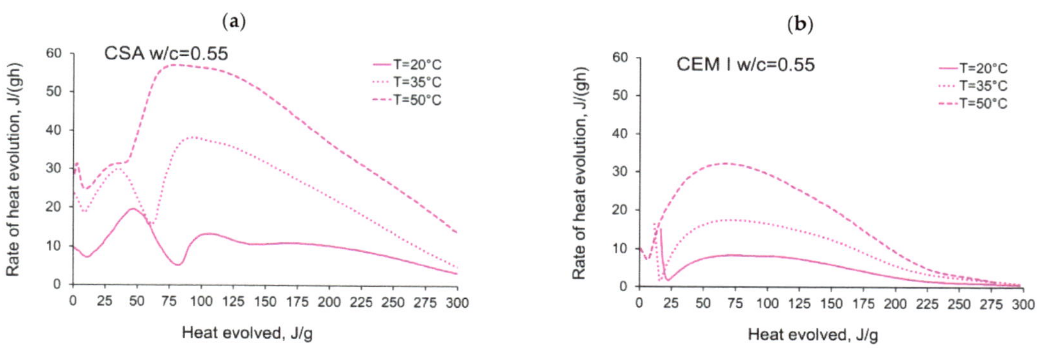

Figure 9. Rate of heat evolution in the function of heat evolved for cement pastes with w/c = 0.55: (**a**) CSA, (**b**) CEM I (OPC).

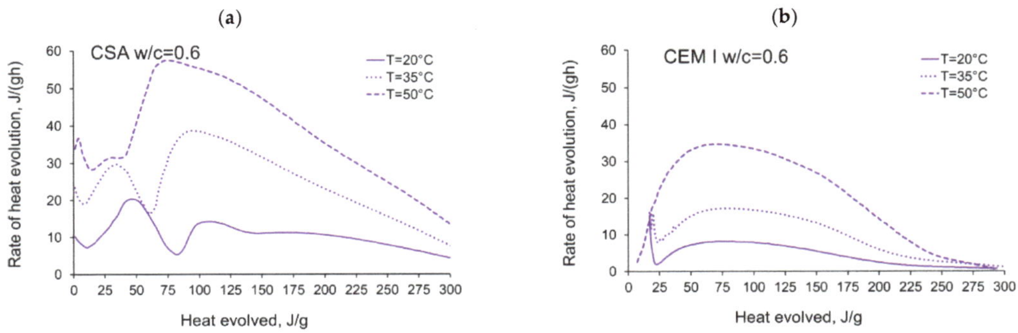

Figure 10. Rate of heat evolution in the function of heat evolved for cement pastes with w/c = 0.6: (**a**) CSA, (**b**) CEM I (OPC).

The computed evolution of activation energy for CSA and CEM I cement pastes with different water/cement ratios can be seen in Figures 11 and 12, respectively. The following observations can be perceived from the apparent activation energy evolution:

- The variations in apparent activation energy of CSA cement pastes are quite substantial, especially compared to the relatively constant values obtained for reference cement pastes made of ordinary Portland cement, CEM I. Moreover, in the case of CSA cement, both water/cement ratio and curing temperature affect the values and the evolution of apparent activation energy.
- For CSA cement, the E_a variations are qualitatively similar considering the two applied sets of curing temperatures (20 °C/35 °C and 20 °C/50 °C); however, they vary quantitatively. In the case of ordinary Portland cement, CEM I, the differences between the two sets of curing temperatures are not so substantial,
- In the case of CSA, for released heat lower than 100 J/g, particularly rapid changes in the values of activation energy are visible, and no stable period is observed. Moreover, these variations are more distinctive for the activation energy calculated using the isothermal tests at 20 °C/35 °C. In this case, the maximum value of E_a reached nearly 100,000 Jmol^{-1} for the hydration heat equal to ~80 J/g, while earlier the activation energy almost dropped to 400–1000 Jmol^{-1} (for heat evolved close to 60 J/g). This effect can already be seen in the graphs presenting the rate of heat released at 20 °C and 35 °C (Figures 4–10), where the overlap of curves is noticeable in the range of hydration heat 50–60 J/g. For hydration heat greater than 100 J/g, the curves describing the activation energy obtained from two sets of curing temperatures (20 °C/35 °C and 20 °C/50 °C) are more congruous. Simultaneously, the influence of the water/cement ratio is perceivable. In this regard, for w/c $\leq$ 0.45, the activation energy successively grows up to the value of 120,000 J/mol^{-1}. For w/c equal to 0.5 and 0.55, the increase of the energy value is insignificant, and for w/c = 0.6, a decrease is even visible.

For CSA cements, with the fairly wide variations observed in the range of hydration heat 0–100 J/g and the previously described subsequent growth, the evolution curves of E_a present a very limited stable period. The mean values of the activation energy obtained for its stable region for CSA cement with different w/c are shown in Table 5. Generally, the greater mean value of activation energy in the considered stable range 101–140 J/g is obtained from the isothermal tests performed at 20 °C/35 °C, except for the cement paste with w/c = 0.3. The values of apparent activation energy for the considered range are also plotted in Figure 13a.

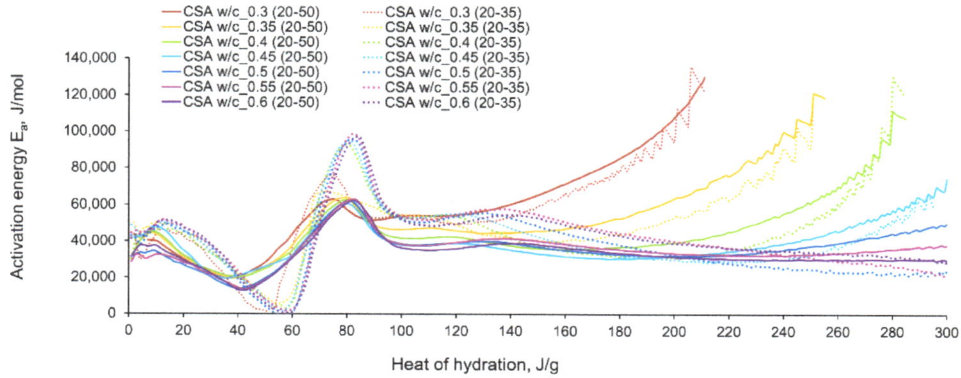

Figure 11. The apparent activation energy of CSA cement pastes with different w/c ratios.

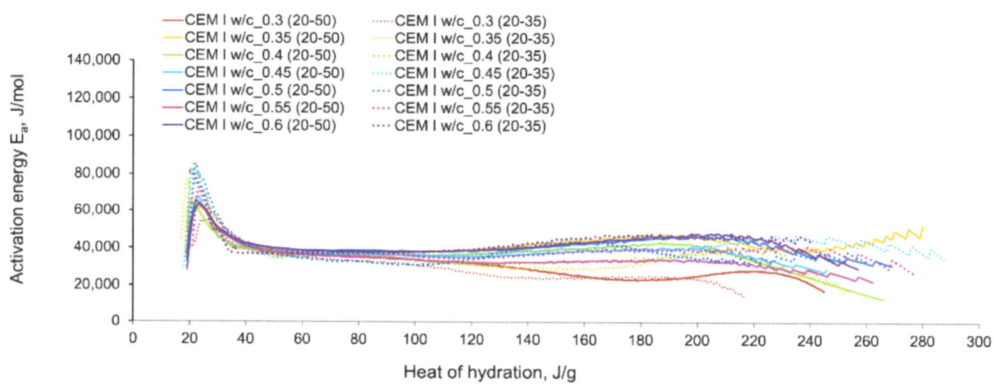

Figure 12. The apparent activation energy of CEM I (OPC) cement pastes with different w/c ratios.

Table 5. The value of E_a calculated for its nearly stable value.

Value	Apparent Activation Energy, J mol^{-1}						
	w/c = 0.3	w/c = 0.35	w/c = 0.4	w/c = 0.45	w/c = 0.5	w/c = 0.55	w/c = 0.6
	Average values calculated from tests 20 °C/35 °C						
CSA—period			in range of evolved heat: 101–136 Jg^{-1}				
CSA—value	50,917	49,131	51,623	53,477	53,230	53,906	51,016
CEM I (OPC)—period			in range of evolved heat: 40–140 Jg^{-1}				
CEM I (OPC)—value	31,211	33,737	34,638	36,667	36,615	36,820	38,282
	Average values calculated from tests 20 °C/50 °C						
CSA—period			in range of evolved heat: 101–136 Jg^{-1}				
CSA—value	54,978	45,968	41,158	38,169	38,915	39,712	36,592
CEM I (OPC)—period			in range of evolved heat: 40–140 Jg^{-1}				
CEM I (OPC)—value	33,922	38,208	37,085	36,805	39,010	36,601	38,967

For ordinary Portland cement, a range with an almost stable value of E_a is between 101–140 J/g (Figure 13b). Contrary to CSA, the slightly greater values of the activation energy are visible from the tests at 20 °C/50 °C. The average values in the stable period for CSA cement are much higher than for ordinary Portland cement. The resulting values are

- for CSA: 51,900 J/mol (tests 20 °C/35 °C) and 42,213 J/mol (tests 20 °C/50 °C),
- for OPC: 35,424 J/mol (tests 20 °C/35 °C) and 37,228 J/mol (tests 20 °C/50 °C).

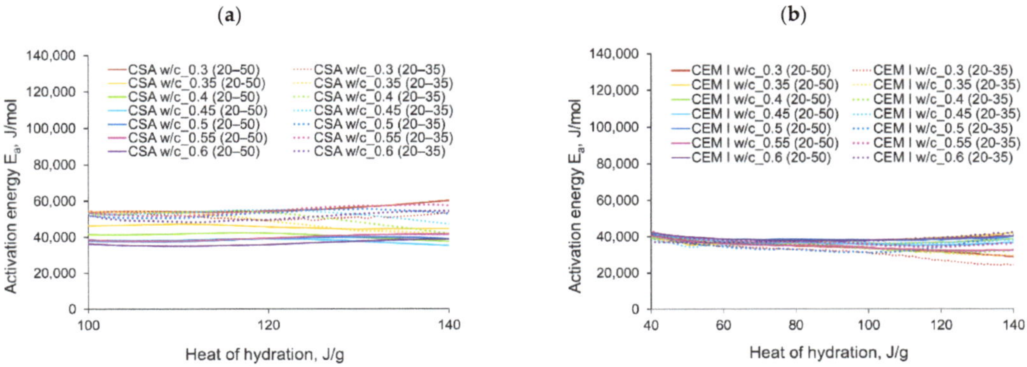

Figure 13. The apparent activation energy in its stable range: (**a**) CSA, (**b**) CEM I (OPC).

4. Conclusions

In this study, the hydration heat of CSA cement pastes with different water/cement ratios has been studied by isothermal tests at different curing temperatures and compared to the ordinary Portland cement. Based on the Arrhenius concept, the coupled effects of temperature and w/c on the apparent activation energy are also examined. The main results can be listed as follows:

- The evolved heat of CSA cement pastes increases with the increase of the water/cement ratio. The largest, 20% increase in heat emission, was recorded for the water/cement change from 0.3 to 0.35. For higher w/c ratios the average difference between consecutive measured w/c ratios was 10%. In the case of OPC, the heat released at the curing temperature of 20 °C is practically the same for different w/c. At a higher curing temperature of 35 °C, the w/c effect is visible but still less pronounced than for CSA cement.
- The temperature of curing affects the evolved heat of CSA cement the most in the first 12 h. After 168 h, there is no significant difference between the evolved heat at 20 °C and 35 °C, while the evolved heat at 50 °C was significantly lower, which is most likely due to water vaporization. Nevertheless, the hydration of CSA pastes is significantly accelerated at 35 °C and 50 °C. The effect of the elevated temperature on the hydration process is much greater than that of ordinary Portland cement.
- The complex thermal behaviour of CSA pastes with several heat flow maxima has been confirmed in the performed isothermal tests at different curing temperatures. Four heat flow peaks in CSA cement are detected for tests at 20 °C, while at 35 °C and 50 °C three peaks are confirmed. Generally, the strength of the hydration peaks decreased with decreasing water/cement ratio. At 35 °C and 50 °C, the end of the induction period and the occurring time of the main peaks are remarkably shortened. Simultaneously, the exothermic peaks have extremely narrow shapes, and the peak value increases greatly compared to tests at 20 °C.
- The variations in apparent activation energy are substantial, and they cannot be considered as a constant value for the CSA type of cement. Only for the relatively small range of the evolved heat (101–140 J/g) did the curves of energy evolution reveal steady behaviour. In this case, the average activation energy depends both on the water/cement ratio and the curing temperature. The average values in the stable period for CSA cement are much higher than those for ordinary Portland cement.
- The hydration region of CSA with especially rapid variations of activation energy covers the initial period of heat release, up to the value of 100 J/g. Using the isothermal results at 20 °C and 35 °C in the calculation procedure, especially sharp changes in energy are obtained. Moreover, both very small (400–1000 J·mol^{-1}) and extremely high (100,000 Jmol^{-1}) values are observed.

To summarize, the conducted tests indicated a different thermal behaviour of CS cement compared to ordinary Portland cement. Calcium sulphoaluminate cement characterized by the rapid course of the hydration process and significant amounts of he being released in the initial period of hydration. These properties should be given speci attention to when choosing the practical application of the CSA cement. The high variabilit of the activation energy and the impossibility of determining its relatively constant valu can lead to problems when using the equivalent age approach for predicting the CS composites' strength development at elevated curing temperatures.

Author Contributions: Conceptualization, J.G. and B.K.; methodology, J.G. and B.K.; software, B.K investigation, M.G.; data curation, M.G.; writing—original draft preparation, M.G. and B.K.; writing review and editing, M.G. and B.K.; visualization, B.K.; supervision, J.G. and B.K.; funding acquisitio J.G. and B.K. All authors have read and agreed to the published version of the manuscript.

Funding: Funding provided by the Silesian University of Technology (project 03/060/RGJ20/008 03/030/RJG21/0118 and BKM-613/RB3/2021).

Institutional Review Board Statement: Not applicable.

Informed Consent Statement: Not applicable.

Data Availability Statement: Data is contained within the article.

Conflicts of Interest: The authors declare no conflict of interest.

References

1. Favier, A.; De Wolf, C.; Schrivener, K.; Habert, G. *A Sustainable Future for the European Cement and Concrete Industry: Technolog Assessment for Full Decarbonisation of the Industry by 2050*; ETH Zurich: Zürich, Switzerland, 2018. [CrossRef]
2. Skjærseth, J.B.; Eikeland, P.O. *Corporate Responses to EU Emissions Trading: Resistance, Innovation or Responsibility*; Taylor & Franci London, UK, 2016.
3. Jain, M. Use and properties of blast furnace slag as a building material—A review. *Int. J. Recent Contrib. Eng. Sci. IT* **2014**, *2*, 5 [CrossRef]
4. Lemougna, P.N.; Wang, K.-T.; Tang, Q.; Nzeukou, A.; Billong, N.; Melo, U.C.; Cui, X.-M. Review on the use of volcanic ashes fc engineering applications. *Resour. Conserv. Recycl.* **2018**, *137*, 177–190. [CrossRef]
5. Giergiczny, Z. Fly ash and slag. *Cem. Concr. Res.* **2019**, *124*, 105826. [CrossRef]
6. Wang, D.H.; Shi, C.J.; Farzadnia, N.; Shi, Z.G.; Jia, H.F.; Ou, Z.H. A review on use of limestone powder in cement-based material Mechanism, hydration and microstructures. *Constr. Build. Mater.* **2018**, *181*, 659–672. [CrossRef]
7. Juenger, M.C.G.; Winnefeld, F.; Provis, J.L.; Ideker, J.H. Advances in alternative cementitious binders. *Cem. Concr. Res.* **2011**, *4* 1232–1243. [CrossRef]
8. Winnefeld, F. Calcium sulfoaluminate cement: An example of a low CO_2—Alternative to portland cement. In Proceedings of th WTA-Colloquium—Effect of Climate Change on Built Heritage, Eindhoven, The Netherlands, 11–12 March 2010; pp. 98–108.
9. Kubissa, W. Air permeability of air-entrained hybrid concrete containing CSA cement. *Buildings* **2020**, *10*, 119. [CrossRef]
10. Nguyen, K.S.; Nguyen-Ngoc, T.H.; Nguyen-Phung, A.T.; Do, Q.M. Preparation of calcium sulfoaluminate cement from bau ite/red mud of Tan Rai—Lam Dong. In Proceedings of the 7th International Conference of Asian Concrete Federation ACF 201 Hanoi, Vietnam, 30 October–2 November 2016; pp. 21–26.
11. Miller, S.A.; Horvath, A.; Monteiro, P.J.M. Readily implementable techniques can cut annual CO_2 emissions from the productio of concrete by over 20%. *Environ. Res. Lett.* **2016**, *11*, 074029. [CrossRef]
12. Chen, I.A.; Hargis, C.W.; Juenger, M.C.G. Understanding expansion in calcium sulfoaluminate–belite cements. *Cem. Concr. Re* **2012**, *42*, 51–60. [CrossRef]
13. Madan Mohan Reddy, K.; Srimurali, M.; Bhaskar, M.; Mohan Reddy, K. Characterization of calcium sulfoaluminate cement. *Int. ChemTech Res.* **2014**, *6*, 1696–1698.
14. Aranda, M.A.G.; De la Torre, A.G. *Sulfoaluminate Cement, Eco-Efficient Concrete*; Woodhead Publishing: Southston, UK, 2013; p 488–522.
15. Winnefeld, F.; Kaufmann, J. Concrete produced with calcium sulfoaluminate cement—A potential system for energy and hea storage. In Proceedings of the First Middle East Conference on Smart Monitoring, Assessment and Rehabilitation of Civ Structures (SMAR 2011), Dubai, United Arab Emirates, 8–10 February 2011; pp. 1–9.
16. Ambroise, J.; Pera, J. Immobilization of calcium sulfate contained in demolition waste. *J. Hazard. Mater.* **2008**, *151*, 840–84 [CrossRef]
17. Hargis, C.W.; Telesca, A.; Monteiro, P.J.M. Calcium sulfoaluminate (Ye'elimite) hydration in the presence of gypsum, calcite an vaterite. *Cem. Concr. Res.* **2014**, *65*, 15–20. [CrossRef]

18. Martin, L.H.; Winnefeld, F.; Müller, C.J.; Lothenbach, B. Contribution of limestone to the hydration of calcium sulfoaluminate cement. *Cem. Concr. Compos.* **2015**, *62*, 204–211. [CrossRef]
19. Winnefeld, F.; Lothenbach, B. Hydration of calcium sulfoaluminate cements—Experimental findings and thermodynamic modelling. *Cem. Concr. Res.* **2010**, *40*, 1239–1247. [CrossRef]
20. Berger, S.; Coumes, C.C.D.; Le Bescop, P.; Damidot, D. Influence of a thermal cycle at early age on the hydration of calcium sulphoaluminate cements with variable gypsum contents. *Cem. Concr. Res.* **2011**, *41*, 149–160. [CrossRef]
21. Li, P.; Gao, X.; Wang, K.; Tam, V.W.Y.; Li, W. Hydration mechanism and early frost resistance of calcium sulfoaluminate cement concrete. *Constr. Build. Mater.* **2020**, *239*, 117862. [CrossRef]
22. Batog, M.; Synowiec, K. *Cement I Spoiwa Specjalne Zawierające Klinkier Siarczanoglinianowy (Cement and Special Binders Containing Sulphoaluminate Clinker)*; Budownictwo Technologie Architektura: Kraków, Poland, 2017; Volume 1, pp. 59–64.
23. Qin, L.; Gao, X.; Zhang, A. Potential application of Portland cement-calcium sulfoaluminate cement blends to avoid early age frost damage. *Constr. Build. Mater.* **2018**, *190*, 363–372. [CrossRef]
24. Ramanathan, S.; Halee, B.; Suraneni, P. Effect of calcium sulfoaluminate cement prehydration on hydration and strength gain of calcium sulfoaluminate cement-ordinary portland cement mixtures. *Cem. Concr. Compos.* **2020**, *112*, 103694. [CrossRef]
25. Li, W. The properties and hydration of portland cement containing calcium sulfoaluminate cement. *Ceram. Silik.* **2018**, *62*, 364–373. [CrossRef]
26. Dovál, M.; Palou, M.; Kovár, V. Heat evolution and mechanism of hydration in $CaO-Al_2O_3-SO_3$ system. *Ceram. Silik.* **2005**, *49*, 104–108.
27. Jeong, Y.; Hargis, C.W.; Kang, H.; Chun, S.C.; Moon, J. The effect of elevated curing temperatures on high Ye'elimite calcium sulfoaluminate cement mortars. *Materials* **2019**, *12*, 1072. [CrossRef] [PubMed]
28. Wang, P.; Li, N.; Xu, L. Hydration evolution and compressive strength of calcium sulphoaluminate cement constantly cured over the temperature range of 0 to 80 °C. *Cem. Concr. Res.* **2017**, *100*, 203–213. [CrossRef]
29. BS EN 197-1:2011 cement. In *Composition, Specifications and Conformity Criteria for Common Cements*; British Standards Institution: London, UK, 2011.
30. Klemczak, B.; Batog, M.; Giergiczny, Z.; Żmij, A. Complex effect of concrete composition on the thermo-mechanical behaviour of mass concrete. *Materials* **2018**, *11*, 2207. [CrossRef] [PubMed]
31. Klemczak, B.; Żmij, A. Reliability of standard methods for evaluating the early-age cracking risk of thermal-shrinkage origin in concrete walls. *Constr. Build. Mater.* **2019**, *226*, 651–661. [CrossRef]
32. Hansen, P.F.; Pedersen, E.J. Maturity computer for controlling curing and hardening of concrete. *Nord. Betongfoerbundet* **1977**, *1*, 21–25.
33. D'Aloia, L.; Chanvillard, G. Determining the "apparent" activation energy of concrete: E_a—Numerical simulations of the heat of hydration of cement. *Cem. Concr. Res.* **2002**, *32*, 1277–1289. [CrossRef]
34. Schindler, A.K.; Folliard, K.J. Heat of hydration models for cementitious materials. *ACI Mater. J.* **2005**, *102*, 24–33.
35. Poole, J.L.; Riding, K.A.; Folliard, K.J.; Juenger, M.C.G.; Schindler, A.K. Methods for calculating activation energy for Portland cement. *ACI Mater. J.* **2007**, *104*, 202–311.
36. Wilińska, I.; Pacewska, B. Influence of selected activating methods on hydration processes of mixtures containing high and very high amount of fly ash. *J. Therm. Anal. Calorim.* **2018**, *133*, 823–843. [CrossRef]
37. CEN, EN 196-11:2018 Methods of testing cement. In *Heat of Hydration. Isothermal Conduction Calorimetry Method*; European Committee for Standardization: Brussels, Belgium, 2018.
38. Glasstone, S.; Laidler, K.; Eyring, H. *The Theory of Rate Processes*; McGraw-Hill Book Company: New York, NY, USA, 1941.
39. Zákoutský, J.; Tydlitát, V.; Černý, R. Effect of temperature on the early-stage hydration characteristics of Portland cement: A large-volume calorimetric study. *Constr. Build. Mater.* **2012**, *36*, 969–976. [CrossRef]
40. Klemczak, B.; Batog, M. Heat of hydration of low-clinker cements. Pt. 1: Semi-adiabatic and isothermal tests at different temperature. *J. Therm. Anal. Calorim.* **2016**, *123*, 1351–1360. [CrossRef]
41. Han, F.; Zhang, Z.; Wang, D.; Yan, P. Hydration kinetics of composite binder containing slag at different temperatures. *J. Therm. Anal. Calorim.* **2015**, *121*, 815–827. [CrossRef]
42. Wirquin, E.; Broda, M.; Duthoit, B. Determination of the apparent activation energy of one concrete by calorimetric and me-chanical means Influence of a superplasticizer. *Cem. Concr. Res.* **2002**, *32*, 1207–1213. [CrossRef]
43. Kada-Benameur, H.; Wirquin, E.; Duthoit, B. Determination of apparent activation energy of concrete by isothermal calorimetry. *Cem. Concr. Res.* **2000**, *30*, 301–305. [CrossRef]
44. Broda, M.; Wirquin, E.; Duthoit, B. Conception of an isothermal calorimeter for concrete—Determination of the apparent activation energy. *Mater. Struct.* **2002**, *35*, 389–394. [CrossRef]
45. Klemczak, B.; Batog, M. Heat of hydration of low-clinker cements. Pt. 2: Determination of apparent activation energy and validity of the equivalent age approach. *J. Therm. Anal. Calorim.* **2016**, *123*, 1361–1369. [CrossRef]
46. Zhang, J.; Ke, G.; Liu, Y. Early hydration heat of calcium sulfoaluminate cement with influences of supplementary cementitious materials and water to binder ratio. *Materials* **2021**, *14*, 642. [CrossRef] [PubMed]
47. Winnefeld, F.; Barlag, S. Calorimetric and thermogravimetric study on the influence of calcium sulfate on the hydration of ye'elimite. *J. Therm. Anal. Calorim.* **2010**, *101*, 949–957. [CrossRef]

48. Bobrowski, A.; Nocuń-Wczelik, W.; Gawlicki, M.; Łagosz, A.K.; Łój, G. *Cement: Metody Badań, Wybrane Kierunki Stosowania (Cement: Testing Methods, Chosen Uses)*; Wydawnictwo AGH: Kraków, Poland, 2015. (In Polish)
49. Neville, A.M. *Properties of Concrete*; Pearson Education: Harlow, UK, 2012.

materials

MDPI

rticle

Microstructure and Durability Performance of Mortars with Volcanic Powder from Calbuco Volcano (Chile) after 4 Hardening Years

Rosa María Tremiño [1], Teresa Real-Herraiz [2], Viviana Letelier [3] and José Marcos Ortega [1,*]

1 Departamento de Ingeniería Civil, Universidad de Alicante, Ap. Correos 99, 03080 Alicante, Spain; rmta2@alu.ua.es
2 Instituto de Matemática Multidisciplinar, Universidad Politécnica de Valencia, Camino de Vera s/n, 46022 Valencia, Spain; tereaher@upv.es
3 Departamento de Obras Civiles, Universidad de la Frontera, Av. Fco. Salazar, Temuco 01145, Chile; viviana.letelier@ufrontera.cl
* Correspondence: jm.ortega@ua.es; Tel.: +34-96-5903-400 (ext. 2470)

Abstract: One of the most popular ways to lessen the impact of the cement industry on the environment consists of substituting clinker by additions. The service life required for real construction elements is generally long, so it would be interesting to obtain information about the effects of new additions after a hardening period of several years. Analyzed here are the effects of the incorporation of volcanic ashes, coming from Calbuco volcano's last eruption (Chile), as clinker replacement, in the durability and pore structure of mortars, after approximately 4 hardening years (1500 days), in comparison with reference specimens without additions. The substitution percentages of clinker by volcanic powder studied were 10% and 20%. The microstructure was characterized with mercury intrusion porosimetry and impedance spectroscopy. In order to evaluate the pozzolanic activity of the volcanic powder after 1500 days, differential thermal analyses were performed. Water absorption after immersion, steady-state diffusion coefficient and length change were also studied. In accordance with the results obtained, the 10% and 20% substitution of clinker by volcanic powder from the Calbuco volcano showed beneficial effects in the mortars after 4 years, especially regarding the microstructure and chloride diffusion, without noticeable influence in their water absorption.

Keywords: volcanic powder; Calbuco volcano; very long term; microstructure; durability

itation: Tremiño, R.M.; eal-Herraiz, T.; Letelier, V.; rtega, J.M. Microstructure and urability Performance of Mortars ith Volcanic Powder from Calbuco olcano (Chile) after 4 Hardening ears. *Materials* **2021**, *14*, 1751. tps://doi.org/10.3390/ma14071751

cademic Editors: Miguel Bravo, Rui asco Silva and António P. C. Duarte

eceived: 23 February 2021
ccepted: 25 March 2021
ublished: 2 April 2021

1. Introduction

At present, the cement industry still constitutes an important pollutant sector [1,2], and it is necessary that this industry contributes to the current worldwide goals regarding global warming reduction. Therefore, for lessening its greenhouse gas emissions and making the cement industry and construction sector in general more sustainable, several strategies have been developed, such as the research on eco-friendly materials, which has experienced great progress in recent years [3–5].

With respect to cement-based materials, one of the most popular ways of reducing the greenhouse gas emissions, due to their manufacture, consists of substituting clinker with supplementary cementitious materials [6–9]. Some of them are residues generated in other industrial processes, so its reuse is also beneficial from an environmental perspective. Among the supplementary cementitious materials available, it is important to highlight silica fume [10,11], ground granulated blast-furnace slag [12,13], and fly ash [14], which have all been commonly used in recent decades. Despite that, developing alternatives to these classical additions is still an important topic of research. Among others, some examples of these new supplementary cementitious materials are brick powder [15], rice husk ash [16], glass powder [17,18], red mud [19], etc.

In relation to the possible use of powders coming from volcanic materials as a addition to cement, research has shown that they would provide adequate service prope ties [20,21]. On the one hand, it has been reported that these volcanic powders can develo pozzolanic activity [20] and improve the durability of cement-based materials [22,2 Nevertheless, other works indicated that the incorporation of volcanic powders as a subs tution of clinker produced a reduction incompressive strength [24,25] and an increase i porosity [21,26], recommending percentages of this addition up to 20%, inwhich an appro imate 23% reduction in strength has been reported [25]. In addition to this, it is importa to highlight that the majority of these studies have analyzed the effects of volcanic powde at relatively short hardening ages [27].

On the other hand, the service life required for construction elements, belonging to re structures, buildings, and other engineering works is generally long [28]. Then, to assess these new additions to cement-based materials, such as volcanic powder, are suitable fc use in real construction elements, it would be very interesting to obtain information abou their behavior in the very long term, after a hardening period of several years. Furthermor this could be particularly relevant in the case of pozzolanic materials.

The main purpose of this research is to study the effects of the incorporation of volcani powder, coming from the last eruption of the Calbuco volcano (Chile), as clinker replac ment, in the durability and pore structure of mortars after approximately 4 hardening year (1500 days), compared to reference mortars without additions. The analyzed percentage of clinker substituted by volcanic powder were 10% and 20%. The pore structure of th mortars was characterized using mercury intrusion porosimetry and the non-destructiv impedance spectroscopy technique. For evaluating the development of the possible po zolanic activity of the volcanic powder at the end of the time period studied, differentia thermal analyses were performed. With respect to the durability of the mortars, the stead state diffusion coefficient, determined from the resistivity of the water-saturated sample and the water absorption after immersion, were obtained. Lastly, the length change in th mortars was measured after 1500 hardening days for assessing the possible effects of th volcanic powder addition in this respect.

2. Materials and Methods

2.1. Materials and Sample Preparation

The volcanic powder (VP) studied in this research came from Calbuco volcano's las eruption (41°20′ S, 72°37′ W, 2003 m.a.s.l.). This volcano is situated in the southern Ande between the cities of Ensenada and Puerto Montt in Chile. The dates of the last subplinia eruption of this volcano were April 22nd–23rd, 2015. Porphyritic-basaltic andesite (~55 by weight of SiO_2) [29] was detected in the erupted debris. This fallout mostly affected th volcano's northeast area, although the finest ashes reached southern Chile and Argentinea Patagonia, where they mainly settled. The distribution of the particle size of the debri showed particle fractions ranging between 3 and 350 μm [30]. In this study, the volcani powder was sieved, and the particles used as addition were those with sizes smaller tha 75 μm. The chemical components of the VP are shown in Table 1 [27], whereas the siz distribution of the powder is represented in Figure 1 [27].

Table 1. Chemical components of volcanic powder.

Components	Value
Fe_2O_3	11.00%
CaO	8.27%
Al_2O_3	14.54%
SiO_2	57.76%
K_2O	2.14%
Na_2O	2.41%
MgO	2.44%
TiO_2	1.42%

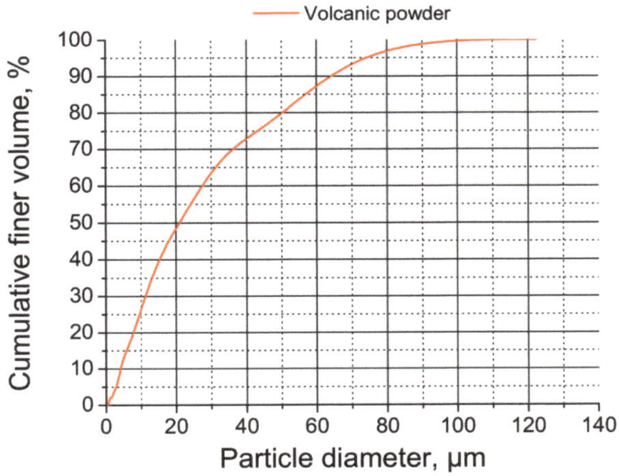

Figure 1. Size distribution of particles of studied powder.

Three kinds of mortar were studied in this research, in which Calbuco's VP and Portland cement were combined. This cement was type CEM I 42.5 R [31]. Firstly, a reference mortar without VP, designed as REF in the results and discussion sections, was prepared. On the other hand, mortars with two different VP binders were also made. These binders incorporated 10% and 20% in weight as cement CEM I 42.5 R substitution, named VP10 and VP20, respectively. A water/cement ratio of 0.5 was used for all of the studied mortars, as well as an aggregate/cement ratio equal to 3. The fine quartz sand accomplished the provisions of standard UNE-EN 196-1 [32] and was provided by the company Normensand (Beckum, Germany) [33–35].

Two types of sample were cast. One of them consisted of cylindrical samples with dimensions 10 cm in diameter, and 15 cm in height. In addition, prismatic specimens with the dimensions of 25 mm × 25 mm × 285 mm were also made. All of them were stored in a chamber at 20°C at a 95% relative humidity (RH) during the initial 24 h. After that curing period, they were de-moulded. In order to optimize the total amount of material required and their storage, cylindrical samples were cut to obtain slices with 1 cm in height. Finally, the samples were kept in an optimum laboratory condition (20 °C and 100% RH) until 1500 hardening days (approximately 4 years), when they were tested. This condition consisted of storing the specimens in hermetically sealed containers, which contained distilled water in their bottom part to achieve a 100% relative humidity environment. The mortar samples were placed into the containers, avoiding contact with water by using a rack above the water level. In addition, these containers were kept in a chamber at a controlled temperature of 20 °C. The dimensions of the hermetically sealed containers used were 325 mm × 265 mm × 150 mm.

2.2. Mercury Intrusion Porosimetry

Mercury intrusion porosimetry is useful for obtaining data regarding the pore network of materials [36–38]. Here, the porosimetry test was performed using a Poremaster-60 GT porosimeter manufactured by Quantachrome Instruments (Boynton Beach, FL, USA). The specimens were oven-dried at 50 °C over 48 h before performing the test. The results analyzed in this research were total porosity, pore size distribution, and percentage of Hg retained at the end of the experiment. The pore size distribution was analyzed, taking into account the next intervals: <10 nm, 10–100 nm, 100 nm to 1µm, 1–10 µm, 10 µm to 0.1 mm, and >0.1 mm [39,40]. Two measurements were made on each type of mortar at the hardening age studied. Pieces taken from the 1 cm disks, with irregular shapes, were tested. The total weight of the pieces tested in each of the measurements was approximately 1.5 g.

2.3. *Impedance Spectroscopy*

The impedance spectroscopy technique is used for obtaining data related to th microstructure of cement-based materials [41–44]. In this research, the impedance spec troscopy measurements were done with an analyzer model Agilent 4294A (Agilent Tech nologies, Kobe, Japan). This device allows measurements between 10^{-14} and 0.1 F. Th electrodes were circular (diameter 8 cm). They were made with flexible graphite, whic was fixed to a copper disk, also measuring 8 cm in diameter. The range of frequencies wa 100 Hz to 100 MHz.

Contacting and non-contacting measurements were conducted [41]. The experimenta records were fit to the equivalent circuits shown in Figure 2a,b [41], which consiste of several resistances and capacitances [41]. The resistance R_1 provides data about th percolating pores in the material [41], the resistance R_2 provides information about th overall pores [41], the capacitance C_1 provides information regarding a sample's soli fraction [41], and the C_2 capacitance is associated with the pores' surface contacting wit the electrolyte present inside the pore network [42]. In this work, regarding the por structure characterization, only the results of the parameters R_2, C_1 and C_2, determine using the non-contacting method have been analyzed, on account of their greater accurac The R_1 resistance, which can be determined using the contacting method, was only use for calculating the steady-state chloride diffusion coefficient in water-saturated specimen as will be explained in Section 2.6. Eight different slices, measuring 1 cm in height, wer analyzed with this technique for each mortar type.

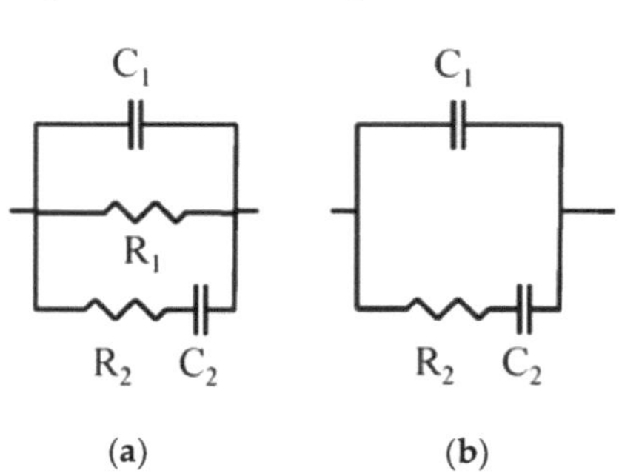

Figure 2. (**a**) Equivalent circuit for the contacting method [18,27]; (**b**) Equivalent circuit for th non-contacting method [18,27]. Reprinted with permission from [18,27] 2018 MDPI.

2.4. *Differential Thermal Analysis*

The differential thermal analyses were conducted with a simultaneous TG-DTA mode TGA/SDTA851e/SF/1100 from Mettler Toledo (Columbus, OH, USA), which allows opera tion from room temperature, up to 1100 °C. The heating ramp selected was 20 °C/min up t 1000 °C in N_2 atmosphere. The curve weight derivate versus temperature was determine Three measurements were made on each type of mortar at the hardening age studied. Th powder samples tested with this technique were obtained from milling pieces taken fron disks of 1 cm in thickness. The total weight of powder in each one the measurements wa approximately 20 mg.

2.5. *Water Absorption*

The absorption after immersion was determined following the procedure describe in the ASTM Standard C642-06 [45]. Six pieces, taken from slices of 1 cm in height, wer tested for each kind of mortar studied.

2.6. Steady-State Diffusion Coefficient

The steady-state chloride diffusion coefficient was determined from the electrical resistivity of the samples. The resistivity was calculated from the impedance resistance R_1 measured in water saturated samples. As explained previously, the resistance R_1 is directly related to the crossing pores of the material [41] and is therefore equivalent to the electrical resistance of the sample [46]. This test has been successfully used in several research works for different types of cement-based materials [46,47]. For each binder, six different samples of 1 cm in thickness were tested. Finally, the steady-state diffusion coefficient was obtained using the following Equation [48]

$$D_S = \frac{2 \times 10^{-10}}{\rho} \quad (1)$$

where D_s is the chloride steady-state diffusion coefficient through the sample (m^2/s), and ρ is the electrical resistivity of the specimen ($\Omega \cdot m$).

2.7. Shrinkage/Expansion

In order to assess if the incorporation of volcanic powder from the Calbuco volcano resulted in the development of the expansion or shrinkage phenomena in the mortars, their length change was measured after1500 hardening days. For each mortar type, six specimens with dimensions measuring 25 mm × 25 mm × 285 mm were tested. The length change in percentage after 1500 hardening days was determined from the measurements performed with a length comparator according to ASTM Standard C596-01 [49].

3. Results

3.1. Mercury Intrusion Porosimetry

The total porosity results after1500 days are represented in Figure 3. This porosity was slightly greater for the VP20 mortars in comparison to the REF and VP10 ones. Despite that, the total porosity was relatively similar for the different series studied.

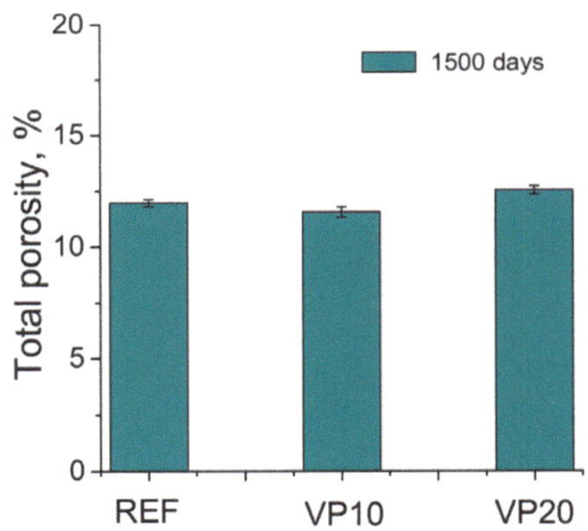

Figure 3. Total porosity results for the analyzed series.

Regarding the pore size distributions of the analyzed mortar series, they are depicted in Figure 4. The proportion of pores with lower sizes was higher for specimens which incorporated volcanic powder in comparison with REF mortars. In this regard, it is important to highlight the greater percentage of pores with diameters in the range of <10 nm, noted for VP10 and VP20 mortars. In view of these results, the addition of volcanic powder gave a more refined pore network to the mortars after 1500 hardening days.

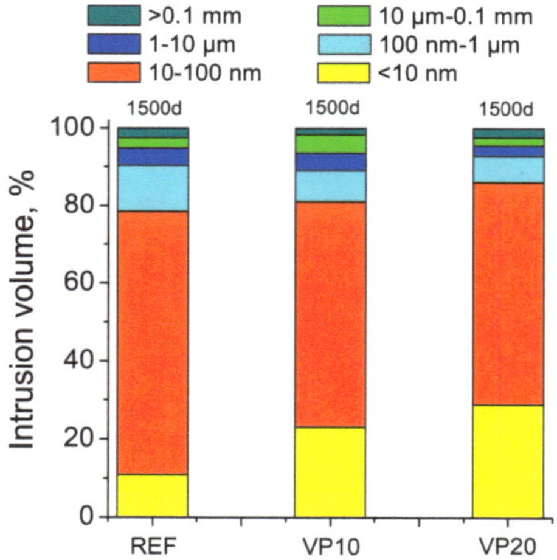

Figure 4. Pore size distributions noted for the series studied.

The results of the percentage of mercury retained at the end of the porosimetry te are depicted in Figure 5. This percentage was greater for specimens with the addition, i comparison to those taken as reference. The highest percentage of mercury retained wa noted for VP20 mortars.

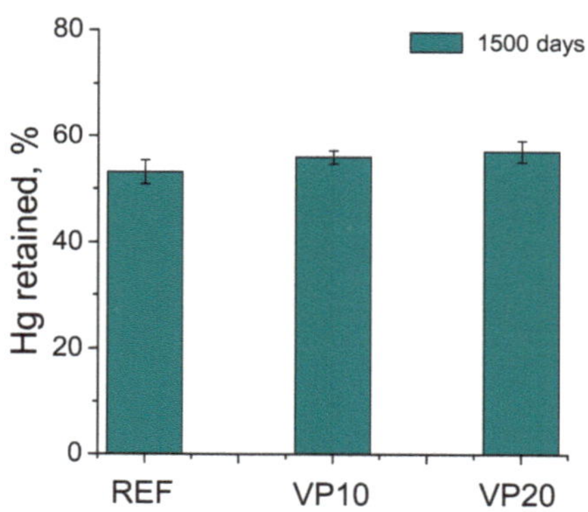

Figure 5. Results of percentage of mercury retained at the end of porosimetry test for the analyze series.

3.2. *Impedance Spectroscopy*

The results of capacitance C_1 can be observed in Figure 6. This capacitance was slightl higher as the percentage of volcanic powder in the mortars rose. Nevertheless, very lo differences between the different binders have been noted in the results of capacitance C

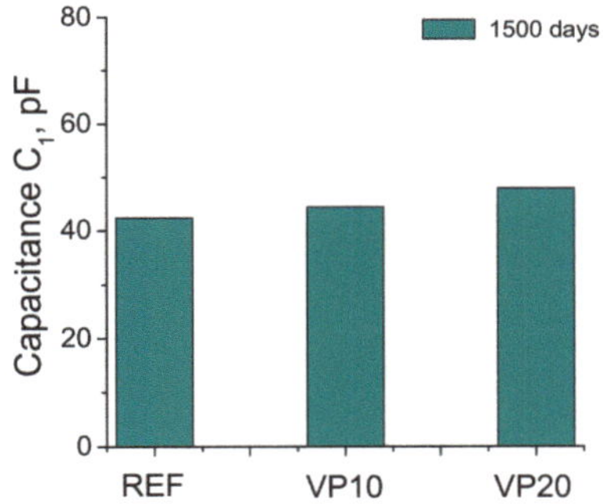

Figure 6. Results of capacitance C_1 for the mortar series studied.

In relation to capacitance C_2, its results have been represented in Figure 7. The greatest value of this capacitance was observed for the VP20 series, followed by VP10. The mortars of the REF series showed the lowest value of this parameter.

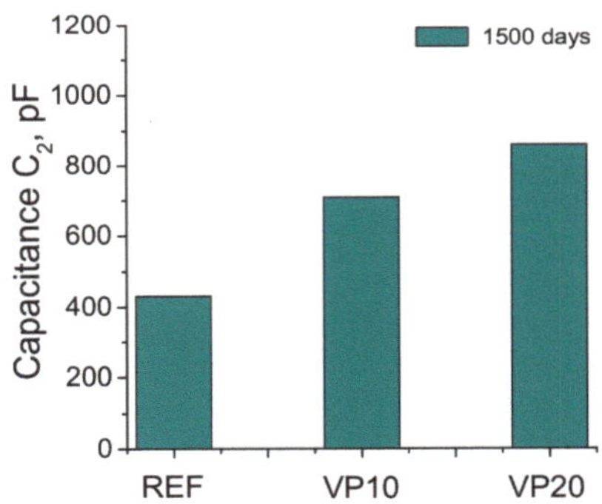

Figure 7. Capacitance C_2 values for the REF, VP10 and VP20 specimens.

The results of impedance resistance R_2 can be observed in Figure 8. This resistance was higher for specimens with volcanic powder after 1500 hardening days, in comparison with reference ones. In addition to this, the value of this parameter increased with the volcanic powder proportion in the binder, showing the VP20 mortars the greatest resistance R_2.

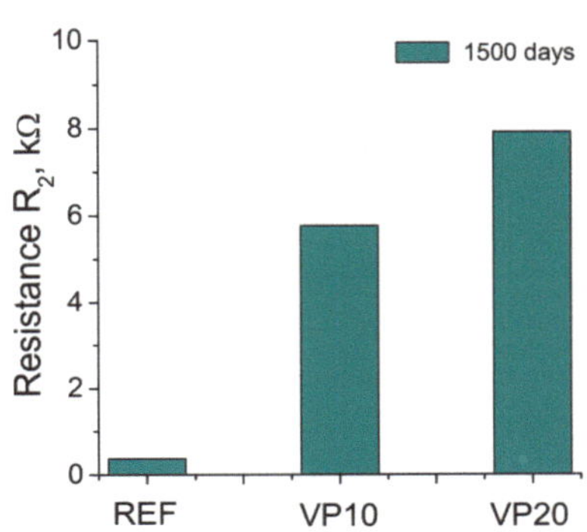

Figure 8. Resistance R_2 results noted for the mortars tested.

3.3. *Differential Thermal Analysis*

The derivate of the weight versus temperature curves obtained for the analyzed morta series after 1500 days are depicted in Figure 9. The portlandite peak area of this curve wa lower as the percentage of clinker replacement by volcanic powder increased.

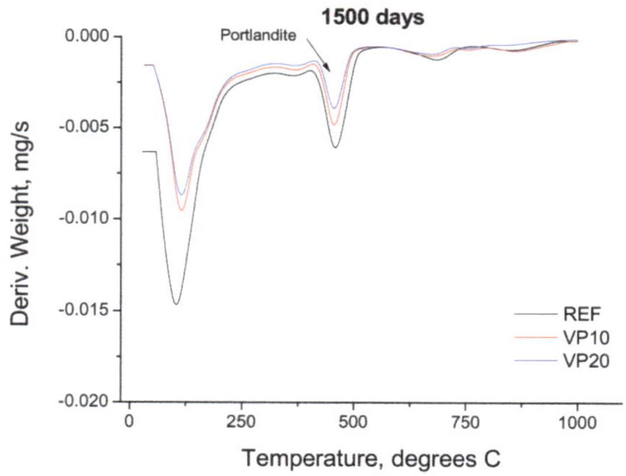

Figure 9. Derivate of weight versus temperature curve obtained for REF, VP10 and VP20 mortar after 1500 hardening days.

3.4. *Steady-State Chloride Diffusion Coefficient*

The results of the steady-state chloride diffusion coefficient obtained from the sample' resistivity for the analyzed series in the very long term have been represented in Figure 1 This coefficient was greater for mortars taken as a reference, in comparison to those whic incorporated volcanic powder. The lowest diffusion coefficient was observed for specimen with the greater percentage of volcanic powder (VP20 mortars).

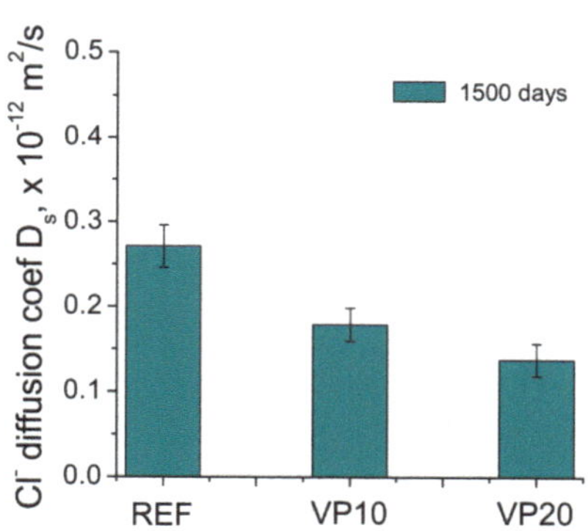

Figure 10. Steady-state chloride diffusion coefficients noted for the binders tested.

3.5. *Water Absorption after Immersion*

The percentages of water absorption after immersion, obtained for the mortar serie analyzed after 1500 hardening days, are represented in Figure 11. This absorption wa slightly greater for VP20 series with respect to REF and VP10 mortars. In spite of this, th differences between the three binders studied were overall scarce.

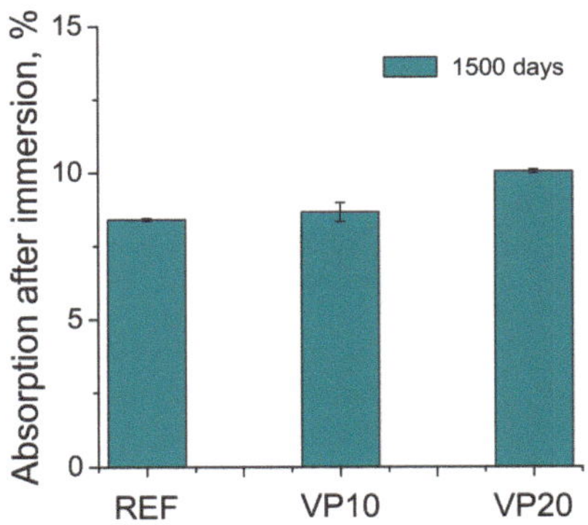

Figure 11. Percentage of absorption after immersion for the series analyzed.

3.6. Shrinkage/Expansion

The percentage of length change noted for the different binders studied is depicted in Figure 12. All the mortars showed an expansion during the analyzed time period. The greatest expansion was obtained for the reference series, while the smallest corresponded to the VP20 binder.

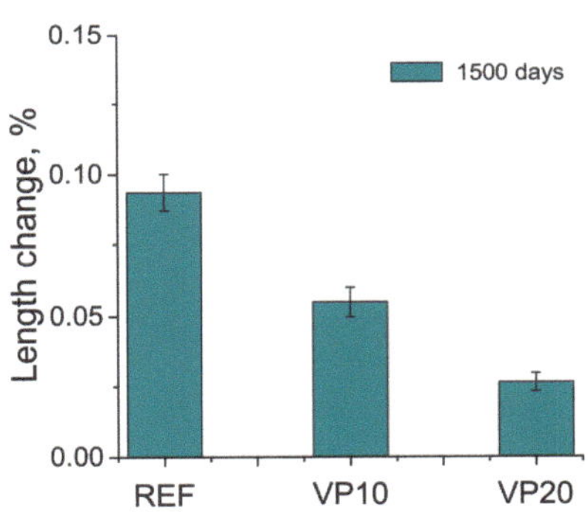

Figure 12. Length change noted for the mortar series analyzed after 1500 hardening days.

4. Discussion

4.1. Microstructure Characterization

Regarding the mercury intrusion porosimetry technique, the similar total porosity in the very long term noted for the mortars tested (see Figure 3) would indicate that they would have a similar fraction of solids and overall volume of pores, independently of the binder used. In this regard, the incorporation of 20% of volcanic powder in the binder would not entail a noticeable increase in the porosity, compared to reference specimens, in keeping with other research [26]. On the other hand, the distributions of pores by size obtained with this technique, after 1500 hardening days (see Figure 4), would reveal a greater pore refinement in the microstructure produced by the volcanic powder addition, as would suggest the greater proportion of finer pores, mainly those with sizes lower than 10 nm, noted for the VP10 and VP20 specimens, in comparison with reference ones. In addition to this, the results of the percentage of Hg at the end of the mercury intrusion porosimetry test (see Figure 5) would corroborate this higher pore refinement of mortars with volcanic powder. This parameter gives data regarding the tortuosity of the pore structure of the material [41]. Therefore, the highest percentages of Hg retained, obtained in the VP10 and VP20 samples, would indicate more tortuosity in their microstructure, directly related to their high pore refinement. The results of pore size distributions and H retained would also show that the microstructure refinement was greater as the proportic of clinker replaced by volcanic powder increased.

The microstructure refinement produced by the addition of volcanic powder afte 1500 hardening days could be a result of the development of its pozzolanic reactions. Thes reactions would form solid phases as products, closing the pore network of the mortar and increasing the relative presence of pores with lower diameters. This pozzolanic activi of the volcanic powder after 4 hardening years would also be suggested by the differenti thermal analyses performed. With this technique, it has been noted lower areas of th portlandite peak in the curves derivate of the weight versus the temperature for series wit volcanic powder, compared to those without an addition (see Figure 9). This pozzolan activity would be in keeping with other research [20,21], in which the effects of additior coming from volcanic materials in the short term have been analyzed.

Furthermore, the results of impedance spectroscopy showed coincidences with th mercury intrusion porosimetry results previously discussed. On one hand, the paramete C_1 gives data regarding global fraction of solids in the sample [41], independently of it pore size distribution. As has been previously described, scarce differences between th different studied mortars were observed in this capacitance (see Figure 6). This woul suggest that global fractions of solids were very similar after 1500 hardening days for th different studied series. This result would agree with that obtained for porosity, whic also revealed slight differences in the total volume of pores, independently of the additio of volcanic powder. The incorporation of volcanic powder produced scarce effects on th global porosity, whereas it increased the refinement of the microstructure, as noted by othe supplementary cementitious materials, such as fly ash [50].

The impedance capacitance C_2 is associated with the internal pores surface contactin with the electrolyte, which fills the material's pore network [42]. Therefore, a rise in thi surface produced by the formation of solids as rough structures on the preexisting wall of the pores, due to the progressive development of pozzolanic and hydration reaction would entail higher values of this capacitance C_2 [43]. Furthermore, for specimens wit similar global porosity and greater pore refinement, the higher presence of finer pore would also produce a higher global pore surface, giving greater capacitance to C_2 value This parameter showed higher values for VP10 and VP20 specimens after 1500 hardenin days compared to reference mortars, which became greater as the percentage of volcani powder in the binder rose (see Figure 7). This result would indicate that the interna pore surface in specimens with this addition was higher than that for referenced one suggesting that the addition of volcanic powder would produce a greater microstructur refinement, as a consequence of the progress of its pozzolanic reactions [20,21]. This woul agree with the pore size distributions and Hg retained results previously discussed.

Finally, the R_2 resistance gives information about the material's pore structure [42 The values of R_2 are higher as greater is the presence of pores with smaller diameters. A previously described, the specimens with volcanic powder showed higher resistances R_2 i comparison with the reference ones (see Figure 8), and the highest values of this resistanc corresponded to VP20 mortars. These results would also coincide with those discussed fo capacitance C_2, pore size distributions and percentage of Hg retained, revealing the por refinement produced by this addition.

4.2. *Durability-Related Parameters*

With regard to the analyzed durability-related parameters of the series with volcani powder in the very long term, the steady-state chloride diffusion coefficient exhibited smaller values for VP10 and VP20 specimens, in comparison with referenced ones (se Figure 10). As happened with some microstructural parameters, this coefficient decrease as the content of volcanic powder in the mortars increased. This result would agree wit those obtained by other authors at short hardening ages [22,23]. It may be related to th high microstructural refinement observed for VP10 and VP20 specimens, which could b a consequence of the development of the volcanic-powder pozzolanic reactions [20,21] already discussed. The presence of a noticeable percentage of pores with lower sizes would entail a more difficult diffusion of aggressive chloride ions throughout the material, which would lead to lower values of the steady-state diffusion coefficient, as previously observed. This result would be relevant, because chlorides constitute one of the most damaging agents which could affect the durability of cementitious materials and, as was observed, the addition of volcanic powder as clinker replacement, improved the performance of the mortars at this point after1500 hardening days.

In relation to the percentage of absorption after immersion, in general, the differences between the studied mortars were scarce (see Figure 11), which would be in keeping with the results of impedance capacitance C_1, as well as total porosity, suggesting that the overall volume of pores in the studied series was relatively similar, irrespective of the binder. In view of the length change results observed for the different series studied (see Figure 12), a length increase was noted for all of them, showing a slight expansion after 1500 hardening days. This could be expected due to the storage of the specimens at an optimum condition during the studied time period. Despite that, it is noticeable that mortars with volcanic powder showed less deformation than referenced ones.

Lastly, considering the results previously discussed, the incorporation of volcanic powder from the Calbuco volcano in mortars, with up to 20% of clinker replacement, would have beneficial effects after 1500 hardening days, in terms of the refinement of pore network and chloride diffusion. Furthermore, this addition would not noticeably worsen other properties, such as total porosity and water absorption by immersion. It is important to highlight that these results, in the very long term, would be interesting for evaluating the possible application of volcanic powder coming from the Calbuco volcano in real cement-based construction elements, which are generally required over a long service life. Furthermore, the use of this volcanic powder as a clinker replacement would also contribute to sustainability, providing additional advantages in relation to the protection of the environment.

5. Conclusions

The main conclusions that can be obtained from the results previously discussed can be summed up as follows:

- Mortars with the addition of volcanic powder from the Calbuco volcano showed a more refined pore network after 1500 hardening days, in comparison with referenced specimens, in view of the results of the impedance C_2 capacitance and R_2 resistance, as well as the pore size distributions obtained by mercury intrusion porosimetry;
- The higher microstructure refinement produced by the incorporation of volcanic powder from the Calbuco volcano could be due to the pozzolanic activity of this addition, as indicated by the differential thermal analyses after 1500 hardening days, which produced a greater proportion of finer pores;
- The global porosity and solid fraction was very similar for all of the studied mortars after 1500 days, irrespective of the presence of volcanic powder in the binder, as showed in the total porosity results, obtained from mercury porosimetry, and the impedance capacitance C_1;
- The use of binders with volcanic powder from the Calbuco volcano produced a reduction of the steady-state chloride diffusion coefficient after 4 years, compared to mortars without this addition. Moreover, this parameter was lower as the percentage of volcanic powder in the binder rose. This improvement in the behaviour regarding the chloride diffusion could be produced by the more refined pore network caused by the addition of volcanic powder in the mortars;
- The incorporation of volcanic powder did not noticeably affect the performance of the mortars with respect to the water absorption after immersion in the very long term, suggesting an overall volume of pores that is fairly similar for the analyzed series;

- In accordance with the results obtained in this work, the 10% and 20% clinker replacement by volcanic powder from the Calbuco volcano would have beneficial effects in the performance of mortars after approximately 4 hardening years, especially regarding their microstructure development and chloride ingress resistance, without considerably affecting their water absorption, with the additional advantage regarding the protection of the environment.

Author Contributions: Conceptualization, V.L. and J.M.O.; methodology, V.L., T.R.-H. and J.M.O.; investigation, R.M.T., T.R.-H., V.L. and J.M.O.; data curation, R.M.T. and J.M.O.; writing—original draft preparation, R.M.T.; writing—review and editing, J.M.O.; supervision, T.R.-H. and J.M.O.; funding acquisition, J.M.O. and V.L. The results included in this paper have been obtained in the PhD thesis carried out by R.M.T. at University of Alicante (Spain), under the supervision of J.M.O. and T.R.-H. All authors have read and agreed to the published version of the manuscript.

Funding: This work was supported by the Conselleria de Educación, Investigación, Cultura y Deporte (at present re-named as Conselleria de Innovación, Universidades, Ciencia y Sociedad Digital) de la Generalitat Valenciana (Spain) (grant code GV/2019/070) and by the Agencia Nacional de Investigación y Desarrollo de Chile (ANID) (grant number FONDECYT REGULAR 1211135).

Institutional Review Board Statement: Not applicable.

Informed Consent Statement: Not applicable.

Data Availability Statement: Data sharing not applicable.

Acknowledgments: Authors wish to thank Cementos Portland Valderrivas S.A. for providing the ordinary Portland cement used in this study.

Conflicts of Interest: The authors declare no conflict of interest.

References

1. Benhelal, E.; Shamsaei, E.; Rashid, M.I. Challenges against CO_2 abatement strategies in cement industry: A review. *J. Environ. Sci. China* **2021**, *104*, 84–101. [CrossRef]
2. Zhang, C.-Y.; Yu, B.; Chen, J.-M.; Wei, Y.-M. Green transition pathways for cement industry in China. *Resour. Conserv. Recycl.* **2021**, *166*, 105355. [CrossRef]
3. Valipour, M.; Shekarchi, M.; Arezoumandi, M. Chlorine diffusion resistivity of sustainable green concrete in harsh marine environments. *J. Clean. Prod.* **2017**, *142*, 4092–4100. [CrossRef]
4. Sobolev, K.; Türker, P.; Soboleva, S.; Iscioglu, G. Utilization of waste glass in ECO-cement: Strength properties and microstructural observations. *Waste Manag.* **2007**, *27*, 971–976. [CrossRef] [PubMed]
5. Palankar, N.; Ravi Shankar, A.U.; Mithun, B.M. Durability studies on eco-friendly concrete mixes incorporating steel slag as coarse aggregates. *J. Clean. Prod.* **2016**, *129*, 437–448. [CrossRef]
6. Liu, Y.; Sidhu, K.S.; Chen, Z.; Yang, E.-H. Alkali-treated incineration bottom ash as supplementary cementitious materials. *Constr. Build. Mater.* **2018**, *179*, 371–378. [CrossRef]
7. Yang, K.-H.; Jung, Y.-B.; Cho, M.-S.; Tae, S.-H. Effect of supplementary cementitious materials on reduction of CO_2 emissions from concrete. *J. Clean. Prod.* **2015**, *103*, 774–783. [CrossRef]
8. Faleschini, F.; Zanini, M.A.; Brunelli, K.; Pellegrino, C. Valorization of co-combustion fly ash in concrete production. *Mater. Des.* **2015**, *85*, 687–694. [CrossRef]
9. Ortega, J.M.; Sánchez, I.; Climent, M.A. Influence of environmental conditions on durability of slag cement mortars. In Proceedings of the 2nd International Conference on Sustainable Construction Materials and Technologies, Ancona, Italy, 28–30 June 2010.
10. Ganjian, E.; Pouya, H.S. Effect of magnesium and sulfate ions on durability of silica fume blended mixes exposed to the seawater tidal zone. *Cem. Concr. Res.* **2005**, *35*, 1332–1343. [CrossRef]
11. Ortega, J.M.; Esteban, M.D.; Rodríguez, R.R.; Pastor, J.L.; Ibanco, F.J.; Sánchez, I.; Climent, M.Á. Influence of silica fume addition in the long-term performance of sustainable cement grouts for micropiles exposed to a sulphate aggressive medium. *Materials* **2017**, *10*, 890. [CrossRef]
12. Çakır, Ö.; Aköz, F. Effect of curing conditions on the mortars with and without GGBFS. *Constr. Build. Mater.* **2008**, *22*, 308–314. [CrossRef]
13. Ortega, J.M.; Pastor, J.L.; Albaladejo, A.; Sánchez, I.; Climent, M.A. Durability and compressive strength of blast furnace slag-based cement grout for special geotechnical applications. *Mater. Constr.* **2014**, *64*. [CrossRef]
14. Berndt, M.L. Properties of sustainable concrete containing fly ash, slag and recycled concrete aggregate. *Constr. Build. Mater.* **2009**, *23*, 2606–2613. [CrossRef]

5. Navrátilová, E.; Rovnaníková, P. Pozzolanic properties of brick powders and their effect on the properties of modified lime mortars. *Constr. Build. Mater.* **2016**, *120*, 530–539. [CrossRef]
6. Xu, W.; Lo, Y.T.; Wang, W.; Ouyang, D.; Wang, P.; Xing, F. Pozzolanic Reactivity of Silica Fume and Ground Rice Husk Ash as Reactive Silica in a Cementitious System: A Comparative Study. *Materials* **2016**, *9*, 146. [CrossRef]
7. Lu, J.-X.; Zhan, B.-J.; Duan, Z.-H.; Poon, C.S. Using glass powder to improve the durability of architectural mortar prepared with glass aggregates. *Mater. Des.* **2017**, *135*, 102–111. [CrossRef]
8. Ortega, J.M.; Letelier, V.; Miró, M.; Moriconi, G.; Climent, M.A.; Sánchez, I. Influence of waste glass powder addition on the pore structure and service properties of cement mortars. *Sustainability* **2018**, *10*, 842. [CrossRef]
9. Ribeiro, D.V.; Labrincha, J.A.; Morelli, M.R. Potential use of natural red mud as pozzolan for Portland cement. *Mater. Res.* **2011**, *14*, 60–66. [CrossRef]
0. Siddique, R. Properties of concrete made with volcanic ash. *Resour. Conserv. Recycl.* **2012**, *66*, 40–44. [CrossRef]
1. Kupwade-Patil, K.; Al-Aibani, A.F.; Abdulsalam, M.F.; Mao, C.; Bumajdad, A.; Palkovic, S.D.; Büyüköztürk, O. Microstructure of cement paste with natural pozzolanic volcanic ash and Portland cement at different stages of curing. *Constr. Build. Mater.* **2016**, *113*, 423–441. [CrossRef]
2. Anwar Hossain, K.M. Performance of volcanic ash based precast and in situ blended cement concretes in marine environment. *J. Mater. Civ. Eng.* **2005**, *17*, 694–702. [CrossRef]
3. Hossain, K.M.A. Blended cement using volcanic ash and pumice. *Cem. Concr. Res.* **2003**, *33*, 1601–1605. [CrossRef]
4. Seddik Meddah, M. Durability performance and engineering properties of shale and volcanic ashes concretes. *Constr. Build. Mater.* **2015**, *79*, 73–82. [CrossRef]
5. Hossain, K.M.A.; Lachemi, M. Strength, durability and micro-structural aspects of high performance volcanic ash concrete. *Cem. Concr. Res.* **2007**, *37*, 759–766. [CrossRef]
6. Kupwade-Patil, K.; Palkovic, S.D.; Bumajdad, A.; Soriano, C.; Büyüköztürk, O. Use of silica fume and natural volcanic ash as a replacement to Portland cement: Micro and pore structural investigation using NMR, XRD, FTIR and X-ray microtomography. *Constr. Build. Mater.* **2018**, *158*, 574–590. [CrossRef]
7. Letelier, V.; Ortega, J.M.; Tremiño, R.M.; Henriquéz-Jara, B.I.; Fustos, I.; Real-Herraiz, T.; Moriconi, G.; Climent, M.A.; Sánchez, I. The use of volcanic powder as a cement replacement for the development of sustainable mortars. *Appl. Sci.* **2020**, *10*, 1460. [CrossRef]
8. European Committee for Standardization. *EN 1992-1-1 Eurocode 2: Design of Concrete Structures—Part. 1-1: General Rules and Rules for Buildings. EN 1992-1-1 Eurocode 2*; European Committee for Standardization (CEN): Brussels, Belgium, 2004.
9. Romero, J.E.; Morgavi, D.; Arzilli, F.; Daga, R.; Caselli, A.; Reckziegel, F.; Viramonte, J.; Díaz-Alvarado, J.; Polacci, M.; Burton, M.; et al. Eruption dynamics of the 22–23 April 2015 Calbuco Volcano (Southern Chile): Analyses of tephra fall deposits. *J. Volcanol. Geotherm. Res.* **2016**, *317*, 15–29. [CrossRef]
0. Reckziegel, F.; Bustos, E.; Mingari, L.; Báez, W.; Villarosa, G.; Folch, A.; Collini, E.; Viramonte, J.; Romero, J.; Osores, S. Forecasting volcanic ash dispersal and coeval resuspension during the April–May 2015 Calbuco eruption. *J. Volcanol. Geotherm. Res.* **2016**, *321*, 44–57. [CrossRef]
1. Asociación Española de Normalización y Certificación. *Composición, Especificaciones y Criterios de Conformidad de Los, Cementos Comunes, AENOR UNE-EN 197-1:2011*; Asociación Española de Normalización y Certificación: Madrid, Spain, 2011.
2. Asociación Española de Normalización y Certificación. *Métodos de Ensayo de Cementos. Parte 1: Determinación de Resistencias Mecánicas, AENOR UNE-EN 196-1:2005*; Asociación Española de Normalización y Certificación: Madrid, Spain, 2005.
3. Normensand GmbH Normensand. Available online: https://www.normensand.de/en/home/ (accessed on 20 March 2021).
4. Normensand GmbH Material Information Sheet Normensand. Available online: https://www.normensand.de/fileadmin/gruppen/Normensand/Downloads/2017-05-09_Normensand_Material-Information-Sheet.pdf (accessed on 20 March 2021).
5. Normensand GmbH Untersuchungsbericht Normensand. Available online: https://www.normensand.de/fileadmin/gruppen/Normensand/Downloads/Untersuchungsbericht_Gesamtstatistik.pdf (accessed on 20 March 2021).
6. Diamond, S. Mercury porosimetry. *Cem. Concr. Res.* **2000**, *30*, 1517–1525. [CrossRef]
7. Ouellet, S.; Bussière, B.; Aubertin, M.; Benzaazoua, M. Microstructural evolution of cemented paste backfill: Mercury intrusion porosimetry test results. *Cem. Concr. Res.* **2007**, *37*, 1654–1665. [CrossRef]
8. Horpibulsuk, S.; Rachan, R.; Chinkulkijniwat, A.; Raksachon, Y.; Suddeepong, A. Analysis of strength development in cement-stabilized silty clay from microstructural considerations. *Constr. Build. Mater.* **2010**, *24*, 2011–2021. [CrossRef]
9. Pedeferri, P.; Bertolini, L. *La Durabilità del Calcestruzzo Armato*; Mc Graw-Hill Education: Milano, Italy, 2000; ISBN 9788838608452.
0. Ho, L.S.; Nakarai, K.; Duc, M.; Kouby, A.L.; Maachi, A.; Sasaki, T. Analysis of strength development in cement-treated soils under different curing conditions through microstructural and chemical investigations. *Constr. Build. Mater.* **2018**, *166*, 634–646. [CrossRef]
1. Cabeza, M.; Merino, P.; Miranda, A.; Nóvoa, X.R.; Sanchez, I. Impedance spectroscopy study of hardened Portland cement paste. *Cem. Concr. Res.* **2002**, *32*, 881–891. [CrossRef]
2. Cabeza, M.; Keddam, M.; Nóvoa, X.R.; Sánchez, I.; Takenouti, H. Impedance spectroscopy to characterize the pore structure during the hardening process of Portland cement paste. *Electrochim. Acta.* **2006**, *51*, 1831–1841. [CrossRef]
3. Ortega, J.M.; Sánchez, I.; Climent, M.A. Impedance spectroscopy study of the effect of environmental conditions in the microstructure development of OPC and slag cement mortars. *Arch. Civ. Mech. Eng.* **2015**, *15*, 569–583. [CrossRef]

44. Tang, S.W.; Cai, X.H.; He, Z.; Zhou, W.; Shao, H.Y.; Li, Z.J.; Wu, T.; Chen, E. The review of pore structure evaluation in cementitiou materials by electrical methods. *Constr. Build. Mater.* **2016**, *117*, 273–284. [CrossRef]
45. ASTM. *Standard Test Method for Density, Absorption, and Voids in Hardened Concrete, ASTM C642-06*; ASTM International: We Conshohocken, PA, USA, 2006; p. 3.
46. Ortega, J.M.; Sánchez, I.; Antón, C.; De Vera, G.; Climent, M.A. Influence of environment on durability of fly ash cement morta *ACI Mater. J.* **2012**, *109*, 647–656.
47. Ortega, J.M.; Sánchez, I.; Climent, M.A. Durability related transport properties of OPC and slag cement mortars hardened und different environmental conditions. *Constr. Build. Mater.* **2012**, *27*. [CrossRef]
48. Andrade, C.; Alonso, C.; Arteaga, A.; Tanner, P. Methodology based on the electrical resistivity for the calculation of reinforceme service life. In Proceedings of the 5th CANMET/ACI International Conference on Durability of Concrete, Supplementary Pape Barcelona, Spain, 4–9 June 2000; Malhotra, V.M., Ed.; American Concrete Institute: Farmington Hills, MI, USA; pp. 899–915.
49. ASTM. *Standard Test Method for Drying Shrinkage of Mortar Containing Hydraulic Cement, ASTM C596-01*; ASTM International: We Conshohocken, PA, USA, 2001; p. 3.
50. Bijen, J. Benefits of slag and fly ash. *Constr. Build. Mater.* **1996**, *10*, 309–314. [CrossRef]

materials

MDPI

Article

Evaluation of Eco-Efficient Concretes Produced with Fly Ash and Uncarbonated Recycled Aggregates

Miren Etxeberria

Department of Civil and Environmental Engineering, Universitat Politècnica de Catalunya·BarcelonaTECH, Campus Nord, 08034 Barcelona, Spain; miren.etxeberria@upc.edu

Abstract: The fabrication of conventional concrete, as well as remains from demolition, has a high environmental impact. This paper assessed the eco-efficiency of concrete made with uncarbonated recycled concrete aggregates (RCA) and fly ash (FA). Two concrete series were produced with an effective water/cement ratio of 0.50 (Series 1) and 0.40 (Series 2). In both series, concretes were produced using 0% and 50% of RCA with 0%, 25% and 50% FA. After analysing the compressive strength, and carbonation and chloride resistance of those concretes, their eco-efficiency based on the binder intensity and CO_2-eq intensity was assessed. We found that the use of 50% uncarbonated RCA improved the properties of concretes produced with FA with respect to using natural aggregates. The concrete made of 25% FA plus RCA was considered the most eco-efficient based on the tests of compressive, carbonation and chloride properties with the values of 4.1 kg CO_2 m^{-3} MPa^{-1}, 76.3 kg CO_2 m^{-3} mm^{-1} $year^{0.5}$ and 0.079 kg CO_2 m^{-3} C^{-1}, respectively. The uncarbonated RCA improved carbonation resistance, and FA improved chloride resistance. It can be concluded that the use of 50% un-carbonated RCA combined with FA considerably enhanced the properties of hardened concrete and their eco-efficiency with respect to concretes produced with natural aggregates.

Keywords: eco-efficiency; recycled aggregate concrete; fly ash; mechanical properties; carbonation; chloride resistance; durability; sustainable construction

Citation: Etxeberria, M. Evaluation of Eco-Efficient Concretes Produced with Fly Ash and Uncarbonated Recycled Aggregates. *Materials* **2021**, *14*, 7499. https://doi.org/10.3390/ma14247499

Academic Editors: Miguel Bravo, Rui Vasco Silva and António P. C. Duarte

Received: 6 November 2021
Accepted: 3 December 2021
Published: 7 December 2021

1. Introduction

Concrete production has a high environmental impact due to the abundant energy consumption and CO_2 emissions of cement production, releasing approximately 5–7% of total global anthropogenic CO_2 emissions and 3% of total greenhouse gases emissions in the atmosphere [1–3]. In general, the impact of concrete is assessed as a functional unit of the amount of cement employed in the concrete on the environment weight or volume [4]. Adequate performance in terms of mechanical and durability properties guarantees the eco-efficiency of concrete [4]. It can be said that the demolition of concrete structures causes a considerable volume of waste that terminates in landfills. In 2018, the construction and demolition waste was 35.4% of the total waste (2277 million tonnes) generated in the EU by all economic activities and households, of which only 54.2% was recovered [5,6]. In this vein, the use of supplementary cementitious materials, such as fly ash (FA) and recycled concrete aggregates (RCA, recycled aggregates produced crushing concrete structures), in the production of green concrete (CRAFA concrete) can help increase the eco-efficiency of concretes [7–10].

RCA has been proven to be suitable for concrete production [11]. In addition, the employment of recycled aggregates concrete (RAC, concrete produced using recycled aggregates) as a structural material has been widely analysed and validated in many applications [12,13]. However, taking into consideration its durability properties, it was found that the RAC concrete had lower chloride ion diffusion resistance than natural aggregate concrete (NAC) when concretes were produced with ordinary Portland cement (OPC) [14–17]. The studies on carbonation resistance tend to be inconsistent. Guo et al. [15],

Adessina et al. [17] and Silva et al. [18], among other researchers, stated that RAC produce with a high percentage of RCA achieved higher carbonation depth than NAC. Zeng [19 stated that the best replacement percentage of natural aggregates with RCA was 50% a this avoided a decrease in carbonation resistance.

However, certain researchers [16,20] have stated that the RAC mixtures had similar c higher resistance to carbonation than NAC due to the old adhered mortar. According t Leemann and Loser [21], there was no systematic difference in the carbonation coefficier independently of the replacement levels of RCA. Although the RCA particles that are eithe porous or already carbonated at the time of concrete production can lead to a local increas of carbonation depth, the impact on the carbonation coefficient is not significant, as there i only an increase of 10% at a given compressive strength [21]. Pedro et al. [22] documente that when the quality of the cement paste in the new concrete was found to be considerabl better than that of the RCA, an increase of the carbonation coefficient with increasing level of NAC replacement was inevitable.

Today, the use of FA in proportions of up to 15–30% [23,24] in concrete production i usual. The use of FA causes low strength in concretes at initial ages [25]. However, at late ages, the strength and durability of concrete increases [24,26,27]. In addition, the chlorid ion penetration resistance of concrete is higher when FA is employed [28,29]. However, th concrete produced with FA achieves lower carbonation resistance than the concrete mad of 100% OPC [30].

The concrete produced with a combination of FA and RCA (CRAFA) could improv durability and eco-efficiency required for new sustainable concretes [31]. According t Faella et al. [32] and Kurda et al. [33], the CRAFA concrete (produced with 50% RCA an 50% FA) exhibited lower water permeability compared with NAC concrete after 28 days achieving a similar compressive strength with the NAC (produced with OPC) at 90 days Moreover, a significant attenuation of the chloride-ion penetration was achieved by addin FA to the RAC mixtures [32]. Kurda et al. [33], Kim et al. [34] and Kou and Poon [35], amon other researchers, determined that the total charge passed through concrete (measured i C) increased with increasing levels of coarse RCA replacement, and the opposite occurre when increasing the incorporation levels of FA. According to Corinaldesi and Moriconi [36 the addition of FA proved to be effective in reducing the chloride ion penetration depth in concrete, even when RCA was used. Moreover, they concluded that the addition of FA to RAC increased the carbonation rate. Similar results were obtained by Faella et al. [32 Jianmin et al. [37] and Poon and Kou [35,38].

However, the CRAFA mixtures were prepared with a lower water/binder ratio [36 as a result of the refinement of the pore system, carbonation did not present risks o reinforcement corrosion. Moreover, Tian et al. [39] concluded that the influence that ha the most impact on the depth of carbonation of the CRAFA was that of the water/binde ratio, signifying that the carbonation age and the FA content had a minor effect. Xia et al. [20] found that although the use of FA in RAC production improved the internal por structure by decreasing the porosity of the concrete, it also resulted in the reduction of th total alkaline content that can be carbonated, resulting in greater carbonation depths.

The eco-efficiency of the cement used in concrete production was first analysed by Damineli et al. [4]. The optimum mix design is calculated by determining the binde intensity (Bi) and CO_2-eq intensity (Ci) of the concrete. The Bi value measures the amoun of binder necessary to deliver a unit of strength and, consequently, the efficiency of the employed binder. It has been suggested that the minimum stabilised Bi value at 28 days after production is 5 kg m^{-3} MPa^{-1} for concretes of >60 MPa. Research suggests tha this value will be an achievable goal for normal strength concrete (<50 MPa) in the nea future and the ready-mix concretes <40 MPa consume the most cement, having a Bi valu of 10–12 kg m^{-3} MPa^{-1}. For 20 MPa concretes, the minimum Bi value is approximately 13 kg m^{-3} MPa^{-1} [4]. A higher Bi value (>20 kg m^{-3} MPa^{-1}) is necessary when larg volumes of weak or low elastic modulus waste materials are employed in concrete pro duction. To achieve a moderate Bi value, needed for higher concrete strengths, the binder consumption per unit of MPa should always be low [40].

The Ci value of concretes is determined by dividing the emitted CO_2-eq in kg per m^3 of concrete by its compressive strength at 28 days, and it is expressed as kg CO_2 m^{-3} MPa^{-1}. The Ci value considers the contribution of concrete mixtures to the global warming potential (GWP) per unit volume and strength, and it has been analysed intensively during recent years [3,41,42]. Damineli et al. [4], studying international data, determined that the average estimated Ci value was 7.1 kg CO_2 m^{-3} MPa^{-1}. All Ci values > 13 kg CO_2 m^{-3} MPa^{-1} presented a high Bi value and no clinker replacement [4]. The data suggest [4] that the mix design and the selection of raw materials allow reducing Ci values by a factor of 10, from 15 kg CO_2 m^{-3} MPa^{-1} to a minimum of 1.5 kg CO_2 m^{-3} MPa^{-1}. The minimum estimated Ci value was 1.5–2 kg CO_2 m^{-3} MPa^{-1}. According to de Matos et al. [41], all concretes with Ci < 5.0 kg CO_2 m^{-3} MPa^{-1} are in the range of 70–80 MPa, while for the range of 50–60 MPa, the average Ci value is 6.2 kg CO_2 m^{-3} MPa^{-1}. Similar values have been obtained by Celic et al. [43]. Randl et al. [42] also confirmed that the use of FA and the improved mix design were an effective solution for reducing Ci values of concretes.

This paper analyses the beneficial application of 50% of uncarbonated RCA as a substitution for natural aggregates together with 0%, 25% and 50% of FA in the replacement of cement for adequate properties of CRAFA concrete production. The compressive strength and durability properties of carbonation and chloride resistance were determined and compared with conventional concrete. The eco-efficiency of the produced concretes was assessed by measuring Bi and Ci values with respect to those obtained for compressive strength, carbonation coefficient and chloride resistance. The results guaranteed that CRAFA concrete (produced with FA and RCA) could maintain high strength and durability properties. Consequently, CRAFA achieved a higher level of eco-efficiency than the concretes made with Portland cement and natural aggregates.

2. Materials and Methods

2.1. Binder

The CEM II A-L 42.5R cement (88% clinker, 10% limestone and 2 minority component; CEM II) was employed in the concrete production. Class F coal-fly ash was used at 0%, 25% and 50% to replace CEM II. The CEM I 52.5R cement (high strength and rapidly hardened, 98% clinker and 2% of minority component; CEM I) was used, replaced by FA by 50%. CEM II and CEM I were produced by the same cement producer using the same type of clinker for their manufacture. The density of both types of cement was 3.1 kg/dm^3. The coal-FA produced by a Spanish coal-fired power plant had a density of 2.16 kg/dm^3. Table 1 shows the composition of the binder materials.

Table 1. Composition of binder materials presented as the percentages of the total weight.

	CaO	SiO_2	Fe_2O_3	Al_2O_3	MgO	SO_3	Na_2O	K_2O	LOI
CEM II	61.8	18.3	3.3	4.1	1.4	2.9	0.1	0.8	5.1
CEM I	63.5	19.4	3.4	4.2	1.4	3.0	0.12	0.53	3.7
FA	2.3	58.4	7.3	21.6	1.9	0.2	0.9	2.1	3.1

Aggregates

Three fractions of natural limestone aggregates were used to produce concrete: fine natural aggregates (NS of 0/4 mm) and two natural coarse aggregates (CA1 of 5/10 mm and CA2 of 8/20 mm). The grading distribution of the aggregates (Figure 1) was determined following EN 933-1 specification. The dry density of NS, CA1 and CA2 was 2.59 kg/dm^3, 2.64 kg/dm^3 and 2.66 kg/dm^3, respectively. The absorption capacity was 1.79%, 0.69% and 0.57%, respectively. The dry density and absorption capacity were determined following the EN 1097-6 specifications.

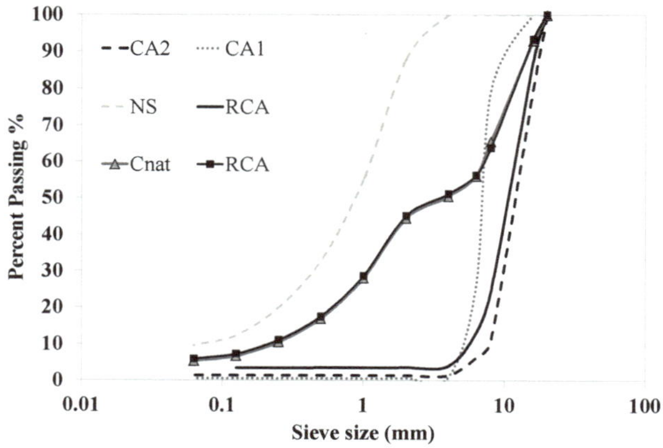

Figure 1. The grading distribution of all the aggregate fractions (NS, CA1, CA2 and RCA) and th corresponding mixtures used for concrete production with 100% of natural aggregates (C) and 50 coarse recycled concrete aggregates (RCA).

The RCA aggregates were produced by crushing a one-year-old parent concret specifically, 30 MPa cylindrical concrete samples (150 mm Ø and 300 mm long) obtaine from a concrete manufacturing company. The component of the parent concrete was CEM II A-L 42.5R cement and limestone aggregates. RCA was used for concrete productio immediately after the parent concrete was crushed. The parent concrete presented carbonation depth of 3–4 mm. In consequence, the RCA aggregates were uncarbonate Figure 2 illustrates that the RCA was uncarbonated: (a) pink colour by phenolphthalei (b) DRX analysis, in which $Ca(OH)_2$ component is shown. The grading distribution c RCA was a 5/20 mm fraction (Figure 1). Its dry density was 2.33 kg/dm^3 and absorptio capacity 5.35%. The RCA had a lower density and higher absorption than that of th limestone aggregates (calcium carbonate).

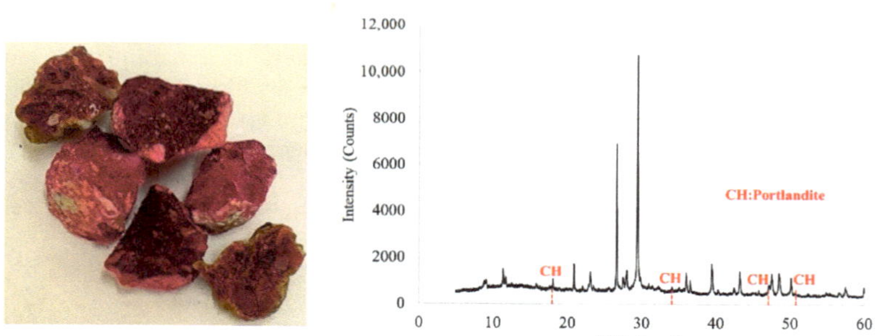

(a) (b)

Figure 2. Recycled aggregates were uncarbonated: (**a**) the phenolphthalein (pink colour) showed the presence of $Ca(OH)_2$; (**b**) DRX analysis shows $Ca(OH)_2$.

2.2. Mix Design and Production Process

Two series of concrete production were carried out, employing an effective wa ter/cement ratio of 0.50 for Series 1 and 0.40 for Series 2 (Table 2). Each concrete wa assigned the numeration of −1 or −2 depending on the manufacturing series 1 or 2 respectively.

Table 2. Mixture proportions of Series 1 and Series 2 concretes. The values are given as weight over volume of concrete production (kg/m^3) (W: water, Ad: admixture). Ad. was used in % with respect to the cement's weight

	CEM	FA	W Total	NS	CA1	CA2	RCA	Ad. (%)	Slump Test (cm)
				Series 1					
CC-1	350	-	192.5	900	275.2	647	-	0.50	20
CFA25-1	262.5	87.5	192.5	867.9	275.2	647	-	0.50	19
CFA50-1	175	175	192.5	835.9	275.2	647	-	0.50	19
CFA50.2-1	175	175	192.5	835.9	275.2	647	-	0.50	19
CRA-1	350	-	213.4	900	137.6	323.5	400	0.50	20
CRAFA25-1	262.5	87.5	213.4	867.9	137.6	323.5	400	0.50	20
CRAFA50-1	175	175	213.5	835.9	137.6	323.5	400	0.50	19
CRAFA50.2-1	175	175	213.5	835.9	137.6	323.5	400	0.50	20
				Series 2					
CC-2	350	-	157.5	945.3	289.1	679.6	-	1	20
CFA25-2	262.5	87.5	157.5	913.3	289.1	679.6	-	1.1	16
CFA50-2	175	175	157.5	881.2	289.1	679.6	-	1.6	21
CFA50.2-2	175	175	157.5	881.2	289.1	679.6	-	1.6	21
CRA-2	350	-	180.4	945.3	144.5	339.8	425.2	1	18
CRAFA25-2	262.6	87.5	180.4	913.3	144.5	339.8	425.2	1.2	16
CRAFA50-2	175	175	180.5	881.2	144.5	339.8	425.2	1.4	16
CRAFA50.2-2	175	175	180.5	881.2	144.5	339.8	425.2	1.4	16

In each series, eight different concretes were produced. Four types of binders were employed: (1) 100% CEM II cement; (2) 75% CEM II cement plus 25% FA (FA25); (3) 50% CEM II cement plus 50% FA (FA50); and (4) 50% CEM I cement plus 50% of FA (FA50.2). Those binders were mixed with two types of aggregate fractions: (1) 100% of natural aggregates and (2) 50% of RCA in replacement (in volume) of natural coarse aggregates. All the concretes were produced employing natural sand. The grading distribution of the different aggregate fraction mixtures is described in Figure 2. The concrete produced with 100% of natural aggregates and the four binders previously mentioned were designated as CC, CFA25, CFA50 and CFA50.2, respectively. The assignation of concretes produced with 50% of RCA and the same four binders were CRA, CRAFA25, CRAFA50 and CRAFA50.2. Figure 3 shows the manufactured concrete and the employed materials.

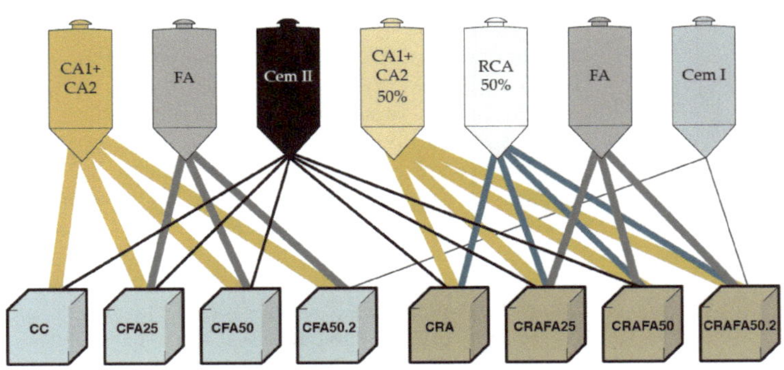

Figure 3. Description of materials employed and produced concrete. The eight concretes were produced in both series.

Following Neville's [44] definition of the effective amount of water in the mixture, which occupies the space outside the aggregate particles, the effective water/cement ratio for Series 1 (0.50) and Series 2 (0.40) was kept constant. The reason for this was to achieve the same conditions with respect to the hydration of the cement paste caused by th high absorption of RCA. RCA was used with high moisture content in nearly saturate surface-dry conditions (80–90% of water absorption capacity), but not totally saturated i order to avoid the influence of water surface layers on the mechanical properties of th concrete [45]. The RCA moisture content was measured before its use and the dosages wer adjusted according to the remaining effective water absorption capacity (the effective wate absorption of the aggregates was determined by submerging them in water for 20 min) c RCA and the natural aggregates. While the effective absorption capacity of NS was 90% and that of CA1 and CA2 50% of their total absorption capacity, the effective absorptio capacity of RCA was 70%.

A superplasticiser based on modified polycarboxylates was employed in all mixture at 0.50% with respect to the cement's weight in all the concrete mixtures produced in Serie 1 (Table 2) to achieve a slump of 190–200 mm. In Series 2, the superplasticiser dosag was increased to 1–1.6% to achieve a slump test of 160–210 mm. The slump value wa determined following the UNE-EN 12350-2:2020 specifications [46].

The concrete samples were produced and cured following UNE-EN 12390-2:200 regulations [47] and manually compacted using a steel rod. The concrete samples wer then covered with a plastic sheet and air-cured for the first 24 h. After 24 h of casting, th samples were demoulded and then stored at a temperature of 21 °C and 95% humidit until reaching the test ages for the physical, mechanical and chloride ion penetration test The samples that were submitted to an accelerated carbonation process were taken fron the humidity room after 28 days and placed in the laboratory under pre-conditioning fo 14 days at a CO_2 concentration of 400 ppm, 21 ± 1 °C and 50–55% humidity before the being placed in the CO_2 chamber.

2.3. Test Procedure

2.3.1. Compressive Strength

The compressive strength of concretes was determined using a compression machin with a loading capacity of 3000 kN. The compressive strength was measured at the ages o 7, 28 and 90 days following the EN 12390-3:2009 specifications [48]. Three cubic sample (100 × 100 × 100 mm) were used for each concrete mixture, at a total of nine samples.

2.3.2. Carbonation and Chloride Resistance

Concretes' carbonation resistance was determined by analysing the carbonation dept presented by concrete specimens after being submitted to accelerated carbonation for (7 and 14 days. Six cylindrical samples of 100 mm Ø and 100 mm long and six cubi samples of 100 × 100 × 100 mm were analysed for each concrete. All the samples wer cured in the humidity room for 28 days and were later exposed to a pre-conditionin period of 2 weeks under laboratory conditions (21 ± 1 °C and 50–55% humidity) befor their exposure to accelerated carbonation. After the pre-conditioning period (0 day), th carbonation depth and compressive strength of the first concrete samples, two cylindrica samples of 100 mm Ø and 100 mm long and two cubic samples of 100 × 100 × 100 mm o each concrete mixture were tested, respectively. The carbonation depth was determined b spraying a colour indicator solution on a freshly broken surface. A solution composed o 1 g of phenolphthalein indicator in 70 mL of ethanol and 30 mL of demineralised wate was applied following EN specifications. The four cylindrical samples of 100 mm Ø anc 100 mm long and four cubic samples (100 × 100 × 100 mm) of each concrete mixture wer placed in a climate chamber with CO_2 control for the accelerated carbonation test at 20% CO_2, 20 °C and 60% humidity, following GB T50082-2009 standards [49]. After 7 anc 14 days of exposure, two cylindrical samples for each concrete mixture were measured fo carbonation depth. Additionally, the compressive strength of the cubic samples submittec to carbonation was determined.

The 20% CO_2 concentration was found to be excessive. It is known that the carbonatior coefficient increases with increased CO_2 concentration [50]. Moreover, at long test periods the high concentrations of CO_2 cause modifications in the microstructure of the cement paste when compared to samples subjected to natural exposure only [51]. However, subjecting the concrete test samples to conditions of high concentration of CO_2 over short periods of testing (for 14 days in this study) allows for the determining of the concrete's resistance to carbonation [52].

The chloride penetrability of the concrete was determined following the American Standard Test Method, ASTM C1202 standards [53], using a 50 mm thick and 100 mm Ø concrete disc cut from the 100 mm Ø and 200 mm length concrete cylinder. The concrete's resistance to chloride-ion penetration is represented by the total charge (measured in C) during a test period of 6 h, maintaining a potential difference of 60 V dc across both ends of the specimens. In this study, the chloride ion penetrability test was carried out on the concrete samples at 28 and 90 days. Each result was the average of three measurements.

2.3.3. Eco-Efficiency Assessment

Bi and Ci values were used to assess the concretes' eco-efficiency as defined by Damineli et al. [4]. Three Bi values, the weight of binder (kg) employed for one m^3 of concrete production to deliver one unit of performance, compressive strength (Bi-comp), performance of carbonation coefficient (Bi-carb) and chloride resistance (Bi-chlor), were determined for each concrete. For each type of concrete, three Ci values were also determined, as the amount of CO_2 emitted per unit of compressive strength (Ci-comp), per carbonation coefficient (Ci-carb) and per chloride resistance (Ci-chlor).

The CO_2 emissions (kg) per weight of CEM II and CEM I were adapted from existing data provided by the cement production company, "Cement Molins Industrial (CMI)", located in Barcelona. The Ci values of CEM I and CEM II were 0.753 kg $CO_{2\text{-}I}$ and 0.674 kg $CO_{2\text{-}II}$, respectively. The supply and transport of raw materials and cement production have been included in the given values. The CO_2 emissions per kg of FA were considered as 0.023 kg $CO_{2\text{-}FA}$ (including transportation), a similar value considered by Matos et al. [41]. Natural and recycled aggregates employed in this study were produced within 30 km of Barcelona, having similar environmental costs [2]. Therefore, the aggregates used were not considered in the analysis. The chemical admixtures or amount of electricity employed for concrete production were also not taken into account. They could be valued as having approximately 10% of the total impact of the global warming potential (CO_2-eq m^{-3}) [1], a value similar to that in all of the produced concretes. The eco-efficiency of RCAs owing to the reduction of landfill use and the low consumption of natural resources was not considered either, as there is no significant difference in direct CO_2 emissions. The use of recycled aggregates significantly reduces the consumption of natural resources [3,54]; however, this was not considered in this work. The absorption of CO_2 during concrete carbonation was also not considered. Consequently, the Ci results probably underestimate the impact on global warming. However, the obtained data are still helpful in analysing the potential detrimental effect on the environment.

The CO_2 emissions per m^3 of concrete are determined by Equation (1).

$$CO_{2\text{-}CONCRETE} = CEM\ (II\ or\ I) \times CO_{2\text{-}I/II} + FA \times CO_{2-FA} \tag{1}$$

Finally, the concretes' Ci value was determined by dividing the CO_2 emission (in kg) per m^3 of concrete by its unit of performance. Compressive strength was expressed as kg $CO_2 \cdot m^{-3}$ MPa^{-1}, carbonation as kg CO_2 m^{-3} mm^{-1} $year^{0.5}$ and chloride resistance performance as kg CO_2 m^{-3} C^{-1}.

The lower Bi-comp and Ci-comp values obtained guarantee a higher efficiency of required strength in the concrete. Whereas, a higher Bi and Ci for carbonation and chloride resistance guarantee a higher efficiency in achieving required durability values within the concrete. Moreover, only the concrete that achieved the durability property values (carbonation and chloride resistance) defined by the Spanish Structural Code [55] for XC3 and XC4 exposition environment and the ASTM C1202 specifications [53], respectively, whether similar or not too high, could be considered eco-efficient concretes.

Figure 4 describes the flowchart of the test procedure and the analysis carried out.

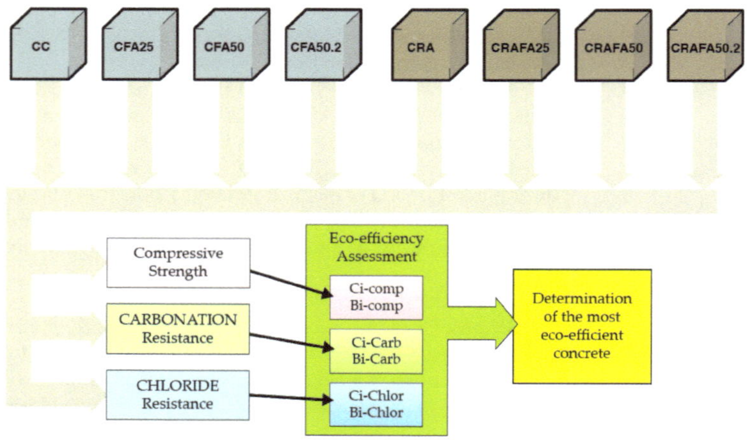

Figure 4. Flowchart of the test procedure carried out.

3. Results

3.1. Compressive Strength

The compressive strength and the standard deviation of all concretes at 7, 28 an 90 days are shown in Table 3. The CC-1 and CC-2 mixtures achieved the highest compre sive strength in Series 1 and Series 2, respectively, at tested ages, 7, 28 and 90 days.

Table 3. Physical properties of concretes produced in Series 1 (with effective w/c of 0.50) and 2 an the compressive strength of concretes at 7, 28 and 90 days. The standard deviation of each averag value is given between brackets.

	Series 1 (Effective w/c 0.50)			Series 2 (Effective w/c 0.40)		
	Compressive Strength (MPa)			Compressive Strength (MPa)		
	7 Days	28 Days	90 Days	7 Days	28 Days	90 Days
CC	36.7 (0.72)	42.0 (1.4)	49.6 (0.8)	45.1 (0.49)	53.7 (0.41)	61.1 (1.7)
CFA25	29.0 (1.9)	39.3 (1.7)	49.4 (1.7)	39.7 (0.90)	49.7 (1.53)	56.7 (2.15)
CFA50	17.5 (0.8)	29.1 (0.0)	39.5 (1.6)	28.6 (0.05)	37.6 (0.95)	54.1 (0.85)
CFA50.2	22.1 (0.6)	32.7 (0.0)	42.2 (0.8)	31.5 (0.28)	42 (0.17)	55.3 (2.2)
CRA	34.5 (2.4)	41.3 (0.3)	47.5 (1.0)	43.0 (0.95)	48.6 (0.65)	57.1 (1.45)
CRAFA25	31.5 (0.13)	38.7 (0.24)	48.1 (1.1)	37.7 (0.39)	46.2 (2.16)	57.6 (0.6)
CRAFA50	19.2 (0.33)	26.9 (0.98)	37.8 (0.2)	24.2 (0.76)	35.4 (0.58)	51.95 (1.05)
CRAFA50.2	23.7 (0.61)	30.5 (1.9)	41.8 (1.2)	29.1 (1.07)	44.01 (0.76)	56.3 (1.9)

The concretes produced with FA had a lower initial strength than the CC concrete In Series 1 (Figure 5a), at 28 and 90 days, CFA25-1 achieved similar strength to the CC- concrete. On the contrary, at 28 and 90 days, the CFA25-2 concrete (Series 2, Figure 5b suffered a reduction of 13% and 7%, respectively, compared to that of the CC-2 concrete. In addition, all the 50% FA concretes tested achieved lower strength values than those of th CC concretes at all the tested ages. Similar results were obtained by Nath and Sarker [56 at 28 and 90 days.

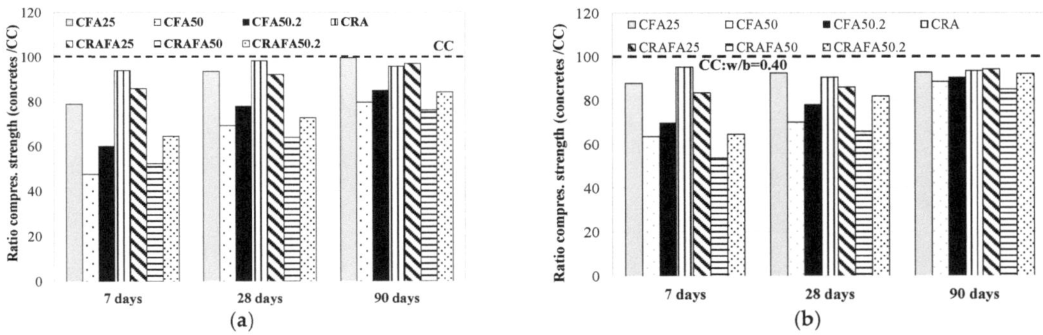

Figure 5. The compressive strength of all concretes with respect to the CC concrete: (**a**) Series 1, (**b**) Series 2.

In addition, the CFA50-1 and CFA50-2 concretes with FA replacing CEM II were found to have lower strength values than the CFA50.2-1 and CFA50.2-2 concretes, when FA was the replacement of CEM I. While it was determined that in Series 2, the CFA50.2-2 concrete achieved a <10% reduction of the compressive strength compared to the CC-2 concrete at 90 days, the CFA50.2-1 concretes produced in Series 1 achieved a decrease of 15% with respect to the CC-1 concrete.

According to the results obtained from the concretes produced with RCA, the CRA-1 and CRA-2 concretes achieved compressive strength levels similar to CC-1 and CC-2, respectively, at all ages. While the CRA-1 achieved a maximum reduction of 6% with respect to CC-1 concrete at 7 days, the CRA-2 achieved a maximum reduction of 9.5% with respect to CC-2 at 28 days. In Series 1, FA in the production of RAC led to a lower reduction of the compressive strength compared to that of the FA concretes produced with natural aggregates at 7 days of curing. Probably, this difference was due to a reaction to the FA caused by the presence of a slightly higher amount of portlandite in the concretes produced with uncarbonated RCA (See Figure 2, uncarbonated RCA). In addition, in both series, the CRAFA25 concretes achieved similar (at 28 days) and higher (at 90 days) strength than the CRA concretes. It must be noted that a minute reaction occurred when a high volume of FA was employed in the concrete production [24,57], due to the reduction of the compressive strength value for the 50% FA replacement. Nonetheless, in Series 2, the influence of 50% FA was more prominent. While CRAFA50.2-1 had lower strength than RCA-1 by 26% at 28 days and 12% at 90 days, the CRAFA50.2-2 had 15% and 1.4% lower strength than the RCA-2 concrete.

3.2. Durability Properties: Carbonation and Chloride Resistance

Due to the lower compressive strength of the concrete produced with 50% FA in the replacement of CEM II, only the CFA50.2 and CRAFA50.2 concrete mixtures were considered for the durability analysis.

3.2.1. Carbonation Resistance

Table 4 shows the carbonation depth and the compressive strength of the samples submitted to accelerated carbonation at day 0 (after the pre-conditioning) and at 7 and 14 days of the accelerated carbonation process.

The CC and CRA concrete mixtures (produced with 100% of CEM II) achieved the lowest carbonation depth in both Series. In addition, the CC-2 and CRA-2 concretes achieved 0 mm of carbonation after 7 days of the carbonation process (Table 4). Furthermore, it was determined that increased quantities of FA resulted in increased carbonation depth. The carbonation depth increased considerably when 50% FA was employed as this had a lower amount of $Ca(OH)_2$ in the cement paste [58], a result which is in line with other research reports [30,38,59].

Table 4. Carbonation depth and compressive strength after the accelerated carbonation process of all concretes produced in Series 1 and Series 2.

Concrete Type	Carbonation Depth (mm)			K_{acc} (mm/day$^{0.5}$)	K_{nat} (mm/year$^{0.5}$)	Compressive Strength (MPa)		
	0 Day	7 Day	14 Day			0 Day	7 Day	14 Day
Series 1 (effective w/c 0.50)								
CC-1	0	5.3	9.1	2.35	2.08	45.0	46.6	53.9
CFA25-1	0	8.9	12.8	3.4	3.01	43.0	41.7	51.8
CFA50.2-1	0	14.9	18.8	5.1	4.49	37.0	38.1	40.3
CRA-1	0	6.9	7.7	2.16	1.9	44	48.6	53.4
CRAFA25-1	0	7.8	9.7	2.65	2.3	46	53.8	54.7
CRAFA50.2-1	0	12.5	15.0	4.13	3.65	38	38.8	36.8
Series 2 (effective w/c 0.40)								
CC-2	0	0.0	3.2	0.702	0.62	58	59.9	61.7
CFA25-2	0	6.34	8.47	2.290	2.03	55	56.7	59.4
CFA50.2-2	0	9.44	13.74	3.651	3.24	47	51.8	58.2
CRA-2	0	0.0	5.63	1.23	1.09	53	55.7	55.1
CRAFA25-2	0	6.19	7.67	2.1025	1.86	55	55.1	55.7
CRAFA50.2-2	0	9.03	14.29	3.7452	3.32	52	50.5	52.5

In Series 1, the concretes produced with RCA suffered lower carbonation depth than those with natural aggregates, probably because RCA concretes had higher amounts of $CaO_{reacted}$ [58]. This lower carbonation depth also occurred in Series 2, where it was found that the carbonation depth of CRAFA25-2 was lower than that of the CFA25-2 concrete. However, when 50% FA was employed, the CFA50.2-040 and CRAFA50.2-2 concretes achieved similar values.

Figure 6a,b show the linear regression of the mean depths as a function of the square root of time (up to 14 days) for the Series 1 and 2 concretes, with the linear regression' slope representing the carbonation coefficient (k_{acc}) of each concrete. The k_{acc} (Table 4) was determined assuming a steady-state condition (i.e., constant carbonation coefficient) defined by Fick's first law of diffusion shown in Equation (2).

$$X_c(t) = k_{acc} \cdot (t)^{0.5} \quad (2)$$

where X_c is the carbonation depth (mm), k_{acc} is the carbonation coefficient and t is the time (day).

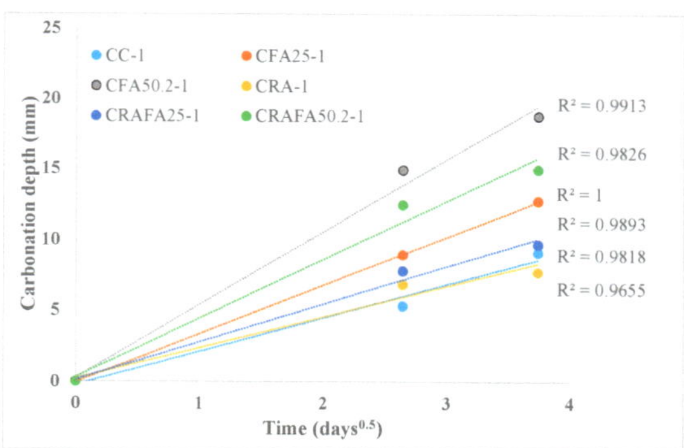

(a)

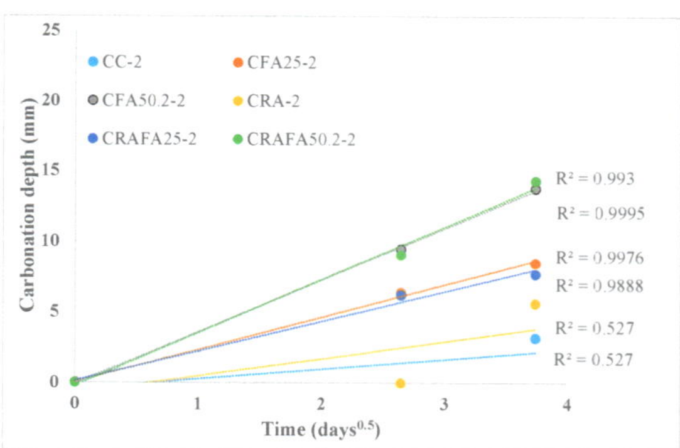

(b)

Figure 6. (**a**) Carbonation depth versus time (days$^{0.5}$) at 14 days of exposure to a 20% CO_2 concentration in Series 1. (**b**) Carbonation depth versus time (days$^{0.5}$) at 14 days of exposure to a 20% CO_2 concentration in Series 2.

In addition, the natural carbonation coefficient, k_{nat}, of each concrete was determined (Table 4) assuming the relationship between k_{acc} and k_{nat}, which has been previously defined [60,61] and is shown in Equation (3).

$$\frac{k_{acc}}{k_{nat}} = \frac{\sqrt{\varnothing_{acc}}}{\sqrt{\varnothing_{nat}}} \tag{3}$$

where $\varnothing_{acc}$ and $\varnothing_{nat}$ are CO_2 concentrations in accelerated carbonation (20%) and the natural carbonation process (400 ppm, in Barcelona), respectively.

Figure 6a and Table 4 (see Series 1) show that the CFA50.2-1 concrete achieved the highest k_{acc} value (5.6 mm/day$^{0.5}$), followed by the CFA25-1 concrete. The CC-1 concrete achieved the lowest k_{acc} (2.3 mm/day$^{0.5}$). However, the use of 50% RCA in concrete production caused a reduction of the k_{acc} coefficient compared to those employing the same binder type and natural aggregates. This effect was more evident in concretes produced with FA.

Here, the uncarbonated RCA aggregates were carbonated during their contact with CO_2, thus causing an increased carbonation resistance. Previous research [16,20] has validated the acceptable behaviour of uncarbonated RCA. Although, up to the current date, the majority of research has concluded that the concrete produced with RCA achieves lower carbonation resistance as a result of the high porosity of the aggregates [17,21,22,30]; moreover, the carbonated or uncarbonated state of RCA could modify this behaviour. The CRA-1, CRAFA25-1 and CRAFA50.2-1 concretes achieved a lower k_{acc} coefficient by 8.3%, 22.1% and 18.7% compared to the CC-1, CFA25-1, and CFA50.2-1 concretes, respectively.

In Series 2 (Figure 6b), the CC-2 concrete achieved the lowest k_{acc} value, followed by the CRA-2 concrete. The CRA-2 concrete achieved a higher carbonation coefficient, probably because the quality of the new cement paste was superior to that of RCA, thus causing an increase in the carbonation coefficient [22]. The fact that the CC-2 and CRA-2 concretes achieved an *R*-squared value of 0.53, much lower than any other concrete, is probably due to the high CO_2 concentration resulting in an acceleration of the carbonation process from 7 to 14 days. However, testing the lower CO_2 concentration would be required to verify the carbonation coefficient value. The *R*-squared value of all the other concrete mixtures was >0.80 (the majority achieving $R = 0.99$), guaranteeing that at 7 and 14 days, the relationship between carbonation depth and time was constant.

Figure 7a,b show the relationship between the compressive strength of the carbonated concrete samples from 0 to 14 days versus carbonation depth for the Series 1 and 2 concretes, respectively. The concretes with higher compressive strength achieved lower carbonation depth (Figure 7a), and this conclusion is supported by previous research [21,36,39]. On day 0 of the carbonation process, the compressive strength of the concretes depended only on the type of binder employed in concrete production and not on RCA. Nevertheless, the carbonation depth of concrete produced with RCA was lower than that produced with natural aggregates when the same type of binder was employed. Similar results were obtained by Thomas et al. [16], whose research revealed that the carbonation rate of RAC was slightly lower than for the CC concrete with respect to similar compressive strength concretes.

The results for Series 2 were similar to those for Series 1 (Figure 7b). In general, the concretes with the higher initial compressive strength achieved the lower carbonation depth. However, in Series 2, the influence of RCA on the carbonation depth was less evident. Nevertheless, due to the low water/cement value of those concretes, they achieved an adequate carbonation resistance, as previously demonstrated [36,39].

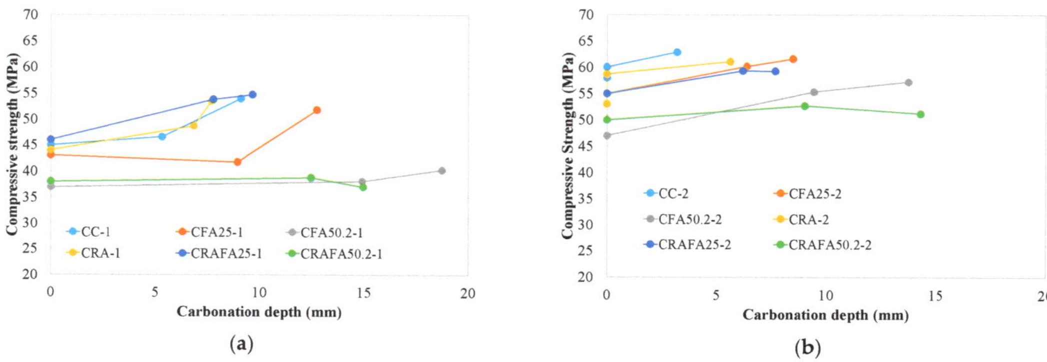

Figure 7. (**a**) Carbonation depth versus compressive strength for the carbonated samples in Series 1. (**b**) Carbonation depth versus compressive strength for the carbonated samples in Series 2.

Figure 8a,b indicate the carbonation depth versus the compressive strength at 28 day for all concretes (Series 1 and 2). The carbonation depth levels depended on the concrete' compressive strength at 28 days. All the concretes produced with RCA in Series 1 achieve similar compressive strengths and a lower carbonation depth than those made with natura aggregates. In contrast, in Series 2, the employment of RCA caused a reduction of th compressive strength, but the carbonation coefficient was similar when FA was employe in concrete production.

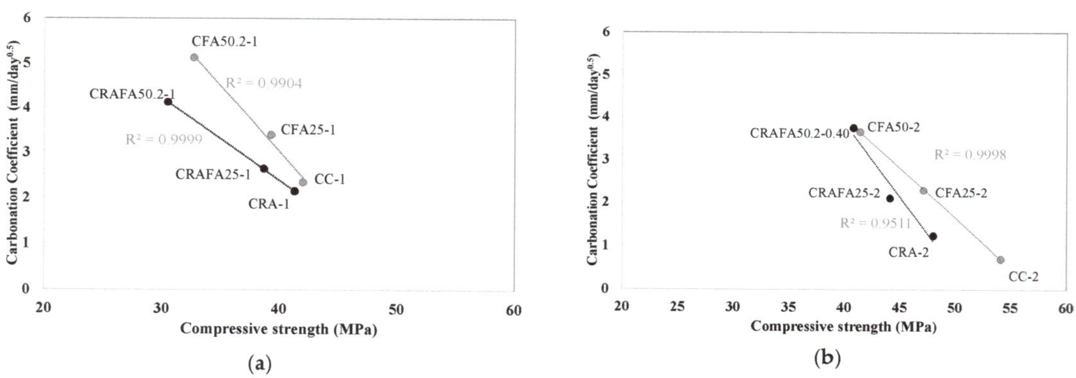

Figure 8. Compressive strength versus carbonation depth for all the concretes. (**a**) Series 1 and (**b**) Series 2. The concretes produced with RCA are represented by the black dots.

According to Silva et al. [18], to avoid corrosion of concretes exposed to enviro mental conditions of the classes XC3 and XC4 (defined in the EN 206-1 specification the maximum accelerated carbonation coefficient should be 35 mm/year$^{0.5}$ for XC3 an 50 mm/year$^{0.5}$ for XC4 for a service life of 50 years. Consequently, the results obtaine in the present paper estimated that the Series 1 CC-1, CRA-1 and CRAFA25-1 concret mixtures would be adequate for use only in an XC4 environment with the k_{acc} values o 45, 41 and 49 mm/year$^{0.5}$, respectively. The XC3 environment, being less restrictive, als allowed a smaller cover width of the concrete structure. In addition, the Series 2 CC- and CRA-2 concretes were adequate for employment in both XC3 and XC4 environment and the CFA25-2 and CRAFA25-2 concretes would also be suitable for use in an XC environment.

According to the Spanish Structural Concrete Code [55], the minimum cover in an XC3 environment for a service life of 50 and 100 years should be 20 mm and 30 mm, respectively. In the case of concrete exposed to an XC4 environment for a service life of 50 and 100 years, the minimum cover should be 25 mm and 35 mm, respectively. Therefore, in accordance with the natural carbonation coefficient, k_{nat}, determined by Equation (3) and described in Table 3, and considering the minimum covers defined by the Structural Code, the Series 1 CC-1, CRA-1 and CRAFA25-1 concretes would be adequate for both the XC3 and XC4 environments. Furthermore, the CFA25-1 concrete could also be used in an XC4 environment for both service life periods. The Series 2 CC-2, CFA25-2, CRA-2 and CRAFA25-2 concrete mixtures would be adequate for an XC3 environment, and the six analysed concretes, including those produced with 50% FA, would be adequate for an XC4 environment for a service life of 50 and 100 years. However, further tests should be carried out to verify those behaviours in an accelerated carbonation process of high CO_2 concentration.

3.2.2. Chloride Resistance

Figure 9 shows the chloride penetration resistance at 28 and 90 days for each concrete. They are categorised as low, moderate and high corrosion risk level in accordance with the value of the total passed charge.

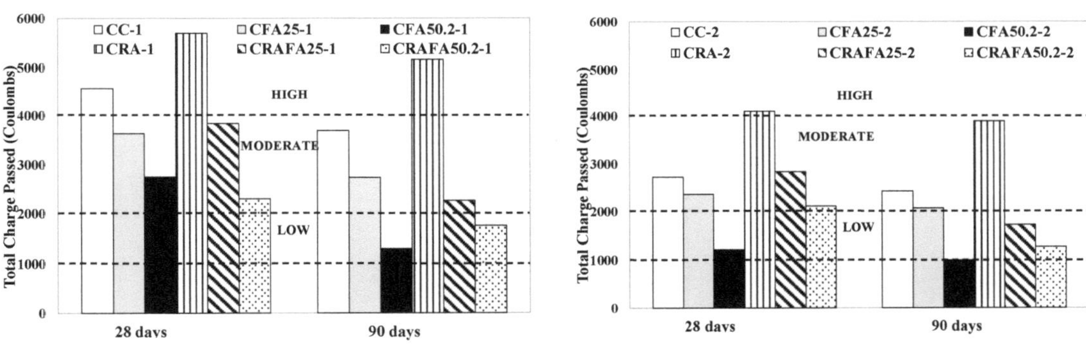

(a) (b)

Figure 9. Chloride-ion penetration of concrete mixtures and ASTM corrosion ranges. (**a**) Series 1, (**b**) Series 2 concrete.

Figure 9a describes Series 1 concretes at 28 days. The concretes CC-1 and CRA-1 were classified as high corrosion risk level. All the concretes produced with FA were classified as moderate risk level. The 50% FA concretes (CFA50.2-1 and CRAFA50.2-1) achieved the lowest values of total passed charge. The use of RCA in concrete production was found not to increase the corrosion risk level. Probably, as mentioned by Xiao et al. [20], the use of FA would have resulted in an improvement of the internal pore structure, not only decreasing the porosity of RAC but also improving the interfacial transition zone (ITZ) between the coarse RCA and the new cement paste. This achievement was also described by Sunayana and Barai [31].

At 90 days, the 50% FA concretes (CFA50.2-1 and CRAFA50.2-1) achieved the lowest values, followed by the CRAFA25-1 concrete. The concretes produced using 25% FA and RCA reduced the corrosion risk with respect to concretes produced with FA and natural aggregates. However, the CRA-1 concrete was found to still have a high corrosion risk level after 90 days of curing. While the reduction of the total passed charge from 28 to 90 days for the CFA50.2-1, CFA25-1 and CC-1 concrete mixtures was 53%, 25% and 19%, respectively, the reduction rate of CRAFA50.2-1, CRAFA25-1 and CRA-1 was 24%, 41% and 9%, respectively. CFA50.2-1 achieved the highest decrease, followed by the CRAFA25-1 and CRAFA50.2-1 concretes.

In contrast, the Series 2 concrete samples (Figure 9b) after 28 consecutive days c curing were classified as moderate corrosion risk level, with the exception of the CRA 2 and CFA50.2-2 concretes which were classified as high and low corrosion risk leve respectively. Testing samples at 90 days demonstrated that while the CC-2, CFA25- and CRA-2 concretes maintained a moderate corrosion risk level, CFA50.2-2 achieved negligible corrosion risk level and the CRAFA25-2 and CRAFA50.2-2 concretes achieve low corrosion risk levels. With respect to the decrease of the total passed charge from 2 to 90 days, while CFA50.2-2, CFA25-2 and CC-2 was 18%, 13% and 11%, respectively, th reduction rate of CRAFA50.2-2, CRAFA25-2 and CRA-2 was 40%, 39% and 5%, respectivel The concretes produced with RCA and FA achieved the highest reduction of corrosion ris

3.3. *Eco-Efficiency of Concretes*

The eco-efficiency of concretes was assessed by measuring the Bi and Ci values, takin into account the mechanical and durability performance. To evaluate the Bi values, 350 k of binder per m^3 of concrete was used for all the produced concretes. As for the Ci value three binder mixtures were considered: 100% CEM II, 75% CEM II plus 25% FA and 50% CEM I plus 50% FA.

3.3.1. Eco-Efficiency for Mechanical Performance

The obtained Bi-comp values at 28 and 90 days are illustrated in Figure 10 for all th concretes (the compressive strength values at 28 and 90 days are shown in Table 3). Th lower Bi-comp value guarantees a higher efficiency of the concrete for a required strengt value.

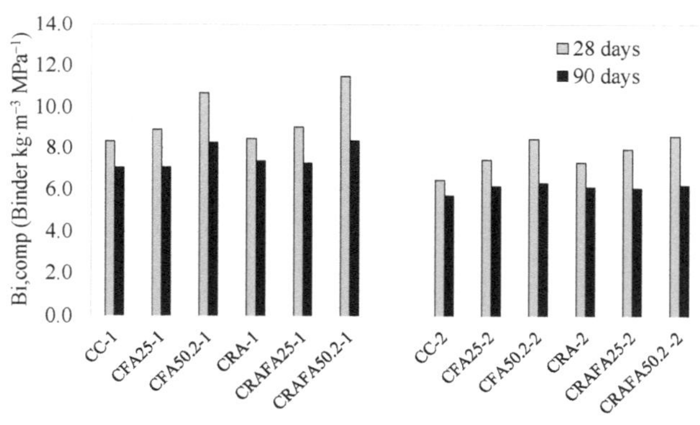

Figure 10. The binder intensity (Bi-comp) for Series 1 and Series 2 concretes at 28 days and 90 day of curing.

In Series 1, at 28 days, the CC-1 and CRA-1 concrete mixtures achieved the lowes Bi-comp values of 8.3 kg m^{-3} MPa^{-1} and 8.5 kg m^{-3} MPa^{-1}, respectively, whereas, to th contrary, the 50% FA concretes CFA50.2-1 and CRAFA50.2-1 achieved the highest values o 10.7 and 11.5 kg m^{-3} MPa^{-1}, respectively. These results are acceptable considering tha FA replaced 50% of the cement, and Damineli et al. [4] stated that ready mix concrete < 40 MPa, which utilise the most cement, have a Bi of 10–12 kg m^{-3} MPa^{-1}. It was als determined that the effect of RCA on the Bi-comp value was almost negligible in concrete produced with 0% and 25% FA, as RCA did not affect the reduction of compressiv strength. After 90 days of curing, the concretes produced with 50% FA achieved th highest reduction of Bi-comp compared to the values at 28 days. The CFA50.2-1 an CRAFA50.2-1 concretes achieved the value of 8.2 kg m^{-3} MPa^{-1}. In addition, the CFA25- and CRAFA25-1 concretes achieved a similar Bi-comp value to that of the CC or CRA concrete (7.0 kg m^{-3} MPa^{-1}).

In Series 2 concretes, the binder achieved higher effectiveness due to the use of a lower water/binder ratio than in Series 1 concretes. This also occurred in Series 1, at 28 days, where it was determined that the CC-2 and CRA-2 concretes achieved the most moderate Bi-comp values (6.5 and 7.3 kg m^{-3} MPa^{-1}, respectively). The Bi-comp value was slightly higher when RCA was employed. The concretes CFA50.2-2 and CRAFA50.2-2 achieved the highest values with 8.5 and 8.6 kg m^{-3} MPa^{-1}, respectively, because of lower compressive strength. After 90 days of curing, all the Series 2 concretes achieved a similar Bi-comp value, approximately 6 kg m^{-3} MPa^{-1}, regardless of the type of binder or aggregates. According to various researchers [4], the Bi-comp values of all concretes were considered acceptable.

As mentioned above, in order to assess the Ci-comp value it is necessary to determine the kg CO_2-eq m^{-3} for the produced concretes, different binder types being only considered. A lower Ci-comp value guarantees higher efficiency. Figure 11a,b illustrate the relationship between each concrete's Ci-comp versus the compressive strength at 28 and 90 days, respectively. The concretes produced with a higher amount of FA and the highest compressive strength [1] were found to achieve the lowest Ci-comp value, and consequently, were considered the most eco-efficient mixtures.

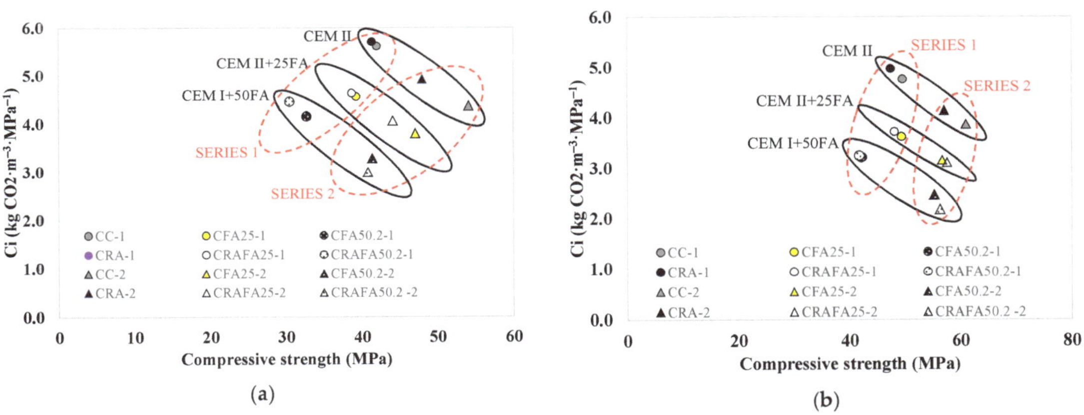

Figure 11. Ci (CO_2 intensity) versus compressive strength for the concretes at (**a**) 28 days and (**b**) 90 days.

Figure 11a shows that the Series 1 CC-1 and CRA-1 concretes achieved a Ci-comp value of 5 kg CO_2 m^{-3} MPa^{-1}. However, the CFA25-1 and CRAFA25-1 concretes achieved a lower Ci-comp value (4.1 kg CO_2 m^{-3} MPa^{-1}) as they achieved compressive strengths similar to CC-1 and CRA-1 concretes. The concrete produced employing 50% FA achieved the lowest Ci values due to the upper range of the FA's cement replacement [42]. However, they had the lowest compressive strength. The values in Series 1 were within the range of the average estimated Ci-comp for international values [4]. In Series 2, RCA negatively affected the compressive strength of concrete [9,62], as a consequence of which the CRA-2 concrete achieved the highest Ci value (4.4 kg CO_2 m^{-3} MPa^{-1}). It was determined that the 50% FA concretes were the most efficient concretes. The CRAFA50.2-2 concrete achieved the lowest value with 2.7 kg CO_2 m^{-3} MPa^{-1} followed by the CFA50.2-2 concrete with the value of 3.0 kg CO_2 m^{-3} MPa^{-1}, representing concretes that were considered very efficient [41].

At 90 days of curing (Figure 9b), the concretes belonging to the same Series achieved similar compressive strength values, with an evident improvement of Ci-comp values in FA-containing concretes [42]. All the concretes achieved a Ci-comp value lower than 5 kg CO_2 m^{-3} MPa^{-1}, which is considered to be efficient [41]. In Series 1, the CRAFA25-1 concrete achieved an adequate relationship between both the Ci-comp and the compressive strength values. In Series 2, the CRAFA50.2-2 concrete achieved a Ci-comp value of 1.9 kg CO_2 m^{-3} MPa^{-1} with a compressive strength of 56 MPa (7% lower than that of CC-2). The concretes produced with RCA, 25% and 50% of FA, achieved similar values to those wit natural aggregates.

3.3.2. Eco-Efficiency for Durability Performance

According to the durability performance, higher values of Bi and Ci define a highe eco-efficiency of the concretes. Table 5 shows the obtained results. With respect to th binder efficiency and CO_2-eq intensity in carbonation resistance in Series 1, the concret with the highest efficiency was CRA-1, followed by the CC-1, CRAFA25-1 and CFA25- concretes. In Series 2, due to the lack of positive influence of RCA, the CC-2 concrete ha the highest efficiency, followed by the CRA-2, CRAFA25-2 and CFA25-2 concretes.

Table 5. Bi and Ci values with respect to carbonation and chloride resistance values of Series 1 an Series 2 concretes. The units of Ci values are expressed as kg CO_2 m^{-3}·mm^{-1}·$year^{0.5}$.

	Bi-Carb (Binder kg m^{-3} mm^{-1} $year^{0.5}$)	Ci-Carb (kg CO_2 m^{-3} mm^{-1} $year^{0.5}$)	Ci-Chlor-28d (kg CO_2 m^{-3} C^{-1})	Ci-Chlor-90d (kg CO_2 m^{-3} C^{-1})
Series 1 (effective w/c 0.50)				
CC-1	167.85	113.13	0.052	0.064
CFA25-1	116.14	59.38	0.049	0.065
CFA50.2-1	77.81	30.19	0.049	0.105
CRA-1	183.12	123.42	0.041	0.046
CRAFA25-1	149.17	76.27	0.047	0.079
CRAFA50.2-1	95.66	37.12	0.059	0.078
Series 2 (effect. w/c 0.40)				
CC-2	564.51	380.48	0.087	0.097
CFA25-2	172.53	88.21	0.076	0.087
CFA50.2-2	108.18	41.98	0.112	0.136
CRA-2	320.52	216.038	0.058	0.061
CRAFA25-2	187.95	96.098	0.063	0.104
CRAFA50.2-2	105.49	36.77	0.058	0.096

Figure 12a,b show the Bi values for the Series 1 and Series 2 concrete mixtures, r spectively, according to chloride resistance (Bi-chlor). The employment of FA improve the binder's efficiency with respect to chloride resistance. In addition, the use of RCA i concrete did not modify the risk level of chloride attack.

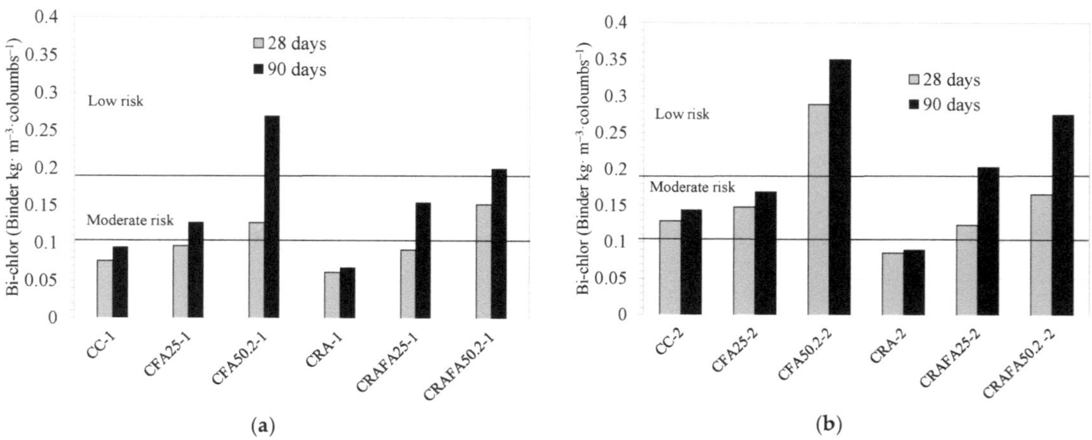

Figure 12. Bi-chlor values for the (**a**) Series 1 and (**b**) Series 2 concretes.

Table 5 shows the Ci-chlor values for the Series 1 and Series 2 concretes at 28 and 90 days. At 28 days, the Series 1 CRAFA50.2-1 concrete achieved the highest eco-efficiency, followed by CC-1, CFA25-1 and CFA50.2-1. At 90 days, although the CFA50.2-1 concrete achieved the highest efficiency, the CRAFA25-1 and CRAFA50.2-1 concretes using RCA also achieved high eco-efficiency. In Series 2, at 28 days, CRA50.2-2 achieved the highest eco-efficiency, followed by the two concretes produced with natural aggregates. At 90 days, the CRAFA25-2 and CRAFA502-2 concretes were also considered to be of high eco-efficiency.

4. Discussion

The concretes produced with the lowest Bi-comp and Ci-comp values were considered the most eco-efficient. Figure 13a,b show the Bi-comp versus Ci-comp values for Series 1 and Series 2 concretes, respectively. In Series 1 (Figure 13a), the concretes produced with 25% FA presented the highest eco-efficiency. The employment of RCA did not decrease the compressive strength of the concrete. Consequently, the CRAFA25-1 was considered the most eco-efficient material. In Series 2 (Figure 13b), at 28 days, the concretes produced with 25% FA (CFA25-2 and CRAFA25-2) were considered the most eco-efficient. However, at 90 days, the CRAFA50.2-2 concrete was the most eco-efficient, followed by CFA50.2-2, CRAFA25-2 and CFA25-2.

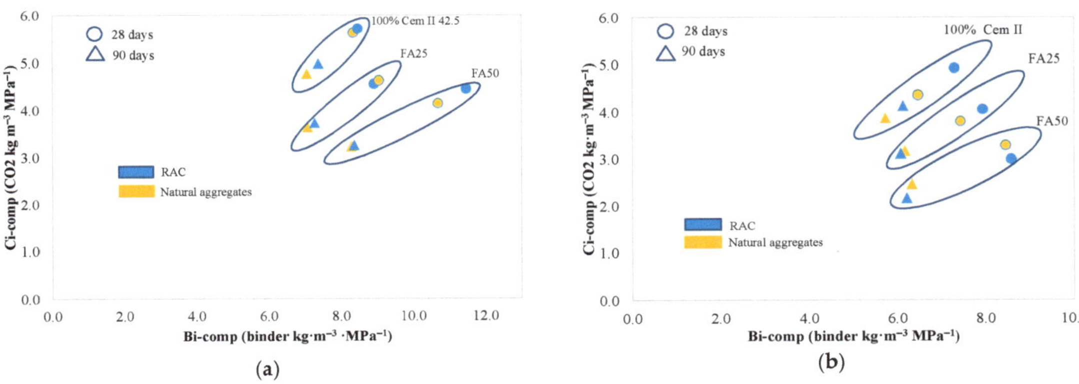

Figure 13. The Bi-comp versus Ci-comp relation for (**a**) Series 1 and (**b**) Series 2.

To assess the concrete's eco-efficiency for carbonation resistance, it was considered that the reinforced concrete should be exposed to an XC4 environment for 100-year service life. According to the Spanish Structural Concrete Code [55], the reinforced concrete requires a minimum of 35 mm of cover in that condition, which we considered in this study. Based on those characteristics, the optimum Bi-carb value was determined to be 100 $kg{\cdot}m^{-3}{\cdot}mm^{-1}{\cdot}year^{0.5}$. Similarly, the optimum Ci-carb values were determined as the maximum kg CO_2-eq emissions of each concrete made with a specific binder and are presented in Table 6. In addition, Table 6 describes the ratio between the experimental results for the Ci-carb values compared to the optimum Ci-carb value.

The Series 1 CC-1 and CRA-1 concretes achieved the highest ratio values. Consequently, they were considered to be the most durable concretes; however, their CO_2 emissions proved to be unnecessarily high. The CFA25-1 and CRAFA25-1 concretes were considered the most efficient concretes with values of 1.17 and 1.49, respectively, achieving a value >1, durable enough for this defined environment exposure. In addition, the employment of RCA increased the eco-efficiency using the same amount and type of binder. In contrast, the CFA50.2-1 and CRAFA50.2-1 concretes achieved a ratio value < 1 and these concretes, which cannot be considered durable, do not guarantee the uncarbonation of the cover, thus causing corrosion of the reinforced bars.

Table 6. The optimum Ci-carb values of each concrete and the ratios of the obtained results wit respect to the optimum results.

	CC	CFA25	CFA50.2	CRA	CRAFA25	CRAFA50.
Optimum Ci-carb	67.4	51.12	38.8	67.4	51.12	38.8
			Series 1 (effective w/b = 0.50)			
Ratio Ci-carb/optimum Ci-carb	1.68	1.16	0.78	1.83	1.49	0.96
			Series 2 (effect. w/b = 0.40)			
Ratio Ci-carb/optimum Ci-carb	5.65	1.73	1.08	3.21	1.88	1.05

All Series 2 concretes were considered durable even though they achieved a ratio valu > 1. However, the CFA50.2-2 and CRAFA50.2-2 concretes had the highest eco-efficienc since their ratio value was the closest to 1. Furthermore, due to the reduction of th water/cement ratio, the CC-2 and RCA-2 concretes achieved very high ratio values an excellent durability properties. Still, the kg CO_2-eq emissions were too high and inadequat for an XC4 environment.

In order to assess the eco-efficiency of the concretes with respect to the chloride io resistance, it was necessary to achieve a moderate corrosion risk (a maximum charge o 4000 C). Therefore, the optimum Ci-chlor (Table 7) was defined as the maximum kg CO_2-e emitted from concretes produced with a specific binder in order to achieve a moderat corrosion risk. The analysis was carried out at 28 days of curing as the values were mor critical at this age than at 90 days.

Table 7. The optimum Ci-chlor values of each concrete and the ratios of the obtained results wit respect to the optimum results.

	CC	CFA25	CFA50.2	CRA	CRAFA25	CRAFA50.2
Optimum Ci-chlor	0.059	0.045	0.034	0.059	0.045	0.034
			Series 1 (effective w/b = 0.50)			
Ratio Ci-chlor/optimum Ci-chlor, 28 days	0.88	1.10	1.45	0.70	1.04	1.74
			Series 2 (effect. w/b = 0.40)			
Ratio Ci-chlor/optimum Ci-chlor, 90 days	1.47	1.69	3.30	0.98	1.41	1.90

Table 7 shows the optimum Ci-chlor values for each concrete and the ratio of th experimentally obtained values with respect to the optimum value. At 28 days, the Serie 1 CC-1 and CRA-1 concretes were not considered durable as the ratio value was <1. Th CRAFA50.2-1 achieved the highest value, being the most durable and the most eco-efficien concrete, followed by CFA50.2-1 and CRAFA25-1.

In Series 2, similar to Series 1, the durability and eco-efficiency increased with increas ing levels of FA. The CFA50.2-2 achieved the highest value, followed by CRAFA50.2- Considering the use of RCA, the CRAFA50.2-1 and the CRAFA50.2-2 concretes had the high est eco-efficiency in Series 1 and Series 2, respectively, making their application possible i this context.

5. Conclusions

The following conclusions can be drawn based on the results presented above:

- In concretes produced with a water/cement ratio of 0.50, the use of 50% uncarbonated RCA together with Portland cement did not influence the reduction of compressive strength. In addition, when the concrete was produced with FA, the use of RCA improved the compressive strength of 41.8 MPa compared to concrete employing natural aggregates.
- In concretes produced with a water/cement ratio of 0.40, RCA negatively affected the compressive strength when Portland cement was used for concrete production. However, combining it with FA resulted in the concrete achieving similar compressive strength of 56 MPa to conventional concrete at 90 days.
- Concretes with higher compressive strength achieved higher carbonation resistance. While the use of RCA in concrete production increased the carbonation resistance, it was determined that the employment of FA reduced it. Additionally, the carbonation resistance of concretes produced with FA increased when RCA was used in their fabrication.
- The employment of FA improved the chloride ion penetration resistance of the concrete. In addition, the use of RCA together with FA improved the resistance to chloride ion penetration compared to that of the concretes produced with natural aggregates, even more evidently at 90 days of curing.
- Considering the eco-efficiency analysis:
- Based on the mechanical performance at 28 days, the concretes produced with 25% FA and 50% RCA (CRAFA25) achieved the highest eco-efficiency when an effective water/cement ratio of 0.50 and 0.40 was employed. At 90 days of curing, the CRAFA25-1 and CRAFA50.2-2 concretes achieved the highest eco-efficiency.
- Based on the conclusions of carbonation resistance, the CC and CRA concretes achieved the highest durability and the highest value of eco-efficient values. However, for reinforced concrete exposed to the XC4 environment for 100 years of service life (required 35 mm of cover), the CRAFA25-1 and CRAFA50.2-2 concretes also achieved the minimum durability required according to standards and consequently the highest eco-efficiency.
- Based on the results of minimum chloride resistance where the concretes were required to achieve a moderate corrosion risk: the CRAFA50.2-1 and CRAFA50.2-2 concretes, produced with an effective water/cement ratio of 0.50 and 0.40, respectively, had the highest eco-efficiency. The eco-efficiency of the CRAFA25-1 and CRAFA25-2 concretes was also considered valid.

It can be concluded that the use of 50% uncarbonated RCA as a replacement for natural aggregates combined with FA not only considerably improved the properties of concrete but also their eco-efficiency with respect to concretes produced with natural aggregates. The concrete produced with 50% RCA and 25% FA (CRAFA25) achieved the highest eco-efficiency in accordance with the mechanical performance and carbonation resistance of the concretes. Furthermore, the concrete produced with 50% RCA and 50% FA (CRAFA50, produced with 50% CEM I) had the highest eco-efficiency concerning chloride resistance. All the concretes produced with an effective water/cement ratio of 0.40 were considered durable with respect to carbonation and chloride resistance, independently of the type of binder and aggregates, and met the required standard specifications.

Funding: This research received no external funding.

Institutional Review Board Statement: Not applicable.

Informed Consent Statement: Not applicable.

Acknowledgments: The author thanks Fernando Álvarez and Carles Arraz for their collaboration in the experimental work and the staff of the Laboratory of Technology of Structures and Materials "Lluis Agulló" of the UPC for their support.

Conflicts of Interest: The authors declare no conflict of interest.

References

1. Schneider, M. The cement industry on the way to a low-carbon future. *Cem. Concr. Res.* **2019**, *124*, 105792. [CrossRef]
2. Meyer, C. The greening of the concrete industry. *Cem. Concr. Compos.* **2009**, *31*, 601–605. [CrossRef]
3. de Souza, A.M.; de Lima, G.E.S.; Nalon, G.H.; Lopes, M.M.S.; Júnior, A.L.D.O.; Lopes, G.J.R.; de Olivier, M.J.A.; Pedroti, L.C Ribeiro, J.C.L.; de Carvalho, J.M.F. Application of the desirability function for the development of new composite eco-efficienc indicators for concrete. *J. Build. Eng.* **2021**, *40*, 102374. [CrossRef]
4. Damineli, B.L.; Kemeid, F.M.; Aguiar, P.S.; John, V.M. Measuring the eco-efficiency of cement use. *Cem. Concr. Compos.* **2010**, 3 555–562. [CrossRef]
5. Tam, V.W.Y.; Soomro, M.; Evangelista, A.C.J. A review of recycled aggregate in concrete applications (2000–2017). *Constr. Buil Mater.* **2018**, *172*, 272–292. [CrossRef]
6. European Statistics-Eurostat. *Energy, Transport and Environment Statistics 2020 Edition*; European Statistics-Eurostat: Luxembour 2020.
7. Marinković, S.; Dragaš, J.; Ignjatović, I.; Tošić, N. Environmental assessment of green concretes for structural use. *J. Clean. Pro* **2017**, *154*, 633–649. [CrossRef]
8. de Carvalho, J.M.F.; Fontes, W.C.; de Azevedo, C.F.; Brigolini, G.J.; Schmidt, W.; Peixoto, R.A.F. Enhancing the eco-efficiency c concrete using engineered recycled mineral admixtures and recycled aggregates. *J. Clean. Prod.* **2020**, *257*, 120530. [CrossRef]
9. Kurda, R.; de Brito, J.; Silvestre, J.D. Water absorption and electrical resistivity of concrete with recycled concrete aggregates an fly ash. *Cem. Concr. Compos.* **2019**, *95*, 169–182. [CrossRef]
10. Kurda, R.; de Brito, J.; Silvestre, J.D. Influence of recycled aggregates and high contents of fly ash on concrete fresh propertie *Cem. Concr. Compos.* **2017**, *84*, 198–213. [CrossRef]
11. Chen, W.; Jin, R.; Xu, Y.; Wanatowski, D.; Li, B.; Yan, L.; Pan, Z.; Yang, Y. Adopting recycled aggregates as sustainable constructio materials: A review of the scientific literature. *Constr. Build. Mater.* **2019**, *218*, 483–496. [CrossRef]
12. Pedro, D.; de Brito, J.; Evangelista, L. Influence of the use of recycled concrete aggregates from different sources on structur concrete. *Constr. Build. Mater.* **2014**, *71*, 141–151. [CrossRef]
13. Etxeberria, M.; Mari, A.; Vázquez, E. Recycled aggregate concrete as structural material. *Mater. Struct.* **2007**, *40*, 529–54 [CrossRef]
14. Otsuki, N.; Miyazato, S.-I.; Yodsudjai, W. Influence of Recycled Aggregate on Interfacial Transition Zone, Strength, Chlorid Penetration and Carbonation of Concrete. *J. Mater. Civ. Eng.* **2003**, *15*, 443–451. [CrossRef]
15. Guo, H.; Shi, C.; Guan, X.; Zhu, J.; Ding, Y.; Ling, T.-C.; Zhang, H.; Wang, Y. Durability of recycled aggregate concrete—A revie *Cem. Concr. Compos.* **2018**, *89*, 251–259. [CrossRef]
16. Thomas, C.; Setién, J.; Polanco, J.A.; Alaejos, P.; de Juan, M.S. Durability of recycled aggregate concrete. *Constr. Build. Mater.* **201** *40*, 1054–1065. [CrossRef]
17. Adessina, A.; Ben Fraj, A.; Barthélémy, J.-F.; Chateau, C.; Garnier, D. Experimental and micromechanical investigation on th mechanical and durability properties of recycled aggregates concrete. *Cem. Concr. Res.* **2019**, *126*, 105900. [CrossRef]
18. Silva, R.; Neves, R.; de Brito, J.; Dhir, R. Carbonation behaviour of recycled aggregate concrete. *Cem. Concr. Compos.* **2015**, *6* 22–32. [CrossRef]
19. Zeng, X. Progress in the research of carbonation resistance of RAC. *Constr. Build. Mater.* **2020**, *230*, 116976. [CrossRef]
20. Xiao, J.; Lei, B.; Zhang, C. On carbonation behavior of recycled aggregate concrete. *Sci. China Ser. E Technol. Sci.* **2012**, *5* 2609–2616. [CrossRef]
21. Leemann, A.; Loser, R. Carbonation resistance of recycled aggregate concrete. *Constr. Build. Mater.* **2019**, *204*, 335–341. [CrossRe
22. Pedro, D.; De Brito, J.; Evangelista, L. Performance of concrete made with aggregates recycled from precasting industry wast Influence of the crushing process. *Mater. Struct.* **2015**, *48*, 3965–3978. [CrossRef]
23. Chandra, S.; Bendapudi, K. Contribution of Fly ash to the properties of Mortar and Concreteb. *Int. J. Earth Sci. Eng.* **2011**, 1017–1023.
24. Xu, G.; Shi, X. Characteristics and applications of fly ash as a sustainable construction material: A state-of-the-art review. *Resou Conserv. Recycl.* **2018**, *136*, 95–109. [CrossRef]
25. Lothenbach, B.; Scrivener, K.; Hooton, R.D. Supplementary cementitious materials. *Cem. Concr. Res.* **2011**, *41*, 1244–125 [CrossRef]
26. Kocak, Y.; Nas, S. The effect of using fly ash on the strength and hydration characteristics of blended cements. *Constr. Build. Mate* **2014**, *73*, 25–32. [CrossRef]
27. Malhotra, V.M. Durability of concrete incorporating high-volume of low-calcium (ASTM Class F) fly ash. *Cem. Concr. Compo* **1990**, *12*, 271–277. [CrossRef]
28. Lam, L.; Wong, Y.L.; Poon, C.S. Effect of Fly Ash and Silica Fume on Compressive Andfracture Behaviors of Concrete. *Cem. Conc Res.* **1998**, *28*, 271–283. [CrossRef]
29. Dinakar, P.; Reddy, M.K.; Sharma, M. Behaviour of self compacting concrete using Portland pozzolana cement with differer levels of fly ash. *Mater. Des.* **2013**, *46*, 609–616. [CrossRef]

0. Kurda, R.; de Brito, J.; Silvestre, J.D. Carbonation of concrete made with high amount of fly ash and recycled concrete aggregates for utilization of CO_2. *J. CO2 Util.* **2019**, *29*, 12–19. [CrossRef]
1. Sunayana, S.; Barai, S.V. Partially fly ash incorporated recycled coarse aggregate based concrete: Microstructure perspectives and critical analysis. *Constr. Build. Mater.* **2021**, *278*, 122322. [CrossRef]
2. Faella, C.; Lima, C.; Martinelli, E.; Pepe, M.; Realfonzo, R. Mechanical and durability performance of sustainable structural concretes: An experimental study. *Cem. Concr. Compos.* **2016**, *71*, 85–96. [CrossRef]
3. Kurda, R.; de Brito, J.; Silvestre, J.D. Combined influence of recycled concrete aggregates and high contents of fly ash on concrete properties. *Constr. Build. Mater.* **2017**, *157*, 554–572. [CrossRef]
4. Kim, K.; Shin, M.; Cha, S. Combined effects of recycled aggregate and fly ash towards concrete sustainability. *Constr. Build. Mater.* **2013**, *48*, 499–507. [CrossRef]
5. Kou, S.-C.; Poon, C.S. Long-term mechanical and durability properties of recycled aggregate concrete prepared with the incorporation of fly ash. *Cem. Concr. Compos.* **2013**, *37*, 12–19. [CrossRef]
6. Corinaldesi, V.; Moriconi, G. Influence of mineral additions on the performance of 100% recycled aggregate concrete. *Constr. Build. Mater.* **2009**, *23*, 2869–2876. [CrossRef]
7. Yang, H.F.; Deng, Z.H.; Li, X.L. Microscopic Mechanism Study on Carbonation Resistance of Recycled Aggregate Concrete. *Adv. Mater. Res.* **2011**, *194–196*, 1001–1006. [CrossRef]
8. Kou, S.; Poon, C.S. Enhancing the durability properties of concrete prepared with coarse recycled aggregate. *Constr. Build. Mater.* **2012**, *35*, 69–76. [CrossRef]
9. Tian, F.; Hu, W.X.; Cheng, H.M.; Sun, Y.L. Carbonation Depth of Recycled Aggregate Concrete Incorporating Fly Ash. *Adv. Mater. Res.* **2011**, *261–263*, 217–222. [CrossRef]
0. Pelisser, F.; Vieira, A.; Bernardin, A.M. Efficient self-compacting concrete with low cement consumption. *J. Clean. Prod.* **2018**, *175*, 324–332. [CrossRef]
1. de Matos, P.R.; Sakata, R.D.; Prudêncio, L.R. Eco-efficient low binder high-performance self-compacting concretes. *Constr. Build. Mater.* **2019**, *225*, 941–955. [CrossRef]
2. Randl, N.; Steiner, T.; Ofner, S.; Baumgartner, E.; Mészöly, T. Development of UHPC mixtures from an ecological point of view. *Constr. Build. Mater.* **2014**, *67*, 373–378. [CrossRef]
3. Celik, K.; Meral, C.; Gursel, A.P.; Mehta, P.K.; Horvath, A.; Monteiro, P.J.M. Mechanical properties, durability, and life-cycle assessment of self-consolidating concrete mixtures made with blended portland cements containing fly ash and limestone powder. *Cem. Concr. Compos.* **2015**, *56*, 59–72. [CrossRef]
4. Neville, A.M. *Properties of Concrete*; Pearson: London, UK, 2011.
5. Poon, C.; Shui, Z.; Lam, L.; Fok, H.; Kou, S. Influence of moisture states of natural and recycled aggregates on the slump and compressive strength of concrete. *Cem. Concr. Res.* **2004**, *34*, 31–36. [CrossRef]
6. UNE-EN 12350-2. *Testing Fresh Concrete—Part 2: Slump Test-Ensayos de Hormigón Fresco. (Parte 2: Ensayo de Asentamiento.)*; AENOR: Madrid, Spain, 2020.
7. UNE-EN 12390-2. *Testing Hardened Concrete–Part 2: Making and Curing Specimens for Strength Tests. (Ensayos de Hormigón Endurecido-Parte 2. Fabricación y Curado de Probetas para Ensayos de Resistencia)*; AENOR: Madrid, Spain, 2020.
8. UNE-EN-12390-3. *Testing Hardened Concrete—Part 3: Compressive Strength of Test Specimens. (Ensayos de Hormigón Endurecido. Parte 3. Determinación de la Resistencia a Compresión de Probetas)*; AENOR: Madrid, Spain, 2020.
9. GB/T50082-2009. *Standard for Test Method of Long Term Performance and Durability of Ordinary Concrete*; China Architecture and Building Press: Beijing, China, 2009.
0. Shah, V.; Bishnoi, S. Carbonation resistance of cements containing supplementary cementitious materials and its relation to various parameters of concrete. *Constr. Build. Mater.* **2018**, *178*, 219–232. [CrossRef]
1. Castellote, M.; Fernandez, L.; Andrade, C.; Alonso, C. Chemical changes and phase analysis of OPC pastes carbonated at different CO_2 concentrations. *Mater. Struct.* **2009**, *42*, 515–525. [CrossRef]
2. Hyvert, N.; Sellier, A.; Duprat, F.; Rougeau, P.; Francisco, P. Dependency of C–S–H carbonation rate on CO_2 pressure to explain transition from accelerated tests to natural carbonation. *Cem. Concr. Res.* **2010**, *40*, 1582–1589. [CrossRef]
3. ASTM C1202. *Standard Test Method for Electrical Indication of Concrete's Ability to Resist Chloride Ion Penetration*; American Society for Testing Materials: West Conshohocken, PA, USA, 2019.
4. Kubissa, W.; Jaskulski, R.; Chen, J.; Ng, P.-L.; Godlewska, W.; Reiterman, P. Evaluation of Ecological Concrete Using Multi-criteria Ecological Index and Performance Index Approach. *Arch. Civ. Eng. Environ.* **2019**, *12*, 97–107. [CrossRef]
5. Ministry of Presindency; Ministerio de la Presidencia. Codigo Estructural (Structural code-Concrete), Boletín Oficial del Estado, Madrid, España. 2021. Available online: https://www.boe.es/eli/es/rd/2021/06/29/470 (accessed on 3 December 2021).
6. Nath, P.; Sarker, P. Effect of Fly Ash on the Durability Properties of High Strength Concrete. *Procedia Eng.* **2011**, *14*, 1149–1156. [CrossRef]
7. Hemalatha, T.; Ramaswamy, A. A review on fly ash characteristics–Towards promoting high volume utilization in developing sustainable concrete. *J. Clean. Prod.* **2017**, *147*, 546–559. [CrossRef]
8. Leemann, A.; Nygaard, P.; Kaufmann, J.; Loser, R. Relation between carbonation resistance, mix design and exposure of mortar and concrete. *Cem. Concr. Compos.* **2015**, *62*, 33–43. [CrossRef]

59. Khunthongkeaw, J.; Tangtermsirikul, S.; Leelawat, T. A study on carbonation depth prediction for fly ash concrete. *Constr. Buil. Mater.* **2006**, *20*, 744–753. [CrossRef]
60. Parrott, L. *A Review of Carbonation in Reinforced Concrete; British Cement Association; BRE/BCA Report C/1*; 1987; Available onlin https://trid.trb.org/view.aspx?id=280156 (accessed on 2 December 2021).
61. Heede, P.V.D.; De Belie, N. A service life based global warming potential for high-volume fly ash concrete exposed to carbonatio *Constr. Build. Mater.* **2014**, *55*, 183–193. [CrossRef]
62. Etxeberria, M.; Vázquez, E.; Mari, A.; Barra, M. Influence of amount of recycled coarse aggregates and production process o properties of recycled aggregate concrete. *Cem. Concr. Res.* **2007**, *37*, 735–742. [CrossRef]

materials

MDPI

rticle

nfluence of the Mix Proportion and Aggregate Features on the ²erformance of Eco-Efficient Fine Recycled Concrete Aggregate Mixtures

iego Jesus De Souza [1,*], Mayra T. de Grazia [1], Hian F. Macedo [1], Leandro F. M. Sanchez [1], abriella P. de Andrade [1], Olga Naboka [2], Gholamreza Fathifazl [2] and Pierre-Claver Nkinamubanzi [2]

1 Department of Civil Engineering, Faculty of Engineering, University of Ottawa, Ottawa, ON K1N 6N5, Canada; mtagl010@uottawa.ca (M.T.d.G.); hdefr095@uottawa.ca (H.F.M.); leandro.sanchez@uottawa.ca (L.F.M.S.); puentedg@gmail.com (G.P.d.A.)

2 National Research Council Canada, Ottawa, ON K1V 1J8, Canada; olga.naboka@nrc-cnrc.gc.ca (O.N.); reza.fathifazl@nrc-cnrc.gc.ca (G.F.); pierre-claver.nkinamubanzi@nrc-cnrc.gc.ca (P.-C.N.)

* Correspondence: dsouz025@uottawa.ca

Abstract: Most of the previous research on recycled concrete aggregates (RCA) has focused on coarse RCA (CRCA), while much less has been accomplished on the use of fine RCA particles (FRCA). Furthermore, most RCA research disregards its unique microstructure, and thus the inferior performance of concrete incorporating RCA is often reported in the fresh and hardened states. To improve the overall behaviour of RCA concrete advanced mix design techniques such as equivalent volume (EV) or particle packing models (PPMs) may be used. However, the efficiency of these procedures to proportion eco-efficient FRCA concrete still requires further investigation. This work evaluates the overall fresh (i.e., slump and rheological characterization) and hardened states (i.e., non-destructive tests, compressive strength and microscopy) performance of sustainable FRCA mixtures proportioned through distinct techniques (i.e., direct replacement, EV and PPMs) and incorporating different types of aggregates (i.e., natural and manufactured sand) and manufacturing processes (i.e., crusher fines and fully ground). Results demonstrate that the aggregate type and crushing process may influence the FRCA particles' features. Yet, the use of advanced mix design techniques, particularly PPMs, may provide FRCA mixes with quite suitable performance in the fresh (i.e., 49% lower yield stress) and hardened states (i.e., 53% higher compressive strength) along with a low carbon footprint.

Keywords: fine recycled concrete aggregate; residual cement paste (RCP); mix design technique of concrete; rheological behaviour; hardened property of concrete

itation: De Souza, D.J.; de Grazia, .T.; Macedo, H.F.; Sanchez, L.F.M.; e Andrade, G.P.; Naboka, O.; athifazl, G.; Nkinamubanzi, P.-C. ıfluence of the Mix Proportion and ggregate Features on the erformance of Eco-Efficient Fine ecycled Concrete Aggregate ixtures. *Materials* **2022**, *15*, 1355. ttps://doi.org/10.3390/ a15041355

cademic Editors: Miguel Bravo, Rui asco Silva and António P.C. Duarte

eceived: 8 December 2021
ccepted: 19 January 2022
ublished: 12 February 2022

1. Introduction

Concrete is a widely used material for critical infrastructure worldwide; over 10 billion tons of concrete are produced annually in the modern industrial society [1]. However, the construction industry faces crucial challenges in finding cost-effective strategies to reduce the carbon footprint and embodied energy associated with producing and using concrete and its ingredients [2–4]. Amongst the viable solutions, the re-use of construction waste, the so-called recycled concrete aggregates (RCA), has been receiving increased attention in the civil industry [5–10].

The intrinsic difference between RCA and natural aggregate (NA) is the fact that CRCA is a two-phase material composed of the original virgin aggregate (OVA) and residual mortar (RM). Likewise, fine recycled concrete aggregates (FRCA) comprise the original virgin aggregate (OVA) and residual cement paste (RCP). The properties of RCA concrete made of CRCA have been extensively investigated [3,10–17]; therefore, numerous documents such as ACI-555R [18], BS EN 1744-6:2006 and BS EN 933-11:2009 along with

RILEM technical reports, such as [19,20], provide guidelines and criteria on how to de with and use CRCA for proportioning RCA concrete. Yet, although RCA's use in concret has grown over the past years, concerns related to its mechanical properties, durabilit and long-term behaviour prevent its widespread use for structural purposes [6,9,14,21,22 especially due to the lack of efficient quality control and proportioning procedures. On th other hand, FRCA is considered a very low-quality material due to its high amount of RC adhered to the fine particles [15,21]; thus, much less has been developed and accomplishe with the use of FRCA. The amount of RCP varies depending upon the type and quality c the OVA (i.e., lithotype, texture, shape, etc.), RCP inner quality (i.e., mechanical properties crusher type and manufacturing process, etc., which leads to substantial variability to th final product [13,23,24]. The few existing studies suggest that FRCA negatively influence the fresh state, mechanical properties and durability of recycled concrete [22]. Otherwise, has been demonstrated that FRCA replacement ratios lower than 30% do not negativel impact the mechanical properties of recycled concrete [23].

Studies have demonstrated that mortar mixtures made of partial or full direct r placement of NA by FRCA displayed more somewhat similar fresh state behaviour tha conventional mortar, once both aggregates (either NA or FRCA) were in saturated surfac dry conditions [24]. Conversely, it has been verified that self-consolidating recycled concret mixtures incorporating 50% and 100% of FRCA demonstrated less self-compacting abilit than conventional mixtures [25]. Moreover, results suggest that recycled mixtures bearin 100% of FRCA may yield 35% higher drying shrinkage [26] along with 30% and 50% lowe compressive strength [27] and modulus of elasticity [28,29], respectively, than convention companion mixtures. In all the above studies, the low performances of the recycled mi tures were deemed to be linked to some of the FRCA features, such as their higher porosit absorption and the presence of important amounts of residual cement paste adhered t the particles.

In this context, advanced mix design techniques have been developed to optimiz recycled concrete mixtures, allowing more efficient re-use of RCA in concrete and enablin better performances in the fresh and hardened states. The literature suggests that th key factor for an efficient proportioning of recycled concrete mixtures bearing CRCA is t account for their unique microstructure (i.e., original virgin aggregate—OVA and residu mortar—RM), while the mix design and proportioning techniques such as the equivalen mortar volume (EMV) [30] and equivalent volume (EV) [31] methods were established wit this concept and showed promising results. Likewise, particle packing models (PPMs) hav demonstrated to successfully proportion recycled mixtures bearing CRCA with suitabl fresh- and hardened-state properties [32]. Nevertheless, to the authors' best knowledge, th literature has not largely explored mix design techniques accounting for the unique FRC microstructure. Thus, this work aims to appraise the influence of the FRCA features an mix-proportion technique on the fresh and hardened performance of sustainable recycle concrete mixtures designed for structural purposes.

2. Background in Mix Proportioning Techniques

2.1. Proportion RCA Concrete

2.1.1. Direct Replacement Methods

One of the earliest attempts to use a mix proportion of RCA concrete was performe through direct replacement methods (DRMs), which can be conducted by weight or volum This approach treats RCA as homogeneous material and partially replaces a certain amoun of natural aggregates accounting for their distinct microstructure and or the presence o RM [33,34]. Moreover, the direct replacement of the NA by RCA (i.e., OVA + RM) withou accounting for the RM makes the total amount of coarse aggregate (i.e., OVA and NA lower than the NA in the companion conventional mixture. Thus, DRM techniques ofte result in recycled concrete mixtures with similar fresh-state behaviour. Still, it is quit inferior hardened-state performance compared to a companion CC mix, especially whe moderate to high amounts of RCA is used [35]. Nevertheless, recent studies have show that suitable fresh- and hardened-state performances of RCA concrete might be achieved using DRM when adjustments are made to the water-to-binder ratio along with high amounts of binder materials (e.g., >400 kg/m^3) and chemical admixtures [36,37]. In other words, this approach offsets the use of RCA since PC is by far the concrete ingredient presenting the most negative impact on sustainability [38]. Hence, "equivalent volume methods" were proposed in the literature, and amongst those, the equivalent volume (EV) method stands out as a promising technique.

2.1.2. Equivalent Volume Methods

The equivalent mortar volume (EMV) method was initially proposed by Fathifazl et al. 2009a and treated RCA as a multi-phase material composed of RM and OVA. Thus, the volume of cementitious materials added to the recycled mix is reduced according to the amount of existing RM content adhered to the RCA particles. It has been suggested that EMV can reach similar or even somewhat higher compressive and tensile splitting strengths than conventional concrete [39,40]. Besides the comparable compressive strength, EMV-proportioned recycled mixtures often yield a similar modulus of elasticity than companion CC (i.e., varying, on average, from −5% to 11%) [41,42]. The enhancement in the modulus of elasticity of EMV-proportioned RCA concrete is due to the equivalent volumetric ratio of natural aggregate (i.e., either OVA or NA) between CC and the recycled mixture, whereas in DRM-proportioned RCA concrete, the overall volume of natural aggregate is decreased in the mixture. Otherwise, EMV-proportioned mixtures often face challenges in the fresh state (e.g., high consistency, low flowability and unsuitable rheological profile) due to the lower availability of fresh mortar in the mixture [43]. Moreover, the RM was found to behave as an aggregate in the fresh state while acting as a mortar in the hardened state, negatively affecting the flowability of recycled mixtures [44,45].

In order to address the issues often faced in the fresh state by EMV-proportioned RCA mixes, especially in eco-efficient mixtures with a reduced binder content, a modification to the preliminary method was proposed by Hayles et al. 2018. The EMV-mod method adds a supplementary factor to the previous strategy to optimize the "cement-to-sand mass ratio" of the RM and thus allows the selection of the amount of cement to be accounted for in the new RCA mix. Hence, EMV-mod-proportioned mixtures show better improvement in fresh-state performance (i.e., lower consistency and higher flowability) than EMV mixes. However, this approach was verified to present an important drawback since the binder content is typically increased for a given targeted strength compared to conventional and EMV-designed mixes [44].

Recently, a new mix design procedure, the so-called equivalent volume (EV) method, was developed by Ahimoghadam et al. 2020 based on the main concepts of the previous EMV and EMV-mod methods. The EV was established to solve the issues faced by EMV and EMV-mod mix-proportioned mixtures in the fresh state, along with providing recycled mixtures with an eco-friendly character and suitable performance in the hardened state [31]. The novelty of this method is that the RCA-proportioned mix is based on a companion conventional concrete displaying the same volumetric amount of cement paste and aggregates (fine and coarse), while in the EMV (and EMV-mod), the same amount of coarse aggregates and mortar was considered for calculations. Furthermore, the RM is considered in this method as the summation of residual paste and residual fine aggregates. Hence, the total volumetric cement paste content in an EV-proportioned mixture is defined as the summation of residual and fresh-paste volumes. As opposed to EMV and EMV-mod methods, EV-proportioned mixes often yielded suitable fresh-state behaviour and cement efficiency (PC contents of about 300 kg/m^3) due to the significant increase in fine aggregates and decrease in coarse particles in the system [31,41].

2.2. Particle Packing Models (PPM)

Particle packing models (PPMs) are advanced mix design techniques used to improve the performance of concrete mixtures in fresh and hardened states. PPMs are divided into discrete and continuous models; either discrete or continuous, PPMs aim to optimize granular systems' particle size distribution (PSD), reducing the material's porosity and consequently increasing the system's packing density [46]. The discrete approach refers to the packing of multimodal distributions containing two or more discrete size classes of particles [47]. In contrast, the continuous model is based on the basic assumption that all possible aggregates are present within the particle distribution of the system [48,49]. Typically, PPM-mix-proportioned mixes present lower porosity and amounts of the binder when compared to conventionally designed concrete.

The first continuous PPM was established by Fuller in 1907, and it is based on a power curve that accounts for two variables: the distribution factor (q) and the largest particle size (D) [50]. Over the years, several authors have developed different models; yet, the newest and most recognized continuous PPM is the Alfred model (also known as modified Andreasen), which Funk and Dinger initially created in 1980. This model calculates the optimum PSD based on a coefficient of distribution (q) along with the largest (D_L) and smallest (D_S) particle sizes (i.e., diameter) present in the system (Equation (1)). The coefficient of distribution (q) is adopted based on the fresh-state requirements of the mix; values ranging from 0.20 to 0.23 are often selected for self-consolidating mixtures, whereas distribution factors of 0.26–0.28 are normally targeted for vibrated and or pumped concrete. It is worth noting that 0.37 is the q value that yields the highest packing density, hence the lowest porosity to granular systems [51].

$$CPFT = 100 \times \left(\frac{D_P{}^q - D_S{}^q}{D_L{}^q - D_S{}^q} \right) \quad (1)$$

where: CPFT is the cumulative (volume) percent finer than D_P, and D_P is the particle size.

PPMs have been widely and successfully used to improve performance, especially in the hardened state of eco-efficient [32,52], conventional [52] and high-performance concrete [53,54]. Nevertheless, very little research has been conducted on the use of continuous PPMs to optimize FRCA concrete, especially to proportion eco-efficient recycled mixtures with full replacement of natural aggregates by FRCA [33,46].

3. Scope of Work

As previously described, there is currently limited research on the influence of the FRCA particle features (i.e., amount of residual cement paste, FRCA type: natural vs. manufactured and manufacturing process) and mix design techniques (i.e., conventional vs. advanced) on the fresh- and hardened-state performance of eco-efficient recycled concrete mixtures made of 100% FRCA. This work is divided into two main phases: (1) production and characterization of FRCA materials and (2) appraisal of the fresh- and hardened-state properties of eco-efficient FRCA concrete mixtures. The first phase evaluates the impact of the FRCA type (i.e., originally fabricated with natural vs. manufactured sand) and manufacturing process (i.e., crushing process) on the physical properties of the granular material. In the second phase, three mix design techniques (i.e., DRM, EV and PPM) are selected to proportion eco-efficient FRCA mixtures made of natural or manufactured sand. Finally, evaluations on the fresh- and hardened-state properties are conducted, and comparisons with a companion conventional concrete (0% FRCA) are performed. It is worth noting that the term eco-efficient used in the context of this work refers to concrete mixtures displaying equal or lower binder contents than their companion conventional mixtures. In other words, there is no need to use a supplementary amount of binder to obtain a performance in the recycled mixture as is conventionally adopted in RCA research.

4. Materials and Methods

The experimental program is divided into two main streams as described in Section 3 (i.e., FRCA characterization and appraisal of mix design techniques). However, the FRCA materials' characterization (i.e., production and characterization of FRCA materials) is displayed in this section since they are initially required to properly create a mix-proportion of recycled concrete mixtures for analyses in the fresh and hardened states.

4.1. FRCA Production Process

The FRCA used in this work was manufactured in the laboratory; it is derived from crushed companion conventional concrete specimens, mix-proportioned through the ACI method (i.e., absolute volume method) and displaying a design compressive strength (i.e., 28-day value) of 35 MPa. Two families of FRCA were fabricated, one containing a natural siliceous sand (ACI-NS) and the other incorporating a limestone-manufactured sand (ACI-MS). A coarse limestone aggregate with a nominal maximum size of 19 mm, an oven-dry specific gravity of 2.78 g/cm^3 and 0.75% absorption was used to produce the conventional concrete mixtures. Neither chemical admixture nor mineral admixtures were employed in the conventional concrete mixtures. Table 1 summarizes the conventional concrete mixture proportions used to fabricate the FRCA material.

Table 1. Companion conventional concrete mix design proportions.

ACI-NS	**Type**	**Mass (kg/m^3)**
Cement	GU	370
Fine aggregate	Natural sand	898
Coarse aggregate	Limestone	1032
Water	-	174
ACI-MS		Mass (kg/m^3)
Cement	GU	370
Fine aggregate (MS)	Manufactured sand	934
Coarse aggregate	Limestone	1032
Water	-	174

Two hundred conventional concrete cylinders (i.e., 100 mm × 200 mm in size) were cast from each of the concrete mixtures used in this research incorporating NA and MS. The specimens were kept in the moulds at laboratory temperature, demoulded after 24 h and moist-cured over 28 days. In order to appraise the influence of the crushing process on the FRCAs' inner quality, two manufacturing methods (i.e., crusher fines—CF and fully ground—FG) were employed to produce the recycled granular material. Figure 1 presents the CF method, where coarse and fine RCA are produced simultaneously. The CF process presents two series of crushing and three stages. First, concrete specimens were crushed into a 19 mm maximum gap opening in the jaw crusher. Then, the produced material followed a second crushing stage with a 5 mm gap opening. Finally, the material was divided into FRCA and CRCA; only the fine fraction (i.e., 150 µm–5 mm) was sieved and used in this work as per the Canadian Standards Association (CSA, [55]).

The second method (FG—full process Figure 1) only produces FRCA by multiples (and continuous) series of crushing (5 mm gap opening in the jaw crusher) and sieving as per CSA [55]. In both manufacturing methods, the particle size distribution of the final FRCA materials ranged from 150 µm to 5 mm. Furthermore, regardless of the crushing process, two FRCA mixtures were produced in this work: FRCA from natural (FRCA NS) and manufactured (FRCA MS) sands. Both were fabricated according to the two methods described above (i.e., CF and FG).

Figure 1. Summary of crusher fines (CF) and fully ground (FG) method of production.

4.2. *Materials Characterization*

4.2.1. FRCA

Residual cement paste (RCP)

The RCP was measured in this work on all FRCA materials through the solubl silica sub-procedure according to ASTM C1084-15 [56] and C114-18 [57]. The procedur involves the selective extraction of silica from Portland cement by cold dilute hydrochlori acid with subsequent gravimetric analysis of soluble silica. The procedure assumes tha only silica from Portland cement is soluble in cold dilute hydrochloric acid, while silic from aggregates is not soluble. Samples of 2.5 g of all FRCA materials were analyzed a per the above-mentioned procedures. They were then weighed to determine the mas difference before and after the test, representing the amount of residual cement paste i the FRCA material (Table 2). Overall, both the distinct raw materials (i.e., NS or MS) an manufacturing processes (i.e., CF or FG) influenced the amount of RCP adhered to th FRCA particles. Yet, the manufacturing process seems to have a greater impact on the RC adhered to the FRCA particles. The RCP values for the various mixtures ranged from 11.4 (FRCA-MS-FG) to 16.8% (FRCA-MS-CF).

Table 2. Properties of FRCA and natural aggregates.

Physical Property	FRCA-NS-CF	FRCA-NS-FG	NS	FRCA-MS-CF	FRCA-MS-FG	MS
RCP content (wt.%)	15.5	11.5	-	16.8	11.4	-
Saturated surface-dry specific gravity (kg/L)	2.47	2.56	2.70	2.51	2.58	2.76
Oven-dry specific gravity (kg/L)	2.29	2.41	2.67	2.33	2.44	2.74
Water absorption (%)	7.87	6.38	0.86	7.76	6.16	0.65

Potential of RCP for further hydration

Isothermal heat conduction calorimetry (IHCC) of the FRCA was determined in a isothermal calorimeter G-563-62-L0071/AD by Controls Group. The test was performe to determine whether the residual cement paste (RCP) adhered to the FRCA particle might have some potential for further hydration in concrete. The test was conducted o original FRCA particles and cement powder for comparison purposes. It is worth notin that IHCC was performed on freshly prepared samples to avoid the possible reaction o unhydrated cement with moisture from the air. Ten grams of each FRCA material wer selected for testing; 3.5 g of double-deionized water was added to the samples to evaluat the cementing capability of the RCP encountered in the FRCA particles. After mixing fo 60 s, the samples were placed in the calorimeter and tested. It is important to note that th plastic scintillation vials were filled with 26.49 ± 0.2 g of glass beads as a reference. All the IHCC results are presented on the same basis (per gram of cement) since the amount of cement in each sample is variable. The IHCC results obtained in this work are presented in the analysis and discussion section.

Specific gravity and water absorption

The specific gravity and water absorption capacity of all FRCA materials were evaluated according to the method proposed by [58]. The conventional procedure from CSA [55] was found to not yield reliable results for FRCA due to its cohesiveness and binding behaviour. As per [58], the samples were first saturated in 0.1% of sodium hexametaphosphate solution, a clay dispersant commonly used in soil analysis for clay suspensions. Then, for a given sample, a concentration of 1 g/L of the dispersant was used to prevent the agglomeration of the particles. Then, the procedure was performed as per CSA [55], except that the samples were left for a further 24 h in a pycnometer to have entrapped air bubbles removed, as proposed by [58]. Table 2 summarizes the FRCAs' physical characterization results and the properties of natural and manufactured fine aggregates [55].

Particle size distribution (PSD)

Figure 2a shows the particle size distribution (PSD) for the FRCA materials derived from natural sand (i.e., NS-CF and NS-FG), while Figure 2b illustrates the results of the FRCA materials derived from manufactured sand (i.e., MS-CF and MS-FG). The square-dotted lines represent the particle size distribution thresholds of CSA [55]. Overall, the PSD of the FRCA can be significantly affected by the crushing process adopted. For instance, FRCA-NS-CF (crusher fines—Figure 2a) predominantly comprises coarser particles, yielding a fineness modulus of 3.27, whereas FRCA-NS-FG (fully ground—Figure 2a) displayed a higher amount of fine particles and thus a fineness modulus of 2.53. Moreover, FRCA-NS-FG presented a PSD similar to the natural sand and fineness modulus. The same trend was observed for the FRCA-MS from crusher fines and fully ground types (Figure 2b). FRCA-MS-CF presented a fineness modulus of 3.17, while FRCA MS-FG and the original manufactured sand displayed fineness moduli values of 2.70 and 2.85, respectively.

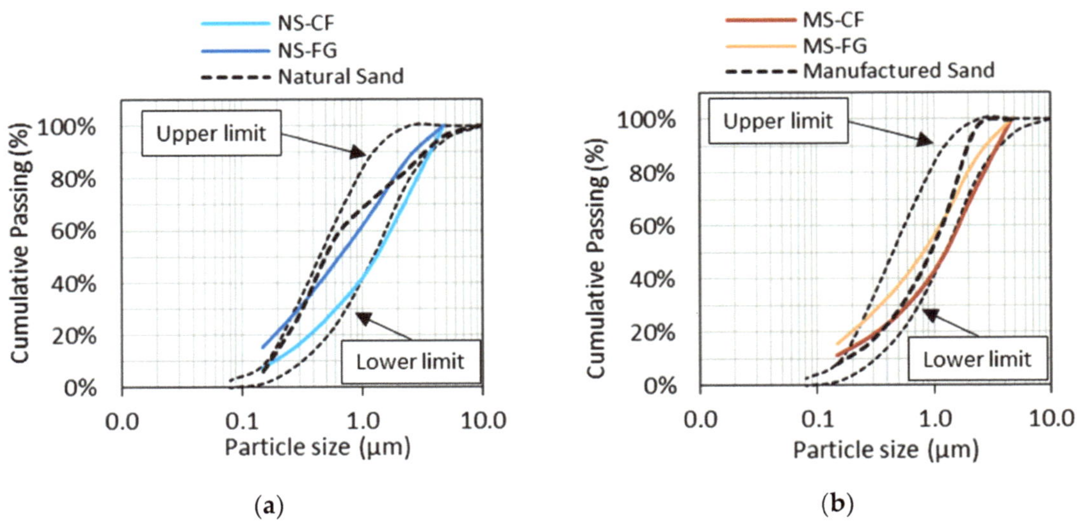

Figure 2. Particle size distribution curves for FRCA derived from natural (**a**) and manufactured sand (**b**) companion conventional concrete.

4.2.2. Binder and Filler

A general-use (equivalent to ASTM Type I) Portland cement was used in all mixes. I addition, a limestone filler with a smaller PSD than PC (Figure 3), known as performanc filler, was employed in the mixtures proportioned through PPM methods. The chemica compositions of the PC and filler, obtained via X-ray fluorescence in a Rigaku Supermini20 WDXRF Spectrometer, and the PC mineralogical phases calculated through Bogue's equa tions [59] are presented in Table 3. Moreover, the specific gravity and specific surface are analyses were performed through a gas pycnometer (AccuPyc II by Micromeritics) an BET (Horiba SA-9600 by Meritics), respectively, and the results are summarized in Figure

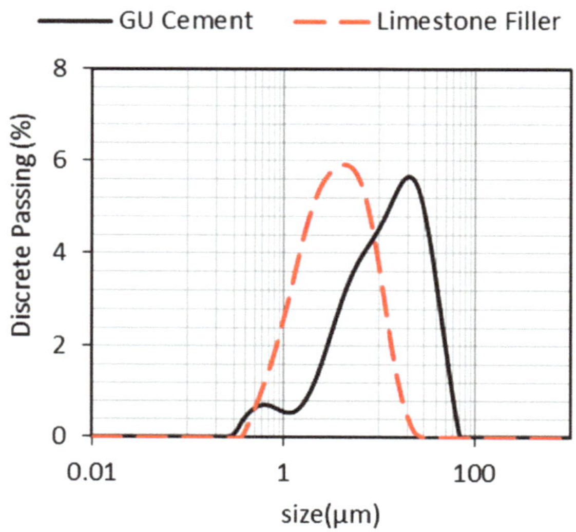

Figure 3. PSD of PC and limestone filler used in the PPM FRCA mixtures.

Table 3. Physicochemical characterization of GU cement and limestone filler.

Chemical Composition	Filler	Cement	Mineralogical Phases		
CaO	52.28	61.4	C_3S	52.9	
SiO_2	4.09	18.9	C_2S	14.3	
Al_2O_3	0.23	4.9	C_3A	6.7	
Fe_2O_3	0.15	3.7	C_4AF	11.3	
MgO	1.69	2.3	Physical Properties	Cement	Filler
K_2O	0.07	0.9			
Na_2O	0.04	0.3	Specific Gravity (g/cm^3)	3.03	1.00
SO_3	-	5.3			
LOI	41.3	2.1	SSA (m^2/g)	2.60	3.70

4.3. Concrete Manufacturing

4.3.1. Mix design Approaches (DRM, EV and PPM)

A total of three distinct concrete mixtures (i.e., DRM, EV and PPM) with 100% replacement of fine NA by FRCA were designed. Moreover, two control mixes (e.g ACI method—0% FRCA) containing NS or MS were also produced for compariso purposes. Concrete mixtures proportioned using the DRM method were designe according to the ACI protocol (i.e., absolute volume method), purposely treating FRC as presenting the same properties of fine NA; the latter is a common practice adopted i the construction industry.

The EV method [31] was adapted to design recycled mixtures made of 100% FRCA and accounting for the RCP content attached to the fine aggregate particles. However, the main concept of the EV technique was kept the same as follows: the proportioned FRCA mixture is based on a conventional mixture, having the same amount (in volume) of cement paste and aggregates. In other words, the total amount of cement paste in the EV mix is considered as the summation of the RCP plus the fresh paste added to the mix. In contrast, the total volume of aggregates is the volume of original virgin aggregates (OVA) plus the volume of natural aggregates (NA). In addition to the above methods, a novel approach incorporating the EV concept (i.e., accounting for the RCP of the FRCA) into a continuous PPM technique (i.e., Alfred's model) was also adopted in this research. A limestone filler was used in the PPM-mix-designed mixtures to lessen the overall interparticle porosity and binder content of the system. The distribution factor (q) selected in this work was 0.28 to ensure a suitable fresh-state behaviour (i.e., targeting a vibrated and/or pumped concrete) with a minimum requirement of chemical admixtures.

4.3.2. Concrete Mixtures Containing FRCA

The DRM-, EV- and PPM-designed recycled mixtures were proportioned with 100% FRCA replacement from the four FRCA sources previously mentioned. The target 28-day compressive strength was selected as 35 MPa as per CSA [60] for the C1 exposure class (i.e., structural concrete exposed to environmental conditions displaying freeze–thaw cycles and de-icing salts). Therefore, an air-entraining agent (AEA) was used in all mixtures to incorporate an air content of 7–8% required for the C1 exposure class as per CSA [55]. A water-to-cement ratio (w/c) of 0.35 was adopted for all FRCA mixes since preliminary trial batches indicated the need to reduce the w/c of the FRCA mixtures to yield close compressive strength results (i.e., 35 MPa) to companion conventional concrete mixes with w/c of 0.47. Moreover, the consistency of the mixture (i.e., slump values) was aimed to lie within 100 ± 20 mm. A summary of all mixtures is presented in Table 4. It worth noting that the FRCA material was added to the mixer in oven-dry conditions since previous studies [61,62] showed that pre-saturating FRCA would negatively affect the fresh- and hardened-state properties of the material.

Table 4. FRCA and control concrete mix design proportions.

Mixture	Cement [b] (kg/m^3)	FRCA (kg/m^3)	Natural FA (kg/m^3)	Natural CA (kg/m^3)	Filler (kg/m^3)	Water (kg/m^3)	Mid-Range + HRWR [a] (%)
ACI-NS	370	-	685	1032	-	174	-
ACI-MS	370	-	704	1032	-	174	-
DRM-NS-CF	497	524		1032	-	174	0.25
DRM-NS-FG	497	546		1032	-	174	0.25
DRM-MS-CF	497	533		1032	-	174	0.25
DRM-MS-FG	497	551		1032	-	174	0.25
EV-NS-CF	373	714	-	1005	-	131	0.32
EV-NS-FG	373	740	-	1014	-	131	0.32
EV-MS-CF	373	732	-	1004	-	131	0.32
EV-MS-FG	373	752	-	1006	-	131	0.32
PPM-NS-CF	308	879	-	806	108	108	0.32
PPM-NS-FG	333	907	-	797	83	117	0.30
PPM-MS-CF	299	898	-	809	118	105	0.40
PPM-MS-FG	332	915	-	798	84	116	0.36

[a] HRWR: high-range water-reducer admixture. [b] Cement content presented in Table 4 refers to the total amount of new cement. The old cement (i.e., RCP) is accounted for in the FRCA portion.

The mixes proportioned using the DRM method (i.e., ACI—absolute volume method) treated the FRCA material as the same as natural fine aggregates, resulting in mixtures with high amounts of PC (e.g., nearly 500 kg/m^3). Conversely, since the EV and PPM methods account for the RCP attached to the FRCA particles, the mixtures proportioned by these procedures presented a much lower PC content. In the case of the EV method, it wa possible to obtain mixtures with Portland cement contents (≈373 kg/m^3) very similar t those of conventional concrete designed with the ACI method (i.e., 370 kg/m^3). Likewis mixtures designed through the continuous PPM technique (distribution factor of 0.28) an with the use of inert fillers yielded final Portland cement contents ranging from 333 t 299 kg/m^3 (using a minimum dosage of chemical admixtures).

4.4. Specimens' Preparation and Testing Procedures

Forty litres of concrete were manufactured for each of the mixtures tested in this wor to appraise their fresh-state behaviour. Furthermore, 100 mm × 200 mm concrete cylinde from all the fourteen mixes were fabricated according to CSA A23.2-3C [55], demoulde after 24 h and moist-cured over 28 days before testing.

4.5. Assessing Performance

4.5.1. Fresh-State Behaviour

The fresh state behaviour of the fourteen mixtures was investigated by slump me surements and rheological characterization using a planetary rheometer (IBB). The IB rheometer (Figure 4a) consists of an H-shape impeller (100 mm height and 130 mm lengt and a bowl with a diameter of 360 mm and 250 mm height and is considered reliable whe assessing concrete with consistencies ranging from 40 mm to 300 mm. The dataset (i.e torque and rotation values) recorded by the IBB over the experimental process was used t plot the rheological profiles of all mixtures. It is worth noting that the IBB results are nc displayed in fundamental units; thus, the yield stress and plastic viscosity are not given i Pa and Pa.s, respectively, but rather in N.m and rpm.

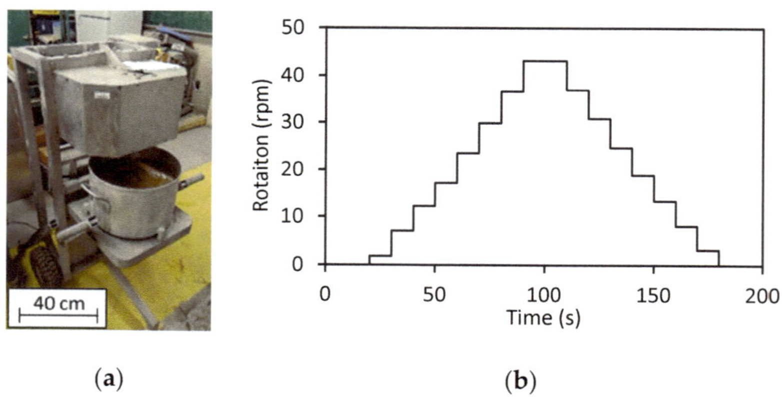

Figure 4. (**a**) IBB rheometer and (**b**) test programme used for analyzing the rheology cycle.

The rheological profiles of the mixtures evaluated were determined with a two-ste process: an increase in shear rate up to approximately 43 RPM, followed by a decreas period at the same stepwise, keeping a constant rotation for roughly 10 s at each ste (Figure 4b). The test was performed three consecutive times for each of the mixture evaluated; the first test was verified to be largely different from the two consecutive one and thus has been considered as pre-shearing. However, the second and third cycles wer mainly identical; hence, to avoid "fluid memory" and mitigate potential segregation, th second cycle was selected for analysis and used in this research.

4.5.2. Hardened-State Behaviour

The evaluation of hardened-state properties was performed through compressiv strength and non-destructive tests. For compressive strength, three samples of each mixtur were tested over time (i.e., 7, 14 and 28 days) as per CSA A23.1-14 [55]. Likewise, non-destructive techniques (i.e., surface electrical resistivity—ER and dynamic modulus of elasticity through ultrasonic pulse velocity test—UPV) were selected for this study to assess the microstructure quality of the distinct recycled concrete mixtures, as per ASTM C1876-19 [63] and ASTM C597-16 [64], respectively. It is worth mentioning that a viscoelastic gel was applied in between the transducers and concrete specimen to enable a suitable connection for the UPV test, and a frequency of 54 kHz was selected. Moreover, all specimens were tested in saturated surface-dry conditions at 23 ± 3 °C.

4.6. Microstructure Analysis

The microstructure quality of the concrete specimens was evaluated using scanning electron microscopy (SEM). The selected samples were first cut using a diamond-bladed masonry saw followed by subsequent polishing using a mechanical polishing table with grits of 30, 60, 140, 280 (80–100 μm), 600 (20–40 μm), 1200 (10–20 μm), 3000 (4–8 μm), 10,000 (2–4 μm) and 20,000 (1–2 μm). After that, the polished sections were dried at 60 °C for 24 h. Once the polishing process was completed, the samples were examined in a Jeol JSM- 6610LV SEM operated under a low vacuum, with an electron acceleration voltage of 20 kV and a spot size of 57. Backscattered electron (BSE) was conducted at different magnifications (i.e., ×300, ×500, ×750, ×1000 and ×1500) to study the mixture's quality, morphology and distribution. The SEM images and the description of the microscopic features observed in the different recycled concrete mixtures are presented in the discussion section of the manuscript.

5. Experimental Results

5.1. Fresh-State Behaviour

The consistency results (i.e., slump test) of ACI, DRM, EV and PPM mixtures are presented in Figure 5. As previously mentioned, all mixtures assessed in the study were designed with a targeted slump range of 100 ± 20 mm. Overall, regardless of the mix design technique, incorporating the natural sand (NS) resulted in a slightly higher slump than MS mixtures (either for control or RCAs mixtures). The control ACI mixtures were manufactured with the same water content (174 kg/m^3); yet, the ACI-NS resulted in a higher slump (110 mm) than ACI-MS (95 mm). Moreover, although the DRM mixtures were produced with the same water content as ACI mixtures, admixtures were required (i.e., 0.25% of mid-range/HRWR) to achieve the target slump (100 ± 20 mm). Otherwise, a lower amount of water (i.e., 131 kg/m^3) was used for all EV mixtures along with a 0.32% of mid-range/HRWR. The PPM mixtures achieved the required slump with the lowest amount of water (i.e., ranging from 105 to 116 kg/m^3) and a somewhat higher mid-range/HRWR (i.e., 0.32 to 0.40%) when compared to the EV mixtures. Finally, it is worth mentioning that mixtures containing FRCA-FG yielded overall higher slump values when compared to FRCA-CF.

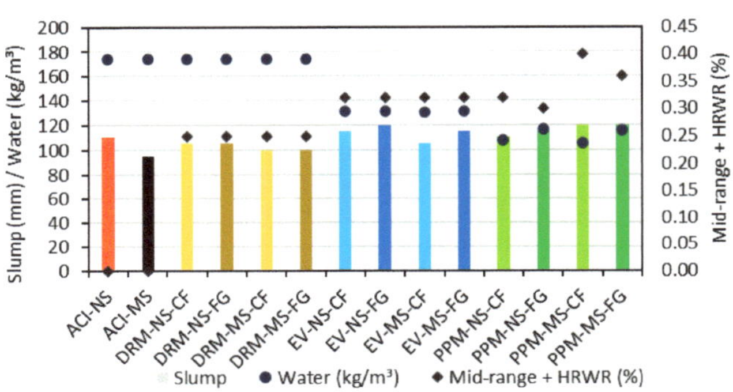

Figure 5. Comparison between slump, water content and admixtures of distinct mixtures.

The rheological profiles for the ACI, EV and PPM mixtures are presented in Figure It is important to notice that due to the torque limitations of the IBB rheometer used for th analysis, the DRM mixtures could not be investigated in this section since they presente high cohesiveness and low flowability.

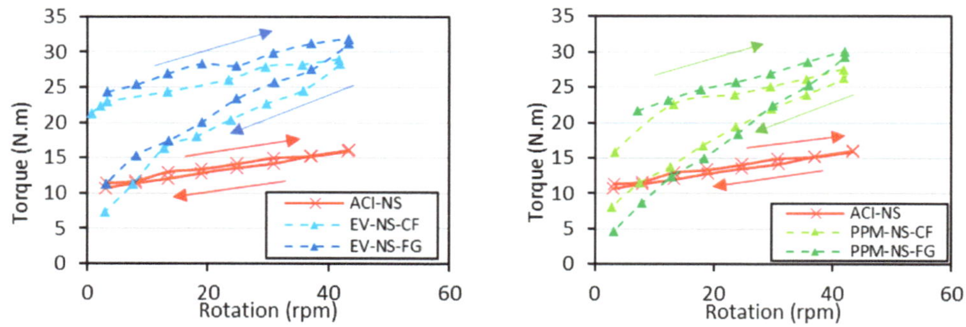

(**a**)

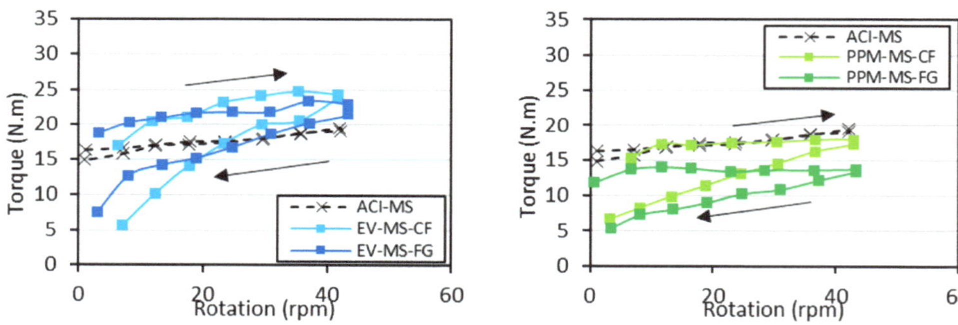

(**b**)

Figure 6. Rheological profiles of concrete mixtures made of NS (**a**) or MS (**b**).

Analyzing the plots, one first verifies that all mixtures yielded a shear-thinning be haviour, i.e., decrease in viscosity (slope of the torque vs. rotation curve) as a function of th torque applied regardless of the mix design procedure adopted. Second, the PPM mixture displayed the lowest yield stress, i.e., requiring a lower amount of torque to enable flov (i.e., 4.74 N.m to 8.17 N.m), followed by EV (7.32 N.m to 11.33 N.m) and ACI (10.70 N.m t 14.81 N.m)-mix-designed mixtures. Moreover, besides the rheological profiles and the min mum torque values mentioned above, two "extra" key rheological parameters can also b extracted (i.e., secant viscosity, the slope of the straight line between the first and the last de celeration points, and hysteresis area). A summary of all these values is displayed in Table One may notice that ACI mixtures reached the lowest secant viscosity (7 N.m/rad/s, o average), followed by the PPM (23.45 N.m/rad/s) and EV (28.88 N.m/rad/s) mixtures, o average. Furthermore, all concrete mixtures fabricated incorporating NS aggregate exhib ited greater secant viscosity (74% higher, on average). Likewise, comparing the crushing process adopted, the recycled mixtures made of FG aggregates yielded about 22% lowe secant viscosity than CF mixtures; the only exception is the PPM-NS FG mixture, which yielded a secant viscosity around 37% higher than PPM-NS-CF.

Table 5. Fresh-state properties of concrete mixtures appraised.

Mixture	ACI-NS	EV-NS-CF	EV-NS-FG	PPM-NS-CF	PPM-NS-FG
Minimum torque (N.m)	10.70	7.33	11.33	8.17	4.74
Secant viscosity (N.m/rad/s)	8.05	32.28	29.73	27.90	38.10
Hysteresis area (N.m.rad/s)	0.59	5.39	4.98	3.63	5.00
Mixture	**ACI-MS**	**EV-MS-CF**	**EV-MS-FG**	**PPM-MS-CF**	**PPM-MS-FG**
Minimum torque (N.m)	14.81	10.12	7.48	6.75	5.45
Secant viscosity (N.m/rad/s)	6.15	31.67	20.83	15.88	11.91
Hysteresis area (N.m.rad/s)	0.36	4.66	4.09	3.09	3.19

Finally, the time dependency (thixotropic vs. rheopexy behaviour) of all mixtures was also evaluated by calculating the hysteresis area under the torque vs. rotation curve. Overall, all mixtures presented positive HA, corresponding to a thixotropic behaviour; in other words, the viscosity of those mixtures decreases over time under constant shear stress or shear rate. ACI mixes presented the lowest HA values, followed by PPM and EV. Furthermore, mixtures made of MS presented lower HA than those with NS, whereas mixtures made of FG aggregates also yielded lower HA values when compared to CF mixes.

5.2. Hardened-State Behaviour

In this section, compressive strength (CS) and non-destructive tests results are presented for all mixtures manufactured in the laboratory. Figure 7 displays the CS data obtained for the control and recycled concrete mixtures. Globally, regardless of the mix design method, concrete mixtures incorporating MS aggregate displayed slightly higher compressive strength (i.e., 5.1%) than NS-made mixtures. Moreover, the RCA production process significantly impacted the CS values once FG mixtures reached 23.5% higher than CF-made mixtures; this behaviour was even more evident for PPM-proportioned mixtures (i.e., PPM-NS-FG and PPM-MS-FG achieved 53% higher values than their respective CF mixtures).

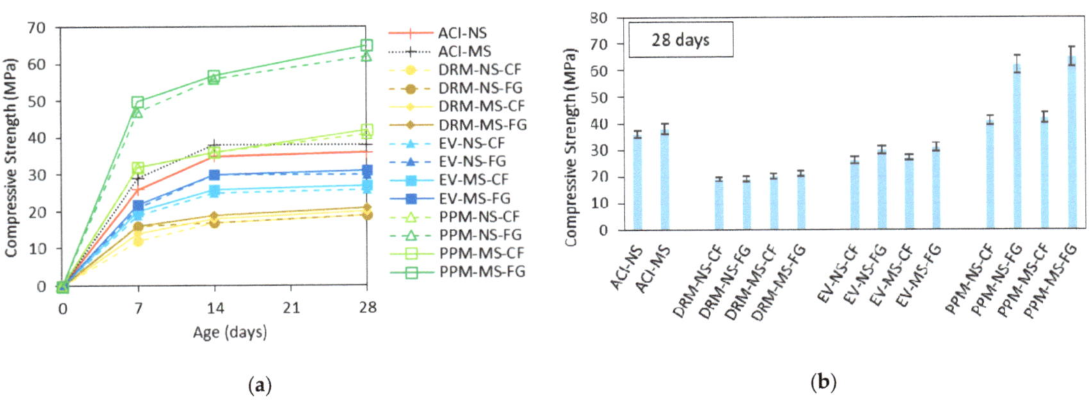

Figure 7. Compressive strength for all mixtures appraised: (**a**) over time and (**b**) at 28 days with the error bars.

Comparing the results, one verifies that the highest compressive strength values were obtained by PPM-proportioned mixtures (i.e., ranging from 41 to 65 MPa at 28 days), whereas the mixtures proportioned via DRM yielded the lowest results (i.e., between 19 and 21 MPa at 28 days). EV-designed mixes' results lay between PPM and DRM mixtures (i.e., 26 to 31 MPa). The control mixes designed with natural (AC-NS) and manufactured

(ACI-MS) sands presented 28-day compressive strength results within the range of 35 t 40 MPa.

The compressive strength results mentioned above correlate quite well with surfac electrical resistivity (Figure 8a) and the dynamic modulus of elasticity (Figure 8b) value obtained. Analyzing the plots, one notices that the mixtures containing MS presente higher electrical resistivity (ER) values (5.9%, on average) than NS-made mixtures; likewis the FG mixtures yielded 17.2% higher ER values, on average than those made of C aggregates. Regarding the mix design techniques, the highest values were obtained fc PPM mixtures (ER values between 10.8 and 14.3 KΩ·cm at 28 days), followed by E proportioned mixes (ranging from 6.5 to 7.3 KΩ·cm) and then DRM mixes (3.9–4.4 KΩ·cm respectively. A somewhat similar trend to the ER values was verified when evaluating th dynamic modulus of elasticity results of the RCA mixtures. In other words, incorporatin MS and NS did not significantly impact the test results. Yet, the crushing process of the RC significantly influenced the dynamic modulus of elasticity values since concrete specimen made of FG aggregates achieved a 16.3% higher dynamic modulus than CF-made mixture Moreover, PPM-proportioned mixtures presented the overall highest results (i.e., an averag of 51 GPa for PPM-FG and 42 GPa for PPM-CF at 28 days) followed by EV mixes (i.e., 35 GP for EV-FG and 27 GPa for EV-CF at 28 days) and finally DRM mixtures (i.e., an averag of 16 GPa for both DRM-FG and DRM-CF, at 28 days). On the other hand, conventiona concrete mixtures (ACI either NS or MS) obtained dynamic modulus of elasticity values c 49 GPa, just below the PPM-FG mixtures.

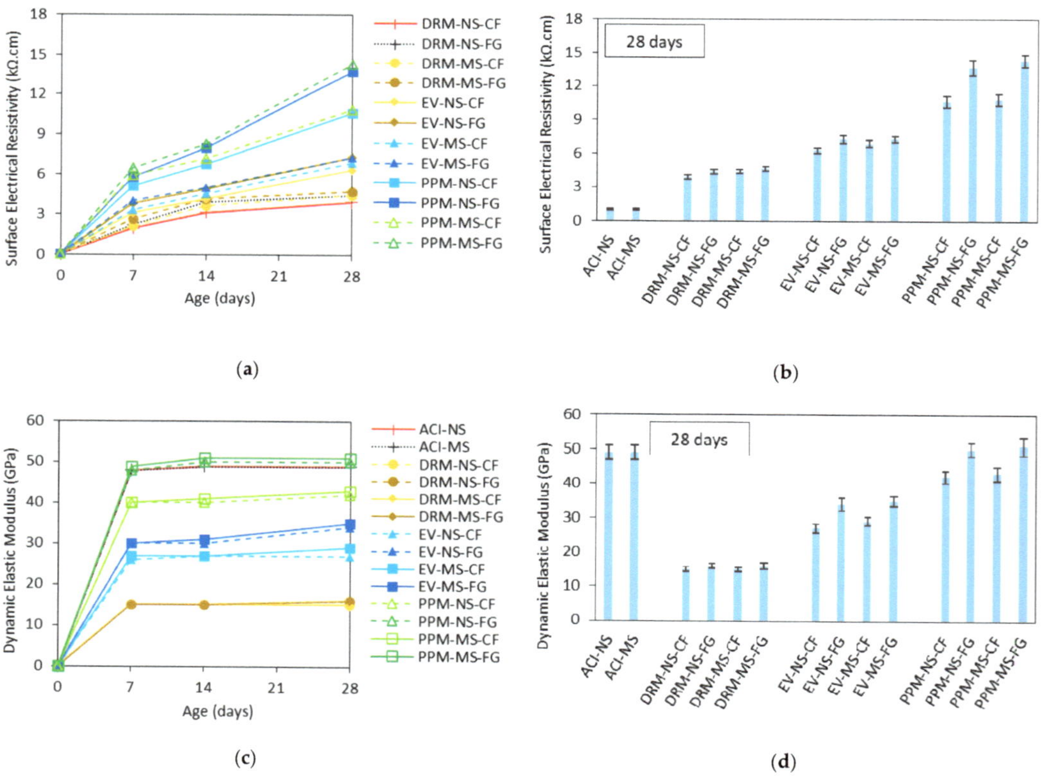

Figure 8. Non-destructive tests results: (**a**) surface electrical resistivity over time, (**b**) surface electrica resistivity at 28 with error bars, (**c**) dynamic elastic modulus over time, and (**d**) dynamic elasti modulus at 28 with error bars.

6. Analysis and Discussion

As previously mentioned, the present work is divided into two main parts: (1) production and characterization of FRCA materials and (2) appraisal of the fresh- and hardened-state properties of FRCA concrete mixtures proportioned using distinct mix design techniques. Although the data from the characterization of the FRCA, such as RCP content, specific gravity and water absorption and particle size distribution, were displayed in the experimental program section (since they were needed to design and manufacture the various recycled concrete mixtures), a global discussion on the obtained data seems appropriate herein. Therefore, the current *analysis and discussion section* will be divided into six sub-sections. First, a discussion of the influence of the raw materials and manufacturing process on the FRCA features is conducted. Second, a volumetric fraction evaluation of the various recycled mixtures components obtained through distinct mix design techniques is performed, followed by their impact on those mixtures' fresh and hardened states. Finally, microstructure analyses of the recycled mixtures are conducted, along with a discussion on the binder efficiency of FRCA mixtures proportioned using distinct mix design techniques.

6.1. Influence of the Raw Materials and Manufacturing Process on the FRCA Features

The analysis of the characterization results demonstrates an important role of the raw materials used (i.e., NS or MS) or the RCA's manufacturing processes (i.e., CF or FG) in the fresh and hardened states of FRCA concrete. Amongst the distinct properties appraised (i.e., specific gravity and water absorption, particle size distribution and RCP content), the RCP content showed an important influence of the raw material since FRCA made of MS presented a higher RCP than FRCA made of NS (i.e., 9% higher). This result is likely due to the rougher texture and more angular surface of the MS material, which may enhance the bond between the cement paste and aggregates. Likewise, comparing the FRCA's manufacturing processes, CF-made materials yielded 23% higher fineness modulus and 30% more RCP attached (on average), which is linked to the number of crushing processes, i.e., the higher the number of crushing cycles, the lower the amount of RCP attached to the fine aggregate particles. Consequently, FRCA-CF presented 4% lower specific gravity and 20% higher water absorption than FRCA manufactured through the FG crushing process.

From the particle size distribution of the FRCAs, it was verified that a higher number of crushing cycles (i.e., FG process) produces a greater amount of fine material (i.e., particles smaller than 150 μm that are often discarded due to lack of quality). In other words, an appropriate and optimized process is required to produce FRCA to minimize waste generation (i.e., particles smaller than 150 μm) along with energy consumption during the manufacturing process. Otherwise, FG manufacturing processes generate FRCAs with 29.1% less RCP content (on average) than CF. Moreover, the greater the amount of RCP adhered to the particles, the lower the quality of the materials, and the more challenging it is for FRCA concrete mixture proportioning to achieve comparable fresh- and hardened-state properties to conventional companion mixtures.

The above discussion indicates that high amounts of RCP adhered to the FRCA particles may lead to a lower-quality material since more inner deterioration such as microcracks and flaws is expected. Conversely, the presence of RCP, depending on the condition of the cement grains (i.e., fully hydrated or not), may stimulate further hydration, which could help seal pre-existing inner cracks and thus enhance the overall quality of the FRCA material. In this regard, isothermal calorimetry was performed to quantify the flow and the amount of the heat released from the FRCA materials as mentioned in Section 4.2.1. Figure 9a,b present the heat flow curves and the cumulative heat curves (respectively) of distinct original FRCA materials (FRCA NS and FRCA MS).

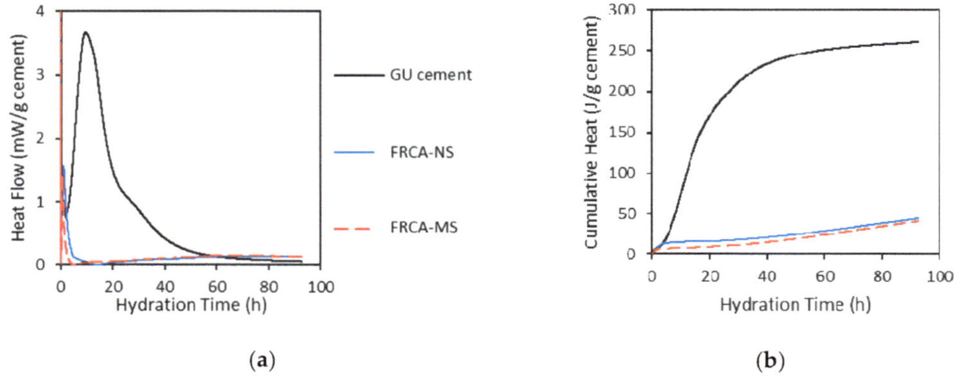

Figure 9. Isothermal calorimetry analysis: (**a**) heat flow curves and (**b**) cumulative calorimetric curve of original and powder FRCA samples.

The calorimetric measurements indicate that the overall acceleration of exothermi reactions from FRCAs was strongly reduced compared to general Portland GU-type cemen samples. The FRCAs' heat flow curves are not very well defined. Yet, two phases wer identified (i.e., strong initial period and very slow induction period), which is a ver different behaviour compared to the traditional GU cement behaviour with clear initia induction, accelerating and decelerating periods. Moreover, the results demonstrate tha the difference in raw materials (i.e., MS versus NS) does not significantly influence th amount of heat released over the first 92 h of the test. Yet, both FRCAs demonstrate some hydraulic activity. The cumulative heat achieved by the FRCA materials after 92 h corresponded to about 15% of the heat released by the GU-type cement. This behaviou dos not significantly change the mechanical properties of the FRCA concrete over time, bu this demonstrates that further hydrates might be theoretically formed in the system, whic could contribute to an increase in the overall quality of the FRCA concrete. Further analysi is still required in future projects to quantify this potential aspect more precisely.

6.2. Volumetric Evaluation of Distinct Mix Design Methods to Proportion FRCA Mixtures

The distinct mix design methods used in this research played a very important rol in the volumetric fraction components of both FRCA and CC mixes, as expected; ye to better visualize this influence, Figure 10a,b are presented. The air percentage in th plots represents the fresh-state air entrainment/entrapment measured in the laborator Analyzing Figure 10, one notices that the overall amount of Portland cement in bot conventional concrete mixtures made of NS and MS represents 11.8% of their total volum whereas for DRM, EV and PPM-proportioned mixtures, the values obtained were 15.9% 11.8% and 10.5% (on average), respectively. These values demonstrate the inefficienc of DRM techniques to proportion eco-efficient FRCA concrete to achieve comparabl mechanical properties to CC. Conversely, the efficiency of PPM methods is conduciv to the design of eco-efficient recycled mixtures, especially with the use of inert filler Furthermore, the type of FRCA manufacturing process and material (i.e., CF or FG, mad of NS or MS) seems to play a role in the binder demand of the FRCA mixtures, since th higher the RCP content, the lower the binder demand, except for DRM mixtures, wher no influence was noticed. Finally, it is worth noting that the volume of RCP is lowe in mixtures containing FG aggregates when compared to CF. In contrast, the volume o residual (i.e., old) fine aggregate (presented as FRCA in Figure 10) is constant in mixture proportioned with the same mix design technique (e.g., DRM-NS-CF has equal FRCA an higher RCP than DRM-NS-FG).

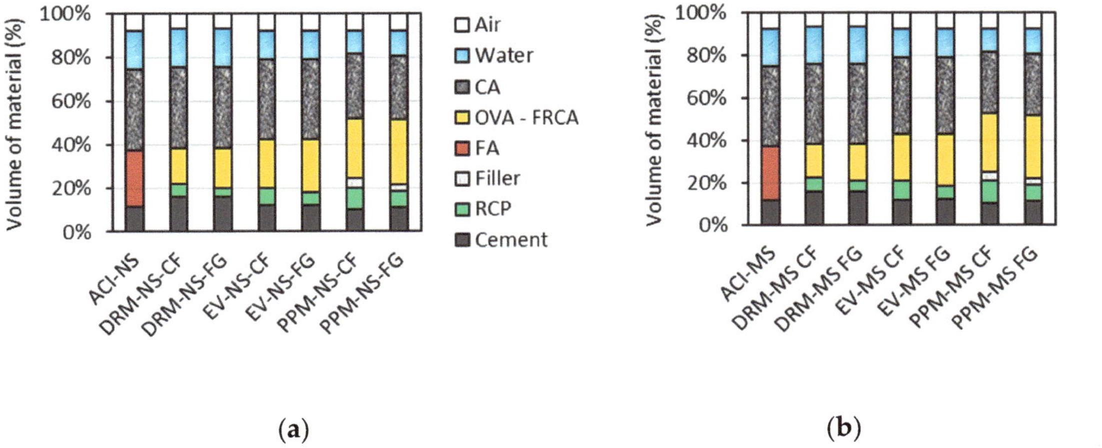

Figure 10. Volumetric fraction of each component for mixtures with FRCA made of NS (**a**) or MS (**b**).

6.3. Fresh-State Behaviour

All mixtures assessed in the study were designed with a targeted slump range of 100 ± 20 mm, being thus considered (in practice) as "workable" mixtures to be used as cast-in-place/vibrated concrete. Yet, although similar consistencies (i.e., slump value) were selected for all mixtures, quite variable minimum torque results, ranging from 4.74 N.m (PPM-NS-FG) to 14.81 N.m (ACI-MS), were obtained. In general, the higher the slump value, the lower the minimum torque required (Figure 11). To attest to the validity of the minimum torque results gathered using the IBB rheometer, an analytical evaluation was conducted as proposed by [65] (Figure 11—τ_c). These authors developed an analytical model to determine concrete mixtures' yield stress (τ_c) based on the slump test results. Analyzing the data below and the experimental comparison between slump vs. minimum torque, the calculated yield stress ($\tau_{c)}$ also decreases with the increase in a slump, which attests to the validity of the minimum torque results gathered.

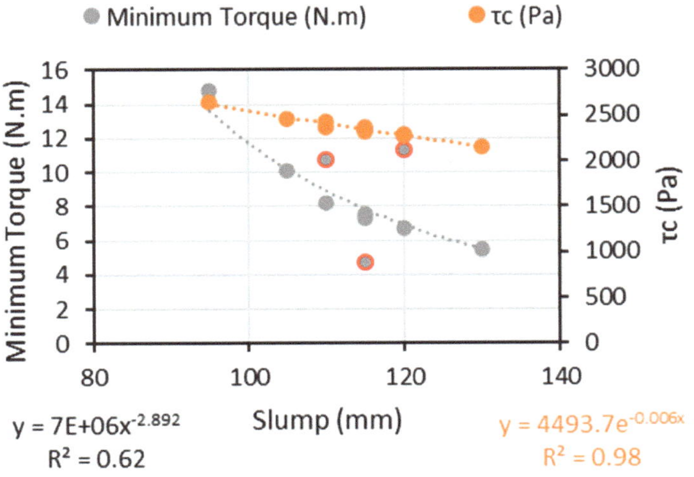

Figure 11. Yield stress experimental versus calculated through Roussel model.

According to Bingham's model, the rheological behaviour of conventional concret mixtures is described very often [66,67] and is represented by a linear relationship b tween shear stress and shear rate. However, granular suspensions such as concrete ca also display different rheological behaviours, as the viscosity changes according to th torque applied. Moreover, the FRCA concrete mixes evaluated in this research were foun to follow a distinct nonlinear trend (i.e., shear-thinning behaviour) where the viscosit lessens with the increase in shear rate. Recent research evaluating the rheological pr file of concrete with reduced PC content [52] and RCA mixtures [68] verified that suc eco-efficient mixtures might be precisely described in the fresh state through the Hersche Bulkley model (Equation (2)). Hence, Figure 12 illustrates the rheological profiles gathere from all a) EV and b) PPM mixtures in comparison to CC mixtures and modelled wit Herschel-Bukley's approach.

$$\tau = \tau_0 + k_{HB}\dot{\gamma}^n \quad (2$$

where τ is the shear stress, τ_0 is the yield stress, $\dot{\gamma}$ is the shear rate, k_{HB} is the viscosit constant of Herschel–Bulkley, and n is the flow behaviour factor, which is $n < 1$ for suspe sions presenting shear-thinning behaviour and $n > 1$ for shear-thickening ones [66]. It worth noting that for the sake of the following analyses, the concrete mixtures appraised i this work were assumed to behave as incompressible and homogenous fluids. Moreove segregation has not been observed during the rheological tests, thus disregarded herein.

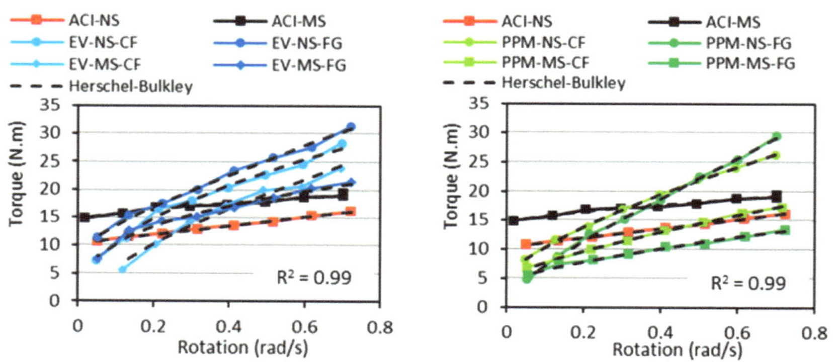

(**a**) (**b**)

Figure 12. Rheological profiles of (**a**) EV and (**b**) PPM mixes compared to Herschel–Bukley's mode

Analyzing the data gathered through Herschel–Bukley's model (Table 6) and plottin them over the results acquired in the laboratory for each mixture (Figure 12) enhance the rheological outcomes obtained in this research once the coefficient of determinatio obtained were averagely equal to 0.99. Moreover, one can verify that all mixtures displaye low-flow behaviour factors (i.e., n < 1), except for the ACI-NS mixture, where the n factc close to 1 indicates an almost linear behaviour (i.e., no change in viscosity as a function c torque). This correlation with the flow behaviour factor validates the shear-thinning aspe of the FRCA mixes since they require less torque applied in the system at higher shear stres regimes. The overall experimental initial torque required by each mix design represent 12.8 N·m, 9.1 N·m and 6.3 N·m on average for CC, EV and PPM mixes, respectivel Furthermore, the k_{HB} constant was also observed to yield the lowest values for PPM mixe compared to EV and CC mixtures. It is important to notice that the higher k_{HB} values, th higher the slope of the rheological profile curve (i.e., higher viscosity).

Table 6. Parameters of Herschel–Bukley's equation and minimum torque required for all mixtures inves-tigated.

Mixture	Minimum Torque—Experimental (N·m)	τ_0 (N·m)	K_{HB} (N·m/rad/s)	n
ACI-NS	10.7	10.4	8.0	1.05
ACI-MS	14.8	14.5	6.0	0.71
EV-NS-CF	7.3	0.2	32.5	0.50
EV-NS-FG	11.3	8.8	28.3	0.78
EV-MS-CF	10.1	0.1	31.2	0.70
EV-MS-FG	7.5	0.1	23.8	0.37
PPM-NS-CF	8.2	4.9	27.7	0.71
PPM-NS-FG	4.7	1.8	36.8	0.84
PPM-MS-CF	6.7	5.1	16.2	0.79
PPM-MS-FG	5.5	4.2	11.3	0.70

The above results suggest that the PPM-proportioned mixtures have the most suitable rheological behaviour for vibrated and or pumped applications. Moreover, following the study on the minimum torque, the derivative of the Herschel–Bulkley model of each concrete mixture (Table 6) was performed to calculate the real viscosity of each mixture at different shear rates (Figure 13). The results show that, as previously discussed, ACI-NS is the only mixture where the viscosity did not change as a function of the torque applied (i.e., viscosity approximately constant and equal to 8 N·m/rad/s). The higher the torque, the lower the viscosity for all the other mixtures.

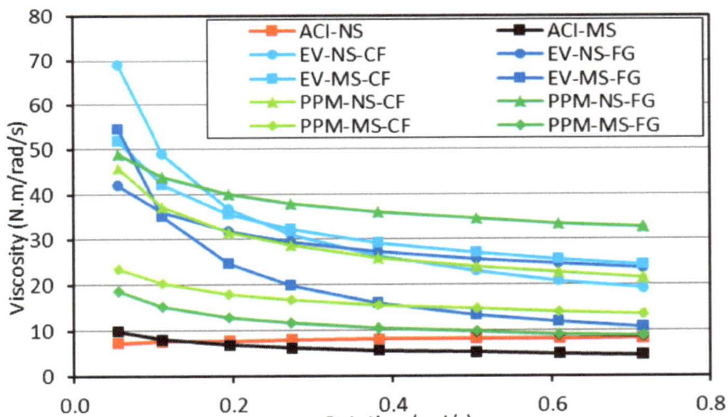

Figure 13. Real viscosity of non-Newtonian mixtures calculated through Herschel–Bulkley model.

6.4. Hardened-State Behaviour

The type of FRCA material (i.e., CF vs. FG) was found to have a low impact on the surface electrical resistivity (ER) and compressive strength (CS) of DRM and EV mixes and a more significant influence on the PPM-mix-proportioned mixtures. It is well known that the mechanical properties of concrete are directly related to the quality of its microstructure. Moreover, the results of CS and ER are quite well correlated; overall, the higher the compressive strength of the concrete, the higher the electrical resistivity. In general, the mixtures proportioned with FRCA-FG aggregates showed higher strength and ER results for the same w/c ratio; these results are due to the lower amount of RCP adhered to the FRCA-FG particles along with their more angular in shape and rougher in texture characters. Moreover, an increase of nearly 53% in CS and 31% in ER was obtained for PPM-designed mixes, whereas a slight average increase of about 16% and 6% was observed for EV and DRM mixes, respectively. The above results suggest that the FRCA particles' features (inner quality, shape and texture) are more important in mixtures with a mor optimized particle size distribution (improved granular skeleton) and thus have lowe porosity. Conversely, the raw-material type (i.e., NS or MS) does not seem to influence th mechanical properties of FRCA concrete significantly.

Another important discussion to be addressed is on the dynamic elastic modulu (DEM) obtained in this research. The DEM results for the FRCA concrete mixtures follow close correlation with the compressive strength, i.e., the higher the CS, the higher the DEM values. However, only the PPM-NS-FG and PPM-MS-FG mixtures could reach simila values to the companion conventional concrete mixtures (i.e., ACI-NS and ACI-MS). Thes findings may be justified by a thorough evaluation of the FRCA mixtures compositions. Fo instance, the replacement of natural fine aggregate by FRCA introduced theoretically mor interfacial transition zones (ITZ) to the system, which may have governed (and lessenec the overall DEM of the recycled mixture. On the other hand, PPM-mix-designed recycle mixtures presented a higher DEM; this is associated with a better microstructure qualit (i.e., optimization of the granular skeleton, lower porosity, etc.) of PPMs systems, enhancin the stiffness response of concrete mixtures as previously reported by [52,69]. Moreove the FRCA manufacturing processes also impacted on the DEM results (i.e., on averag FRCA-FG displayed a 16% greater DEM than FRCA-CF). The latter is related to the amour of RCP in the FRCA, which introduces more porosity and defects/flaws to the material' microstructure and thus slows the ultrasonic propagation through the material.

The above discussion confirms both the influence of the manufacturing process (i.e FG displays less RCP when compared to CF) and the mix-proportioning technique (i.e PPM mixes showed improved behaviour compared to DRM and EV mixtures) on th mechanical properties of FRCA concrete.

6.5. Microstructure Analysis of FRCA Mixes

It is well established that the mechanical behaviour of CC depends on the quality of it compounds, such as hydrated products and the ITZ, which in turn is directly related to th porosity (or w/c ratio) of the mixture [70]. As previously discussed, FRCA is a multi-phas material comprised of OVA and RCP; hence FRCA concrete is expected to display a highe amount (and distinct types) of ITZs than conventional concrete (i.e., between OVA an RCP—old ITZ, and between FRCA and new cement paste—new ITZ) [71,72]. To furthe understand the hardened-state responses gathered in this research, the microstructure o the various recycled mixes proportioned using distinct mix design methods (i.e., DRM EV and PPMs) were qualitatively analyzed through the scanning electron microscop (SEM). Thus, three SEM samples per mix were saw cut, polished and prepared as per 4. Figure 14 illustrates the SEM observations made at 750× magnification for FRCA mixture incorporating NS.

The micrographs revealed that PPM-designed mixtures (i.e., both PPM-FG and PPM CF) presented a much-enhanced microstructure when compared to EV or DRM propo tioned mixes since suitable (and better) adherence between the residual cement paste (RCP and the new cement paste (CP) was observed in all samples. Moreover, minor microcrack were identified within the RCP in both PPM-proportioned mixtures (highlighted areas i Figure 14a,b), being slightly more pronounced for the CF mix. The latter agrees with th mechanical results obtained in this work (i.e., PPM made of FG aggregates yielded bette mechanical responses). These observations emphasize the suitability of PPMs to proportio FRCA concrete with an enhanced microstructure and mechanical response. Finally, th inclusion of inert fillers in PPM systems (i.e., limestone fillers with smaller particle siz distribution than Portland cement) might also have contributed to the improvement of th new ITZ (Figure 14a,b) as previously reported [73–75], hence, enhancing the mechanica properties of FRCA concrete as discussed in Section 5.2.

Figure 14c,d indicate that EV and DRM-mix-designed mixtures display lower qualit in ITZs (i.e., old and new) than PPM mixes. Yet, EV samples showed an enhanced qualit of their new cement paste compared to DRM mixes (Figure 14c,d), which agrees wit their better mechanical responses, especially compressive strength. Moreover, it has been found that the quality of the new ITZ and cement paste of FRCA concrete is influenced by the material's pre-saturation, which is released over time to the new cement paste and generates further porosity in the mix [54]. The latter occurs especially in recycled mixtures that disregard the RCP in the mix design, such as DRM, and are indeed noticed in Figure 14d. However, it is worth mentioning that recent research [76] performed on cementitious mortars made of FRCA shows that the water may be transferred within the first 6 min and remains constant for up to 90 min. Therefore, further investigation is still required to better understand water transfer in cementitious materials made of FRCA since this behaviour may depend on the recycled material's features such as type (i.e., mineralogy), shape and texture and porosity and absorption of the aggregate, along with the amount and quality of the residual cement paste.

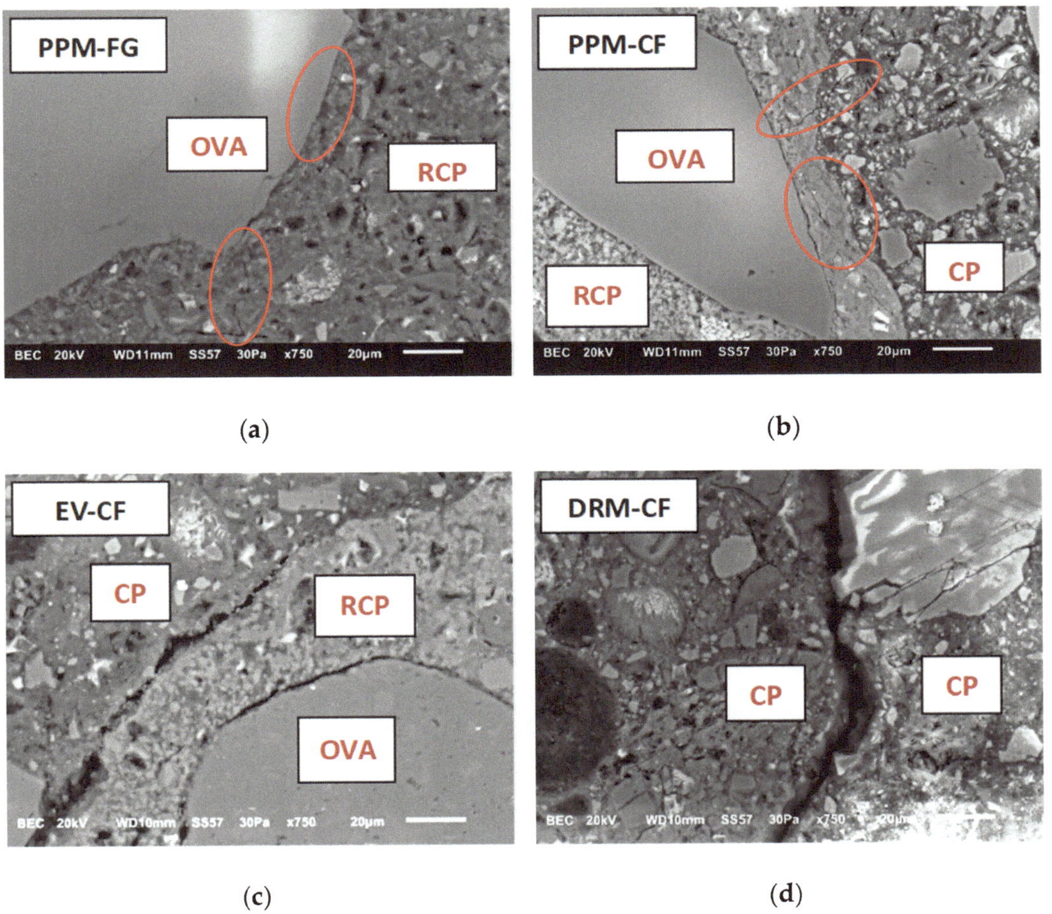

Figure 14. SEM micrographs of the old and new ITZs present in different RCA mixtures designed by (**a**) PPM-NS-CF, (**b**) PPM-NS-FG, (**c**) EV-NS-CF and (**d**) DRM-NS-CF—BSE images at a magnification of 750×.

6.6. Binder Efficiency

In order to evaluate the eco-efficiency of conventional concrete mixtures, Damineli et al. [77] developed an index, the so-called binder intensity (bi), which accounts for the amount of binder (kg/m^3) required to obtain one unit of a given mechanical property; fo example, 1 MPa of compressive strength at 28 days. Figure 15a displays the amount o new Portland cement (presented in Table 4) used to reach the 28-day compressive strengt results gathered for each of the three mix design methods utilized in this work. Analyzin the plot, it is clear that the overall performance of recycled mixtures is not directly linked t the amount of Portland cement used but rather to the microstructure quality of the materi (related to the efficiency of the mix-proportioning technique). For instance, among a FRCA mixes, the PPM-MS-FG yielded the highest 28-day compressive strength (i.e., nearl 65 MPa) while presenting a moderate binder content (332 kg/m^3).

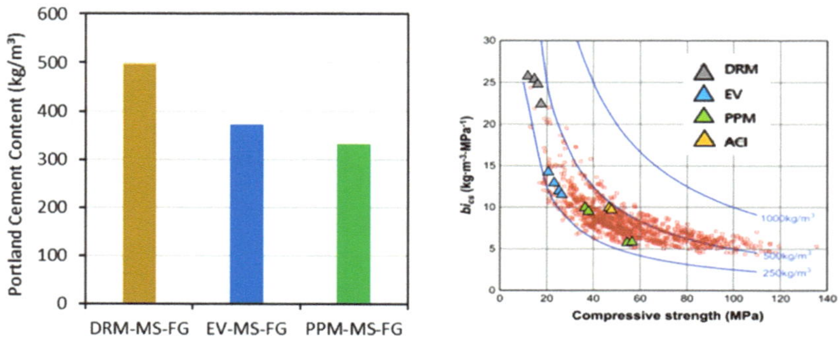

(a) (b)

Figure 15. (**a**) Portland cement (PC) content levels for the mixtures in analysis; (**b**) relationshi between bi and compressive strength at 28 days (adapted from [77]).

To evaluate the binder efficiency of all conventional concrete and FRCA mixture studied in this experimental program, the *bi* factor was calculated for all of them, and th results are illustrated in Figure 15b. It can be observed from the plot that PPM-mix-designe mixtures (green triangles) were the most eco-efficient FRCA mixes designed in this researc since they yielded the lowest *bi* values (8.3 kg·m^3·MPa^{-1} with an approximate average 2 day compressive strength of 53 MPa). Furthermore, one may notice that some PPM mixture are quite close to the bottom line of the plot (i.e., Portland cement content = 250 kg/m^3 which represents the lowest amount of Portland cement conventionally used in the mark for conventional concrete. Otherwise, conventional concrete mixes presented an overa *bi* factor of 10 kg·m^3·MPa^{-1} for an average 28-day compressive strength of about 37 MP while DRM and EV mixes displayed *bi* factors of approximately 25 and 13.1 kg·m^3·MPa^{-} for average compressive strengths of 20 and 29 MPa, respectively. These results demonstrat the promising character of PPM techniques in creating mix proportions of eco-efficien FRCA concrete with suitable fresh- and hardened-state properties.

7. Conclusions

The current research aimed to understand the overall behaviour of FRCA concret made of distinct raw materials (i.e., natural sand—NS or manufactured sand—MS), man facturing processes (i.e., CF vs. FG) and mix design techniques (DRM, EV and PPMs). Th main findings of this research are presented hereafter:

- It is clear that the manufacturing (i.e., crushing, sieving, etc.) process influences th overall quality of the FRCA material. With only two consecutive series of crushin the so-called CF generates recycled particles with coarser PSD, and high amoun of adhered RCP. Yet, multiple crushing series, the so-called FG, were observed t produce FRCA particles with similar PSD to natural aggregates and displayed les RCP attached to the particles due to further separation of mortar during crushin

Moreover, CF particles seem to be more rounded in shape and smoother in texture when compared to FG particles.

- The fresh-state behaviour of FRCA mixtures proportioned through both EV and PPM methods demonstrated an interesting performance, showing a shear-thinning profile (i.e., a decrease in viscosity as a function of the torque). Yet, PPM-mix-designed FRCA concrete demonstrated the best rheological variables (i.e., minimum torque required and lowest secant viscosity) among all mixtures; the latter indicates its suitability for applications under high-torque regimes such as pumping and or vibrating processes
- Interesting hardened-state properties (i.e., compressive strength and dynamic elastic modulus) were achieved using both EV and PPM techniques. However, in general, the DRM method yielded the worst mechanical performance obtained in this work. Yet, PPM-mix-designed FRCA proved to be the most promising mix design technique to proportion FRCA concrete mixes with suitable mechanical properties, binder efficiency and an enhanced microstructure (i.e., SEM analysis).
- It seems that both the aggregate types (NS vs. MS) and crushing processes (CF vs. FG) play an important role in the overall FRCA quality (i.e., RCP content, particle shape, texture, etc.) along with fresh- and hardened-state performances. Yet, the results obtained in this experimental campaign demonstrated that the mix design procedure adopted to produce FRCA concrete is even more important than the FRCA features and inner quality.
- Finally, microscopic analysis needs to be assessed considering distinct FRCA concrete mixes through the use of both EV and PPM methods in order to investigate the different failure mechanisms of FRCA concrete, especially when subjected to harsh environmental conditions. This would propose valuable remarks on the mechanical and durability-related properties of FRCA mixes against CC mixes as well as potential improvements on the EV mixes.

Author Contributions: Conceptualization, L.F.M.S., M.T.d.G., H.F.M. and G.F.; methodology, D.J.D.S., M.T.d.G., H.F.M., L.F.M.S., G.P.d.A., O.N., G.F. and P.-C.N.; validation, D.J.D.S., M.T.d.G., H.F.M., L.F.M.S. and G.P.d.A.; formal analysis, D.J.D.S., M.T.d.G. and H.F.M.; investigation, M.T.d.G. and H.F.M.; data curation, D.J.D.S., M.T.d.G. and H.F.M.; writing—original draft preparation, D.J.D.S., M.T.d.G. and H.F.M.; writing—review and editing, D.J.D.S., M.T.d.G., H.F.M., L.F.M.S., G.P.d.A., O.N., G.F. and P.-C.N.; visualization, L.F.M.S.; supervision, L.F.M.S. and G.F. All authors have read and agreed to the published version of the manuscript.

Funding: This research received no external funding.

Informed Consent Statement: Not applicable.

Acknowledgments: The authors gratefully acknowledge the financial support from the National Research Council Canada (NRC) and the University of Ottawa for taking part in this collaboration research project, which was funded by the CEI R&D unit. The authors would like to acknowledge that the characterization of FRCA was performed at the CEI Materials Laboratory from NRC, and a report was prepared for the CEI R&D unit. They would also like to thank CANMET Mining and André Demers for supporting with the necessary equipment for aggregate crushing and Tomlinson Group for generously providing aggregates used in this work. Moreover, De Souza and De Grazia benefits from the prestigious Vanier scholarship funded by NSERC.

Conflicts of Interest: The authors declare no conflict of interest.

References

1. Meyer, C. The greening of the concrete industry. *Cem. Concr. Compos.* **2009**, *31*, 601–605. [CrossRef]
2. Worrell, E.; Price, L.; Martin, N.; Hendriks, C.; Meida, L.O. Carbon dioxide emissions from the global cement industry. *Annu. Rev. Energy Environ.* **2001**, *26*, 303–329. [CrossRef]
3. Pacheco, J.; de Brito, J. Recycled Aggregates Produced from Construction and Demolition Waste for Structural Concrete: Constituents, Properties and Production. *Materials* **2021**, *14*, 5748. [CrossRef]
4. Corinaldesi, V.; Moriconi, G. Influence of mineral additions on the performance of 100% recycled aggregate concrete. *Constr. Build. Mater.* **2009**, *23*, 2869–2876. [CrossRef]

5. Huang, B.; Wang, X.; Kua, H.; Geng, Y.; Bleischwitz, R.; Ren, J. Construction and demolition waste management in China throug the 3R principle. *Resour. Conserv. Recycl.* **2018**, *129*, 36–44. [CrossRef]
6. Cabrera-Covarrubias, F.G.; Gómez-Soberón, J.M.; Almaral-Sánchez, J.L.; Arredondo-Rea, S.P.; Gómez-Soberón, M.C.; Corra Higuera, R. An Experimental Study of Mortars with Recycled Ceramic Aggregates: Deduction and Prediction of the Stress-Strai *Materials* **2016**, *9*, 1029. [CrossRef] [PubMed]
7. Mendivil-Escalante, J.M.; Gómez-Soberón, J.M.; Almaral-Sánchez, J.L.; Cabrera-Covarrubias, F.G. Metamorphosis in the Porosit of Recycled Concretes Through the Use of a Recycled Polyethylene Terephthalate (PET) Additive. Correlations between th Porous Network and Concrete Properties. *Materials* **2017**, *10*, 176. [CrossRef]
8. Letelier, V.; Tarela, E.; Muñoz, P.; Moriconi, G. Assessment of the mechanical properties of a concrete made by reusing bot Brewery spent diatomite and recycled aggregates. *Constr. Build. Mater.* **2016**, *114*, 492–498. [CrossRef]
9. Letelier, V.; Tarela, E.; Muñoz, P.; Moriconi, G. Combined effects of recycled hydrated cement and recycled aggregates on th mechanical properties of concrete. *Constr. Build. Mater.* **2017**, *132*, 365–375. [CrossRef]
10. Gonzalez, V.C.L.; Moriconi, G. The influence of recycled concrete aggregates on the behavior of beam–column joints under cycli loading. *Eng. Struct.* **2014**, *60*, 148–154. [CrossRef]
11. Yunchao, T.; Zheng, C.; Wanhui, F.; Yumei, N.; Cong, L.; Jieming, C. Combined effects of nano-silica and silica fume on th mechanical behavior of recycled aggregate concrete. *Nanotechnol. Rev.* **2021**, *10*, 819–838. [CrossRef]
12. Tang, Y.; Feng, W.; Chen, Z.; Nong, Y.; Guan, S.; Sun, J. Fracture behavior of a sustainable material: Recycled concrete with wast crumb rubber subjected to elevated temperatures. *J. Clean. Prod.* **2021**, *318*, 128553. [CrossRef]
13. Etxeberria, M. Evaluation of Eco-Efficient Concretes Produced with Fly Ash and Uncarbonated Recycled Aggregates. *Materia* **2021**, *14*, 7499. [CrossRef] [PubMed]
14. Feng, J.; Dong, C.; Chen, C.; Wang, X.; Qian, Z. Effect of Manufactured Sand with Different Quality on Chloride Penetratio Resistance of High–Strength Recycled Concrete. *Materials* **2021**, *14*, 7101. [CrossRef]
15. Zang, X.; Zhu, P.; Chen, C.; Yan, X.; Wang, X. Time-Dependent Shrinkage Model for Recycled Fine Aggregate Thermal Insulatio Concrete. *Materials* **2021**, *14*, 5581. [CrossRef]
16. Bravo, M.; Duarte, A.P.C.; De Brito, J.; Evangelista, L.; Pedro, D. On the Development of a Technical Specification for the Use c Fine Recycled Aggregates from Construction and Demolition Waste in Concrete Production. *Materials* **2020**, *13*, 4228. [CrossRe [PubMed]
17. Corinaldesi, V.; Letelier, V.; Moriconi, G. Behaviour of beam–column joints made of recycled-aggregate concrete under cycl loading. *Constr. Build. Mater.* **2011**, *25*, 1877–1882. [CrossRef]
18. Rao, A.; Jha, K.N.; Misra, S. Use of aggregates from recycled construction and demolition waste in concrete. *Resour. Conser Recycl.* **2007**, *50*, 71–81. [CrossRef]
19. Cyr, M.; Lawrence, P.; Ringot, E. Efficiency of mineral admixtures in mortars: Quantification of the physical and chemical effect of fine admixtures in relation with compressive strength. *Cem. Concr. Res.* **2006**, *36*, 264–277. [CrossRef]
20. RILEM Technical Committee. Progress of Recycling in the Built Environment. Available online: http://www.springer.com/civil engineering/book/978-94-007-4907-8 (accessed on 17 October 2021).
21. Fan, C.-C.; Huang, R.; Hwang, H.; Chao, S.-J. The Effects of Different Fine Recycled Concrete Aggregates on the Properties c Mortar. *Materials* **2015**, *8*, 2658–2672. [CrossRef]
22. Li, L.; Zhan, B.J.; Lu, J.; Poon, C.S. Systematic evaluation of the effect of replacing river sand by different particle size ranges o fine recycled concrete aggregates (FRCA) in cement mortars. *Constr. Build. Mater.* **2019**, *209*, 147–155. [CrossRef]
23. Evangelista, L.; de Brito, J. Mechanical behaviour of concrete made with fine recycled concrete aggregates. *Cem. Concr. Compo* **2007**, *29*, 397–401. [CrossRef]
24. Bouarroudj, M.E.; Remond, S.; Michel, F.; Zhao, Z.; Bulteel, D.; Courard, L. Use of a reference limestone fine aggregate to stud the fresh and hard behavior of mortar made with recycled fine aggregate. *Mater. Struct.* **2019**, *52*, 18. [CrossRef]
25. Carro-López, D.; González-Fonteboa, B.; Martínez-Abella, F.; González-Taboada, I.; De Brito, J.; Varela-Puga, F. Proportionin fresh-state properties and rheology of self-compacting concrete with fine recycled aggregates. *Hormigón y Acero* **2018**, *69*, 213–22 [CrossRef]
26. Kou, S.-C.; Poon, C.S. Properties of concrete prepared with crushed fine stone, furnace bottom ash and fine recycled aggregate a fine aggregates. *Constr. Build. Mater.* **2009**, *23*, 2877–2886. [CrossRef]
27. Khatib, J.M. Properties of concrete incorporating fine recycled aggregate. *Cem. Concr. Res.* **2005**, *35*, 763–769. [CrossRef]
28. Hassan, R.Y.; Faroun, G.A.; Mohammed, S.K. Mechanical properties of concrete made with coarse and fine recycled aggregate *Mater. Today Proc.* **2021**. [CrossRef]
29. Akono, A.-T.; Chen, J.; Zhan, M.; Shah, S.P. Basic creep and fracture response of fine recycled aggregate concrete. *Constr. Buil Mater.* **2021**, *266*, 121107. [CrossRef]
30. Fathifazl, G.; Abbas, A.; Razaqpur, A.G.; Isgor, O.B.; Fournier, B.; Foo, S. New Mixture Proportioning Method for Concrete Mad with Coarse Recycled Concrete Aggregate. *J. Mater. Civ. Eng.* **2009**, *21*, 601–611. [CrossRef]
31. Ahimoghadam, F.; Sanchez, L.F.M.; De Souza, D.J.; Andrade, G.P.; Noël, M.; Demers, A. Influence of the Recycled Concret Aggregate Features on the Behavior of Eco-Efficient Mixtures. *J. Mater. Civ. Eng.* **2020**, *32*, 04020252. [CrossRef]
32. Li, L.G.; Lin, C.J.; Chen, G.M.; Kwan, A.K.H.; Jiang, T. Effects of packing on compressive behaviour of recycled aggregate concret *Constr. Build. Mater.* **2017**, *157*, 757–777. [CrossRef]

3. Pradhan, S.; Kumar, S.; Barai, S.V. Recycled aggregate concrete: Particle Packing Method (PPM) of mix design approach. *Constr. Build. Mater.* **2017**, *152*, 269–284. [CrossRef]
4. Gómez-Soberón, J.M.V. Shrinkage of Concrete with Replacement of Aggregate with Recycled Concrete Aggregate. In *SP-209: ACI Fifth Int Conf on Innovation in Design with Emphasis on Seismic, Wind and Environmental Loading, Quality Control, and Innovation in Materials/Hot Weather Concreting;* Malhotra, V.M., Ed.; Universitat Politècnica De Catalunya · Barcelona Tech: Barcelona, Spain, 2002; pp. 475–495.
5. Topçu, I.B.; Şengel, S. Properties of concretes produced with waste concrete aggregate. *Cem. Concr. Res.* **2004**, *34*, 1307–1312. [CrossRef]
6. Silva, R.V.; De Brito, J.; Dhir, R.K. Properties and composition of recycled aggregates from construction and demolition waste suitable for concrete production. *Constr. Build. Mater.* **2014**, *65*, 201–217. [CrossRef]
7. Knaack, A.M.; Kurama, Y.C. Behavior of Reinforced Concrete Beams with Recycled Concrete Coarse Aggregates. *J. Struct. Eng.* **2015**, *141*, 4014009. [CrossRef]
8. Van Oss, H. CEMENT (Data in Thousand Metric Tons Unless Otherwise Noted). 2013. Available online: https://prd-wret.s3-us-west-2.amazonaws.com/assets/palladium/production/atoms/files/myb1-2013-cemen.pdf (accessed on 6 July 2019).
9. Jiménez, C.; Barra, M.; Josa, A.; Valls, S. LCA of recycled and conventional concretes designed using the Equivalent Mortar Volume and classic methods. *Constr. Build. Mater.* **2015**, *84*, 245–252. [CrossRef]
0. Anike, E.E.; Saidani, M.; Ganjian, E.; Tyrer, M.; Olubanwo, A.O. Evaluation of conventional and equivalent mortar volume mix design methods for recycled aggregate concrete. *Mater. Struct.* **2020**, *53*, 22. [CrossRef]
1. De Souza, D.J.; Sanchez, L.; Ahimoghadam, F.; Fathifazl, G. Comparison among Different Mix-Design Procedures for RCA Concrete. *ACI Spec. Publ.* **2019**, *334*, 99–121. Available online: https://www.concrete.org/publications/internationalconcreteabstractsportal.aspx?m=details&ID=51720255 (accessed on 2 February 2021).
2. Kim, J. Properties of recycled aggregate concrete designed with equivalent mortar volume mix design. *Constr. Build. Mater.* **2021**, *301*, 124091. [CrossRef]
3. Bizinotto, M.B.; Faleschini, F.; Fernández, C.G.J.; Hernández, D.F.A. Effects of chemical admixtures on the rheology of fresh recycled aggregate concretes. *Constr. Build. Mater.* **2017**, *151*, 353–362. [CrossRef]
4. Hayles, M.; Sanchez, L.F.M.; Noël, M. Eco-efficient low cement recycled concrete aggregate mixtures for structural applications. *Constr. Build. Mater.* **2018**, *169*, 724–732. [CrossRef]
5. Ahimoghadam, F. Effect of the Source of Recycled Concrete Aggregate (RCA) on the Behaviour of New Recycled Concrete Mixtures. Master's Thesis, University of Ottawa, Ottawa, ON, Canada, 2019.
6. Sunayana, S.; Barai, S.V. Recycled aggregate concrete incorporating fly ash: Comparative study on particle packing and conventional method. *Constr. Build. Mater.* **2017**, *156*, 376–386. [CrossRef]
7. Dinger, D.R.; Funk, J.E. Particle-Packing Phenomena and Their Application in Materials Processing. *MRS Bull.* **1997**, *22*, 19–23. [CrossRef]
8. Fennis, S.A.A.M.; Walraven, J.C. Using particle packing technology for sustainable concrete mixture design. *Heron* **2012**, *57*, 73–102.
9. Koutný, O.; Kratochvíl, J.; Švec, J.; Bednárek, J. Modelling of Packing Density for Particle Composites Design. *Procedia Eng.* **2016**, *151*, 198–205. [CrossRef]
0. Mangulkar, M.N.; Jamkar, S.S. Review of particle packing theories used for concrete mix proportioning. *Int. J. Sci. Eng. Res.* **2013**, *4*, 143–148.
1. Vogt, C. Ultrafine Particles in Concrete Influence of Ultrafine Particles on Concrete Properties and Application to concrete Mix Design. Ph.D. Thesis, KTH, Stockholm, Sweden, 2010.
2. de Grazia, M.T.; Sanchez, L.F.M.; Romano, R.C.O.; Pileggi, R.G. Investigation of the use of continuous particle packing models (PPMs) on the fresh and hardened properties of low-cement concrete (LCC) systems. *Constr. Build. Mater.* **2019**, *195*, 524–536. [CrossRef]
3. Amario, M.; Rangel, C.S.; Pepe, M.; Filho, R.D.T. Optimization of normal and high strength recycled aggregate concrete mixtures by using packing model. *Cem. Concr. Compos.* **2017**, *84*, 83–92. [CrossRef]
4. Sun, Y.; Wang, Z.; Gao, Q.; Liu, C. A new mixture design methodology based on the Packing Density Theory for high performance concrete in bridge engineering. *Constr. Build. Mater.* **2018**, *182*, 80–93. [CrossRef]
5. CSA A23.1-14/A23.2-14. Concrete Materials and Methods of Concrete Construction/Methods of Test and Standard Practices for Concrete. 2014. Available online: https://www.scc.ca/en/standardsdb/standards/28327 (accessed on 17 October 2021).
6. ASTM C1084. Standard Test Method for Portland-Cement Content of Hardened Hydraulic-Cement. Available online: https://www.astm.org/c1084-19.html (accessed on 17 October 2021).
7. ASTM C114-18. Standard Test Methods for Chemical Analysis of Hydraulic Cement. Available online: https://standards.globalspec.com/std/4472896/astm-c114-18 (accessed on 17 October 2021).
8. Rodrigues, F.; Evangelista, L.; De Brito, J. A new method to determine the density and water absorption of fine recycled aggregates. *Mater. Res.* **2013**, *16*, 1045–1051. [CrossRef]
9. Taylor, H.F.W. *Cement Chemistry*, 2nd ed.; Thomas Telford Publishing: London, UK, 1997; pp. 271–272.

60. CSA A23.1:19/CSA A23.2:19. Concrete Materials and Methods of Concrete Construction/Test Methods and Standard Pra tices for Concrete. 2019. Available online: https://civilnode.com/download-standard/10644237618004/csa-a23119a232 -concrete-materials-and-methods-of-concrete-constructiontest-methods-and-standard-practices-for-concrete (accessed on 17 C tober 2021).
61. Ferreira, L.; de Brito, J.; Barra, M. Influence of the pre-saturation of recycled coarse concrete aggregates on concrete properti *Mag. Concr. Res.* **2011**, *63*, 617–627. [CrossRef]
62. Cartuxo, F.; de Brito, J.; Evangelista, L.; Jiménez, J.; Ledesma, E. Rheological behaviour of concrete made with fine recycle concrete aggregates—Influence of the superplasticizer. *Constr. Build. Mater.* **2015**, *89*, 36–47. [CrossRef]
63. C09 Committee Standard Test Method for Bulk Electrical Resistivity or Bulk Conductivity of Concrete. *ASTM Int.* **2019**, 1- [CrossRef]
64. ASTM C 597-02. Standard Test Method for Pulse Velocity through Concrete. Available online: https://afzir.com/knowledge/w content/uploads/2018/02/ASTM-C-597.pdf (accessed on 17 October 2021).
65. Roussela, N.; Coussot, P. "Fifty-cent rheometer" for yield stress measurements: From slump to spreading flow. *J. Rheol.* **2005**, 4 705–718. [CrossRef]
66. Bao, K.; Lavrov, A.; Nilsen, H.M. Numerical modeling of non-Newtonian fluid flow in fractures and porous media. *Compu Geosci.* **2017**, *21*, 1313–1324. [CrossRef]
67. Ferraris, C.F.; Billberg, P.; Ferron, R.; Feys, D.; Hu, J.; Kawashima, S.; Koehler, E. Role of Rheology in Achieving Successf Concrete Performance. *Concr. Int.* **2017**, *39*, 43–51.
68. Singh, R.B.; Singh, B. Rheological behaviour of different grades of self-compacting concrete containing recycled aggregate *Constr. Build. Mater.* **2018**, *161*, 354–364. [CrossRef]
69. Yousuf, S.; Sanchez, L.; Shammeh, S. The use of particle packing models (PPMs) to design structural low cement concrete as a alternative for construction industry. *J. Build. Eng.* **2019**, *25*, 100815. [CrossRef]
70. Djerbi, A. Effect of recycled coarse aggregate on the new interfacial transition zone concrete. *Constr. Build. Mater.* **2018**, *19* 1023–1033. [CrossRef]
71. Poon, C.; Shui, Z.; Lam, L. Effect of microstructure of ITZ on compressive strength of concrete prepared with recycled aggregat *Constr. Build. Mater.* **2004**, *18*, 461–468. [CrossRef]
72. Otsuki, N.; Miyazato, S.-I.; Yodsudjai, W. Influence of Recycled Aggregate on Interfacial Transition Zone, Strength, Chloric Penetration and Carbonation of Concrete. *J. Mater. Civ. Eng.* **2003**, *15*, 443–451. [CrossRef]
73. Dobiszewska, M.; Schindler, A.K.; Pichór, W. Mechanical properties and interfacial transition zone microstructure of concre with waste basalt powder addition. *Constr. Build. Mater.* **2018**, *177*, 222–229. [CrossRef]
74. Wu, K.; Shi, H.; Xu, L.; Ye, G.; De Schutter, G. Microstructural characterization of ITZ in blended cement concretes and its relatio to transport properties. *Cem. Concr. Res.* **2016**, *79*, 243–256. [CrossRef]
75. Mohammed, M.; Dawson, A.; Thom, N.H. Macro/micro-pore structure characteristics and the chloride penetration of se compacting concrete incorporating different types of filler and mineral admixture. *Constr. Build. Mater.* **2014**, *72*, 83–9 [CrossRef]
76. Maimouni, H.; Remond, S.; Huchet, F.; Richard, P.; Thiery, V.; Descantes, Y. Quantitative assessment of the saturation degree model fine recycled concrete aggregates immersed in a filler or cement paste. *Constr. Build. Mater.* **2018**, *175*, 496–507. [CrossRe
77. Damineli, B.L.; Kemeid, F.M.; Aguiar, P.S.; John, V.M. Measuring the eco-efficiency of cement use. *Cem. Concr. Compos.* **2010**, 3 555–562. [CrossRef]

materials

MDPI

Article

Freezing and Thawing Resistance of Fine Recycled Concrete Aggregate (FRCA) Mixtures Designed with Distinct Techniques

Cassandra Trottier [1,*], Mayra T. de Grazia [1], Hian F. Macedo [1], Leandro F. M. Sanchez [1,*], Gabriella P. de Andrade [1], Diego J. de Souza [1], Olga Naboka [2], Gholamreza Fathifazl [2], Pierre-Claver Nkinamubanzi [2] and André Demers [3]

1 Department of Civil Engineering, Faculty of Engineering, University of Ottawa, Ottawa, ON K1N 6N5, Canada; mayra.grazia@uottawa.ca (M.T.d.G.); hdefr095@uottawa.ca (H.F.M.); gandr070@uottawa.ca (G.P.d.A.); dsouz025@uottawa.ca (D.J.d.S.)
2 National Research Council Canada (NRCC), Ottawa, ON K1A 0R6, Canada; olga.naboka@nrc-cnrc.gc.ca (O.N.); gholam-reza.fathi-fazl@nrc-cnrc.gc.ca (G.F.); pierre-claver.nkinamubanzi@nrc-cnrc.gc.ca (P.-C.N.)
3 CANMET Energy (NRCAN), Ottawa, ON K1V 1E1, Canada; andre.demers@canada.ca
* Correspondence: ctrot059@uottawa.ca (C.T.); leandro.sanchez@uottawa.ca (L.F.M.S.)

Abstract: The pressure to use sustainable materials and adopt practices reducing the carbon footprint of the construction industry has risen. Such materials include recycled concrete aggregates (RCA) made from waste concrete. However, concrete made with RCA often presents poor fresh and hardened properties along with a decrease in its durability performance, especially when using its fine fraction (i.e., FRCA). Most studies involving FRCA use direct replacement methods (DRM) to proportion concrete although other techniques are available such as the Equivalent Volume (EV) and Particle Packing Models (PPMs); yet their impact on the durability performance, especially its performance against freezing and thawing (F/T), remains unknown. This work, therefore, appraises the F/T resistance of FRCA mixtures proportioned through various mix proportioning techniques (i.e., DRM, EV and PPMs), produced with distinct crushing processes (i.e., crusher's fines vs. finely ground). The results show that the mix design technique has a significant influence on the FRCA mixture's F/T resistance where PPM-proportioned mixtures demonstrate the best overall performance, exceeding the specified requirements while DRM-proportioned mixtures failed F/T resistance requirements. Moreover, the crushing process plays an important role in the recycled mixtures' cracking behavior under F/T cycles, where less processing leads to fewer cracks while remaining the most sustainable option overall.

Keywords: fine recycled concrete aggregates; low cement concrete; eco-friendly concrete; the durability of FRCA concrete; mix design; particle packing model; freezing and thawing; microscopic assessment

Citation: Trottier, C.; de Grazia, M.T.; Macedo, H.F.; Sanchez, L.F.M.; de Andrade, G.P.; de Souza, D.J.; Naboka, O.; Fathifazl, G.; Nkinamubanzi, P.-C.; Demers, A. Freezing and Thawing Resistance of Fine Recycled Concrete Aggregate (FRCA) Mixtures Designed with Distinct Techniques. *Materials* **2022**, *15*, 1342. https://doi.org/10.3390/ma15041342

Academic Editors: Rui Vasco Silva, Miguel Bravo and António C. Duarte

Received: 30 December 2021
Accepted: 8 February 2022
Published: 11 February 2022

1. Introduction

Tackling climate change is among the top priorities worldwide, with emphasis on reducing CO_2 emissions. Canada has set its target of reaching the so-called net-zero by 2050 [1]. There is indeed an urgency to adopt alternative solutions to reduce the carbon footprint in the construction industry. The production of Portland cement alone amounts to approximately 5% of global CO_2 emissions [2]. Moreover, sources of the non-renewable natural aggregates are becoming depleted nearby major continuously expanding urban centers resulting in longer transportation distances and, thus, an increased carbon footprint [3–5]. Therefore, the use of recycled concrete aggregates (RCA) has been progressively considered as a sustainable alternative to the non-renewable natural aggregate. RCA can help decrease the carbon footprint of new concrete (i.e., lessening the amount of new Portland cement required when accounting for the residual cement paste into the mixture), limiting the use of natural resources and reducing landfill's usage, provided that the RCA is in close proximity to the concrete plant/construction site [6–9].

RCA is produced by crushing returned or demolished concrete into aggregate-size fractions. Generally, the coarse RCA fraction is used in new concrete construction, ofte limited to non-structural applications due to concerns over the material's variability, qualit and presence of impurities (i.e., other waste debris), whereas the fine RCA (FRCA) fractio is rejected, being considered as a by-product or waste [4,5].

Much of the research on RCA implicitly treats the material as being homogeneous an directly replaces certain proportions of natural aggregate by RCA through partial or fu direct replacement methods (DRM) in the concrete mixture without further consideratio of its multi-phase nature, especially for FRCA. Indeed, the multi-phase nature of FRC consisting of residual cement paste (RCP) adhered to residual sand (RS) and/or fracture coarse original virgin aggregate (OVA), influences the performance of recycled concret mixtures, often resulting in inferior mechanical and durability properties. Moreover, give the high variability of RCA, accurately estimating and/or predicting the fresh and hardene state performance of recycled mixtures designed through the DRM method remains a cha lenge [10–13]. However, recent studies demonstrated that when the unique microstructur of the RCA is accounted for in the mix proportioning of recycled mixtures [7–9,14], adequat fresh and hardened state behavior might be obtained. Amongst the mix design technique successfully implemented in this regard, the Equivalent Volume (EV) [9] method showe to be quite promising, enabling a replacement ratio of 100% while achieving the targete properties. On the other hand, several advanced mixture proportioning techniques base on particle packing models (PPMs) were used for both coarse RCA [15,16] and FRCA [17,18 mixtures, yielding interesting results, especially in the hardened state (i.e., compressiv strength). Nevertheless, the impact of using distinct mix design techniques on the durabilit and long-term performance of recycled mixtures incorporating FRCA is yet to be dete mined. This work appraises the durability performance against freeze and thawing (F/T cycles of recycled concrete mixtures designed through various proportioning technique (i.e., DRM, EV and PPM) and crushing techniques while incorporating 100% of FRCA.

2. Background

2.1. Multi-Phase Nature of Recycled Concrete Aggregates (RCA)

Concrete waste is crushed then sieved to obtain certain particle sizes. Fine recycle concrete aggregates (FRCA) are normally considered as a by-product of coarse RCA (CRCA production. Consequently, CRCA is composed of mainly original virgin aggregate (OVA and residual mortar (RM); the latter can even represent up to 60% of the CRCA volume [19 Meanwhile, FRCA is generally composed of RM, which can be further divided into resi ual cement paste (RCP) and residual sand (RS). Some fractured OVA may detach durin the crushing process and thus can be part of the FRCA particles. Therefore, FRCA ca be composed of only RM, only OVA or RS and a combination of RM and OVA as high lighted in Figure 1 through the yellow, green and blue contours, respectively. In additio FRCA contains original interfacial transition zones (ITZ) between the OVA/RS and RC enhancing the multi-phase nature of the aggregate.

Physical properties of the FRCA particles are influenced by several factors such as th amount of RCP found in FRCA particles that can be influenced by the type and qualit of the OVA and RS (i.e., lithotype, texture, shape, etc.), the RCP's quality (i.e., porosit and mechanical properties) and the FRCA crushing process (i.e., crusher type, crushin series and energy, etc.), leading to important variability among the final product [20,21 Moreover, when RCA is used in concrete, new ITZs are formed between the OVA/R to new cement paste and residual to new cement paste interfaces; thus, adding anothe component to the distinct microstructure of concrete made of FRCA.

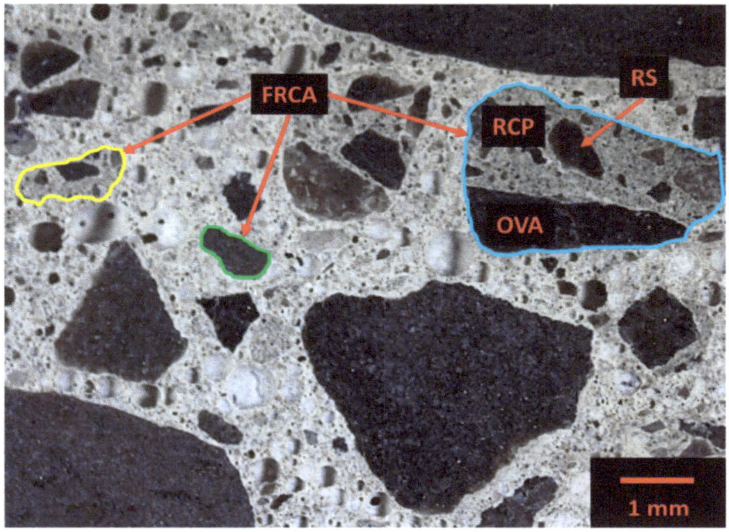

Figure 1. Fine recycled concrete aggregate (FRCA) concrete multi-phase nature including FRCA particles with original virgin aggregate (OVA), residual cement paste (RCP) and residual sand (RS).

FRCA is generally considered a very low-quality material due to its high amount of RM, resulting in low specific gravity, high porosity and absorption, the latter of which corresponds to the increased water demand of the FRCA concrete. Some studies were performed to quantify the RM of FRCA [22–24]; however, the quantification of the RCP is more relevant as FRCA may be composed of only RM, which makes the results obtained on those studies somewhat inconsistent [25,26]. De Andrade et al. [18] used a discrete PPM approach, the so-called Compressible Packing Model (CPM) developed by De Larrard [27], to proportion recycled concrete mixtures made of 100% FRCA and found that better control over the properties of the material might be achieved despite the FRCA features; however, the RCP in the FRCA was quantified in this study only for comparative purposes.

Despite all recent advancements in recycled concrete using FRCA, when compared to conventional concrete (CC), recycled concrete mixtures designed through conventional methods using partial (over 30%) or full FRCA replacement were observed to often yield lower performance in the fresh (i.e., high consistency and low flowability) and hardened (i.e., low compressive strength and stiffness) states along with durability [11,28–41].

2.2. *Available Mixture Proportioning Techniques for RCA Concrete*

2.2.1. Direct Replacement Method (DRM)

Direct replacement methods (DRM), either partial/full or by weight/volume, are one of the earliest attempts to mixture proportion concrete using FRCA. Studies show that a replacement ratio of less than 30% provides optimal results [29,31–33,38,42–44], although suitable properties were achieved using 100% replacement in some cases. Moreover, results of recycled concrete mixtures designed through DRM were verified to be extremely variable. As such, it was suggested by Nedeljković et al. [45] in a critical review of the use of FRCA to develop new approaches when proportioning recycled mixtures.

2.2.2. Equivalent Volume (EV)

The Equivalent Volume (EV) [9] is the most recent mix design technique created that accounts for RCP content adhered to the CRCA particles to proportion concrete made of CRCA, allowing up to 100% of CRCA replacement while achieving the targeted properties. This method was developed based on the Equivalent Mortar Volume (EMV [7]) and EMV modified (EMV-mod [8]), overcoming the issues presented in the fresh state and ec efficiency, respectively. The EV mix design technique develops a CRCA concrete that bea the same total volume of cement paste (i.e., residual and new) and the same total volum of aggregates (i.e., OVA and residual and new sand) as its companion CC. This techniqu was used and adapted for FRCA in a feasibility study [46].

2.2.3. Particle Packing Models (PPMs)

Particle packing models (PPMs) are advanced mix design techniques used to improv the performance of concrete mixtures in both fresh and hardened states while optimizing th granular system, resulting in an improved packing density and reduced porosity [47–51 Hence, concrete mix proportioned through PPMs often present lower Portland cemen content and enhanced durability aspects when compared to CC. Although PPMs and oth similar techniques are rarely used for RCA, previous studies [16–18] showed promisin results. PPMs are divided into discrete and continuous models, the former of which a sumes the existence of a given number of discrete particle sizes that are rearranged to reac the maximum packing density [16,49,51], whereas the latter assumes a continued distrib tion of particles within the system [49,51]. The Modified Andreasen model (also know as the Alfred model) is the most recent continuous PPM, which calculates the optimu particle size distribution based on a coefficient of distribution (q) along with the largest (D and smallest (D_S) particle diameter present in the system. The coefficient of distributio is normally adopted based on the fresh state requirements of the mixture. Distributio coefficients ranging between 0.20 and 0.23 are often selected for self-consolidating mixture while values between 0.26 and 0.28 are normally targeted for vibrated and/or pumpe concrete. It is worth noting that 0.37 is the coefficient of distribution that yields the highe packing density and thus lowest porosity to the granular system as per [51]. The optimu particle size distribution is calculated as per Equation (1):

$$CPFT = 100 \times \left(\frac{D_P{}^q - D_S{}^q}{D_L{}^q - D_S{}^q} \right), \quad (1)$$

where $CPFT$ is the cumulative (volume) percent finer than D_P and D_P is the partic diameter.

2.3. Durability and Long-Term Performance of FRCA Concrete

2.3.1. Non-Destructive Tests (NDT) and Stiffness Damage Test (SDT)

A convenient evaluation of the inner quality of concrete mixtures may be performe using non-destructive techniques (NDT), such as ultrasonic pulse velocity (UPV) an surface electrical resistivity (ER). Such techniques were verified to be suitable to apprais the inner quality of CC mixtures subjected to aggressive/harsh environments [52] alon with the effects of RCA replacement [9,42,53–55]. Studies show that the higher the RC replacement proportioned with DRM, the lower the NDT results due to the lower inn quality of the RCA [42,56,57].

Moreover, regardless of the mixture proportioning technique used, RCA concrete mi tures generally present lower stiffness when compared to CC [39,58,59], likely associate with the presence of microcracks within the RCA. The Stiffness Damage Test (SDT-[60–62 was shown to be a very reliable tool to quantify physical damage in concrete through th stiffness damage index (SDI) and plastic deformation index (PDI), which relate the numb of inner cracks with either the energy dissipated over the loading–unloading cycles or th unrecoverable deformation after five loading cycles, respectively. Moreover, the modulu of elasticity of the material can be calculated using the average of the secant modulus of th second and third cycles [61,62] through the SDT approach. However, the SDT has nev been used to evaluate the inner quality and flaws (i.e., microcracks) of FRCA concrete.

2.3.2. Freeze-Thaw Resistance

Freeze-thaw (F/T) resistance of CC is required in Canada's harsh environments and is linked to various factors such as the aggregate features, cement paste porosity and free water availability in the system. These parameters may be even more critical for recycled concrete since its higher porosity may contribute to an increase in the material's absorption and permeability [63,64]. Some studies have shown promising results regarding the F/T resistance of concrete made with FRCA [29,30,54]. However, the effect of highly porous aggregates such as RCA on permeability is not yet fully understood as it may play a dual role: highly porous aggregates are normally weaker and easily saturated, whereas increased permeability leads to dissipation of internal hydraulic pressure that may improve F/T resistance [65].

2.3.3. The Damage Rating Index (DRI)

The Damage Rating Index (DRI) is a microscopic tool developed to assess the damage and its extent in CC [66–70]. Using a stereomicroscope at 16× magnification, distress features (i.e., cracks) are counted in 1 cm × 1 cm squares drawn on the surface of a polished concrete section after which weighting factors [71] are applied to those features to balance their relative importance towards the distress mechanism and normalized to 100 cm^2 for comparative purposes. Moreover, the extended version of the DRI (i.e., without applying weighting factors) can be used to evaluate the crack propagation and distribution within a CC [66,68] as well as recycled concrete [72–74].

Nevertheless, the durability performance of concrete made of FRCA, namely its freezing and thawing resistance, while using several distinct mixture proportioning techniques (i.e., DRM, EV and PPMs) and crushing processes (i.e., crushing fines vs. fully ground) remains mostly unknown. Moreover, techniques such as the SDT and DRI are novel test procedures used to evaluate the inner quality and or damage degree of concrete and have never been used to evaluate concrete made of FRCA before.

3. Scope of the Work

Currently, there is a lack of studies on the durability and long-term performance of RCA, especially FRCA, which hinders its use in the industry. Moreover, the impact of distinct mix design procedures on the durability performance of recycled concrete mixtures made of FRCA is mostly unknown. Thus, this work aims to evaluate the durability related properties, namely the resistance against freezing and thawing (F/T) cycles, of concrete made of 100% FRCA replacement proportioned through various mix proportion techniques (i.e., DRM, EV and PPM), produced with distinct crushing processes (i.e., crusher's fines (CF) vs. finely ground (FG)) and derived from conventional concrete mixtures incorporating natural and manufactured sand. By using non-destructive techniques (NDT), the compressive strength and the Stiffness Damage Test (SDT), the FRCA concrete mixtures' inner quality and flaws were appraised before being subjected to F/T cycles as per ASTM C666 [75]. Furthermore, the FRCA concrete mixtures subjected to F/T cycles were assessed using a microscopic technique, the Damage Rating Index (DRI) and its extended version, to better understand the F/T-induced damage development in the recycled concrete mixtures over time.

4. Materials and Methods

4.1. FRCA Production and Raw Materials Characterization

Two-hundred 35 MPa CC cylinders (100 mm × 200 mm, w/c of 0.47) were fabricated according to ASTM C39 [76] while varying the fine aggregate type (i.e., natural (NS) and manufactured (MS)). A Portland cement (General Use-GU, similar to ASTM Type I) was used along with a coarse limestone aggregate with a nominal maximum size of 19 mm.

Following the CC production, the specimens were de-molded after 24 h and moist cured (100% RH and 20 °C) for 28 days before being processed to produce the FRCA. Figure 2 displays a) the crusher's fines (CF) and b) fully ground (FG) crushing procedures used to produce the FRCA. For the CF–FRCA, concrete specimens were jaw crushed twic on a 19 mm maximum gap opening then sieved to divide the particles sizes; only the fin fraction of the crushed material was used. Meanwhile, the FG-FRCA was produced b crushing the CRCA material obtained from sieving greater than 4.75 mm through multip series of crushing (5 mm gap opening in the jaw crusher). Figure 3 shows the particle siz distribution of all fine aggregate used in this study, including the specified limits as per [4

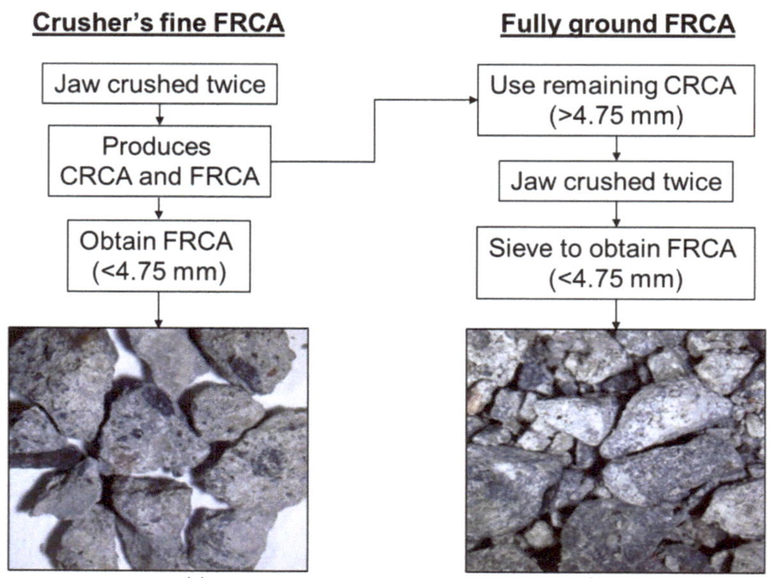

(a) (b)

Figure 2. Summary of the two methods of FRCA production adopted in this experimental wor (**a**) crusher's fines (CF) and (**b**) fully ground (FG).

Further characterization of the residual cement paste (RCP) content was performed fc each type of FRCA through the soluble silica sub-procedure according to ASTM C1084 [77 and C114 [78]. This procedure, as detailed in [46], was selected due to the use of coars limestone aggregate thus, preventing the selection of other chemical procedures such a maleic acid digestion. Moreover, specific gravity and water absorption (Table 1) wer evaluated according to the new method proposed by Rodrigues et al. [79], in which FRC samples are first saturated in a 0.1% dispersant solution to prevent FRCA cohesion befor continuing with a similar procedure as per [4].

Table 1. Properties of FRCA and natural aggregates.

Physical Property	FRCA NS–CF	FRCA NS–FG	NS	FRCA MS–CF	FRCA MS–FG	MS	Coarse Limestone
RCP content (wt.%)	15.5	11.5	-	16.8	11.4	-	-
SSD specific gravity (kg/L)	2.47	2.56	2.70	2.51	2.58	2.76	2.79
OD specific gravity (kg/L)	2.32	2.42	2.67	2.36	2.44	2.74	2.78
Water absorption (%)	7.87	6.38	0.86	7.76	6.16	0.65	0.42
Fineness modulus	3.27	2.53	2.59	3.17	2.70	2.85	-

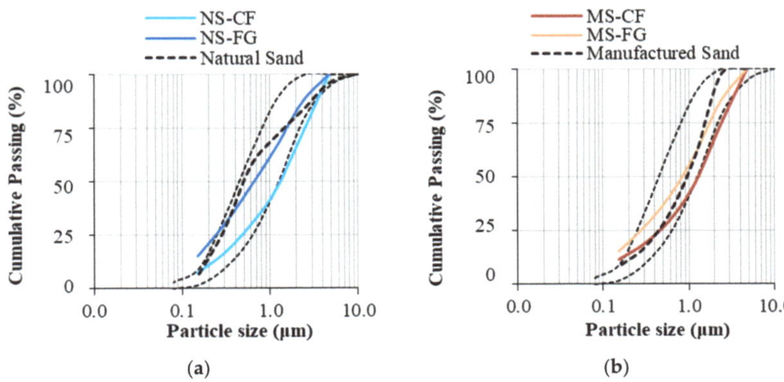

Figure 3. Particle size distribution curves for FRCA derived from (**a**) natural sand (NS) and (**b**) manufactured sand (MS).

In order to produce the FRCA concrete mixtures, the same General Use-GU Portland cement and coarse limestone aggregate (19 mm of maximum nominal size) were selected for use. Their physical properties were determined as shown in Table 2. Moreover, a limestone filler with a particle size distribution (Figure 4) smaller than the Portland cement (i.e., performance filler) was added to the mixtures proportioned through the PPM method to enhance its eco-efficiency; this is further discussed in the following section.

Table 2. Physical properties characterization.

Material	Mass (g)	Volume (cm^3)	Specific Gravity (g/cm^3)	Specific Surface Area (m^2/g)
Portland cement	31.9	10.49	3.03	1.00
Limestone filler	19.5	7.56	2.60	3.70

4.2. Mix Design Procedures and Proportions

Three distinct FRCA concrete mixtures proportioned with distinct techniques (i.e., DRM, EV and PPM) and containing 100% replacement of the fine aggregate by FRCA were designed (Table 3). Moreover, two control mixtures (e.g., 0% FRCA) incorporating natural (NS) and manufactured (MS) sand were also mix-proportioned for comparative purposes. The water-to-cement (w/c) ratio of all FRCA mixtures was selected as 0.35. In addition, an air-entraining agent (AEA) was used in all mixtures (i.e., control and recycled) to incorporate an amount of air of 7–8% according to [4], as required for structural concrete exposed to freeze-thaw cycles. Lastly, a combination of polycarboxylate-based high range and mid-range water reducer was used in all mixtures at different amounts to achieve consistency (i.e., slump value) of 100 ± 20 mm. Note that the EV and PPM methods account for the RCP attached to the FRCA particles; thus, the mixtures proportioned by these procedures presented much lower Portland cement content.

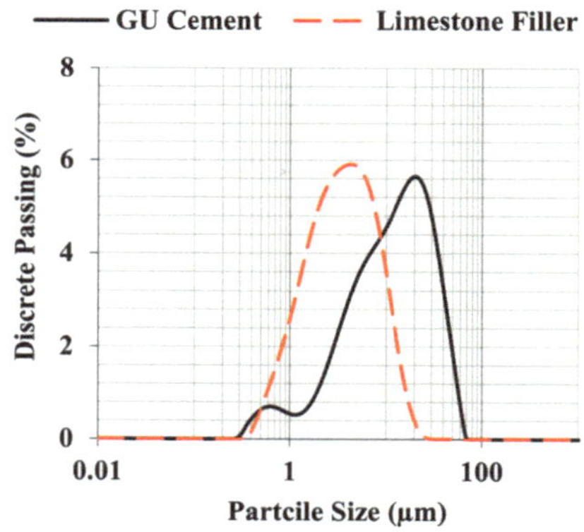

Figure 4. Particle size distribution of Portland cement and limestone filler used in the particle packing model (PPM) mixtures.

Table 3. FRCA and control concrete mix design proportions.

Mixture	Portland Cement (kg/m^3)	FRCA (kg/m^3)	Natural Fine Aggregate (kg/m^3)	Natural Coarse Aggregate (kg/m^3)	Limestone Filler (kg/m^3)	Water (kg/m^3)	w/c	AEA (%)	Water Reducer (kg/m^3)
ACI–NS	370	-	738	1032	-	174	0.47	0.65	-
ACI–MS	370	-	759	1032	-	174	0.47	0.65	-
DRM–NS CF	497	524		1032	-	174	0.35	0.45	-
DRM–NS FG	497	546		1032	-	174	0.35		-
DRM–MS CF	497	533		1032	-	174	0.35		-
DRM–MS FG	497	551		1032	-	174	0.35		-
EV–NS CF	374	714	-	1005	-	131	0.35	0.50	1.2
EV–NS FG	373	740	-	1014	-	131	0.35		1.2
EV–MS CF	372	732	-	1004	-	130	0.35		1.2
EV–MS FG	373	752	-	1006	-	131	0.35		1.2
PPM–NS CF	308	879	-	806	108	108	0.35	0.50	1.0
PPM–NS FG	333	907	-	797	83	117	0.35		1.0
PPM–MS CF	299	898	-	809	118	105	0.35		1.2
PPM–MS FG	332	915	-	798	84	116	0.35		1.2

4.3. FRCA Concrete Manufacturing

Twenty-seven cylindrical (100 mm × 200 mm) specimens were produced for the C and FRCA concrete mixtures. Likewise, three 75 mm × 75 mm × 300 mm prisms pe concrete mixture (recycled or conventional) were fabricated for the accelerated freeze-thaw tests as per ASTM C666 [75]. All specimens were fabricated according to CSA A23.2-3C [4 de-molded after 24 h and moist-cured (100% RH and 20 °C) for 28 days before testing.

4.4. Hardened Properties

Non-destructive techniques (i.e., electrical resistivity (ER) and dynamic modulus o elasticity through ultrasonic pulse velocity test (UPV)) and compressive strength tests wer performed on three specimens from each concrete mixture at 28 days. Furthermore, th stress–strain relationship of selected mixtures before being subjected to F/T was evaluate at 28 days through the use of the Stiffness Damage Test (SDT) as per Sanchez et al. [61,62 For the SDT, concrete cylinders were subjected to five compressive loading–unloadin cycles at a controlled loading rate of 0.10 MPa/s up to 40% of the 28-day compressiv strength.

4.5. Resistance to Freeze-Thaw Cycles

All recycled mixtures fabricated were tested against F/T cycles through the use o the accelerated laboratory test procedure as per ASTM C666 [75]. The procedure consist of storing the prisms in an F/T chamber, with temperatures ranging from −18 to 4 °

at time intervals not less than two nor more than five hours and producing a total of 300 cycles. Before beginning the F/T cycles, the mass, average length and fundamental transverse frequency as per [80] were recorded. The specimens were removed from the F/T chamber every 36 cycles of exposure (approximately one week) for further mass, length and frequency evaluations and were continuously monitored over time until 300 cycles were reached.

The relative dynamic modulus of elasticity of each sample was calculated based on the fundamental transverse frequency measurements, as shown in Equation (2):

$$P_c = \frac{{n_1}^2}{n^2} \times 100 \tag{2}$$

where the percental relative dynamic modulus of elasticity after "c" cycles of F/T is represented by P_c; n is the fundamental transverse frequency (Hz) at 0 cycles and n_1 is the fundamental frequency after "c" cycles of F/T. It is worth noting that the test was performed up until 300 cycles or until the sample reached a 40% decrease in the initial relative dynamic modulus of elasticity value.

The length changes were measured according to ASTM C490 [81] and calculated as per Equation (3):

$$L_c = \frac{l_2 - l_1}{L_g} \times 100 \tag{3}$$

where L_c (%) is the length change after "c" cycles of F/T; l_1 and l_2 are the comparator length readings at 0 and "c" cycles, respectively, and L_g is the effective gauge inner length measured between the gauge studs. As per ASTM C 666 [75], an expansion of less than 0.10% is the limit distinguishing F/T resistance.

Finally, the overall performance of the mixtures against F/T action was evaluated based on the durability factor (*DF*) shown in Equation (4):

$$DF = \frac{P_N N}{M} \tag{4}$$

where *DF* is the durability factor of the specimen (0–100%), which corresponds to the reduction in dynamic ME over freeze-thaw cycles; P_N (%) the dynamic ME after the Nth F/T cycle; N refers to the number of cycles at which the dynamic ME reaches the specified minimum value before discontinuing the test or the maximum number of cycles for the analysis (the lower value); and M is the specified number of cycles at which the exposure is to be terminated (i.e., 300 cycles).

4.6. Microscopic Assessment

The Damage Rating Index (DRI) was used to assess the F/T damage in FRCA concrete mixtures. Concrete prisms subjected to 300 cycles of F/T were sawn in half longitudinally using a masonry saw equipped with a diamond blade followed by successive grinding and polishing (i.e., 60, 140, 280, 600, 1200 and 3000 grits) until a flat reflective surface was achieved. A grid of 1 cm × 1 cm squares was drawn on this surface in which distinct types of cracks (i.e., closed crack in the aggregate (CCA), open crack in the aggregate (OCA), cracks in the residual and new cement paste (CCP, RCP and CCP, NCP, respectively)) were counted using a stereomicroscope at 16× magnification. Those cracks were then weighted using the weighting factors proposed by [71] and normalized to 100 cm^2 for comparative purposes. Moreover, the extended version of the DRI accounts for the cracks without any weighting factors in absolute and relative counts [66,68].

5. Results

5.1. Non-Destructive Techniques (NDT), Hardened Properties and Stress–Strain Relationship

The inner quality of the concrete before being subjected to freezing and thawing (F/T) cycles was evaluated through non-destructive techniques (NDT), hardened state properties and the stress–strain relationship. Figure 5a–c display the average surface ER, compressiv strength, dynamic modulus elasticity (obtained from the UPV) as well as the static modulu of elasticity (acquired from the stiffness damage test—SDT), respectively, for the FRC concrete mixtures. Generally, the DRM-proportioned mixtures show the lowest result followed by the EV-proportioned mixtures, with the PPM-proportioned mixture showin the highest results. The highest surface ER result obtained from the PPM-proportione mixtures ranged from 10.8 to 14.3 kΩ·cm with a standard deviation of 3.5 kΩ·cm for the CF FRCA and 5.1 kΩ·cm for the FG-FRCA, while the EV-proportioned mixtures yielded value ranging from 6.5 to 7.3 kΩ·cm. The DRM-proportioned mixtures show the lowest ER (3. to 4.4 kΩ·cm). Similarly, the dynamic moduli of elasticity of 42 to 51 GPa, 27 to 35 GPa an 15 to 16 GPa were obtained for the PPM, EV and DRM-proportioned mixtures, respectivel while 49 GPa was achieved for both CC mixtures (i.e., natural sand (NS) and manufacture sand (MS)). Likewise, targeting a compressive strength of 35 MPa, values ranging fror 65 to 41 MPa (standard deviation of 0.85 MPa for the CF–FRCA and 2.13 MPa for th FG–FRCA) were achieved with PPM-proportioned FRCA mixtures whereas 31 to 26 MP (overall standard deviation of 1.02 MPa) were achieved for the EV-proportioned mixture followed by the DRM-proportioned mixtures with an average of 20 MPa (overall standar deviation of 0.8 MPa). The influence of the aggregate type (i.e., NS vs. MS) within th FRCA was not captured through the ER, dynamic modulus of elasticity and compressiv strength; however, the results were indeed influenced by the crushing procedure especiall captured by the PPM-proportioned mixtures, showing higher results for the fully groun (FG)–FRCA mixtures compared to those of the crusher's fines (CF)–FRCA.

The influence of the different mix designs used to proportion the FRCA mixtures wa also evaluated using the SDT before being subjected to F/T to appraise the inner quality an flaws (i.e., porosity and presence of microcracks within the FRCA) of the recycled mixture The SDI, an output parameter obtained through the SDT, represents the amount of energ required to close inner flaws, while the PDI represents the plastic deformation of concret subjected to loading. Due to testing limitations, only the natural sand (NS) was selecte for evaluation along with the CF crushing procedure since the compressive strengths fo PPM-proportioned mixtures made with CF–FRCA were closer to the targeted 35 MPa (i.e 26 MPa and 41–42 MPa, respectively) compared to the FG–FRCA. Moreover, the PPM proportioned FRCA concrete mixtures were compared by the crushing procedure due t the significant difference observed through the NDT and compressive strength results. Th SDT appraises three important parameters SDI, PDI and static modulus of elasticity, whic are presented in Table 4. Interestingly, the DRM-proportioned FRCA concrete yields an SD of 0.10 (standard deviation of 0.02), similar to the SDI obtained for the PPM-proportione mixtures (i.e., 0.11 with a standard deviation of 0.002 for the CF–FRCA and 0.10 with standard deviation of 0.01 for the FG-FRCA). The EV-proportioned mixture, on the othe hand, shows an SDI of 0.20 (standard deviation of 0.04). The same trend is generall followed for the PDI and static modulus of elasticity, where the EV-proportioned mixtur presents the poorest performance with a PDI of 0.16 (standard deviation of 0.05) and static modulus of elasticity of 21 GPa (standard deviation of 4.1). Meanwhile, the DRM proportioned mixture presents a better performance with a PDI of 0.09 (standard deviatio of 0.01) and a static modulus of elasticity of 26 GPa, (standard deviation of 3.6), yet th PPM-proportioned mixture presents a PDI of 0.07 and 0.08 (standard deviation of 0.001 an 0.02, respectively) for the CF and FG crushing procedure, respectively. The static elasti modulus of the PPM-proportioned mixtures was found to be 29 GPa and 38 GPa (standar deviation of 0.13 and 0.21, respectively) for the CF and FG crushing procedure, respectivel In addition, the SDI, PDI and static modulus of elasticity of CC were found to be 0.12, 0.1 and 45 GPa, respectively, representing the values obtained for a sound concrete.

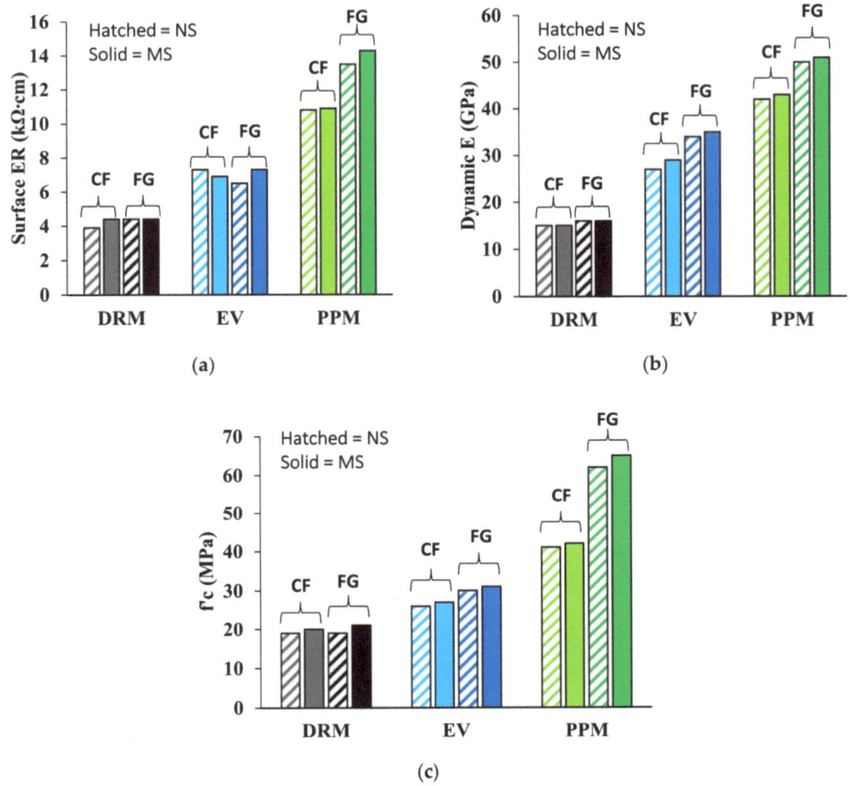

Figure 5. (**a**) Surface electrical resistivity (ER), (**b**) dynamic E and (**c**) compressive strength for all FRCA concrete mixtures.

Table 4. Stiffness damade test (SDT) output.

Mixture	SDI	PDI	Static Modulus of Elasticity (GPa)
DRM–NS–CF	0.10	0.09	26
EV–NS–CF	0.20	0.16	21
PPM–NS–CF	0.11	0.07	29
PPM–NS–FG	0.10	0.08	38

5.2. Freeze-Thaw (F/T) Resistance

5.2.1. Mass Losses

After being subjected to the F/T cycles, the mass loss of the concrete specimens was measured. Figure 6 presents the average mass loss (i.e., residual mass) results for all FRCA mixtures as a function of the number of cycles showing a decrease in the mass as the number of cycles increases. The PPM-proportioned mixtures have the best performance against F/T damage with less than 20% mass loss after 300 cycles (standard deviation ranging from 0.23 to 0.31% throughout the cycles), whereas the EV and DRM-proportioned mixtures show mass losses of up to 45% (standard deviation ranging from 0.08 to 0.24% and 0.17 to 0.39%, respectively, throughout the cycles). Although similar, the DRM-proportioned mixtures following the 300 cycles appeared the most damaged (Figure 7), where excessive spalling on the surfaces along with rounded-edges were observed compared to the EV-proportioned mixtures, with the PPM-proportioned mixtures showing the least amount

of surface damage. The DRM-proportioned FG-FRCA mixture incorporating NS did n
attain 300 cycles, losing more than 40% of its initial mass at 150 cycles.

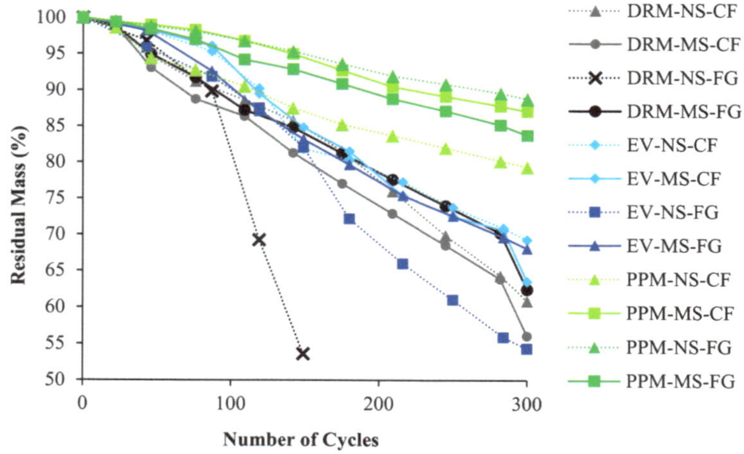

Figure 6. Mass loss as a function of freeze-thaw cycles for all FRCA mixtures.

(a) (b) (c)

Figure 7. Degradation of FRCA mixes at the end of 300 freezing and thawing cycles using (**a**) dire replacement method (DRM); (**b**) equivalent volume (EV) and (**c**) PPM mixture proportions.

5.2.2. Length Changes

The average changes in length for all FRCA concrete mixtures subjected to F/T cycle are displayed in Figure 8, including the 0.10% limit proposed by ASTM C666 [75]. All DRM proportioned mixtures exceed the proposed limit at the end of the 300 cycles (standar deviation ranging from 0.07 to 0.35% throughout all cycles), while both EV- and PPM proportioned mixtures remain below this limit at 0.10% (standard deviation ranging fror 0.01 to 0.04%) therefore, deemed resistant to F/T.

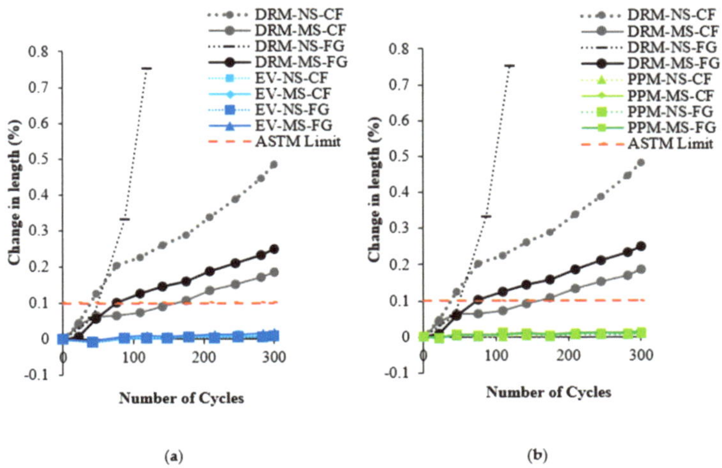

Figure 8. Length change over 300 cycles for (**a**) EV and (**b**) PPM in comparison to DRM mixtures.

5.2.3. Residual Dynamic Modulus of Elasticity

The average residual dynamic modulus of elasticity results for all FRCA mixtures are illustrated in Figure 9, including the ASTM C666 [75] limit considering concrete to be resistant to F/T when the reduction in dynamic modulus of elasticity should be below 40% at 300 cycles. It may be observed that, once again, the DRM mixes yielded the lowest results among the different mixtures, while PPM mixtures showed the best performance (standard deviation ranging from 12 to 35.5% and 4.9 to 21.6%, respectively, throughout the cycles). The EV mixtures also demonstrated acceptable behavior (i.e., about 38% damage at 300 cycles with a standard deviation ranging from 7.3 to 23.4%). Finally, no significant trend was observed for different types of FRCA (i.e., crusher's fines/fully ground (CF/FG) and manufactured sand/natural sand (MS/NS)); hence, the mix design procedure poses a greater influence on the F/T resistance of FRCA.

5.2.4. Durability Factor

The durability factors for each FRCA mixture were calculated following the F/T cycles using Equation (4) and are presented in Figure 10. By analyzing the results, one sees that PPM-proportioned mixtures represent the best performance, with all mixtures having a durability factor above 70%, followed by EV and DRM-proportioned mixtures with above 60% and from 20 to 55%, respectively. No apparent trend was observed for the different types of original fine aggregate used (i.e., NS vs. MS) and the crushing procedure (i.e., CF or FG).

5.3. Damage Rating Index (DRI)

The Damage Rating Index (DRI) was used to complement the above results obtained and better understand the cracking behavior of recycled mixtures using FRCA. Figure 11 displays the cracking behavior of concrete mixtures made of FRCA (i.e., MS-FRCA) subjected to 300 cycles of F/T through the DRI bar chart (Figure 11a), the counts per square centimeters (Figure 11b) and proportions (Figure 11c) as per the extended version of the DRI. In general, the FRCA concrete mixtures follow the same trends as previously discussed, where the PPM-proportioned mixtures show a higher resistance to cracking while the DRM-proportioned mixture produced the most cracks after the 300 cycles. Interestingly, the EV-proportioned mixtures showed less damage through the DRI and its extended version, although its F/T performance (i.e., mass loss and residual dynamic modulus of elasticity)

was similar to that of the DRM-proportioned mixture highlighting the enhancement of th
FRCA concrete when using EV as a mix design method.

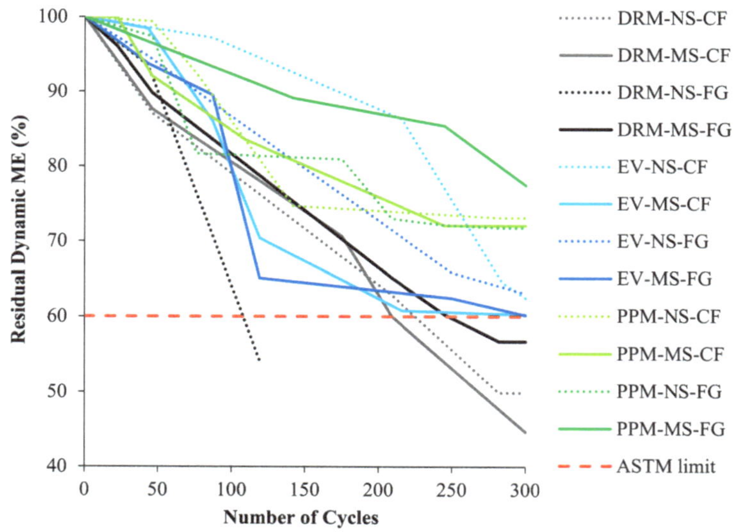

Figure 9. Residual dynamic modulus of elasticity over 300 cycles for PPM and EV in comparison t
DRM mixtures.

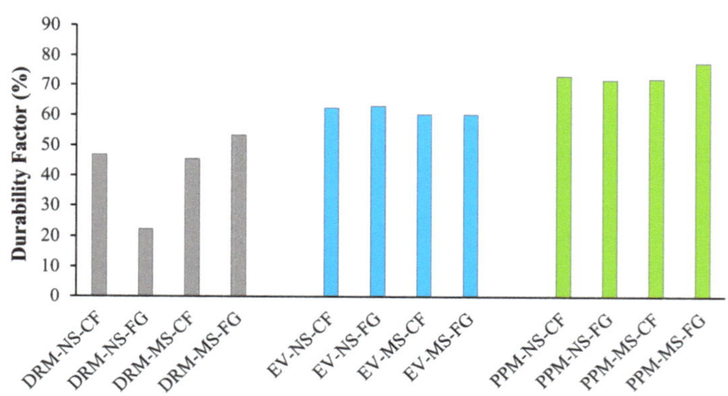

Figure 10. Durability factors of FRCA concrete mixtures subjected to accelerated freezing and thawir
cycles.

Overall, more cracks are observed in the new cement paste (i.e., CCP, NCP) for a
FRCA concrete mixtures, while very few cracks in the residual cement paste (i.e., CCP, RCI
were observed. Although the total number of cracks in the DRM-proportioned mixtur
(i.e., DRI number of 1090 and 693 counts/100 cm^2) is higher than other mixtures (i.e
DRI numbers and counts cm^2 ranging from 370 to 883 and from 267 to 436, respectively
its number of cracks in the cement paste and its proportions (i.e., 232 counts/cm^2 an
35%, respectively) are lower than the EV-proportioned mixtures, which range from 2
to 257 counts/cm^2 and 60 to 70% for the crusher's fines (CF) and finely ground (FC
FRCA, respectively. Otherwise, a lower number of cracks in the cement paste is observe
in the PPM-proportioned mixtures at 91 and 231 counts/cm^2 for the CF and FG FRC
respectively.

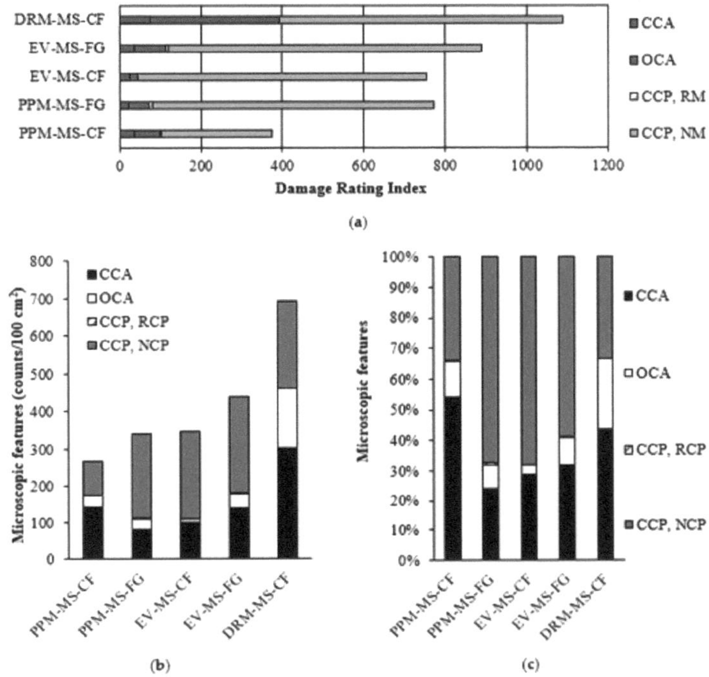

Figure 11. Distress features in FRCA mixtures shown as (**a**) Damage Rating Index (DRI) bar chart, (**b**) counts per 100 cm^2 and (**c**) percentages.

6. Discussion

6.1. Effect of Materials and Mix Design on FRCA Concrete

Several mixture design procedures were used to proportion recycled concrete made of 100% FRCA replacement incorporating two different types of original fine aggregates (i.e., natural sand and manufactured sand) and crushed using two distinct processes (i.e., crusher's fines and fully ground) to evaluate their resistance to freezing and thawing further.

The electrical resistivity and dynamic modulus of elasticity (i.e., NDTs) provide insight into the inner quality (i.e., porosity) and flaws (i.e., presence of microcracks within the FRCA) of the concrete before being subjected to freezing and thawing cycles. As presented in Section 5.1, the DRM-proportioned mixture showed the lowest surface electrical resistivity and dynamic modulus of elasticity (Figure 5a,b, respectively) due to its higher porosity since the residual cement paste is not considered in this mixture design technique, thus increasing the amount of total cement paste in the concrete. Conversely, EV and PPM mixtures were designed to account for the residual cement paste, increasing the inner quality as described by the surface electrical resistivity and dynamic modulus of elasticity. However, when the PPM mix design is used, the mixtures' porosity is further reduced due to the enhancement of the particle distribution [51]. Moreover, the type of original fine aggregate (i.e., natural and manufactured sand) within the FRCA was not captured through the surface electrical resistivity; however, the FRCA made of manufactured sand provided slightly higher results when compared to FRCA made of natural sand through the dynamic modulus of elasticity for the EV- and PPM-proportioned mixtures. Although the EV- and PPM-proportioned mixtures account for the residual cement paste adhered to the original aggregate particles, the crushing procedure influences the inner quality of the FRCA particles and consequently of the concrete. The fully ground crushing process exposes more original aggregate (i.e., residual sand and fragments of original coarse a gregate) due to its excessive processing. Therefore, a lower residual cement paste conte will result in FRCA particles having more original aggregates exposed at the surface of th FRCA (Figure 12), which in turn reduces the number of interfaces formed within the FRC concrete and reduces its porosity.

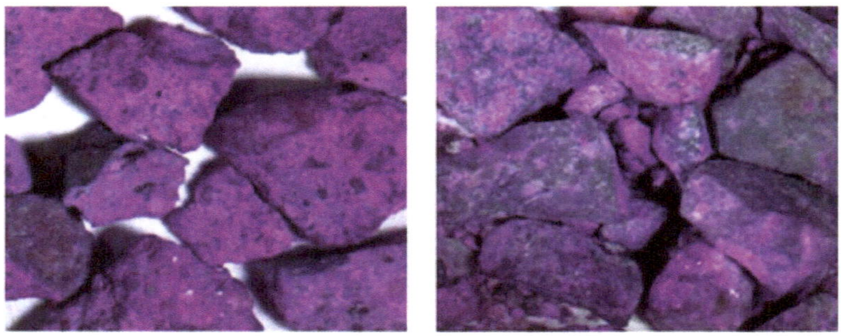

(a) (b)

Figure 12. Phenolphthalein application on FRCA produced by (**a**) CF and (**b**) FG.

The same trend is observed through the compressive strength; however, the influen of the crushing procedure is apparent for the PPM-proportioned mixtures. The fully groun FRCA produced significantly higher results at 62 MPa and 65 MPa for the natural an manufactured sand, respectively, although a 35 MPa concrete was targeted, yet not achieve for the mixtures proportioned through the EV and DRM (i.e., 26–31 MPa and 19–21 MP respectively). Nevertheless, PPM-proportioned FRCA mixtures, especially the fully groun FRCA, present better results when compared to EV- and DRM-proportioned mixtures, thu producing FRCA concrete with a better overall inner quality (i.e., porosity). Besides th lower amount of residual cement paste of the fully ground FRCA, the improvement of th FRCA–PPM–FG mixture may also be explained due to the content of the inert filler. Sin the volume of filler added in PPM mixtures was equivalent to the volume of the residu cement paste of FRCA, FRCA–PPM–FG mixtures resulted in higher cement content an lower filler content than FRCA–PPM–CF mixtures. This slightly higher cement conte might have improved the bond between the FRCA particles, and the new cement pas is fully ground mixtures due to the natural self-healing mechanism developed by th hydration of Portland cement, a phenomenon widely reported in the literature.

6.2. Mechanical Response of FRCA Concrete Subjected to Cyclical Loading

Despite the results obtained through non-destructive techniques (NDTs) and th compressive strength, the EV-proportioned mixtures displayed the lowest performan (i.e., SDI, PDI and modulus of elasticity) out of all mixture proportions used for FRC concrete. As such, the SDIs of the DRM-proportioned mixtures were found to be similar those of the PPM-proportioned mixtures, both of which resemble sound concrete. Howeve the same constituents were used in all mixtures, yet at different proportions, which we captured by the SDT through the SDI and PDI (Table 4). Although the SDI and PD indicate the inner quality of the concrete mixtures without the influence of their strengt differences and corresponding loading, the slopes of the curves and hysteresis area (i. area under the stress–strain curve) obtained through cyclic loading may provide addition insight. Figure 13 shows the stress–strain relationship under cyclic loading for each mi design incorporating natural sand and crusher's fines FRCA including the fully groun FRCA for the PPM mix design due to its apparent difference observed in the NDTs an compressive strength results. The CC (i.e., ACI–NS) represents sound concrete. The slop of the PPM-proportioned mixtures are steeper compared to the DRM and EV mixture therefore, representing a higher stiffness; however, the CC displayed the steepest slope. The EV-proportioned mixture presents the greatest energy dissipation (i.e., largest hysteresis area) and plastic deformation followed by the PPM then the DRM proportioned mixtures. Interestingly, the initial loading slope of the PPM mixture incorporating the fully ground FRCA is similar to that of the CC. Moreover, the fully ground FRCA presents a steeper slope, thus stiffer than that of the crusher's fines FRCA, which was not evident through the SDI, PDI and modulus of elasticity obtained through the SDT.

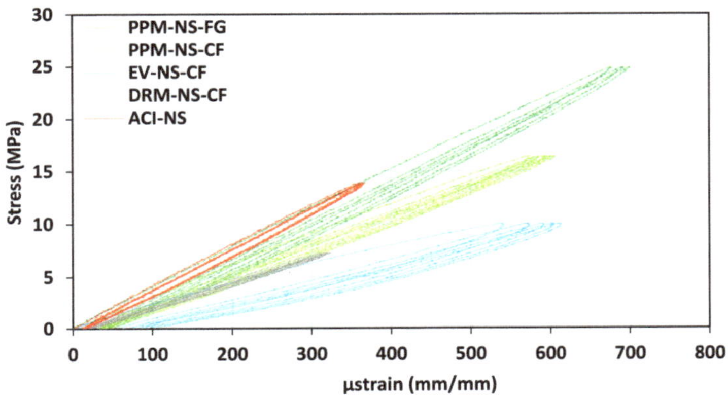

Figure 13. Stress–strain curves obtained from the SDT comparing FRCA–CF and CC mixtures from NS source, and PPM with FRCA of CF and FG types.

Since the residual cement paste was accounted for in EV-proportioned mixtures and disregarded in DRM, the difference in its performance may be justified by the higher FRCA content present in EV mixtures (714–752 kg/m^3) when compared to DRM mixtures (524–551 kg/m^3). Hence, the overall stiffness of the concrete may be governed by the higher number of interfaces found within the FRCA concrete. On the other hand, the PPM-proportioned mixtures contain the most FRCA (879–915 kg/m^3), thus comprised of an even higher number of interfaces; however, all particles were more efficiently distributed/packed in a system with lower porosity.

6.3. Damage Propagation in FRCA Concrete after 300 Freezing and Thawing Cycles

The resistance to freezing and thawing cycles of concrete is crucial in harsh environments such as those experienced in Canada. The freezing and thawing resistance was therefore appraised for each type of mixture proportioning technique, as well as the nature of the original fine aggregate and crushing procedure. The most apparent difference was between the mixture proportioning techniques while, interestingly, the nature of the original fine aggregate (i.e., natural and manufactured sand) and the crushing procedure (i.e., crusher's fines and fully ground) did not show any trends. Similar to the non-destructive techniques and compressive strength results, the DRM-proportioned FRCA concrete showed the poorest performance in mass loss, length change and residual modulus of elasticity and exceeded ASTM C666 [75] limits while both PPM and EV-proportioned mixtures conformed to the standard (Figures 8 and 9, respectively). The PPM-proportioned mixtures displayed the highest resistance to freezing and thawing, while the EV-proportioned mixtures were only slightly above the residual modulus of elasticity requirements, although it did not record any change in length and its mass loss was similar to that of the DRM-proportioned mixtures.

In order to obtain a more thorough understanding of the resistance to freezing and thawing of the various concrete mixtures made of FRCA, a thorough microscopic analysis through the Damage Rating Index (DRI) method was performed. Consequently, the DRM-proportioned mixture evaluated in the microscopic analysis showed a significant numb of open cracks within the aggregates, usually associated with distress as opposed to close cracks in the aggregates (Figure 11), and propagating into the new cement paste (Figure 14 Meanwhile, the cracks tend to propagate through the bulk, new cement paste for FRC concrete mixtures proportioned with EV and PPM and do not propagate through th FRCA particles; however, they may propagate in the interfaces formed between the FRC particles and new mortar (Figure 14b).

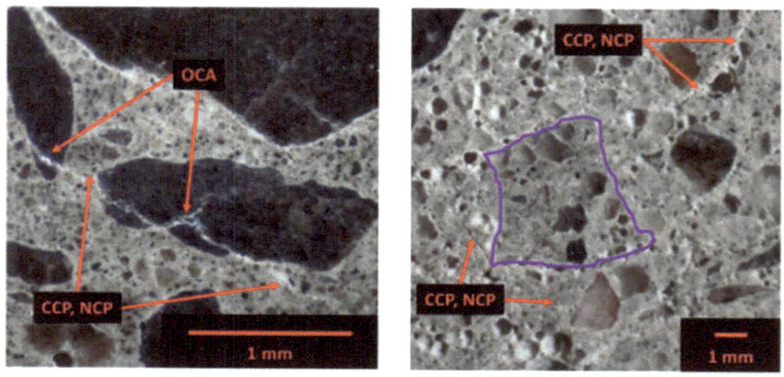

(**a**) (**b**)

Figure 14. Crack propagation through (**a**) coarse aggregate as open cracks in the aggregate (OC and (**b**) in new cement paste (CCP, NCP).

Noticeably, a similar trend is observed for the PPM- and EV-proportioned mixture where fully ground FRCA was more susceptible to cracking than the crusher's fines FRC Before being subjected to freezing and thawing, the fully ground FRCA exhibited bette results than the crusher's fines. A fully ground FRCA is subjected to a more intensiv crushing procedure, hence the higher removal of residual cement paste, exposure of resi ual sand and original virgin aggregate (OVA) and potentially the introduction of a highe number of microcracks; however, cracks produced by freezing and thawing were uncon monly observed within the FRCA and mostly present within the new cement paste. Th crushing procedure, therefore, influenced the cracking behavior of the FRCA concret such that the formation of new interfaces may govern this behavior where residual t new mortar interfaces show more resistance to cracking than residual sand/OVA to ne mortar interfaces due to the difference between the stiffness of the components [82]. Th literature suggests that the strength of concrete made of RCA is governed by its weake ITZ [83–85], while a stronger bond is observed between the residual to new mortar i terface [64,83,86,87]. In addition, more FRCA was used in mixtures containing the full ground FRCA, thus creating more interfaces and weaker zones within the concrete. Sim larly, the PPM-proportioned mixtures contain higher amounts of FRCA and lower coars aggregate content when compared to the EV-proportioned mixtures, thus reducing th differential stresses within the recycled concrete proportioned through PPM and, therefor the number of cracks. Moreover, PPM–FG mixtures contain lower residual cement past and filler content, which may also contribute to the higher number of cracks in the ne cement paste when compared to PPM mixtures made with crusher's fines FRCA. Th additional content of fillers in the PPM mixtures made with crusher's fines FRCA may als have contributed to the quality of its new interface. Nevertheless, crusher's fines FRCA a a by-product of coarse RCA production, which require less processing, hence energy t produce compared to the fully ground FRCA.

7. Conclusions

Three types of mix proportioning techniques (DRM, EV and PPM) were used in this work and assessed by evaluating the inner quality and the resistance to freezing and thawing of FRCA concrete designed with 100% replacement. The main findings can be highlighted hereafter:

- The PPM-proportioned mixtures showed the highest inner quality before being subjected to freezing and thawing, followed by the EV and DRM-proportioned mixtures. The influence of the type of residual sand was not as significant as the crushing procedure, indicating that FRCA subjected to a more rigorous crushing sequence presented a better inner quality;
- The EV and PPM mixture proportioning techniques produced a concrete made of FRCA having adequate freezing and thawing resistance, while the DRM-proportioned mixtures were not considered resistant to freezing and thawing. The PPM mixtures showed the best performance followed by EV then DRM-proportioned FRCA concrete mixtures while no apparent trend was observed between the type of residual sand and crushing procedure;
- The overall durability factor of all FRCA mixtures subjected to freeze-thaw cycles was observed to have considerable variation between the mix design methods. A higher durability factor was observed for PPM mixtures, followed by EV then DRM. This demonstrates that the mix design procedure adopted to design FRCA concrete is more important than the material's quality; PPM and EV-proportioning techniques being capable of reducing the variability presented while using the DRM;
- The DRI captured the differences in the damage propagation of FRCA concrete subjected to freezing and thawing. The highest level of damage was observed in DRM-proportioned mixtures, whereas the EV- and PPM-proportioned mixtures showed less damage. Despite showing a better performance before being subjected to freezing and thawing, the fully ground FRCA concrete presented more damage compared to the crusher's fines FRCA. The crushing procedure significantly influences the mechanical properties, inner quality and crack generation and propagation in FRCA concrete. Further research is therefore required to understand the cracking behavior of RCA (i.e., coarse and fine) concerning its multi-phase nature.

The freezing and thawing resistance of concrete made of FRCA was significantly improved when using mixture proportioning techniques different than the direct replacement method, which failed to comply with the specified requirements. Although it is the most used mix design technique in practice, it is evident that direct replacement methods not only show the poorest performance overall but also require a much higher cement content resulting in an unsustainable approach. Furthermore, it is clear that the inner quality of FRCA manufactured through multiple series of crushing (i.e., FG) is enhanced when compared to only two series (i.e., CF); However, when the proportion was properly mixed, recycled concrete mixtures made of both FG and CF particles displayed quite suitable results. Therefore, it is recommended that the CF method be adopted since it requires less energy and thus provides a more sustainable material with less generation of waste.

Author Contributions: Conceptualization, L.F.M.S., G.F. and P.-C.N.; methodology, L.F.M.S., G.F. and P.-C.N.; validation, L.F.M.S., G.F. and P.-C.N.; formal analysis, C.T., M.T.d.G., H.F.M., G.P.d.A., D.J.d.S., O.N. and A.D.; investigation, C.T., M.T.d.G. and H.F.M.; resources, O.N. and A.D.; data curation, C.T., M.T.d.G. and H.F.M.; writing—original draft preparation, C.T., M.T.d.G., H.F.M., G.P.d.A., D.J.d.S. and L.F.M.S.; writing—review and editing, L.F.M.S.; supervision, L.F.M.S. and G.F.; project administration, L.F.M.S. and G.F.; funding acquisition, L.F.M.S. and G.F. All authors have read and agreed to the published version of the manuscript.

Funding: This research was funded by the National Research Council Canada (NRCC) through a research contract with the University of Ottawa (Principal Investigators—Leandro F. M. Sanchez; uOttawa and Gholamreza Fathifazl; NRCC).

Institutional Review Board Statement: Not applicable.

Informed Consent Statement: Not applicable.

Data Availability Statement: All the data are available within the manuscript.

Acknowledgments: The authors gratefully acknowledge the financial support from the Nation Research Council Canada (NRCC) for taking part in this collaborative research project with th University of Ottawa and Tomlinson Group for generously providing aggregates used in this wo They would also like to thank CANMET Mining for supporting with the necessary equipment f aggregate crushing.

Conflicts of Interest: The authors declare no conflict of interest.

References

1. *Summary Report 2021: Canada's Net-Zero Future*; Canadian Institute for Climate Choices: Ottawa, ON, Canada; Available onlir https://climatechoices.ca/reports/canadas-net-zero-future/ (accessed on 21 January 2022).
2. World Business Council for Sustainable Development. *Cement Industry Energy and CO_2 Performance: Getting the Numbers Rig (GNR)*; World Business Council for Sustainable Development: Geneva, Switzerland, 2016.
3. Ministry of Natural Resources. *The State of the Aggregate Resource in Ontario Study*; Ministry of Northern Development, Min Natural Resources and Forestry: Peterborough, ON, Canada, 2010. Available online: https://files.ontario.ca/environment-an energy/aggregates/aggregate-resource-in-ontario-study/286996.pdf (accessed on 21 January 2022).
4. Canadian Standard Association (CSA). *Concrete Materials and Methods of Concrete Construction/Test Methods and Standard Practic for Concrete. CSA A23.1:19/CSA A23.2:19*; Canadian Standard Association (CSA): Toronto, ON, Canada, 2019.
5. Lamond, J.F.; Campbell, R.L.; Campbell, T.R.; Cazares, J.A.; Giraldi, A.; Halczak, W.; Hale, H.C., Jr.; Jenkins, N.J.T.; Miller, Seabrook, P.T. *Removal and Reuse of Hardened Concrete (ACI 555R)*; American Concrete Institute (ACI): Fermington Hills, MI, US 2001.
6. Marinković, S.; Ignjatović, I.; Radonjanin, V. Life-cycle assessment (LCA) of concrete with recycled aggregates (RAs). In *Handbo of Recycled Concrete and Demolition Waste*; Elsevier: Amsterdam, The Netherlands, 2013; pp. 569–604.
7. Fathifazl, G.; Abbas, A.; Razaqpur, A.G.; Isgor, O.B.; Fournier, B.; Foo, S. New Mixture Proportioning Method for Concrete Mac with Coarse Recycled Concrete Aggregate. *J. Mater. Civ. Eng.* **2009**, *21*, 601–611. [CrossRef]
8. Hayles, M.; Sanchez, L.F.M.; Noël, M. Eco-efficient low cement recycled concrete aggregate mixtures for structural applicatior *Constr. Build. Mater.* **2018**, *169*, 724–732. [CrossRef]
9. Ahimoghadam, F.; Sanchez, L.F.M.; De Souza, D.J.; Andrade, G.P.; Noël, M.; Demers, A. Influence of the Recycled Concre Aggregate Features on the Behavior of Eco-Efficient Mixtures. *J. Mater. Civ. Eng.* **2020**, *32*, 04020252. [CrossRef]
10. Abbas, A.; Fathifazl, G.; Fournier, B.; Isgor, B.; Zavadil, R.; Razaqpur, A.; Foo, S. Quantification of the residual mortar content recycled concrete aggregates by image analysis. *Mater. Charact.* **2009**, *60*, 716–728. [CrossRef]
11. Dean, S.W.; Abbas, A.; Fathifazl, G.; Isgor, O.B.; Razaqpur, A.G.; Fournier, B.; Foo, S. Proposed Method for Determining th Residual Mortar Content of Recycled Concrete Aggregates. *J. ASTM Int.* **2008**, *5*, 1–12. [CrossRef]
12. Guedes, M.; Evangelista, L.; de Brito, J.; Ferro, A.C. Microstructural Characterization of Concrete Prepared with Recycle Aggregates. *Microsc. Microanal.* **2013**, *19*, 1222–1230. [CrossRef]
13. Evangelista, L.; Guedes, M. Microstructural studies on recycled aggregate concrete. In *New Trends in Eco-Efficient and Recycl Concrete*; Woodhead Publishing: Sawston, UK, 2018; Volume 24, pp. 425–451. [CrossRef]
14. Pepe, M.; Toledo Filho, R.D.; Koenders, E.A.B.; Martinelli, E. A novel mix design methodology for Recycled Aggregate Concre *Constr. Build. Mater.* **2016**, *122*, 362–372. [CrossRef]
15. Amario, M.; Rangel, C.S.; Pepe, M.; Filho, R.D.T. Optimization of normal and high strength recycled aggregate concrete mixtur by using packing model. *Cem. Concr. Compos.* **2017**, *84*, 83–92. [CrossRef]
16. Pradhan, S.; Kumar, S.; Barai, S.V. Recycled aggregate concrete: Particle Packing Method (PPM) of mix design approach. *Cons Build. Mater.* **2017**, *152*, 269–284. [CrossRef]
17. Kirthika, S.K.; Singh, S.K.; Chourasia, A. Performance of Recycled Fine-Aggregate Concrete Using Novel Mix-Proportionin Method. *J. Mater. Civ. Eng.* **2020**, *32*, 04020216. [CrossRef]
18. de Andrade, G.P.; Polisseni, G.D.C.; Pepe, M.; Filho, R.D.T. Design of structural concrete mixtures containing fine recycle concrete aggregate using packing model. *Constr. Build. Mater.* **2020**, *252*, 119091. [CrossRef]
19. Behera, M.; Bhattacharyya, S.; Minocha, A.; Deoliya, R.; Maiti, S. Recycled aggregate from C&D waste & its use in concrete— breakthrough towards sustainability in construction sector: A review. *Constr. Build. Mater.* **2014**, *68*, 501–516. [CrossRef]
20. Fan, C.-C.; Huang, R.; Hwang, H.; Chao, S.-J. The Effects of Different Fine Recycled Concrete Aggregates on the Properties Mortar. *Materials* **2015**, *8*, 2658–2672. [CrossRef]
21. Florea, M.V.A.; Brouwers, H.J.H. Properties of various size fractions of crushed concrete related to process conditions and re-us *Cem. Concr. Res.* **2013**, *52*, 11–21. [CrossRef]

2. Beauchemin, S.; Fournier, B. Petrographic Analysis of Aggregate Particles Used in the Accelerated Mortar Bar Test for Evaluating the Potential Alkali-Reactivity of Recycled Concrete Aggregates (RCA). In Proceedings of the 14th International Conference of Alkali Aggregate Reaction, Austin, TX, USA, 20–25 May 2012.
3. Braymand, S.; Roux, S.; Fares, H.; Déodonne, K.; Feugeas, F. Separation and Quantification of Attached Mortar in Recycled Concrete Aggregates. *Waste Biomass Valori.* **2016**, *8*, 1393–1407. [CrossRef]
4. Locati, F.; Zega, C.; dos Santos, G.C.; Marfil, S.; Falcone, D. Petrographic method to semi-quantify the content of particles with reactive components and residual mortar in ASR-affected fine recycled concrete aggregates. *Cem. Concr. Compos.* **2021**, *119*, 104003. [CrossRef]
5. Sosa, M.E.; Lamnek, A.; Benito, D.E.; Zega, C.J.; Di Maio, A.A. Composición y propiedades del agregado fino reciclado en función del tamaño de partícula. In Proceedings of the VII Congreso Internacional—21 Reunión Técnica de la AATH, Salta, Argentina, 28–30 September 2016; pp. 627–633. (In Spanish).
6. Zhao, Z.; Damidot, D.; Remond, S.; Courard, L. Toward the quantification of the cement paste content of fine recycled concrete aggregates by salicylic acid dissolution corrected by a theoretical approach. In Proceedings of the 14th International Congress on the Chemistry of Cement, Beijing, China, 13–16 October 2015.
7. De Larrard, F. *Concrete Mixture Proportioning: A Scientific Approach*, 1st ed.; CRC Press: Boca Raton, FL, USA, 2014.
8. Khatib, J.M. Properties of concrete incorporating fine recycled aggregate. *Cem. Concr. Res.* **2005**, *35*, 763–769. [CrossRef]
9. Guo, Z.; Chen, C.; Lehman, D.E.; Xiao, W.; Zheng, S.; Fan, B. Mechanical and durability behaviours of concrete made with recycled coarse and fine aggregates. *Eur. J. Environ. Civ. Eng.* **2017**, *24*, 171–189. [CrossRef]
0. Kapoor, K.; Singh, S.P.; Singh, B. Evaluating the durability properties of self compacting concrete made with coarse and fine recycled concrete aggregates. *Eur. J. Environ. Civ. Eng.* **2020**, *24*, 2383–2399. [CrossRef]
1. Khalid, F.S.; Saaidin, S.H.; Shahidan, S.; Othman, N.H.; Irwan, J.M.; Guntor, N. Performance of Concrete Containing Fine Recycled Concrete Aggregate (FRCA) and Fine Crumb Rubber (FCR) As Partial Sand Replacement. *IOP Conf. Ser. Mater. Sci. Eng.* **2020**, *917*, 12017. [CrossRef]
2. Revilla-Cuesta, V.; Ortega-López, V.; Skaf, M.; Manso, J.M. Effect of fine recycled concrete aggregate on the mechanical behavior of self-compacting concrete. *Constr. Build. Mater.* **2020**, *263*, 120671. [CrossRef]
3. Zaki, Z.; Khalid, F.S.; Guntor, N.A.A. The Optimum Replacement of Fine Recycled Concrete Aggregate on the Compressive and Splitting Tensile Strength of the Concrete. *Int. J. Integr. Eng.* **2020**, *12*, 18–26. [CrossRef]
4. Rao, A.; Jha, K.N.; Misra, S. Use of aggregates from recycled construction and demolition waste in concrete. *Resour. Conserv. Recycl.* **2007**, *50*, 71–81. [CrossRef]
5. Cartuxo, F.; de Brito, J.; Evangelista, L.; Jiménez, J.; Ledesma, E. Rheological behaviour of concrete made with fine recycled concrete aggregates—Influence of the superplasticizer. *Constr. Build. Mater.* **2015**, *89*, 36–47. [CrossRef]
6. Pereira, P.; Evangelista, L.; de Brito, J. The effect of superplasticisers on the workability and compressive strength of concrete made with fine recycled concrete aggregates. *Constr. Build. Mater.* **2012**, *28*, 722–729. [CrossRef]
7. Zhao, Z.; Rémond, S.; Damidot, D.; Xu, W. Influence of fine recycled concrete aggregates on the properties of mortars. *Constr. Build. Mater.* **2015**, *81*, 179–186. [CrossRef]
8. Kumar, B.V.; Ananthan, H.; Balaji, K. Experimental studies on utilization of recycled coarse and fine aggregates in high performance concrete mixes. *Alex. Eng. J.* **2018**, *57*, 1749–1759. [CrossRef]
9. Pedro, D.; de Brito, J.; Evangelista, L. Structural concrete with simultaneous incorporation of fine and coarse recycled concrete aggregates: Mechanical, durability and long-term properties. *Constr. Build. Mater.* **2017**, *154*, 294–309. [CrossRef]
0. Johnson, C.V.; Chen, J.; Hasparyk, N.P.; Monteiro, P.J.; Akono, A.T. Fracture properties of the alkali silicate gel using microscopic scratch testing. *Cem. Concr. Compos.* **2017**, *79*, 71–75. [CrossRef]
1. Akono, A.-T.; Chen, J.; Zhan, M.; Shah, S.P. Basic creep and fracture response of fine recycled aggregate concrete. *Constr. Build. Mater.* **2021**, *266*, 121107. [CrossRef]
2. Kirthika, S.; Singh, S. Durability studies on recycled fine aggregate concrete. *Constr. Build. Mater.* **2020**, *250*, 118850. [CrossRef]
3. Martínez-García, R.; Rojas, M.; Pozo, J.; Fraile-Fernández, F.; Juan-Valdés, A. Evaluation of Mechanical Characteristics of Cement Mortar with Fine Recycled Concrete Aggregates (FRCA). *Sustainability* **2021**, *13*, 414. [CrossRef]
4. Evangelista, L.; de Brito, J. Mechanical behaviour of concrete made with fine recycled concrete aggregates. *Cem. Concr. Compos.* **2007**, *29*, 397–401. [CrossRef]
5. Nedeljković, M.; Visser, J.; Šavija, B.; Valcke, S.; Schlangen, E. Use of fine recycled concrete aggregates in concrete: A critical review. *J. Build. Eng.* **2021**, *38*, 102196. [CrossRef]
6. Macedo, H.F. Concrete Made with Fine Recycled Concrete Aggregate (FRCA): A Feasibility Study. Master's Thesis, University of Ottawa, Ottawa, ON, Canada, 2019.
7. Sunayana, S.; Barai, S.V. Recycled aggregate concrete incorporating fly ash: Comparative study on particle packing and conventional method. *Constr. Build. Mater.* **2017**, *156*, 376–386. [CrossRef]
8. Fennis, S.A.A.M.; Walraven, J.C.; Uijl, J.A.D. Compaction-interaction packing model: Regarding the effect of fillers in concrete mixture design. *Mater. Struct.* **2012**, *46*, 463–478. [CrossRef]
9. Fennis, S.A.A.M.; Walraven, J.C. Using particle packing technology for sustainable concrete mixture design. *Heron* **2012**, *57*, 73–101.

50. Damineli, B.L.; John, V.M.; Lagerblad, B.; Pileggi, R.G. Viscosity prediction of cement-filler suspensions using interference mod A route for binder efficiency enhancement. *Cem. Concr. Res.* **2016**, *84*, 8–19. [CrossRef]
51. de Grazia, M.T.; Sanchez, L.F.M.; Romano, R.C.O.; Pileggi, R.G. Investigation of the use of continuous particle packing mode (PPMs) on the fresh and hardened properties of low-cement concrete (LCC) systems. *Constr. Build. Mater.* **2019**, *195*, 524–53 [CrossRef]
52. Azarsa, P.; Gupta, R. Electrical Resistivity of Concrete for Durability Evaluation: A Review. *Adv. Mater. Sci. Eng.* **2017**, *201* 8453095. [CrossRef]
53. Topçu, I.B.; Şengel, S. Properties of concretes produced with waste concrete aggregate. *Cem. Concr. Res.* **2004**, *34*, 1307–131 [CrossRef]
54. Bogas, J.A.; de Brito, J.; Ramos, D. Freeze–thaw resistance of concrete produced with fine recycled concrete aggregates. *J. Clea Prod.* **2016**, *115*, 294–306. [CrossRef]
55. Abbas, A.; Fathifazl, G.; Isgor, B.; Razaqpur, A.; Fournier, B.; Foo, S. Durability of recycled aggregate concrete designed wit equivalent mortar volume method. *Cem. Concr. Compos.* **2009**, *31*, 555–563. [CrossRef]
56. Fan, C.-C.; Huang, R.; Hwang, H.; Chao, S.-J. Properties of concrete incorporating fine recycled aggregates from crushed concre wastes. *Constr. Build. Mater.* **2016**, *112*, 708–715. [CrossRef]
57. Marie, I. Optimal Allowable Residual Mortar Content in Recycled Aggregates Crushed from Parent Concrete Implementin Different Waste Materials. *Am. J. Earth Environ. Sci.* **2018**, *1*, 26–33.
58. Knaack, A.M.; Kurama, Y.C. Behavior of Reinforced Concrete Beams with Recycled Concrete Coarse Aggregates. *J. Struct. En* **2015**, *141*, B4014009. [CrossRef]
59. Pedro, D.; de Brito, J.; Evangelista, L. Influence of the use of recycled concrete aggregates from different sources on structur concrete. *Constr. Build. Mater.* **2014**, *71*, 141–151. [CrossRef]
60. Sanchez, L.; Fournier, B.; Jolin, M.; Bastien, J. Evaluation of the stiffness damage test (SDT) as a tool for assessing damage i concrete due to ASR: Test loading and output responses for concretes incorporating fine or coarse reactive aggregates. *Cem. Con Res.* **2014**, *56*, 213–229. [CrossRef]
61. Sanchez, L.; Fournier, B.; Jolin, M.; Bastien, J. Evaluation of the Stiffness Damage Test (SDT) as a tool for assessing damage i concrete due to alkali-silica reaction (ASR): Input parameters and variability of the test responses. *Constr. Build. Mater.* **2015**, *7* 20–32. [CrossRef]
62. Sanchez, L.; Fournier, B.; Jolin, M.; Bastien, J.; Mitchell, D. Practical use of the Stiffness Damage Test (SDT) for assessing damag in concrete infrastructure affected by alkali-silica reaction. *Constr. Build. Mater.* **2016**, *125*, 1178–1188. [CrossRef]
63. Bogas, J.A.; Gomes, A.; Gomes, M.G. Estimation of water absorbed by expanding clay aggregates during structural lightweig concrete production. *Mater. Struct.* **2012**, *45*, 1565–1576. [CrossRef]
64. Yildirim, S.T.; Meyer, C.; Herfellner, S. Effects of internal curing on the strength, drying shrinkage and freeze–thaw resistance concrete containing recycled concrete aggregates. *Constr. Build. Mater.* **2015**, *91*, 288–296. [CrossRef]
65. Mehta, P.K.; Monteiro, P.J.M. *Concrete: Microstructure, Properties, and Materials*, 4th ed.; McGraw-Hill: New York, NY, USA, 201
66. Sanchez, L.; Fournier, B.; Jolin, M.; Bedoya, M.A.B.; Bastien, J.; Duchesne, J. Use of Damage Rating Index to Quantify Alkali-Sili Reaction Damage in Concrete: Fine versus Coarse Aggregate. *ACI Mater. J.* **2016**, *113*, 395–407. [CrossRef]
67. Hooton, R.; Rivard, P.; Fournier, B.; Ballivy, G. The Damage Rating Index Method for ASR Affected Concrete—A Critical Revie of Petrographic Features of Deterioration and Evaluation Criteria. *Cem. Concr. Aggregates* **2002**, *24*, 11228. [CrossRef]
68. Sanchez, L.; Fournier, B.; Jolin, M.; Duchesne, J. Reliable quantification of AAR damage through assessment of the Damage Ratin Index (DRI). *Cem. Concr. Res.* **2015**, *67*, 74–92. [CrossRef]
69. Grattan-Bellew, P.E.; Mitchell, L.D. Quantitative Petrographic Analysis of Concrete—The Damage Rating Index (DRI) Metho A Review. In Proceedings of the Marc-Andre Berube Symposium on Alkali-Aggregate Reactivity in Concrete, Montréal, Q Canada, 31 May–3 June 2006; pp. 321–334.
70. Dunbar, P.A.; Grattan-Bellew, P.E. Results of damage rating evaluation of condition of concrete from a number of structur affected by AAR. In Proceedings of the CANMET/ACI International Workshop On Alkali-Aggregate Reactions In Concret Dartmouth, NS, Canada, 1–4 October 1995; pp. 257–265.
71. Villeneuve, V.; Fournier, B.; Duchesne, J. Determination of the Damage in Concrete Affected by ASR—the Damage Rating Inde (DRI). In Proceedings of the 14th International Conference of Alkali-Aggregate Reaction in Concrete, Austin, TX, USA, 20–25 Ma 2012.
72. Zhu, Y.; Zahedi, A.; Sanchez, L.F.; Fournier, B.; Beauchemin, S. Overall assessment of alkali-silica reaction affected recycle concrete aggregate mixtures derived from construction and demolition waste. *Cem. Concr. Res.* **2021**, *142*, 106350. [CrossRef]
73. Trottier, C.; Zahedi, A.; Ziapour, R.; Sanchez, L.; Locati, F. Microscopic assessment of recycled concrete aggregate (RCA) mixtur affected by alkali-silica reaction (ASR). *Constr. Build. Mater.* **2021**, *269*, 121250. [CrossRef]
74. Trottier, C.; Ziapour, R.; Zahedi, A.; Sanchez, L.; Locati, F. Microscopic characterization of alkali-silica reaction (ASR) affecte recycled concrete mixtures induced by reactive coarse and fine aggregates. *Cem. Concr. Res.* **2021**, *144*, 106426. [CrossRef]
75. ASTM International. *Standard Test Method for Resistance of Concrete to Rapid Freezing and Thawing. ASTM C666*; ASTM Inter-nation West Conshohocken, PA, USA, 2015. [CrossRef]
76. ASTM International. *Standard Test Method for Compressive Strength of Cylindrical Concrete Specimens. ASTM C39*; ASTM Internation West Conshohocken, PA, USA, 2018. [CrossRef]

7. ASTM International. *Standard Test Method for Portland-Cement Content of Hardened Hydraulic-Cement. ASTM C1084*; ASTM International: West Conshohocken, PA, USA, 2015. [CrossRef]
8. ASTM International. *Standard Test Methods for Chemical Analysis of Hydraulic Cement. ASTM C114*; ASTM International: West Conshohocken, PA, USA, 2018. [CrossRef]
9. Rodrigues, F.; Evangelista, L.; De Brito, J. A new method to determine the density and water absorption of fine recycled aggregates. *Mater. Res.* **2013**, *16*, 1045–1051. [CrossRef]
0. ASTM International. *Test Method for Fundamental Transverse, Longitudinal, and Torsional Resonant Frequencies of Concrete Specimens. ASTM C215-14*; ASTM International: West Conshohocken, PA, USA, 2015.
1. ASTM International. *Standard Practice for Use of Apparatus for the Determination of Length Change of Hardened Cement Paste, Mortar, and Concrete. ASTM C490/C490M-21*; ASTM International: West Conshohocken, PA, USA, 2015.
2. Adams, M.P.; Fu, T.; Cabrera, A.G.; Morales, M.; Ideker, J.H.; Isgor, O.B. Cracking susceptibility of concrete made with coarse recycled concrete aggregates. *Constr. Build. Mater.* **2016**, *102*, 802–810. [CrossRef]
3. Poon, C.; Shui, Z.; Lam, L. Effect of microstructure of ITZ on compressive strength of concrete prepared with recycled aggregates. *Constr. Build. Mater.* **2004**, *18*, 461–468. [CrossRef]
4. Qi, B.; Gao, J.M. Fracture Properties and Microstructure of Concrete Made with Recycled Aggregate. *Adv. Mater. Res.* **2013**, *773*, 693–699. [CrossRef]
5. Otsuki, N.; Miyazato, S.-I.; Yodsudjai, W. Influence of Recycled Aggregate on Interfacial Transition Zone, Strength, Chloride Penetration and Carbonation of Concrete. *J. Mater. Civ. Eng.* **2003**, *15*, 443–451. [CrossRef]
6. Evangelista, L.; Guedes, M.; Ferro, A.; de Brito, J. Microstructure of concrete prepared with construction recycled aggregates. *Microsc. Microanal.* **2013**, *19*, 147–148. [CrossRef]
7. Etxeberria, M.; Vázquez, E.; Marí, A. Microstructure analysis of hardened recycled aggregate concrete. *Mag. Concr. Res.* **2006**, *58*, 683–690. [CrossRef]

materials

MDPI

Article

An Experimental Study of Possible Post-War Ferronickel Slag Waste Disposal in Szklary (Lower Silesian, Poland) as Partial Aggregate Substitute in Concrete: Characterization of Physical, Mechanical, and Thermal Properties

Marcin Małek [1], Mateusz Jackowski [1,*], Waldemar Łasica [1], Kamil Dydek [2] and Anna Boczkowska [2]

[1] Faculty of Civil Engineering and Geodesy, Military University of Technology, ul. Gen. Sylwestra Kaliskiego 00-908 Warsaw, Poland; marcin.malek@wat.edu.pl (M.M.); waldemar.lasica@wat.edu.pl (W.Ł.)
[2] Faculty of Materials Science and Engineering, Warsaw University of Technology, ul. Wołoska 141, 02-507 Warsaw, Poland; kamil.dydek@pw.edu.pl (K.D.); anna.boczkowska@pw.edu.pl (A.B.)
* Correspondence: mateusz.jackowski@wat.edu.pl; Tel.: +48-511-06-05-77

Abstract: Aggregates derived from waste, due to the growing awareness of global warming, are mo and more often used in the concrete production process. This way, their disposal not only reduces th pollution of the Earth but also lowers the consumption of natural aggregates, which are limited. Or of the new "eco" aggregates may be a ferronickel slag waste (FNSW), which was generated in pos war metallurgical processes and stored in Szklary (Lower Silesian, Poland). In order to determine th possibility of using ferronickel slag waste aggregate (FNSWA) in the concrete production proces new concrete mixtures were designed and tested. Physical properties (cone slump, air content, p and density), mechanical properties (compressive strength, flexural strength, and tensile strengt and thermal properties (thermal conductivity) were assessed for all new laboratory recipes. Moreov the modulus of elasticity and Poisson's ratio were determined. This study includes five differe contents of FNSWA in the amount of 5%, 10%, 15%, 20%, and 25% of the mass of natural aggregate its partial substitute. The final results were compared to the base sample (BS) containing 100% natur aggregate, which was granite.

Keywords: ferronickel slag waste; by-products; waste disposal; recycling; eco-efficient concret mixture; concrete modifications

Citation: Małek, M.; Jackowski, M.; Łasica, W.; Dydek, K.; Boczkowska, A. An Experimental Study of Possible Post-War Ferronickel Slag Waste Disposal in Szklary (Lower Silesian, Poland) as Partial Aggregate Substitute in Concrete: Characterization of Physical, Mechanical, and Thermal Properties. *Materials* **2021**, *14*, 2552. https://doi.org/10.3390/ma14102552

Academic Editors: Miguel Bravo, Rui Vasco Silva and António P.C. Duarte

Received: 5 April 2021
Accepted: 11 May 2021
Published: 14 May 2021

1. Introduction

The war period was particularly severe for Poland. A significant number of th country's population died, defending their homeland, some lands were taken away, an the losses in the infrastructure were difficult to accurately quantify. It is estimated that th war cost that one resident had to pay was around USD 626 [1]. Despite the initial difficultie however, the reconstruction of the country has begun, thus implementing recovery plan One of them, the Six-Year Plan, assumed the development of the metallurgical industr due to the large amount of natural resources and minerals present in Poland. The priorit of the plan was to increase investment and the policy of intensive induction of the countr on the soviet model. According to the Six-Year Plan, the industrial production alone was increase by 85–95% [2–4], which could not have had a good environmental impact.

The post-war development of the polish industry has left its mark on the environme by leaving soil and groundwater contamination, and a large amount of waste that remain to this day [5,6]. This phenomena is especially visible in Silesian Districts of Southern an South-Western Poland where coal-waste dumps are stored. The enormous amounts material from coal mining and coal production, due to the intense coal exploitation that ha begun in the 19th century, is accumulated on dumps [7]. These materials are exposed t processes of self-heating, water-washing, and biodegradation, which can lead to significa environmental problems around coal-waste dumps, such as high concentrations of toxic compounds in soil or acidic gases and chemicals in the air [8]. Therefore, not only the environment but also human and animal life may be in danger. However, coal-waste is not the only type of waste left after the post-war development of the polish industry. Another big problem lies in South-Western Poland in Lower Silesian voivodeship. A small village, Szklary, within Ząbkowice Śląskie County, is a place where up until 1982 the only nickel ore mine in Poland was located. It has been officially closed due to the unprofitability of further exploitation but huge amounts of ferronickel slag waste are accumulated there to this day [9–11]. As it is stored in dumps, similar processes to coal-waste dumps occur. This work is a proposal of ferronickel slag waste utilization in concrete mixtures to prevent further contamination of groundwater, soil, and air spreading across Ząbkowice Śląskie County, especially Szklary.

Waste and by-products from other industry branches can be re-used by the construction industry as a component of concrete and mortar [12–15]. According to current studies, they can be used as a replacement of natural aggregate [16] and cement [17], as dispersed reinforcement [18–21] to lower surface stress and prevent microcracks or as a filler to increase thermal properties of final composite [22]. Ferronickel slag waste has a potential to substitute natural aggregate as it is a solid discharged from the smelting process of metal nickel or nickel-iron alloy. What is worth mentioning is that it has a high density and durability. Such properties are obtained in the heating and calcining process of nickel, where it is bonded with iron oxide. Moreover, this fusion allows high resistance to corrosion, heat shock, and fatigue, as well as the ferromagnetic effect of the final material [23].

The possibility of ferronickel slag waste utilization in concrete has been studied during recent years by Saha et al. [24–27]. As they reported, a 50% replacement of natural sand with FNSWA increases compressive strength from 38 to 57 MPa in concrete [26]. What was unexpected, other mechanical properties also showed higher values compared to plain concrete. Saha et al. noted about 7% and 13% increase in split tensile strength and flexural strength, respectively. Similar trends were observed by Nguyen et al. [28] as they reported an increase in concrete performance containing ferronickel sand as a fine aggregate. They pointed out, that both, mechanical and durability properties increased, due to the better interfacial transition zone between ferronickel aggregate and cement paste. Nuruzzaman et al. [29] also tested ferronickel slag waste as an aggregate replacement. With the density of 2920 kg/m^3 and substitution in range from 20% to 60% they achieved an increase in concrete mixture workability and a decrease in water absorption of hardened samples. Furthermore, concretes containing 20%, 40%, and 60% showed an increase by 1%, 34%, and 31% compressive strength compared to the base mix (37 MPa), and an increase by about 2%, 15%, and 8% split tensile strength compared to the base mix (4.3 MPa), respectively. The same trend after replacing coarse aggregate with ferronickel was reported by Sun et al. [30] and Qi et al. [31]. However, not only mechanical and psychical properties changes have been reported after FNSWA incorporation in concrete mixture. As presented by Saha [32], ferronickel slag can lower the thermal conductivity, as well. A concrete designed by them, with 100% FNSWA, showed 1.16 W/mK thermal conductivity, which was about two times lower compared to the base mix with only natural aggregate (2.34 W/mK).

This study aims to contribute to this growing area of research by exploring the potential usage of FNSWA mined in Szklary (Lower Silesia, Poland) as a replacement of natural aggregate in concrete. The problem of FNSW in Lower Silesia in Poland is very serious, so far, however, no possible method of utilization of this waste has been efficiently introduced to the Polish industry. To do so, psychical, mechanical, and thermal properties on new concretes were tested. Findings presented in this study make an important contribution to the effective utilization of ferronickel slag waste accumulated in Szklary (Lower Silesia, Poland) and contribute to the environment harmed by ferronickel slag waste stored there. As it has a different mineral composition proportion compared to the ones presented in current studies it may also show different results of concrete properties, especially its mechanical strength. Moreover, such a replacement is consistent with the principle o sustainable development and the circular economy.

2. Materials

2.1. Specimen Preparation

Table 1 shows chemical composition and strength parameters of the Portland cemen (CEM I 42.5 R [33]) used for each concrete mixture, which were determined according t EN 196-6:2019-01 [34] and PN EN 196-1:2016-07 [35]. Moreover, all of the recipes wer based on tap water (chloride content 28 mg/l) and natural granite aggregate, used as filler, with a fraction up to 8.0 mm. Its grain size index (C_U = 7.2 and C_C = 1.1) sugges that aggregate is well compacted [36] and, as presented in Figure 1, fits between upper an lower curves determined in accordance with the EN 12620+A1:2010 [37] standard.

Table 1. Chemical composition and strength parameters of the cement [38].

Compositions	SiO_2	Al_2O_3	Fe_2O_3	CaO	MgO	SO_3	Na_2O	K_2O	Cl
Unit (vol%)	19.5	4.9	2.9	63.3	1.3	2.8	0.1	0.9	0.05
Specific surface area (m^2/kg)					376.3				
Initial setting time (min)					227				
Compressive strength after 2 days (MPa)					28.4				
Compressive strength after 28 days (MPa)					60.8				

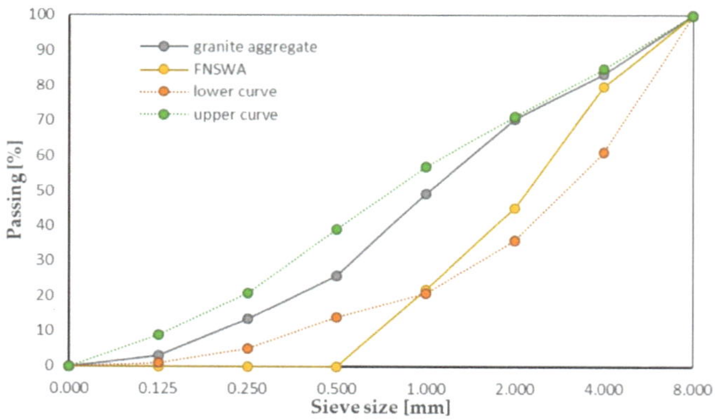

Figure 1. Gradation of crushed granite aggregate and FNSWA.

In order to increase the workability of all mixtures, but at the same time maintain th water-cement ratio w/c at 0.44, the chemical admixture was used. It was a low-alkalin liquid, free of chlorine, and based on an aqueous solution of modified polycarboxyli ethers (melamine and silanes/siloxanes), which helped reduce the accumulated water o the surface of concrete after compaction, as well [39]. The chemical composition of th chemical admixture is given in Table 2.

Table 2. Chemical composition of the admixture.

Compositions	O	Na	Si	K
Unit (vol%)	77.7	14.9	4.8	2.6

2.2. Ferronickel Slag Waste

As a partial substitute to natural granite aggregate, ferronickel slag waste was used (Figure 2). FNSW was mined from the ferronickel waste dump stored in Szklary, within Ząbkowice Śląskie County (Lower Silesia, Poland) and transported to the laboratory.

Figure 2. Ferronickel slag waste aggregate.

Table 3 presents the chemical composition of FNSW used as an aggregate substitute, the density of which is about 2750 kg/m^3. Compared to cement, its chemical composition is similar due to the fact that it contains a significant part of the compounds present in the binder. The FNSWA sieve curve is given in Figure 1.

Table 3. Chemical composition of FNSW.

Compositions	SiO_2	Fe_2O_3	Al_2O_3	CaO	MgO
Unit (vol%)	49.8	24.4	11.9	9.6	4.3

2.3. Mix Composition

Six different types of mortar mixtures were produced, see Table 4. The FNSWA content, which was added rather than the same amount of natural aggregate, was 5%, 10%, 15%, 20%, and 25% of granite mass for samples marked in the article as 5FNSW, 10FNSW, 15FNSW, 20FNSW, and 25FNSW, respectively. The sample with 100% of natural granite aggregate was marked as BM and used to compare test results.

Table 4. Mix proportions (1 m^3).

Mix Symbol	Cement [kg]	Water [kg]	Chemical Admixture [kg]	Granite Aggregate [kg]	FNSWA [kg]
BM	468	207	4.7	1600	0
5FNSW				1520	80
10FNSW				1440	160
15FNSW				1360	240
20FNSW				1280	320
25FNSW				1200	400

2.4. Mix Production

After mixing dry components for 1 min, the liquid components were added to the concrete mixture, and blended together for 4 min. Next, fresh concrete was poured into molds and compacted on a vibrating table. All of the samples were manufactured in 21 °C temperature and 50% humidity inside a laboratory and stored in water according to EN 12390-2:2019-07 [40]. After 28 days, the samples were tested for each property.

From each mix, the following samples were produced: Ten cubes with dimensior of 150 × 150 × 150 mm, five cuboids with dimensions of 40 × 40 × 160 mm, five rolle with a base diameter of 150 mm and a height of 300 mm, and one tile with dimensions 30 × 30 × 6 mm. In total, 60 cubic samples, 30 cuboid samples, 30 cylinders, and six til were made to conduct this study.

3. Methodology

3.1. Fresh Concrete Tests

The slump cone test was carried out according to EN 12350-2:2019-07 [41], when tł air content and pH value of the mix were determined according to EN 12350-7:2019-08 [4 and PN-B-01810:1986 [43], respectively. Five samples for each mixture were investigate These tests were performed in a listed order and right after the mixing process was dor The presented values for the conducted tests are the average values of five samples mac in pursuance of each concrete recipe.

3.2. Hardened Concrete Tests

The density of concrete samples, was determined according to EN 12390-7:2019-08 [4 on 150 × 150 × 150 cubes. Furthermore, mechanical properties were assessed, such a compressive strength, flexural strength, and tensile strength. To do so, a Zwick machir (Zwick, Ulm, Germany) with a force range of 0–5000 kN and maximum stress increas of 0.5 MPa/s was used. The compressive strength was measured according to EN 1239 3:2019-07 [45] on 150 × 150 × 150 mm samples, which were placed in the Zwick machir Next, the compressive strength of each test was determined by dividing the maximu sample load value by the sample cross-sectional area. For the flexural strength test, in three-point bending set-up, 40 × 40 × 160 mm concrete beams were prepared according EN 12390-5:2019-08 [46]. The Zwick machine used for the test enabled loading with a stat force and keeping it in a vertical configuration at a constant assumed level thanks to th supports that allowed only the horizontal movement (the distance between supports wa 100 mm). The last mechanical property, the split tensile strength, was tested on cylinders 150 mm diameter and 300 mm height according to EN 12390-6:2011 [47]. In order to perfor the test, the sample was placed in the Zwick machine on sliding supports, immobilize in the guides of the testing machine. The distance between the supports was greater tha 10 diameters of the sample and the pressure head and supports were radially rounded the point of contact with the sample. The cylinders of 150 mm diameter and 300 mm we also used to define the Modulus of elasticity and Poisson's ratio according to the EN 1239 13:2014-02 [48] standard. Every sample was loaded and the load was removed in the low and upper stress range to determine these material properties of each hardened concre mixture. The modulus of elasticity and Poisson's ratio were determined thanks to line displacements and measuring base lengths noted with Epsilon extensometers (Epsilo Jackson, WY, USA). The scope of the research also included thermal conductivity. Th temperature responses to the material heat flow pulses were the basis for all measuremen done with the ISOMET2114 analyzer (Applied Precision Ltd., Bratislava, Slovakia). Th analyzer probe, containing the resistor heater through which the heat flow was induce had a direct contact with the tested sample. A probe with a diameter of 60 mm an the tested material in the form of cubes (150 × 150 × 150 mm) from each concrete mi were used.

All of the specified above tests were carried out after 28 days of the curing proces Each of them was performed on five samples of every mix and the values presented are th average values (Figure 3).

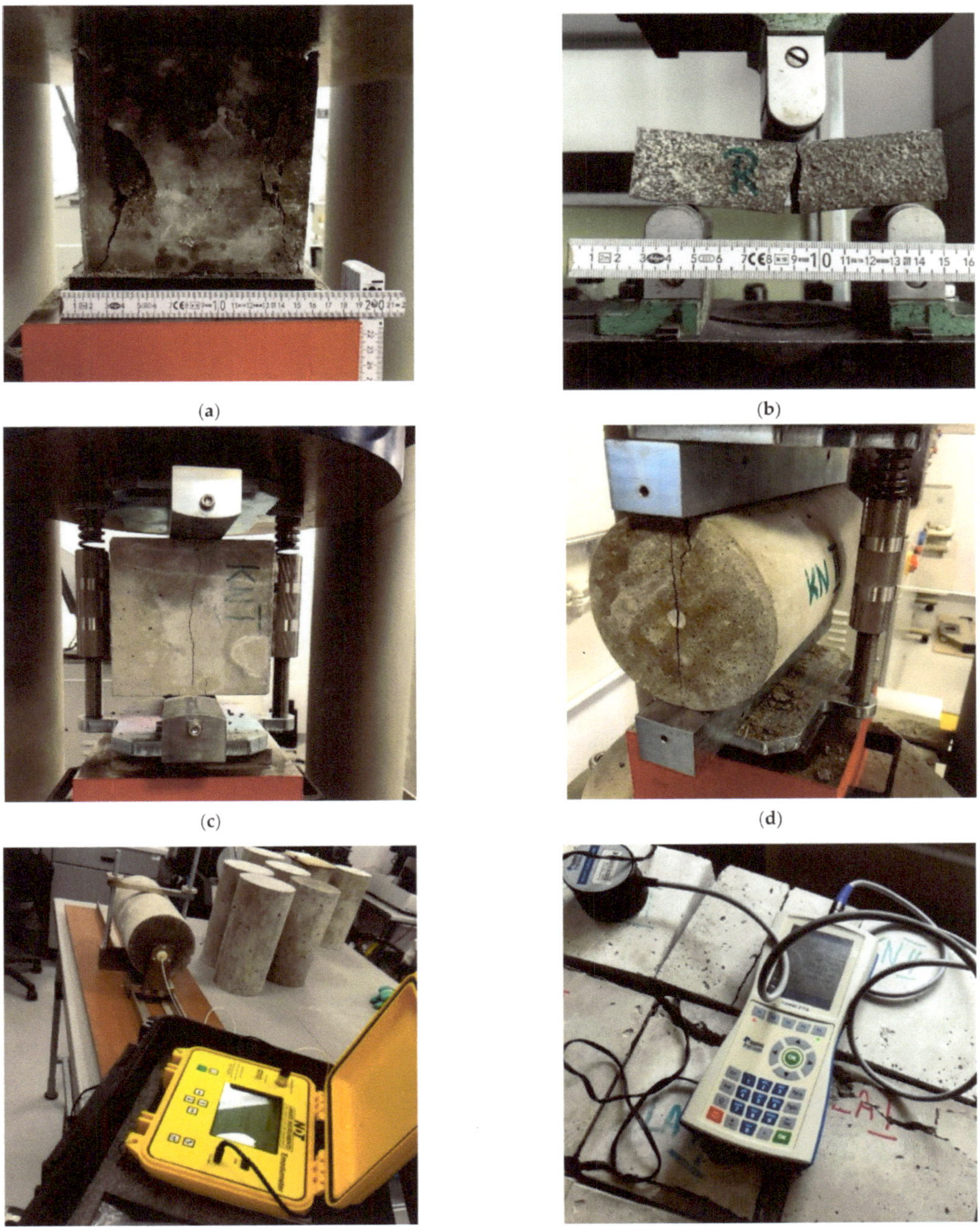

Figure 3. Photographic documentation of tests carried out: (**a**) Compressive strength test, (**b**) flexural strength test, (**c**) split tensile strength test—cubes, (**d**) split tensile strength test—cylinder, (**e**) elastic modulus test, (**f**) thermal conductivity test.

4. Results and Discussion

4.1. The Slump Cone Test

Table 5 shows the slump cone test results of all concrete recipes. According to EI 12350-2:2019-07 [41], they could be classified as a S1 class. Analyzing the obtained result it can be seen that the modification with ferronickel slag did not affect the cone slag at an percentage share of the natural aggregate substitute, and thus the fresh concrete consistenc All of the presented values of the slump cone fall test are in the error range. The slum cone test was also conducted by Saha et al. in their studies [24,26]. They reported howeve a much higher fall as the values fluctuated in the range of 200 to 230 mm and in the rang of 120 to 150 mm for mortars and concretes with ferronickel slag addition, respective This difference in the results may be due to the higher water/cement ratio used by Sah and the smaller size of ferronickel slag waste compared to our study.

Table 5. Fresh concrete test results.

Mix Symbol	Slump Cone [mm]	Consistency Class [41]	Air Content [%]	pH [-]
BM	2 ± 1	S1	2.1 ± 0.1	12.61 ± 0.03
5FNSW	2 ± 1	S1	2.2 ± 0.1	12.65 ± 0.03
10FNSW	2 ± 1	S1	2.4 ± 0.1	12.68 ± 0.04
15FNSW	2 ± 1	S1	2.7 ± 0.1	12.69 ± 0.03
20FNSW	2 ± 1	S1	3.2 ± 0.1	12.71 ± 0.03
25FNSW	1 ± 1	S1	3.6 ± 0.1	12.72 ± 0.04

4.2. Air Content

The air content presented in Table 5 shows that the peak value of this property wa reported for the 25FNSW sample (modified with 25% of ferronickel slag waste). It showe an increment of 1.5% compared to BM. Previous substitutions of granite aggregate als indicated a higher air content in concrete mixtures compared to BM. This phenomena ma be a result of the higher grain size of FNSWA compared to the granite aggregate used i the concrete manufacturing process. It can also indicate more air voids both in fresh an hardened concrete.

4.3. The pH Test

As presented in Table 5, the pH of all concrete mixtures exhibits the alkaline reactio The highest value of pH was reported for samples with 25% ferronickel slag waste aggrega (12.72). Compared to the base mix values (12.61) an increase of about 1% can be note Such a small change may indicate that FNSWA did not affect the pH of the concret mixture significantly.

4.4. Density

The density of concrete samples is presented in Table 6. The substitution equal to 5% 10%, 15%, 20%, and 25% of natural aggregate done with FNSWA, resulted in a densit increase by 1.6%, 2.6%, 3.2%, 6.6%, and 8.7% compared to the BM, respectively. All of th obtained values were in the range between 2187 ± 2 kg/m^3 (BM) and 2378 ± 3 kg/m (25FNSW), and, after analyzing them, an overall trend can be noted as: The higher th FNSWA amount, the higher the density of concrete, see Figure 4. This may be a result of th higher density of FNSWA compared to the natural granite aggregate. Similar to this stud Nuruzzaman et al. [29] also reported an increase in density of concrete containing FNSW As they reported, the density increased by 1.58%, 3.15%, and 4.73% due to the use of 20% 40%, and 60% FNSWA. On the other hand, the opposite trend was reported by Bouasri et al. [49]. They, however, replaced not the aggregate but the cement with ferronickel sla and noted a decrease in concrete density from 2390 to 2330 kg/m^3 after a 10% substitutio Further modifications also showed decreased values—2310 and 2300 kg/m^3 for 15% an

30% replacement of cement, respectively. This may be connected to a higher density of cement powder compared to ferronickel slag.

Table 6. Hardened concrete test results.

Mix Symbol	Density [kg/m^3]	Compressive Strength [MPa]	Split Tensile Strength [MPa]	Flexural Strength [MPa]	Modulus of Elasticity [GPa]	Poisson Coefficient [GPa]	Thermal Conductivity [W/mK]
BM	2187 ± 2	45 ± 1	2.87 ± 0.03	5.9 ± 0.1	31.5 ± 0.4	0.123 ± 0.03	1.88 ± 0.04
5FNSW	2222 ± 2	49 ± 1	2.99 ± 0.05	6.2 ± 0.1	31.6 ± 0.3	0.124 ± 0.03	1.68 ± 0.04
10FNSW	2244 ± 3	53 ± 1	3.60 ± 0.03	6.5 ± 0.1	31.8 ± 0.3	0.129 ± 0.04	1.58 ± 0.04
15FNSW	2258 ± 2	54 ± 1	3.95 ± 0.03	7.4 ± 0.1	32.1 ± 0.4	0.123 ± 0.03	1.56 ± 0.03
20FNSW	2332 ± 2	56 ± 1	4.02 ± 0.03	9.5 ± 0.2	32.4 ± 0.3	0.124 ± 0.04	1.54 ± 0.04
25FNSW	2378 ± 3	59 ± 1	4.11 ± 0.04	9.8 ± 0.1	33.6 ± 0.3	0.123 ± 0.03	1.53 ± 0.03

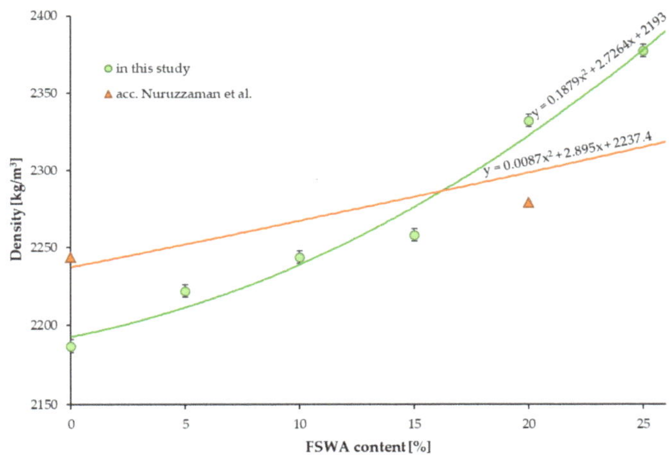

Figure 4. The impact of FNSWA on concrete density.

4.5. Compressive Strength

The results of compressive strength for samples with different contents of FNSWA are presented in Table 6 and in Figure 5. The results were compared to the base mix (without FNSWA), which showed a compressive strength of 45 ± 1 MPa. All of the substitution of natural aggregate, equal to 5%, 10%, 15%, 20%, and 25%, indicated an increase in compressive strength by 8.8%, 17.8%, 20.0%, 24.4%, and 31.1%, respectively. For this case, the highest value of compressive strength (59 ± 1 MPa) was obtained for samples with the highest amount of FNSWA (400 kg/m^3). The same phenomena was reported by Qi et al. [31] as they noted the peak values of compressive strength for samples modified with 50% of FNSWA (44.6 MPa). Another study in line with this trend was published by Nuruzzaman et al. [29]. The authors tested the FNSW concrete with ferronickel slag grading up to 9.50 mm and density of 2920 kg/m^3 with its addition equal to 20%, 40%, and 60%. They reported similar values of compressive strength (52.15 MPa) for samples containing 20% of ferronickel slag compared to 20FNSW samples (56 ± 1 MPa). Moreover, Sun et al. [30] used two types of ferronickel slag for concrete manufacturing—blast furnace slag (BS) and electric furnace slag (ES). They showed that BS indicates an increase in compressive strength by about 6%, 9%, and 11% for samples with 25%, 50%, and 75% of ferronickel used as an aggregate. In the opposition to this were however results of ES concretes, as they presented a decrease in compressive strength with an increase in ferronickel slag content. The opposite trend noted for samples containing ES may be du to the type of ferronickel used, which showed a smooth surface. In turn, it can affect th interactions between components and lower the results of mechanical tests. The lowe compressive strength was reported for samples with 75% and 100% electric furnace sla (58 MPa) and compared to the reference sample (65 MPa), it was a decrease by about 11

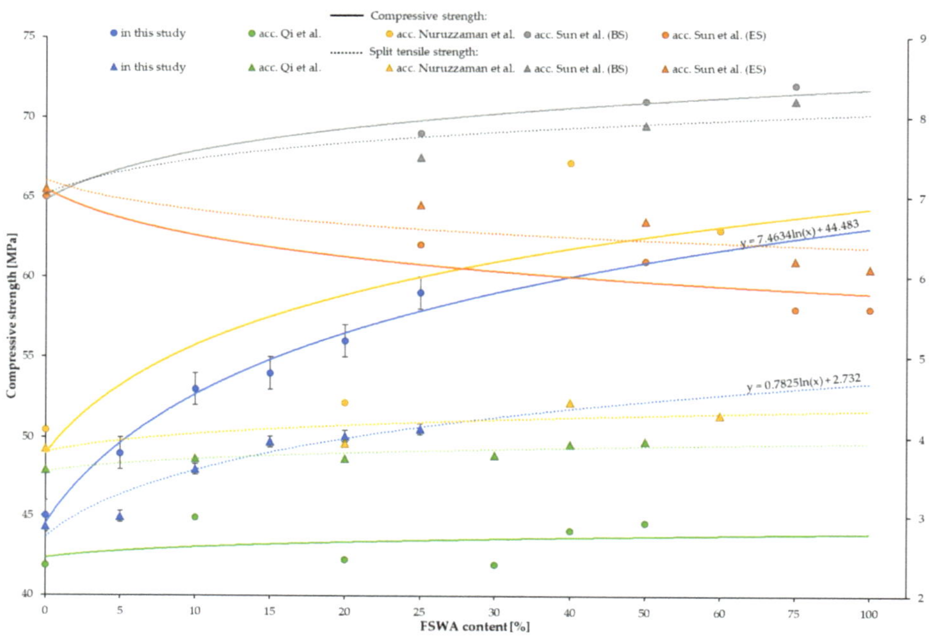

Figure 5. The impact of FNSWA on compressive strength and split tensile strength.

4.6. Split Tensile Strength

Table 6 shows the split tensile strength test results of concrete with a natural aggrega replacement done with FNSWA. After substitution of granite aggregate with 5%, 10%, 15 20%, and 25% of FNSWA, the linear increase in split tensile strength was reported, se Figure 6. Compared to BM, the samples modified with FNSWA showed higher values b around 4%, 25%, 38%, 40%, and 43% values, respectively for 5FNSW, 10FNSW, 15FNSW 20FNSW, and 25FNSW samples. As observed, the difference in split tensile strength for th reference sample and the sample with the lowest amount of FNSWA (80 kg/m^3) was no significant. However, a further increase in FNSWA used in concrete resulted in a visib increase in the split tensile strength, what also was reported by Nuruzzaman et al. [2 Their FNSW concrete showed a 2%, 15%, and 11% increase in split tensile strength for 20 40%, and 60% ferronickel used as an aggregate. Furthermore, Qi et al. [31] tested FNSW as a substitute to natural aggregate. They used five different contents of ferronickel equ to 10%, 20%, 30%, 40%, and 50%. As presented in their study, the highest split tensi strength (3.96 MPa) showed samples with 50% of FNSWA, which was about 10% great than plain concrete. An increasing trend was also reported for BS concretes tested b Sun et al. [30], however their ES concrete showed an opposite phenomena. This may be result of ES particles that contain a number of smooth glass surfaces, which can serious weaken the cohesion between cement paste and aggregates, and mechanical properties o hardened concrete.

4.7. Flexural Strength

Table 6 presents the results of the flexural strength test of concrete samples with FNSWA. For the 5%, 10%, 15%, 20%, and 25% substitution of natural granite aggregate, the obtained flexural strength was about average 5%, 10%, 25%, 61%, and 66% higher than for the plain concrete (5.9 ± 0.1 MPa), respectively. The flexural strength is directly proportional to the FNSWA amount used in the concrete mixture, which is shown by a linear increase in values. Furthermore, the determined R^2 coefficient equal to 0.9, means a good fit. This trend was also reported by Saha et al. [24], as well. They showed however, that a peak in flexural strength is achievable for 50% FNSWA used in concrete and a further increase in the amount of natural aggregate substitute decreases this property. Their study complies with the Sun et al. research [30] as they reported after 28 days a constant value of flexural strength for samples modified with 50% and 75% of ferronickel slag (8.2 MPa). What is worth mentioning is that Sun et al. who used two types of the ferronickel slag described above reported an increase for BS concretes, when the electric furnace slag showed a decrease in flexural strength by about 4%, 6%, 10%, and 12% for 25%, 50%, 75%, and 100% natural aggregate replacement, respectively.

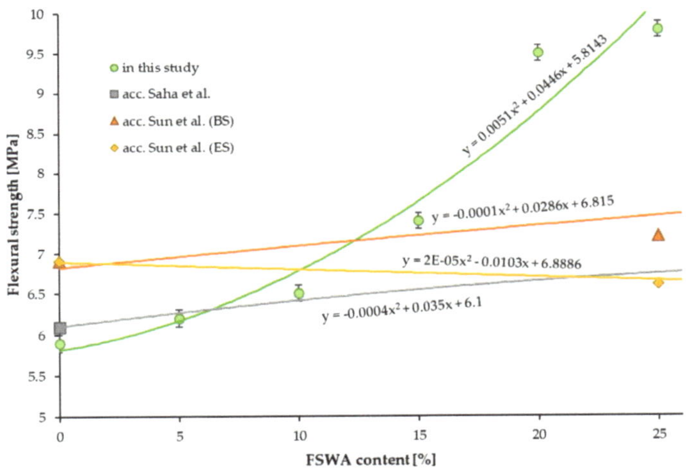

Figure 6. The impact of FNSWA on flexural strength.

4.8. Modulus of Elasticity and Poisson's Coefficient

The modulus of elasticity was slightly influenced by the aggregate substitution done with FNSWA. Results given in Table 6, range from 31.5 ± 0.4 to 33.6 ± 0.3 GPa. The highest values of modulus of elasticity were reported for 25FNSW samples, which compared to BM showed a 6.7% increase in value. This trend was reported for the rest of the samples as well, as the noted values were higher than the plain concrete (0.3%, 1.0%, 1.9%, and 2.9% increase for 5FNSW, 10FNSW, 15FNSW, and 20FNSW samples, respectively). An increasing trend, after replacing 50% of natural sand aggregate with FNSWA, was reported by Sakoi et al. [50] and Saha and Sarker [26], as well. However, their increase was about 14% and 12%, respectively, which is almost two times higher compared to this study. Such a difference was caused by the greater FNSWA amount used by other researchers, which could result in an increase of modulus of elasticity as it is related with the density of concrete components.

On the other hand, the results of the Poisson's ratio, presented in Table 6, indicate no influence of FNSWA to this concrete property, as they are in the measurement error range. Therefore, it can be said that the substitution of natural granite aggregate done with FNSWA in the range from 5% to 25% did not affect the Poisson's ratio. A similar conclusion was made by Qi et al. [31] as they stated that Poisson's ratio of concrete containing ferronick slag is consistent with that of conventional concrete.

4.9. Thermal Conductivity

The obtained results of the thermal conductivity test ranged from 1.53 to 1.88 W/m (Table 6). The concrete modified with 25% FNSWA showed the lowest values of therm conductivity, which were 18.6% less than the plain concrete. Moreover, a linear decreasin trend can be reported as thermal conductivity drops with an increase in the amount c the aggregate substitute. The same phenomena was reported by Saha et al. [32]. The plain concrete reduced its thermal conductivity from 2.34 to 1.58 W/mK, 1.65 W/m 1.36 W/mK, and 1.16 W/mK, with 25%, 50%, 75%, and 100% FNSWA replacement, respe tively. As presented in Figure 7, the thermal conductivity values reported for concrete with 25% ferronickel slag waste in this study and by Saha et al. are almost equal.

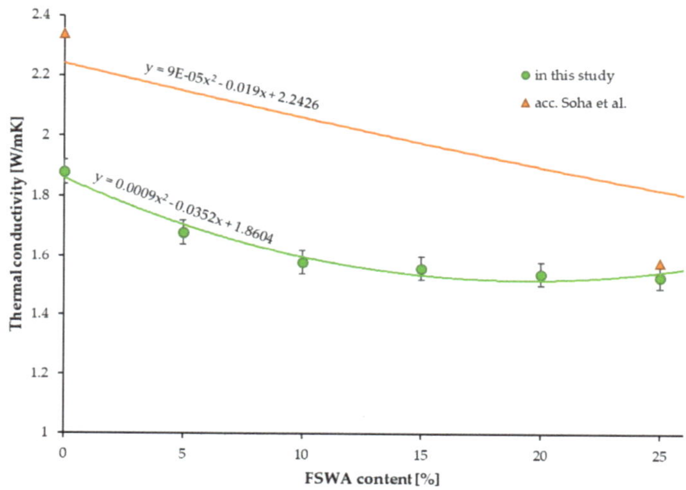

Figure 7. The impact of FNSWA on thermal conductivity.

5. Conclusions

The aim of this study was to evaluate the possibility of using the ferronickel slag wast aggregate stored in dumps in Szklary (Lower Silesian, Poland) as an aggregate substitut in concrete production. To do so, tests on new laboratory concrete mixtures were carrie out, from which final key conclusions can be made:

- The highest air content was reported for 25FNSW samples (3.6%);
- The density of hardened concretes were proportional to the amount of FNSW used. The density increased with the increasing FNSWA and the highest densit (2378 kg/m^3) was found for concrete with 25% FNSWA and the lowest densit (2187 kg/m^3) for plain concrete;
- The substitution of 80, 160, 240, 320, and 400 kg/m^3 natural granite aggregate wit FNSWA caused an increase in the compressive strength by about 1.09, 1.18, 1.20, 1.2 and 1.31 times and in flexural strength by about 1.05, 1.10, 1.25, 1.61, and 1.66 time respectively compared to plain concrete, while the obtained split tensile strength wa increased compared to BM by 4.0%, 25%, 38%, 40%, and 43%, respectively
- A slight influence of FNSWA on the modulus of elasticity was noted as reporte values showed an increase in the range of 0.3% up to 6.7%, depending on the amour of substitute used;
- All of the concretes with FNSWA showed lower thermal conductivity values compar to BM. The highest decrease was reported for 25FNSW samples (18.6%);

- The substitution of natural aggregate done with FNSWA did not affect consistency, the pH value, and Poisson's ratio of concrete.

This research presents a solution to air, soil, and groundwater pollution caused by the remains of post-war mining activities in Szklary (Lower Silesia, Poland). The set goal of this case study, which was to determine the physical, mechanical, and thermal properties of concretes containing FSWA, was successfully achieved. All of the new concretes showed increased strength parameters, crucial in building materials, which can indicate a great application capacity in the current construction industry.

Author Contributions: Conceptualization and methodology, M.M. and M.J.; investigation, M.M., M.J., W.Ł., K.D. and A.B.; data curation, M.J. and M.M.; formal analysis, M.J.; funding acquisition, M.M.; project administration, M.M.; resources, M.M., M.J. and W.Ł.; supervision, M.M.; validation, M.M. and M.J.; visualization, M.J.; writing—original draft preparation, M.J. and M.M.; writing—review and editing, M.M., M.J. and A.B. All authors have read and agreed to the published version of the manuscript.

Funding: This research received no external funding besides the statutory research of particular scientific units.

Institutional Review Board Statement: Not applicable.

Informed Consent Statement: Not applicable.

Data Availability Statement: Data is contained within the article.

Acknowledgments: This work was financially supported by the Dean of Faculty of Civil Engineering and Geodesy of the Military University of Technology as part of a scholarship no. 1/DHP/2020. The authors thank PROFIMA Sp. z o.o., ul. Wojciecha Żywnego 12/30, 02-701 Warsaw for providing the FSWA mined from ferronickel slag dump stored in Szklary (Lower Silesia, Poland) to conduct this research, and the fifth-year students, Bartosz Niewczas and Daniel Moch, for their help in preparing this article.

Conflicts of Interest: The authors declare no conflict of interest.

eferences

Technical Review, Supreme Technical Organization (Poland), Central Institute of Scientific, Technical and Economic Information (Poland), Warsaw, Poland; Central Office of Measures: Warsaw, Poland, 1984.

The Act of July 21, 1950 on the 6-Year Plan for Economic Development and Building the Foundations of Socialism for the Years. *J. Laws* **1950**, *37*, 344.

Tighe, C. Forward to battle for the Six-Year Plan! Polish writers 1945–56. *J. Eur. Stud.* **2015**, *45*, 189–219. [CrossRef]

Pohoski, M. Interrelation between Social Mobility of Individuals and Groups in the Process of Economic Growth in Poland. *Pol. Sociol. Bull.* **1964**, *10*, 17–33.

Malina, G. Ecotoxicological and environmental problems associated with the former chemical plant in Tarnowskie Gory, Poland. *Toxicology* **2004**, *205*, 157–172. [CrossRef] [PubMed]

Berto, C.; Krajcarz, M.T.; Moskal-del Hoyo, M.; Komar, M.; Sinet-Mathiot, V.; Zarzecka-Szubińska, K.; Krajcarz, M.; Szymanek, M.; Wertz, K.; Marciszak, A.; et al. Environment changes during Middle to Upper Palaeolithic transition in southern Poland (Central Europe). A multiproxy approach for the MIS 3 sequence of Koziarnia Cave (Kraków-Częstochowa Upland). *J. Archaeol. Sci. Rep.* **2021**, *35*, 102723. [CrossRef]

dám Nádudvari, Á.; Fabiańska, M.J. Use of geochemical analysis and vitrinite reflectance to assess different self-heating processes in coal-waste dumps (Upper Silesia, Poland). *Fuel* **2016**, *181*, 102–119. [CrossRef]

Carras, J.N.; Day, S.J.; Saghafi, A.; Williams, D.J. Greenhouse gas emissions from low-temperature oxidation and spontaneous combustion at open-cut coal mines in Australia. *Int. J. Coal. Geol.* **2009**, *78*, 161–168. [CrossRef]

Central Statistical Office (GUS)—TERYT (National Register of Territorial Land Apportionment Journal). Available online: https://eteryt.stat.gov.pl/eTeryt/rejestr_teryt/udostepnianie_danych/baza_teryt/uzytkownicy_indywidualni/wyszukiwanie/wyszukiwanie.aspx?contrast=default (accessed on 17 March 2021).

Institute of Mineral and Energy Economy of the Polish Academy of Sciences. Part I: Searching and recognition of defenses planning and organization of geological works. In *Methodology for Documenting the Fold Solid Mines*; Institute of Mineral and Energy Economy of the Polish Academy of Sciences: Cracow, Poland, 2012.

Polish Geological Institute-National Research Institute. *Balance of Prospective Resources of Poland's Minerals at December 31, 2018*; Polish Geological Institute-National Research Institute: Warsaw, Poland, 2020.

12. Małek, M.; Łasica, W.; Jackowski, M.; Kadela, M. Effect of Waste Glass Addition as a Replacement for Fine Aggregate on Properti of Mortar. *Materials* **2020**, *13*, 3189. [CrossRef]
13. Szcześniak, A.; Zychowicz, J.; Stolarski, A. Influence of Fly Ash Additive on the Properties of Concrete with Slag Ceme *Materials* **2020**, *13*, 3265. [CrossRef]
14. Isa, M.; Pilakoutas, K.; Guadagnini, M.; Angelakopoulos, H. Mechanical performance of affordable and eco-efficient ultra-hi performance concrete (UHPC) containing recycled tyre steel fibres. *Constr. Build. Mater.* **2020**, *255*, 119272. [CrossRef]
15. Jurczak, R.; Szmatuła, F.; Rudnicki, T.; Korentz, J. Effect of Ground Waste Glass Addition on the Strength and Durability of Lo Strength Concrete Mixes. *Materials* **2021**, *14*, 190. [CrossRef]
16. Khankhaje, E.; Salim, M.R.; Mirza, J.; Hussin, M.W.; Rafieizonooz, M. Properties of sustainable lightweight pervious concre containing oil palm kernel shell as coarse aggregate. *Constr. Build. Mater.* **2016**, *126*, 1054–1065. [CrossRef]
17. Bhat, J.A. Mechanical behaviour of self compacting concrete: Effect of wood ash and coal ash as partial cement replaceme *Mater. Today Proc.* **2021**, *42*, 1470–1476. [CrossRef]
18. Małek, M.; Jackowski, M.; Łasica, W.; Kadela, M. Characteristics of Recycled Polypropylene Fibers as an Addition to Concre Fabrication Based on Portland Cement. *Materials* **2020**, *13*, 1827. [CrossRef]
19. Kim, S.B.; Yi, N.H.; Kim, H.Y.; Kim, J.H.J.; Song, Y.C. Material and structural performance evaluation of recycled PET fib reinforced concrete. *Cement. Concr. Comp.* **2010**, *32*, 232–240. [CrossRef]
20. Małek, M.; Jackowski, M.; Łasica, W.; Kadela, M.; Wachowski, M. Mechanical and Materials Properties of Mortar Reinforced wi Glass Fiber: An Experimental Study. *Materials* **2021**, *14*, 698. [CrossRef]
21. Małek, M.; Łasica, W.; Kadela, M.; Kluczyński, J.; Dudek, D. Physical and Mechanical Properties of Polypropylene Fibre-Reinforc Cement–Glass Composite. *Materials* **2021**, *14*, 637. [CrossRef]
22. Jackowski, M.; Małek, M.; Życiński, W.; Łasica, W.; Owczarek, M. Characterization of new recycled polymers shots addition f the mechanical strength of concrete. *Mater. Technol.* **2020**, *54*, 355–358. [CrossRef]
23. Kōmura, S.; Takeda, T.; Ōhara, S. Magnetic Diffuse Scattering of Neutrons from Fe-Ni Invar Alloy. *J. Phys. Soc. Jpn.* **197** *35*, 706–711. [CrossRef]
24. Saha, A.K.; Sarker, P.K. Sustainable use of ferronickel slag fine aggregate and fly ash in structural concrete: Mechanical properti and leaching study. *J. Clean. Prod.* **2017**, *162*, 438–448. [CrossRef]
25. Saha, A.K.; Khan, M.N.N.; Sarker, P.K. Value added utilization of by-product electric furnace ferronickel slag as constructic materials: A review. *Resour. Conserv. Recycl.* **2018**, *134*, 10–24. [CrossRef]
26. Saha, A.K.; Sarker, P.K. Compressive strength of mortar containing ferronickel slag as replacement of natural sand. *Proc. En* **2017**, *171*, 689–694. [CrossRef]
27. Saha, A.K.; Sarker, P.K. Durability of mortar incorporating ferronickel slag aggregate and supplementary cementitious materia subjected to wet-dry cycles. *Inter. J. Concr. Struct. Mater.* **2018**, *12*, 29. [CrossRef]
28. Nguyen, Q.D.; Castel, A.; Kim, T.; Khan, M.S.H. Performance of fly ash concrete with ferronickel slag fine aggregate again alkali-silica reaction and chloride diffusion. *Cem. Concr. Res.* **2021**, *139*, 106265. [CrossRef]
29. Nuruzzaman, M.; Casimiro, J.O.C.; Sarker, P.K. Fresh and hardened properties of high strength self-compacting concrete usir by-product ferronickel slag fine aggregate. *J. Build. Eng.* **2020**, *32*, 101686. [CrossRef]
30. Sun, J.; Feng, J.; Chen, Z. Effect of ferronickel slag as fine aggregate on properties of concrete. *Constr. Build. Mater.* **2019**, *2C* 201–209. [CrossRef]
31. Qi, A.; Liu, X.; Wang, Z.; Chen, Z. Mechanical properties of the concrete containing ferronickel slag and blast furnace slag powd *Constr. Build. Mater.* **2020**, *231*, 117120. [CrossRef]
32. Saha, A.K.; Sarker, P.K.; Golovanevskiy, V. Thermal properties and residual strength after high temperature exposure of ceme mortar using ferronickel slag aggregate. *Constr. Build. Mater.* **2019**, *199*, 601–612. [CrossRef]
33. European Committee for Standardization. *EN 197-1:2012 Cement—Part 1: Composition, Specifications and Conformity Criteria f Common Cements*; European Committee for Standardization: Brussels, Belgium, 2012.
34. European Committee for Standardization. *EN 196-6:2019-01 Methods of Testing Cement—Part 6: Determination of Fineness*; Europe Committee for Standardization: Brussels, Belgium, 2019.
35. European Committee for Standardization. *EN 196-1:2016-07 Methods of Testing Cement—Part 1: Determination of Strength*; Europe Committee for Standardization: Brussels, Belgium, 2016.
36. Rudnicki, T. The method of aggregate skeleton in self compacting concrete designing with segment regression. *Cem. Wapno Betc* **2016**, *1*, 10–19. [CrossRef]
37. European Committee for Standardization. *EN 12620+A1:2010 Aggregates for Concrete*; European Committee for Standardizatic Brussels, Belgium, 2010.
38. Górażdże Group: Cement, Concrete, Aggregate. Technical Data Sheet CEM I 42.5 R. Available online: http://www.gorazdze. (accessed on 14 December 2020).
39. Rudnicki, T. Functional Method of Designing Self-Compacting Concrete. *Materials* **2021**, *14*, 267. [CrossRef]
40. European Committee for Standardization. *EN 12390-2:2019-07 Testing Hardened Concrete—Part 2: Making and Curing Specimens f Strength Tests*; European Committee for Standardization: Brussels, Belgium, 2019.
41. European Committee for Standardization. *EN 12350-2:2019-07 Testing Fresh Concrete—Part 2: Slump Test*; European Committee f Standardization: Brussels, Belgium, 2019.

2. European Committee for Standardization. *EN 12350-7:2019-08 Testing Fresh Concrete—Part 7: Air Content—Pressure Method*; European Committee for Standardization: Brussels, Belgium, 2019.
3. Polish Committee for Standardization. *PN-B-01810:1986 Protection Against Corrosion in Building—Protective Properties of Concrete Referring to Reinforcing Steel—Electrochemical Tests*; Polish Committee for Standardization: Brussels, Belgium, 1986.
4. European Committee for Standardization. *EN 12390-7:2019-08 Testing Hardened Concrete—Part 7: Density of Hardened Concrete*; European Committee for Standardization: Brussels, Belgium, 2008.
5. European Committee for Standardization. *N 12390-3:2019-07 Testing Hardened Concrete—Part 3: Compressive Strength of Test Specimens*; European Committee for Standardization: Brussels, Belgium, 2019.
6. European Committee for Standardization. *EN 12390-5:2019-08 Testing Hardened Concrete—Part 5: Flexural Strength of Test Specimens*; European Committee for Stand-ardization: Brussels, Belgium, 2019.
7. European Committee for Standardization. *EN 12390-6:2011 Testing Hardened Concrete—Part 6: Tensile Splitting Strength of Test Specimens*; European Committee for Standardization: Brussels, Belgium, 2011.
8. European Committee for Standardization. *EN 12390-13:2014-02 Testing Hardened Concrete—Part 13: Determination of Secant Modulus of Elasticity in Compression*; European Committee for Standardization: Brussels, Belgium, 2014.
9. Bouasria, M.; Khadraoui, F.; Benzaama, M.-H.; Touati, K.; Chateigner, D.; Gascoin, S.; Pralong, V.; Orberger, B.; Babouri, L.; El Mendili, Y. Partial substitution of cement by the association of Ferronickel slags and Crepidula fornicata shells. *J. Build. Eng.* **2021**, *33*, 101587. [CrossRef]
0. Sakoi, Y.; Aba, M.; Tsukinaga, Y.; Nagataki, S. Properties of concrete used in ferronickel slag aggregate. In Proceedings of the 3rd International Conference on Sustainable Construction Materials and Technologies, Tokyo, Japan, 19–21 August 2013; pp. 1–6.

materials

MDPI

Article

Recycled Mortars with Ceramic Aggregates. Pore Network Transmutation and Its Relationship with Physical and Mechanical Properties

Francisca Guadalupe Cabrera-Covarrubias [1], José Manuel Gómez-Soberón [2,*], Carlos Antonio Rosas-Casarez [1], Jorge Luis Almaral-Sánchez [1,*] and Jesús Manuel Bernal-Camacho [3]

[1] Faculty of Engineering Mochis, Autonomous University of Sinaloa, Fuente de Poseidón y Ángel Flores s/r Col. Jiquilpan, Module B2, Los Mochis, Sinaloa 81210, Mexico; guadalupe.cabrera@uas.edu.mx (F.G.C.-C.) carlos.arc@uas.edu.mx (C.A.R.-C.)

[2] Barcelona School of Building Construction, Polytechnic University of Catalonia, Av. Doctor Marañón 44-5(08028 Barcelona, Spain

[3] School of Engineering Mazatlan, Autonomous University of Sinaloa, Ave. Ejército Mexicano esq. Ave. Universidad s/n., Ciudad Universitaria, Fracc. Antiguo Aeropuerto, Mazatlán, Sinaloa 82017, Mexico; jmbernalc@uas.edu.mx

* Correspondence: josemanuel.gomez@upc.edu (J.M.G.-S.); jalmaral@uas.edu.mx (J.L.A.-S.); Tel.: +34-93-401-6242 (J.M.G.-S.)

Citation: Cabrera-Covarrubias, F.G.; Gómez-Soberón, J.M.; Rosas-Casarez, C.A.; Almaral-Sánchez, J.L.; Bernal-Camacho, J.M. Recycled Mortars with Ceramic Aggregates. Pore Network Transmutation and Its Relationship with Physical and Mechanical Properties. *Materials* **2021**, *14*, 1543. https://doi.org/10.3390/ma14061543

Academic Editors: Miguel Bravo, Rui Vasco Silva and António P.C. Duarte

Received: 18 February 2021
Accepted: 15 March 2021
Published: 21 March 2021

Abstract: The porosity of mortars with recycled ceramic aggregates (10, 20, 30, 50, and 100% as replacement of natural aggregate) was evaluated and analyzed using three different techniques. Th results of gas adsorption (N_2), Scanning Electron Microscopy (SEM) image analysis and open porosi allowed establishing the relationship between the recycled aggregate content and the porosity of thes mortars, as well as the relationship between porosity and the physical and mechanical properties c the mortars: absorption, density, compressive strength, modulus of elasticity, and drying shrinkag Using the R^2 coefficient and the equation typology as criteria, additional data such as Brunaue Emmett, and Teller (BET) surface area (N_2 adsorption) established significant correlations with th mentioned properties; with SEM image analysis, no explanatory relationships could be establishe and with open porosity, revealing relationships were established ($R^2 > 0.9$). With the three technique it was confirmed that the increase in porosity is related to the increase in the amount of cerami aggregate; in particular with gas adsorption (N_2) and open porosity. It was concluded that th open porosity technique can explain the behavior of these recycled mortars with more reliable dat in a simple and direct way, linked to its establishment with a more representative sample of th mortar matrix.

Keywords: recycled mortars; ceramic wastes; recycled aggregates; porosity; adsorption N_2; imag analysis; open porosity

1. Introduction

In recent years, some of the most commonly used terms in the construction secto include the life cycle of materials, waste management, economic sustainability, sustainab development, environmental care, closed-cycle materials, etc.; these terms are related t the increase in the consumption of materials and energy, as well as the currently use incorrect and unsustainable waste management system. This has influenced government industries, and researchers to seek solutions for the implementation of new alternatives t achieve responsible solutions related to resource recovery and waste management [1–6].

It is known that the construction sector generates a large amount of Constructio and Demolition (C&D) waste, which has further increased the interest of researche to find methods to reduce, reuse or recycle it [7]. In the specific case of recycling, th is feasible through selection, grinding, crushing and screening processes, to obtain a a result aggregates of different sizes and properties, which can be used as substitutes for sand or gravel in mortars and concretes [8]. Although previous studies reported that the use of such waste in concrete is detrimental to its mechanical performance and durability—up to certain replacement percentages, its use in mortars (either in the form of fine aggregate as a substitute for natural sand, or in the form of powder as a partial substitute for cement—with an impact on pozzolanic reaction properties) has also been considered to be an alternative [7–13]. For all of the above, the need to use 100% of all resources again as new resources is evident, integrating the paradigm of a sustainable city; and at the same time, there is a need to significantly increase the efficiency of the use of resources, energy, and materials [4].

In the specific case of studies concerning recycled mortars, one of the materials of interest as separated waste is ceramics; for example, in China (2017), 1800 million tons of construction and demolition waste (CDW) were produced, of which 87% are concrete and clay bricks [14]. In general, ceramic waste has its origin from masonry rubble waste, and is represented by clay bricks (solid, hollow, or domed bricks that have been sorted and crushed) [1,3,7,14–17], tiles [1], rubble partition walls [5], roofing tiles [3], or from the ceramic industry for not meeting quality standards (broken, distorted, burned, etc.) [1,18]; these wastes also include ceramic sanitary wastes [6].

At present, numerous investigations have been carried out on recycled ceramic mortars (CRM), within which different variables were studied for their use; one of the main ones has been to discern the role of the recycled aggregate within the mortar matrix, using it as a partial or total replacement of the natural aggregate, or as a partial replacement of the cement.

CRM studies using aggregate replacement established that the particle size of the ceramic aggregate is another important factor in mortar properties (particle size is a function of the crushing method). These researchers established that fine ceramic aggregate produced by rotating ball mill (finer particles) produces more porous mortars than those crushed by jaw crusher (coarser particles) [17]; however, excellent results in terms of improvement of the mortar-brick interfacial transition zone (ITZ) were achieved with the use of the finer ceramic aggregates (even replacing 100% natural aggregate). This improved adhesion of the ITZ was identified as the contribution of the fine powder in the fresh rheology of the CRM, allowing better penetration into the brick surface [16,17]. In another investigation, using brick residues in ranges of particle sizes (0.15–0.3, 0.3–0.6 and 0.6–4.75 mm), they obtained those mortars with finer sand (0.15–0.3 mm) and reduced their workability with respect to the control mortar, due to the fact that they tend to absorb water faster (higher specific surface area); on the other hand, the compressive strength was notably higher, and their elastic modulus and shrinkage were similar to the control mortar [7]. On the other hand, using recycled ceramic aggregate with sizes similar to those of natural sand (10% and 20% replacements), do not cause negative effects on the compressive strength of CRMs, but they do affect the consistency of the mortar; as for shrinkage, it is reduced when 20% of the aggregate is replaced and increased when 10% is replaced (in both cases with respect to the control mix) [1].

Another trend of researchers has been to perform studies on different replacement percentages—including total aggregates. In one study, replacements of 0, 5, 10, 20, and 40% were carried out, concluding that the use of up to 40% of ceramic aggregate does not significantly affect the properties of the CRM, with the only exception being the density (fresh) and its workability [5]. In another investigation that replaced 20, 35, 50, 50, 70, and 100% by ceramic recycled aggregate (ratio 1:6), improvements in flexural behavior and increase in compressive strength were observed; however, the water absorption coefficient was reduced, and the density decreased as the substitution increased. As for shrinkage, it was considered acceptable up to 50% replacements [19].

Continuing with the shrinkage property, in another study, 30, 60, and 100% replacements were performed, and it was reported that the drying shrinkage improves when the recycled sand content increases (due to the release of water previously stored in the porous aggregate); however, the compressive and flexural strength decrease with increasing aggr gate replacement [14]. For extreme cases of replacement (100% of the ceramic aggregate), was shown that CRM does not perform well with respect to the reference mortar; howeve with lower replacement percentages (20 and 50%) it does [3].

In another study, it was concluded that the use of 10% of ceramic aggregate (this tim in powder form), leads in general terms, to improve its performance in comparison wit its reference mortar (contribution of hydraulic capacity) [15].

In any of the previous ways in which the use of ceramic waste has been investigate the property of porosity in CRMs (conferred by the porosity of the ceramic waste itself concomitate as one of the significant causes of variable mechanical behavior and dur bility [16], and as an example, brick can have a porosity of up to 40% [1]. The use such replacement materials interferes with the porosity and pore size distribution of th mortars that include them, and consequently with properties such as compressive strengt shrinkage, and durability (e.g., chloride and sulfate diffusion) [18].

Previous research on porosity in CRM [17], using crushed bricks with two partic size distributions (fine and coarse) and total replacement of natural sand, reported resul of total open porosity, and pore size distribution by Mercury Intrusion Porosimetry (MIP concluding that these aggregates influenced the microstructure of CRMs; specifically, thos that used the fine fraction of ceramic aggregates presented higher pore content in genera as well as higher percentage of micropores and lower percentage of macropores. Simila results were established in another research [20], in which four types of ground clay brick were used to be used as partial replacement of cement (0, 10, 20, and 30%). In this wor MIP was also used, and the results indicated that the pore volume increases with increasin replacement percentage—in turn decreases with increasing curing period; concluding th the resistance increases with the decrease of the total pore volume and with the increase i the proportion of small pores, mentioning that the production of C-S-H gel additional to th reaction of the ground brick with CH, refines the pore size distribution of the CRMs, whic ends up producing the increase in compressive strength. In contrast, reduction in porosit is what was exposed in another study [21], in which ceramic aggregate (product of electric insulators) was used in powder form to replace cement or as an addition (percentages 10, 20 and 30%); the MIP tests performed indicated that the pore volume of each variabl varied from 0.03 to 1.0 m pore diameter (mortars with brick powder), and this being low than the one that did not contain ceramic aggregate. Using histograms of accumulated po volume (variability from 0.05 to 2.0 μm), it was established that the samples with ceram powder presented porosity reductions of 30, 28, and 43%, respectively, with respect to th control mortar, relating this to the gain in compressive strength obtained.

In another study that studied the performance of CRMs in aggressive environmen (chloride and sulfate attack) and their relationship with porosity [18], researchers use ceramic aggregate from tile waste to replace the aggregates and also as a cement subs tute (40%); with the Scanning Electron Microscopy (SEM) technique, they identified th existence of a greater number of voids in the reference mortar than in the CRMs. With th X-Ray Diffraction (EDX) technique, they confirmed that CRMs may have better durabilit behavior, since they have a denser and less porous microstructure matrix, as the ceram powder is the one that densifies the matrix when filling the cavities and reduces the amou and size of the pores. Conclusions that were also supported by a different study [22].

Taking as a reference what has been reported in the literature on research related the properties of CRMs and their incidence with their porosity, the information is shown scarce and also sometimes controversial. Moreover, these studies largely focused on th use of ceramic wastes as a cement substitute. For this reason, this research proposes th approach of using ceramic aggregates as a replacement for natural aggregates (10, 20, 3 50, and 100%), and validating the relationship between their mechanical properties an the porosity of CRMs (using open porosity tests, N_2 adsorption and SEM image analysis seeking to validate viable alternatives in the production of mortars.

2. Materials and Methods

2.1. Materials

Recycled ceramic aggregate (CA) from defective tiles (size and/or geometry) was used; it was acquired from a local waste treatment plant (Reciclàrids SL, located in Sant Joan de Vilatorrada, Barcelona, Spain) certified for its processing. The initial aggregate had a size of 0–5 mm, so the coarse fraction was separated by sieve No. 4 (4.75 mm) and thus only the fine fraction was used. Natural aggregate (NA) purchased from a local construction aggregate company (Arids Anton SL, located in Molins de Rei, Barcelona, Spain) was used; the sand size was 0–4 mm, and of the siliceous type. Portland cement CEM I 42.5 N/SR (Cementos Portland Valderrivas SA, Vallcarca–Sitges, Spain), according to EN 197-1 [23], was used as binder for the manufacture of mortars; and tap water was used as water.

2.2. Properties of Study Aggregates

To obtain the granulometry of the study aggregates, the material was passed through the sieves specified by ASTM C144 [24]; once the procedure was completed, the corresponding curves were plotted, as well as the limits established by the standard. As a result, non-compliance with the established limits was observed; therefore, it was necessary to make a granulometric adjustment to achieve similar profiles between the two aggregates, thus avoiding possible attributable variables in the behavior of the CRMs. For the adjustment, it was proposed to separate each aggregate into two fractions by means of sieve No. 30 (0.59 mm); then, combinations were made between them (each type of aggregate separately) substituting size fractions with progressive increments of 10% of one of the fractions. The common criterion that established the fractions in both types of aggregate was their compactness (combination of fractions that presented the maximum value); which is affected by the shape of the aggregates (sometimes linked to the size of the aggregate). It is worth mentioning that the shape of the aggregate possibly influences aspects such as the adherence of the aggregates with the cement, and foreseeably, it will also influence aspects of workability, fluidity of the mixture, etc. [3,15]. This will be reflected in the density, absorption and finally the porosity of the hardened mortar. The ideal combination for this study was: for CA, 60% of the material retained on the No. 30 sieve, and 40% of the material passing, and for NA, 50% of the material both retained and passing through the No. 30 sieve; Figure 1 shows the particle size curves without adjustment (before), as well as adjusted (after) of the two aggregates.

Physical properties of the aggregates are presented in Table 1; in this table it can be observed in general that the density [25] and bulk density [26] both in oven-dry condition (OD) and in saturation-surface-dry condition (SSD) of the CA is lower than that of the NA; on the contrary, the NA has a lower percentage of voids, a notable difference in absorption, a lower amount of fines [27] and both have similar fineness modulus [28].

Table 1. Physical properties of aggregates [29].

Property	Unit	Recycled Ceramic Aggregate (CA)	Natural Aggregate (NA)
Density (OD)	Kg/cm^3	1820.9	2581.6
Density (SSD)		2155.4	2623.6
Bulk density (OD)		1182.0	1735.1
Bulk density (SSD)		1399.1	1860.8
Void content	%	35.3	32.9
Water absorption coefficient		18.4	1.6
Particles < 75 μm (sieve No. 200)		8.2	2.9
Fineness modulus materials	-	2.8	2.4

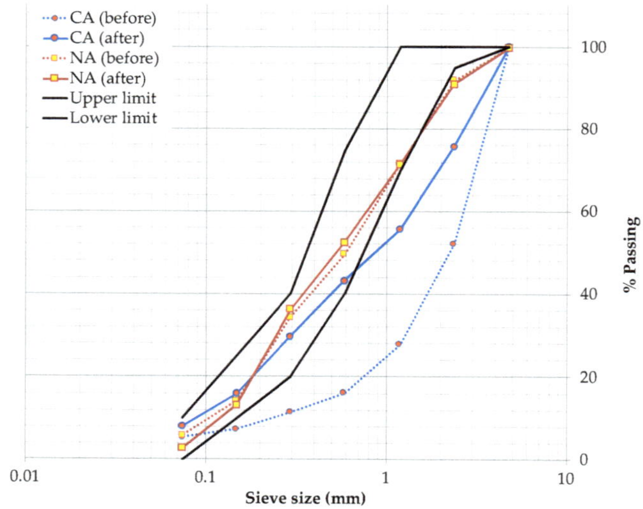

Figure 1. Recycled Ceramic Aggregate (CA) and Natural Aggregate (NA) granulometries.

2.3. Mortar Mixes

CRM specimens were manufactured with mixtures with different percentages of C (10, 20, 30, 50 and 100%) replacing the NA, these were identified as CRMXX (XX represent the percentage of CA). In addition, a usual mortar (UM) with 100% NA was manufacture to be used as a reference mortar. All mortars were designed with a cement:aggregate rati (c:a) = 1:4, as well as an initial water:cement ratio (w/c) = 0.5. Due to the high absorptio of the CA (see Table 1), the amount of water needed for mixing increased to achieve th target workability for all the mixtures (110 ± 5%)—in accordance with ASTM C109 [30] b means of the flow table test ASTM C230 [31].

The mixing procedure consisted of a previous saturation of the aggregates in a perio of one minute with the initial amount of water (result of the w/c ratio), in order to avoi water mobility necessary for the hydration process; then, the cement was added to th mixer container (Mod. E93, Matest brand) which contained the aggregates to start th mixing sequence at medium speed for 60 s, then continued at high speed for 30 s, then th mixture was left to rest for 90 s, and finally the high speed was selected for 60 s more. Onc the process was completed, the flow table test was continued, in which low consistencie were presented due to the lack of water; for this reason, small quantities were graduall added until the aforementioned workability was reached. Table 2 shows the quantities c each of the materials to manufacture one dm^3 of the different CRMs.

2.4. Physical and Mechanical Properties Tests

To compare the behavior of the CRMs studied with respect to their porosity, physic properties (*density*, *absorption*, and open porosity), and mechanical properties (compressiv strength [*fm*], modulus of elasticity [*E*], and *drying shrinkage*) were determined; since th objective of this work focuses on the incidence of porosity, these properties are presented i the Section 3. The specimens used in these tests were 4 × 4 × 16 cm, and for *drying shrinkag* 2.5 × 2.5 × 28.5 cm. The standards used were the Spanish UNE (UNE-EN 1015-10 [32 UNE-EN 1936 [33], and ASTM standards, respectively. More details of the tests for thes properties can be found in previous publications by the authors [29,34].

Table 2. Dosage of the study mortars.

Materials (g)	Mixtures Used for Mortars					
	UM	CRM10	CRM20	CRM30	CRM50	CRM100
Cement	400	433	381	372	348	323
NA [1]	800	780	610	521	348	0
NA [2]	800	780	610	521	348	0
CA [1]	0	70	122	178	278	517
CA [2]	0	104	183	268	417	775
Water	334	390	355	373	397	476
w/c	0.84	0.90	0.93	1.00	1.14	1.48

[1] Size passing through sieve No. 30. [2] Size retained on sieve No. 30. NA = Natural Aggregate. CA = Recycled Ceramic Aggregate. UM = Usual Mortar. CRM = Recycled Ceramic Mortars (where 10, 20, 30, 50 y 100, represents the percentage of CA).

2.5. Nitrogen (N_2) Adsorption Porosimetry

Mortar specimens of 4 × 4 × 16 cm were cured in water for 90 days after manufacture. Then, using a concrete disk cutter, four cuts were made, selecting the central part of the specimen resulting from the cuts (volume of approximately 36 cm^3) and discarding the outer faces (discarding the border effect); then, the central part of each specimen was placed in an oven at 60 °C for drying. Once the central parts of the samples were dried, they were ground in an agate mortar (trying to provoke a diversity of particle sizes), so that each sample under study contained sizes representative of the sieves used in the particle size established by ASTM C144 (with the exception of mesh No. 4 (4.75 mm), since the opening of the sample holder of the porosimetry equipment is 4 mm). Due to the different densities of the CRMs studied, and as a standardization procedure proposed to be performed on all the samples, it was decided to introduce in the sample holder flask for the test the same amount (in weight) of material corresponding to each sieve size (used in aggregate granulometry), resulting finally in 3.5 g samples to be tested for each study variable.

Once the samples were obtained, they were prepared using a VacPrep 061 micromeritics degasser (Micromeritics Instrument Corporation, Norcross, GA, USA), which offers two methods to remove contaminants such as water vapor and adsorbed gases from samples (avoiding interference with surface area measurements). The N_2 adsorption test was performed on a micromeritics TriStar—surface area and porosity analyzer, which uses TriStar 3000 V6.04 A software (Micromeritics Instrument Corporation, Norcross, GA, USA). This equipment is an automated gas adsorption analyzer that contains three ports and uses the principles of physical adsorption and capillary condensation to obtain information on the surface area and porosity of a solid material—it can measure surface areas down to 0.01 m^2/g using nitrogen as adsorbate (Figure 2).

The gas adsorption procedure involves bringing the study sample (cleaned by the degassing method) to a constant temperature (by means of an external bath). Then, the gas (adsorbate) is introduced into the pipette containing the sample in small doses, causing the gas molecules to adhere to the surface of the solid (adsorbent) and form a thin layer that coats the surface of the adsorbent [35]. According to the theory by Brunauer, Emmett, and Teller (BET) [36], it is feasible to determine the number of molecules to form a monolayer of adsorbed gas, which allows the surface area of the sample to be established. As the gas molecules continue to adsorb on the solid (after the first monolayer), multilayers are formed (deposited sequentially one on top of the other); this together with capillary condensation (a process feasible to establish using the Kelvin equation [36], which establishes the ratio of the equilibrium vapor pressure of a curved surface, such as that of a liquid in a capillary or pore, to the equilibrium pressure of the same liquid on a flat surface [37]). In order for the equilibrium pressure of the gas to approach saturation, the pores are filled mainly with adsorbate; therefore, knowing the density of the adsorbate, the volume it occupies and, consequently, the total pore volume can be determined. This adsorption process establishes as a result the so-called "isotherms" (experimental curves of adsorbed gas volumes with respect to relative pressures), which by performing the process inversel (removing known amounts of gas) will establish the desorption isotherms; both curve form a hysteresis that provides information about the different pore shapes [37,38]. Th Barrett, Joyner, and Halenda (BJH) calculation method allows determining the cumulativ or differential pore size distribution by taking the isotherms as data [38]; such distributio are represented by the plot of pore volume versus pore size, usually used to describe th porosity of materials [39].

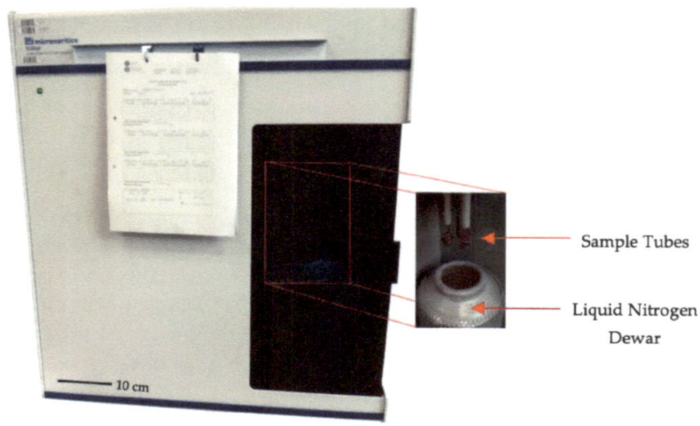

Figure 2. Gas adsorption test equipment.

2.6. Porosity by SEM and Microanalysis of Chemical Elements with an Energy Dispersive Spectrometer (SEM-EDS)

From the 4 × 4 × 16 cm specimens, and after curing in water for a period of 90 day small samples were taken from the central part of the specimens—a cube of approximatel 1 cm^3, taking care that the transversal face of the specimen was the one that was expose for obtaining the images in the microscope. Four samples of the mortars were include inside a circular mold, then covered with transparent epoxy resin (EpoFix resin, Struer Hørsholm, Denmark). The mold with the samples was placed inside an oven at 40 ° for a period of 40 min (hardening of the resin causing the samples to be encapsulate in the resin). The circular sample was then "sliced" to achieve a flat face and with th samples exposed outside the resin (a precision cutter was used); next, the exposed fac was subjected to a progressive roughing-polishing process with different sandpaper (unt the gloss was achieved on the study face). Once the tablet with the different samples wa ready (Figure 3), they were analyzed in a Scanning Electron Microscope (SEM), visuall selecting the possible areas of interest that would allow detecting the natural and ceram aggregates, respectively.

A JEOL JSM-6510 microscope (Jeol Ltd., Tokyo, Japan) was used for image acquisitio obtaining them in two different ways: (a) high resolution secondary electron detector "SE (secondary electron image), and (b) backscattered electron detector "BEI" (backscattere electron image); the latter, with higher resolution than the former, but with greater contra to examine the surface topography of the samples. Two different "zooms" were used obtain the images: 200× in the areas of interest (identification of the types of aggregate interfaces between them and the ITZ), and 500× for the mapping; the latter consiste of determining the chemical composition of the samples, detecting by microanalysis th basic elements that make it up. On this occasion, the equipment used was an energ dispersive spectrometer (EDS) (Inca 200, Oxford Instruments, Abingdon, United Kingdo connected to the SEM, which separates the characteristic X-rays of different elements in an energy spectrum, and by means of the EDS system software, the energy spectrum analyzed in order to determine the abundance of specific elements. By means of scans o the sample, the different basic chemical elements requested are obtained, which taking into consideration the chemistry of the usual reactions of a mortar and its constituent materials, the following were established: Si, Ca, Fe, Al, Mg, Na, Cl, N, and K. Table 3 shows a representative relationship of each of the elements in the CRM samples, which were obtained through EDS analysis.

Figure 3. Recycled Ceramic Mortars (CRM) samples used for SEM analysis: (**a**) UM, (**b**) CRM10, (**c**) CRM20, (**d**) CRM30, (**e**) CRM50 and (**f**) CRM100.

Table 3. Content of each element in the Recycled Ceramic Mortars (CRM) samples obtained by Energy Dispersive Spectrometer (EDS) analysis.

Variable	Percentage Amount of the Elements (%)								
	Si	Ca	Fe	Al	Mg	Na	Cl	N	K
UM	2.78	0.00	2.08	0.69	1.39	2.08	65.28	25.00	0.69
CRM10	28.04	50.15	4.59	7.47	4.80	0.18	0.38	0.00	4.39
CRM20	2.78	0.00	2.08	0.69	1.39	2.08	65.28	25.00	0.69
CRM30	47.77	34.03	5.47	7.94	1.86	0.51	0.13	0.00	2.29
CRM50	23.10	40.82	8.36	7.95	2.30	0.00	0.81	14.55	2.11
CRM100	3.90	-	2.60	1.30	1.95	3.25	87.01	0.00	-

2.7. Image Analysis (Obtained with Scanning Electron Microscope—SEM), Using NI Vision Assistant

The analysis carried out consisted of examining an image by separating the components, and by means of a specific procedure calculating the porosity. For the analysis of the images obtained from the SEM, NI Vision Assistant 2018 software from National Instrument (LabWindows/CVI Version 7.1, Austin Texas, TX, USA) was used. This allows a series of integrated operations with which a "script" (script file, containing processing functions, and relevant parameters, to be used on individual images or in a batch to analyze a collection of images) is created, which allows separating, highlighting, processing, and analyzing the images where the paste and ceramic aggregates of the CRMs under study are shown [40].

Finally, the images selected for porosity analysis were processed from the conjunction of preliminary images obtained in two ways: the one obtained through the main Scanning Electron Microscope-Backscattered Electron Image (SEM-BEI) image of each sample, and the one established through the images corresponding to the mapping of chemical elements. Once entered into the software, the main images were treated by applying a "threshold filter" (pixel color ranges between 0 and 75), to then obtain their histogram (equivaler porosity values in the dark areas). Next, the mapping images were used to establish th elements corresponding to the aggregates (silicon, magnesium, and aluminum), whic were processed with the "advances morphology" and "basic morphology" command The final result of each image process was stored in different buffers and then the "adc operation was applied for all the aggregates, and again the "add" operation was applie to incorporate the porosity values of the main SEM-BEI image. Finally, after completin the procedure, the percentages of aggregates (white pixels), paste (red pixels) and porosit (black pixels) of each processed image were quantified using the "histogram" comman Figure 4 shows graphically the above-mentioned procedure on the image analysis.

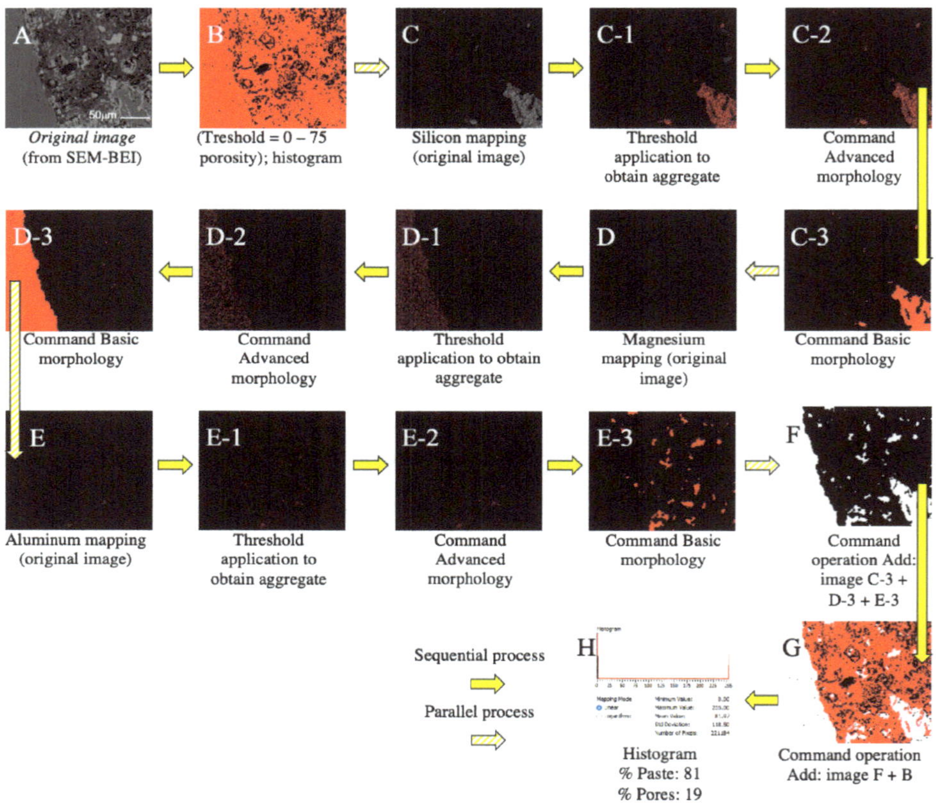

Figure 4. Example of image analysis procedure; CRM10 sample. The steps that have a sequence are linked with yellow arrows, and the steps that present a parallel process (i.e., the same, but for each of the different elements), are linked with arrows with yellow stripes. The image is obtained by SEM (**A**); "Threshold filter" is applied to the original image, and then its histogram is obtained (**B**); The image corresponding to the Silicon mapping is processed (**C**); the same for the image of the Magnesium (**D**) and Aluminum (**E**) mappings. The threshold filter is applied to the mappings to obtain the image representing the aggregates in the sample (**C-1**). The image of the element is processed using the "advances morphology" and "basic morphology" commands (**C-2**,**C-3**); The same steps are performed in the Magnesium and Aluminum mappings (**D-1**,**D-2**,**D-3**,**E-1**,**E-2**,**E-3**). The "add" operation was applied to the processed images of the elements (images **C-3**,**D-3**,**E-3**); and the image (**F**) was obtained. The "add" operation is applied again, to incorporate the porosity values of the original image; image (**G**) [result of (**F**) + (**B**)] is obtained. A histogram is obtained to quantify the percentage of aggregate, paste and pores (**H**).

3. Results and Discussion

The following sections present the data on the physical and mechanical properties of the CRMs studied, followed by the porosity results obtained (gas adsorption test (N_2) and by means of the SEM image analysis process), exploring for each case their correlation, linkage, or general incidence.

3.1. Physical and Mechanical Properties of CRMs

In Table 4 the results corresponding to the physical and mechanical properties of the CRM and UM studied, respectively, are shown. These results (open porosity and *absorption* [29]; *density*, *fm*, *E* and *drying shrinkage* [34]) were analyzed in previous works. The objective of the current work is to establish their relationship with porosity.

Table 4. Mechanical y physical properties of mortars.

Variable	Physical Properties			Mechanical Properties		
	Density (g/cm^3)	Open Porosity (%)	*Absorption* (%)	*fm* (MPa)	*E* (MPa)	*Shrinkage* (%)
		60 días			90 días	
UM	1.95	19.09	11.87	28.77	26,252	−0.0793
CRM10	1.95	22.68	14.18	30.58	26,515	−0.0704
CRM20	1.89	24.96	15.57	30.95	26,251	−0.0847
CRM30	1.86	27.04	16.90	30.10	24,065	−0.0709
CRM50	1.80	32.00	19.91	26.76	21,731	−0.0986
CRM100	1.53	41.17	25.56	20.56	14,194	−0.1141

A simplified and general synthesis of the behavior of these CRMs could be established as follows: *density* decreases with the increase of replacement, reaching a maximum difference of 0.42 g/cm^3 (CRM100 w/r UM); open porosity and *absorption* present similar behavior, since the values increase with the increase in the percentage of CA, with a difference between CRM100 w/r UM of 22.08 and 13.69%, respectively; *fm* presents close values between UM and mortars with replacement of up to 30%, and for replacements of 50 and 100% the values decrease notably (10. 39 MPa between CRM20 w/r CRM100); *E* presents similar behavior to *fm* (decrease with the increase of CA replacement), with difference between the highest and lowest value of 12,321 MPa; *drying shrinkage* established an opposite behavior to *fm* and *E*, by increasing with the increase of CA in mortars, particularly with CRM50 and CRM100 (0.0437% in difference between the lowest and highest value of shrinkage).

3.2. Porosimetry by N_2 Adsorption

Gas adsorption (N_2) is an appropriate technique to characterize pore sizes and distribution mainly in the range of mesopores (from 2 to 200 nm) [37], gel pores (<10 nm), and capillary pores (10 to 50 nm) [41,42]; Therefore, for an investigation in mortars, this can be considered the most approximate, accurate, and adequate. Gas adsorption measurements have been extensively used to determine the surface area and pore size distribution of an important variety of solid materials—among which construction materials are also included [36]. On the other hand, total pore volume and pore size distribution are also common techniques used to describe the porosity of materials; the latter being a distribution of pore volume with respect to pore size [39]. The results of the different variables (besides porosity) obtained by this technique are presented below.

The N_2 adsorption isotherms for the samples studied (ratio of the amount of adsorbed molecules with respect to the pressure at constant temperature) are shown in Figure 5 (the letter "A" at the end of the name corresponds to the adsorption phase; and the letter "D" to desorption, respectively)—sample CRM30 could not be included due to contamination of the sample. It can be seen that CRM100 reports the highest adsorbed amount for all pressures with respect to the rest of the samples, and in the same figure (upper le enlargement) it can also be appreciated that for low pressures (beginning of the curve and in general, the adsorbed amount presents an increase related to the highest amou of CA in the CRMs, where the UM (0% CA) is the one that presents the lowest value With regard to high pressures (end of the curves), which is observed in the upper rig enlargement of the same figure, the amount adsorbed continues to be higher for CRM1 (65.9 cm^3/g); for the rest of the mortars, adsorption caused similar amounts (values fro 32.35 to 35.84 cm^3/g)—the UM being the one with the highest amount. According to th way the hysteresis was presented, it can be said that to remove the gas from the samp lower pressure was necessary; however, for relative pressures of 0.5 (approx.) in all sample the amounts of gases decreased, so desorption becomes slower. From the data provided the isotherms, the surface area of the solid, the size and its descriptive pore parameters, i distribution, etc. were calculated.

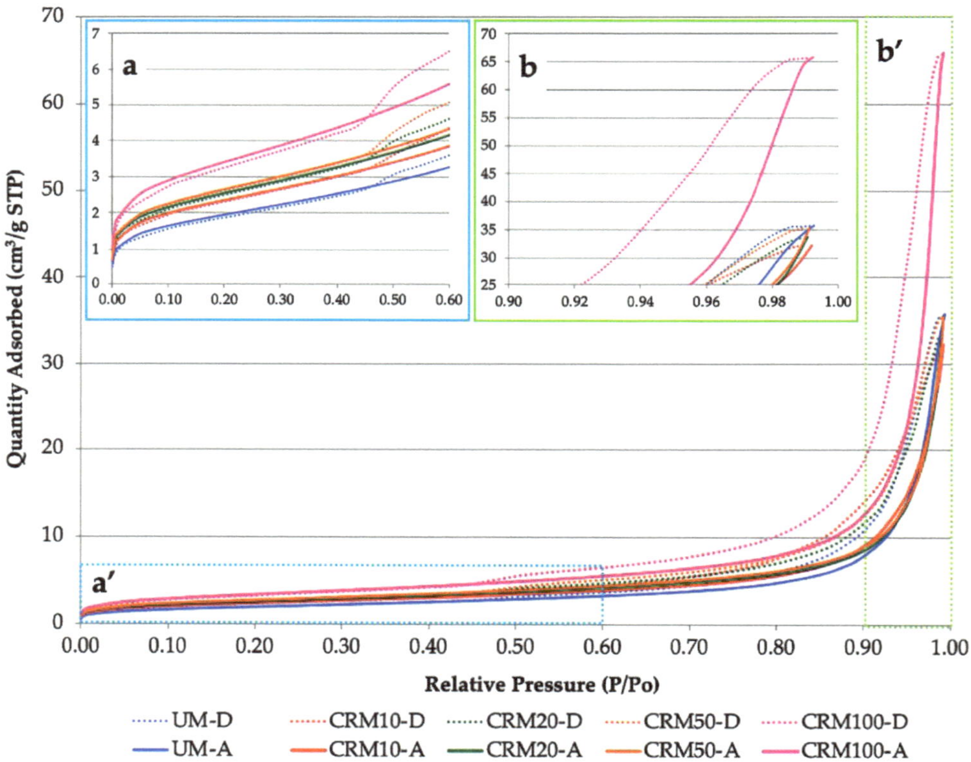

Figure 5. Adsorption-desorption isotherms of CRMs: (**a**) the beginning of the adsorption-desorption isotherm curves is shown; specifically, the amount of molecules adsorbed and desorbed at low pressures (blue box a′); (**b**) the final part of the isotherm curves is shown; specifically, the amount of adsorbed and desorbed molecules at high pressures (green box b′).

With this technique, it is feasible to measure the surface area of solids; and for whic the data in Figure 5 [43] are used, in particular those of the low relative pressure regic (P/P_o from 0 to 0.30). The calculation is carried out using the BET method, since it is a use and established approach to determine the surface area of solid materials, especially thos with different open porosity [37].

The results obtained are shown in Figure 6. In this figure it can be observed th the surface area values present, in general terms, an increase related to the increase the percentage of CA in the CRMs. CRM100 has the highest surface area with respect to UM, with a difference of 5.05 m^2/g (73% higher). According to the above, and due to its greater surface area, it can be affirmed that CRMs have greater porosity, which increases as a function of the CA content [36]. In a previous study, similar behavior was observed (case of concrete with recycled concrete aggregates); in which the surface area increased in relation to the increase in the percentage of recycled aggregate in the mixes (and higher than the reference concrete) [41].

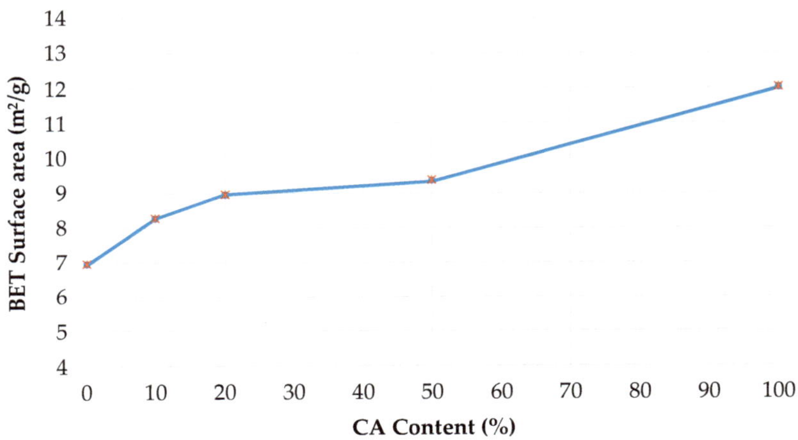

Figure 6. BET surface area according to the Recycled Ceramic Aggregate (CA) content in the Recycled Ceramic Mortars (RCM).

Total pore volume and pore size distribution are two common techniques also used to describe the porosity of materials [39]. Figure 7 shows the pore size distribution curves, which indicate the accumulated pore volume as a function of the average pore diameter of the different CRMs studied and the reference UM (adsorption and desorption phase, respectively). Figure 7a shows that CRM100 is the one that contains significantly higher pore volume, with an accelerated increase of pore volume in the pores with diameters between 20 and 150 nm; while the rest of the mortars contain about 48% lower volume; the general trend of pore volume increase is a function of the percentage of CA in the CRMs. In the desorption phase (Figure 7b), the same order in the behavior of the CRMs is presented; however, the decrease in volume is reflected in smaller pore diameters included between 10 and 70 nm.

The above graphs provide the porosity of the CRMs, which is expressed as cumulative pore volume (cm^3/g) and are presented in Table 5. The above porosity values have a similar trend to the values obtained by means of the open porosity test (Table 4). By the N_2 adsorption technique, the average superior difference of CRM100 with respect to the rest of the mortars is 48% (both in the adsorption and desorption phase), while that obtained by open porosity is 40%.

Table 5. Porosity measured in pore volume (N_2 adsorption and desorption phases).

Mortar	V_{ads} (cm^3/g)	V_{des} (cm^3/g)
UM	0.054321	0.054842
CRM10	0.048443	0.049154
CRM20	0.050341	0.050890
CRM50	0.052853	0.053522
CRM100	0.099844	0.101067

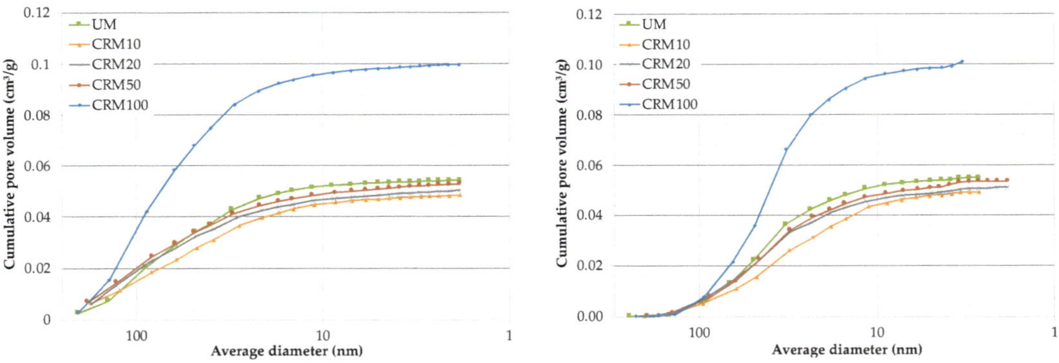

(a) (b)

Figure 7. Pore size distribution with respect to cumulative pore volume: (**a**) adsorption phase; (**b**) desorption phase.

Regarding the size distribution (diameter) of pores defined by IUPAC (Internationa Union of Pure and Applied Chemistry), pores can be classified into: micropores (pore < 2 nm mesopores (between 2–50 nm) and macropores (pore > 50 nm) [36,37]. Figure 8a shows tha the highest porosity percentages of all mortars are found in the macropore zone, with a average difference of 25%, with respect to mesopores and 62% of micropores; in which th rest of the gas was introduced. However, in the desorption phase (Figure 8b), it is indicate that the highest porosity is represented by mesopores, with an average difference with respe to macropores of 27%, and 64% with respect to micropores.

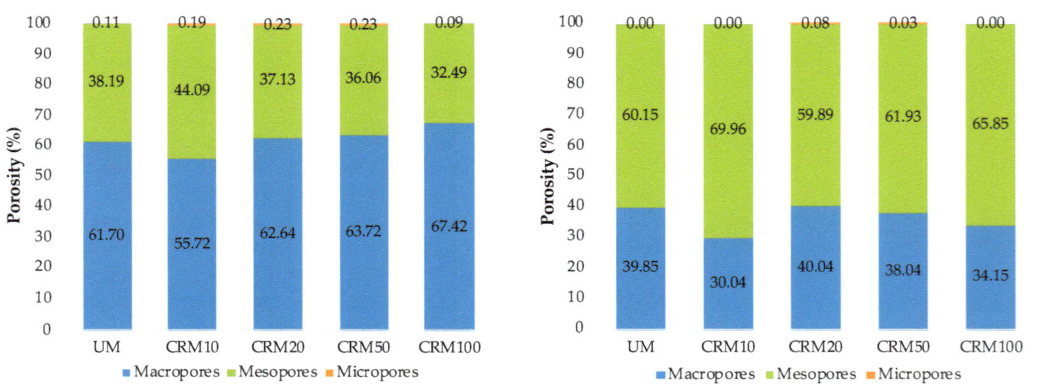

(a) (b)

Figure 8. Pore size distribution according to IUPAC: (**a**) adsorption phase; (**b**) desorption phase.

Research on two types of mortars with 100% CA (with fine aggregate and coars aggregate) reported that the mortar with fine aggregate (similar particle sizes to this study showed higher porosity in general than the mortar with coarse aggregate (particles large than 150 μm). Regarding the distribution of pore sizes, the mortar with fine aggrega presented a lower amount of macropores and a higher amount of mesopores with respect the reference mortar. The above was attributed to the own porosity of the CAs (these infl ence the microstructure of the mortar when it is finely ground) [17]. Therefore, this researc presented similar behaviors to those obtained here—in particular with desorption data.

From complementary information obtained from the gas adsorption technique, and i particular from the pore size distribution curves (Figure 7), several categories of pore radiu can be identified. In Table 6 (adsorption and desorption phase, respectively), the results of the maximum radius (r_{max}), minimum radius (r_{min}), average radius (r_{ave}), medium radius (r_{med}) and critical radius (r_{cri}) are provided; a description of each is given below.

Table 6. Pore radius categories (N_2 adsorption and desorption phases).

Mortar	r_{max} (nm)	r_{min} (nm)	r_{ave} (nm)	r_{med} (nm)	r_{cri} (nm)
		Adsorption de N_2			
UM	103.57	0.91	14.98	33.84	70.50
CRM10	87.80	0.91	12.28	29.22	61.71
CRM20	87.74	0.92	12.52	36.31	87.74
CRM50	91.64	0.92	12.69	37.61	63.89
CRM100	102.64	0.92	16.75	37.89	69.79
		Desorption de N_2			
UM	122.47	1.32	11.69	21.52	24.41
CRM10	101.48	1.32	9.14	16.88	23.61
CRM20	91.92	0.92	9.92	21.09	24.08
CRM50	97.82	0.91	9.58	20.18	22.78
CRM100	113.23	1.63	11.52	20.32	24.26

The r_{max} and r_{min} establish the maximum and minimum radius corresponding to each variable; these were obtained directly from the reports provided by the test equipment. The r_{max} values are in the range of 87.74 to 103.57 nm (adsorption phase), and 91.92 to 122.47 nm (desorption phase); while those of r_{min}, established radius in the range of 0.91 to 0.92 nm (adsorption phase), and 0.91 to 1.63 nm (desorption phase). As can be seen in the table, these did not show any relationship with respect to the CA content.

The r_{ave} was obtained by dividing by two the average diameter value calculated with the equation $4V/A$ (where V = cumulative pore volume, and A = cumulative pore area). The data are obtained with the BJH method (diameters between 17 and 3000 Å). The values obtained show similar behavior to the porosity calculated by N_2 adsorption, which is an increase with the percentage of CA (without considering those of the UM sample), and with similar values in mortars with CA substitutions of up to 50%, as well as a notable increase in the mortar with total replacement of the NA.

The r_{med} corresponds to the radius determined by a Lagrange interpolation of the closest points corresponding to 50% of the total volume of each sample [44,45]; obtaining these from the pore size distribution of the test. Table 6 shows that the r_{med} varies from 29.22 to 37.89 nm, and from 16.88 to 21.52 nm (adsorption and desorption phase, respectively) for the different study variables.

The r_{cri}, or critical pore, is the term given to the corresponding radius that causes the onset of the maximum slope in the curve of pore radius versus the pore volume differential ($dV/dlogD$). This pore radius is usually an indicator of the microstructure of the material and is used to detect a variety of materials [44]. The method for its determination consists of detecting the maximum peak in the pore size versus $dV/dlogD$ curve [14,45,46]. This also indicates the minimum radius of continuous pores within the material [47]; that is, it establishes the space capable of being filled without forming any other adsorption pathway [45,46]. Figure 9 shows the graphs—for adsorption and desorption stages—in which the maximum peak of the curve (with respect to the ordinate axis) is observed; when crossed with the abscissa axis, it establishes the r_{cri}. As an example, that of the variable MCR100 is shown in Figure 9a).

Using the data of the mechanical properties of the CRMs (Table 4), a correlation analysis and equation fitting were performed in order to establish the dependencies of these with respect to the different porosity variables obtained by the N_2 adsorption technique (Table 6). The criteria established to define the adjusted equation were as follows: (a) A coefficient R^2 as close to 1 ($R^2 \cong 1$), (b) The type of equation selected should generate a curve plot coincident to the points of the related variables.

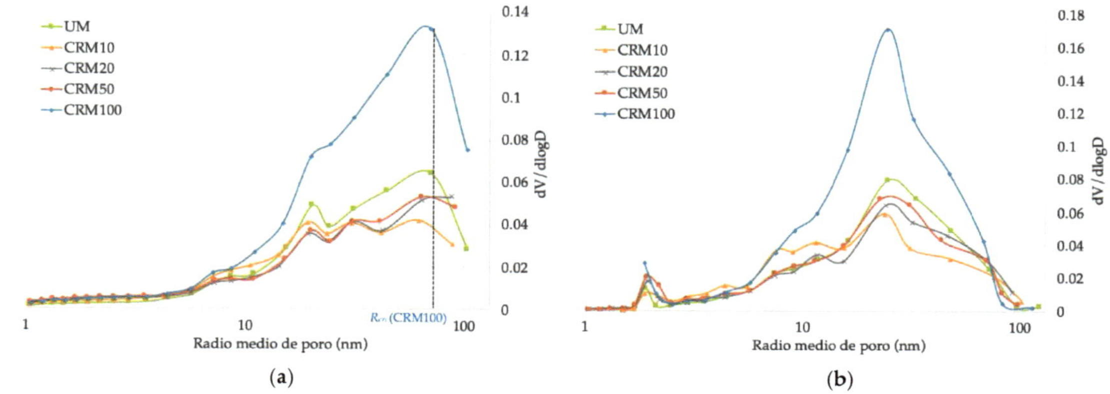

Figure 9. Pore radius with respect to pore volume differential: (**a**) adsorption phase; (**b**) desorption phase.

For the properties of *fm*, *E* and *density* of the CRMs, it was established that the best f
was with the variable of rave porosity (in the adsorption phase); the type of fitting equatic
was that of a second-degree polynomial equation (see Figure 10). In this equation, the F
values and the equations for fitting the data for the three CRM properties are indicate
with a range of validity of application established between CRM10 and CRM100. In th
three equations, the UM sample is isolated from the analysis, as it establishes inconclusiv
values for the study, which may have their origin in aspects inherent to the test or to i
degree of precision (nm).

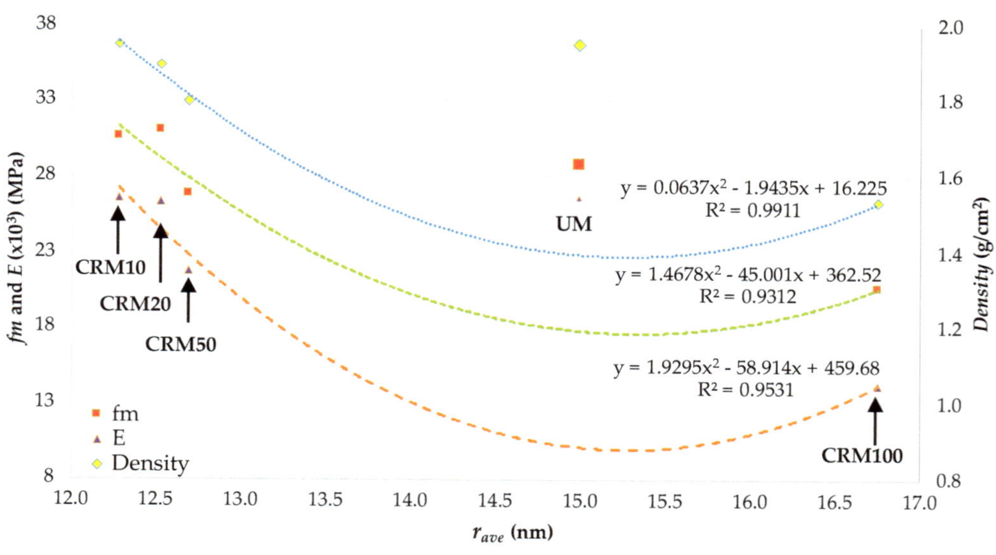

Figure 10. Correlation between *fm*, *E* and *density*, with respect to r_{ave}.

The r_{ave}, as an evident representative parameter (central tendency) of the porosity of the CRMs, meets the expectations of its capacity to link the physical-mechanical behavior of the mortars under study to their porosity. Therefore, an increase in r_{ave} is related to losses in *fm* and *density*, with CA causing the increase in r_{ave}, which in turn produces a weaker CR matrix structure. On the other hand, the nature of the type of equation established (secor order polynomial) may suggest the existence of another variable not included in this stuc which could be the cause of the nonlinearity in the equation. Finally, the existence of a significant change in the plotting of the curves between the CRM50 and CRM100 variables is notorious, and is worthy of future studies to establish the continuity of behavior for the percentages of CA between them.

As for the adsorption and *drying shrinkage* properties, these established a better linkage with the r_{max} porosity variable (adsorption phase), obtaining significant R^2 with second-degree polynomial type equation curves as shown in Figure 11. As in the previous case, their application is valid for the range between CRM10 to CRM100, as well as the assumption of another variable not included in the study.

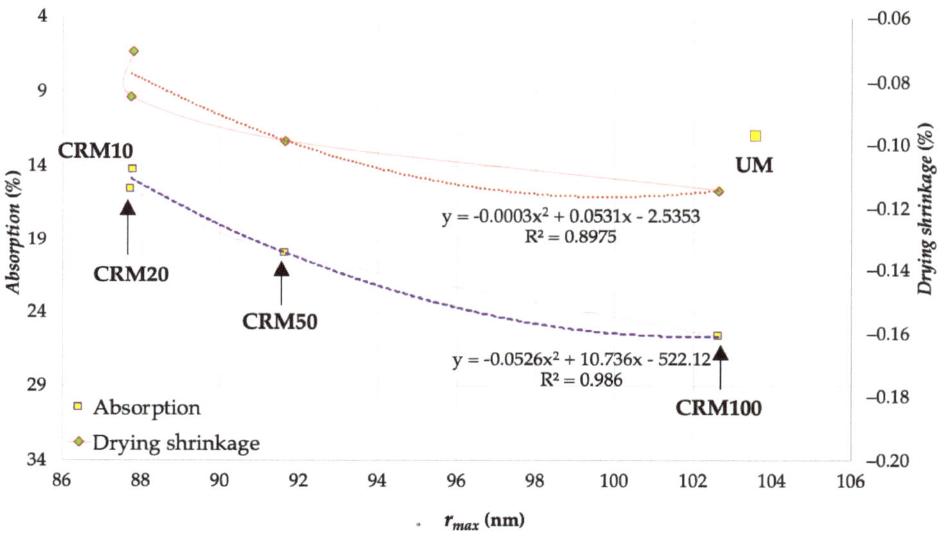

Figure 11. Correlation between *absorption* and *drying shrinkage* with respect to r_{max}.

In this case, it seems that the existence of large pores in the CRM matrix is what best explains the behavior of these two properties—both intrinsically involve, in their respective behavioral phenomena, the ability of water mobility between the exterior and interior of the matrix; therefore, the existence of these large pores seems to provide the conditions that promote these behaviors, the increases in the CA in the CRM being what favors the increase in their size.

From the results of the BET surface area property with respect to the physical and mechanical properties of CRMs, *density*, *E*, and *absorption* obtained $R^2 > 0.9$; while for *fm* and *drying shrinkage* were 0.89 and 0.79, respectively. All the fitting equations were of the second-degree polynomial type. On this occasion they are valid for all the research variables—dispersion of variables according to expectations (Figure 12).

3.3. Porosimetry by SEM Image Analysis

The results of the porosity determined by SEM image analysis are presented in Figure 13; in this, two important aspects are highlighted: (a) a significant increase in the percentage of porosity of the CRMs with respect to the reference UM—especially with the CRM100 and (b) for the CRM10, CRM20, CRM30, and CRM50 variables, with values close to or without evident order—attributable to the complex analysis process (prompted by aspects such as the selection of the capture image by the microscope, or by the quality of the image to be treated).

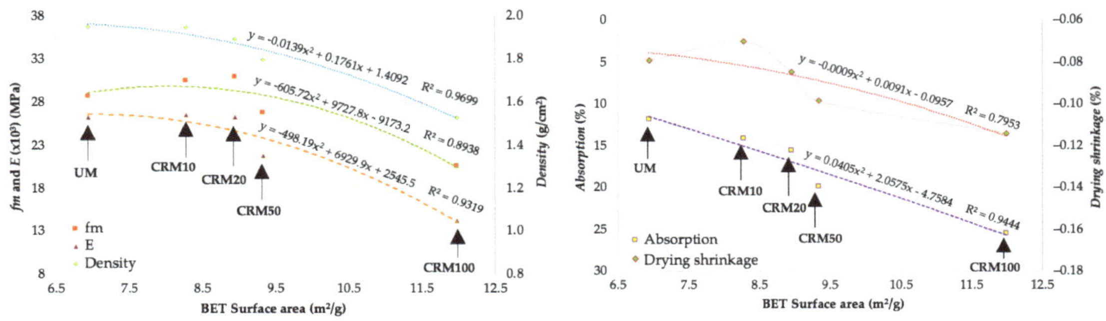

(a) (b)

Figure 12. Correlation of BET surface area and: (**a**) *fm*, *E* and *density*; (**b**) *absorption* and *drying shrinkage*.

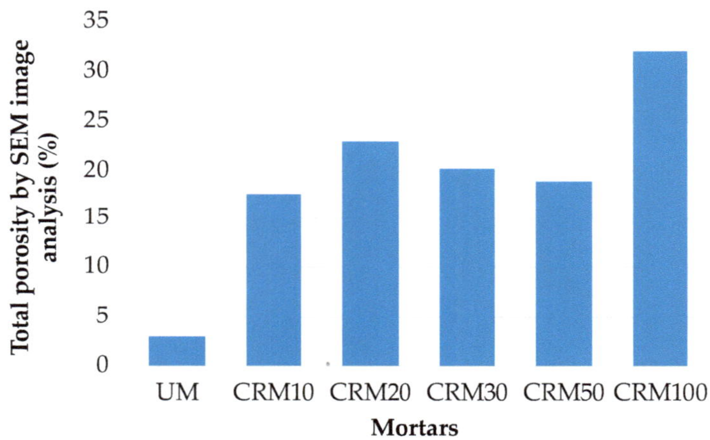

Figure 13. Percentage of total porosity of CRMs by SEM image analysis.

With respect to the porosity values obtained by this technique, a correlation analys was also performed to compare the physical and mechanical properties of the CRMs. I Figure 14 we present the results of *fm*, *E* and *density* with respect to the total porosit obtained by SEM image analysis. The best-fit regressions were with a second-degree pol nomial trend, with application to all the study variables—including CRM30; of these, th relationship with respect to *density* is the most outstanding. Comparatively, the adjus ments achieved with this technique to establish the porosity with respect to the physic and mechanical properties of CRMs are less satisfactory than those obtained with N adsorption—not because of its calculation accuracy, but because of the degree of represe tativeness of the test sample.

3.4. Total Open Porosity

The results of open porosity presented in this study were determined in previou work [29]; however, a brief abstract of their determination would be the use of dry, su merged, and saturated state weights, applying the formulation of the standards [33]. Wit the data obtained for the CRMs under study, an analysis was performed to establish th relationship with the physical and mechanical properties. Regarding the properties of *f* *E* and *density*, R^2 values higher than 0.9 were obtained, with a second-degree polynomi trend (see Figure 15a), in the case of *drying shrinkage* it was 0.8101 (see Figure 15b); final *absorption* was not included in the study of correlations because its determination used data identical to those used to establish open porosity (numerical redundancy that produces identity). In any of the properties correlated with open porosity, it is evident that open porosity establishes a satisfactory relationship, with an ordered, equidistant, and continuous dispersion distribution of the studied variables.

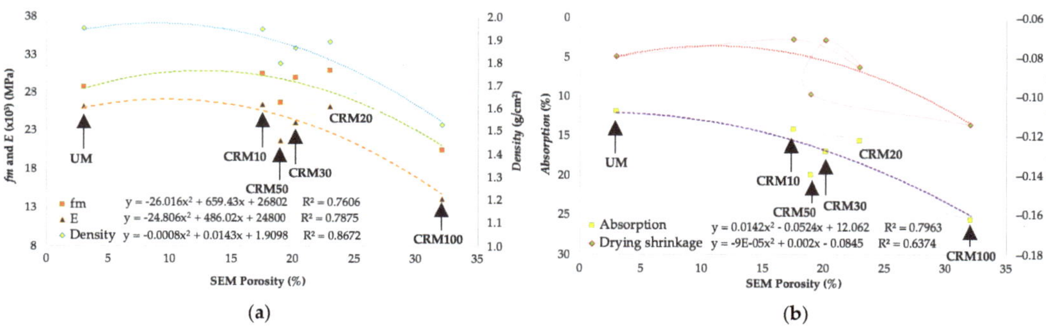

Figure 14. Correlation of Porosity by SEM and: (**a**) *fm*, *E* and *density*; (**b**) *absorption* and *drying shrinkage*.

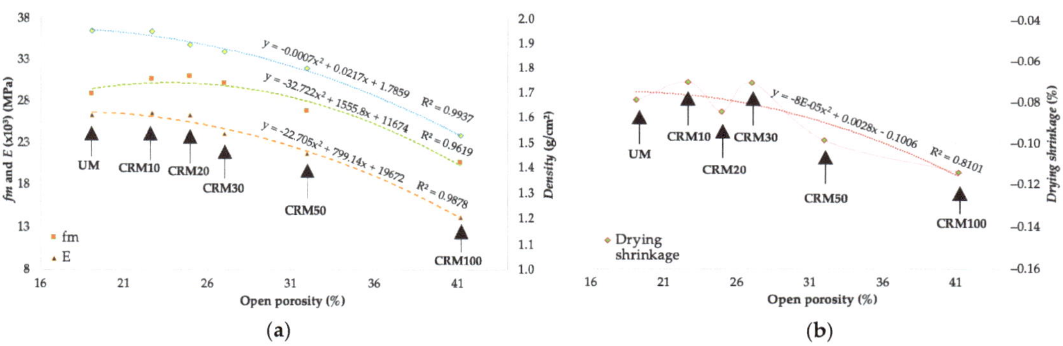

Figure 15. Correlation Open porosity and: (**a**) *fm*, *E* and *density*; (**b**) *drying shrinkage*.

3.5. Correlations between Porosimetry Techniques

With the results obtained from the three different porosimetry techniques, it was determined which of these offered a better correlation with the physical and mechanical properties of the CRMs. From the previous results, it was observed that the most significant correlations between the different porosity values were presented with the data obtained from the open porosity—the simplest, most direct method and the one that establishes a more representative sample of the matrix of a mortar. The results of total porosity—the only common result comparable among the study techniques for the three techniques—are presented in Figure 16 (porosity by N_2 is included in the adsorption and desorption phase—both very similar). In this graph, it has been chosen to present the results in percentage to simplify the comparison; thus, allowing making visible the same behavior of porosity increase between the study variables (CA content) and the techniques used.

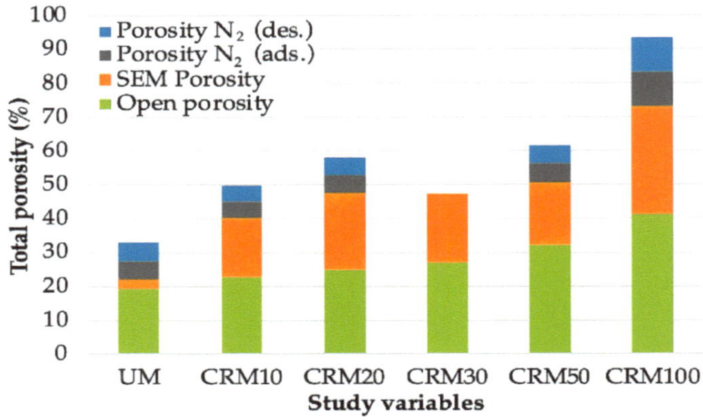

Figure 16. Percentages of total porosity obtained by the techniques used.

To numerically validate the previous results of total porosity of the CRMs, takir into account the different experimental techniques of their determination, the correlatio with the data obtained with the same were analyzed. It was observed that the compa son between open porosity and N_2 porosity (desorption phase) resulted in a coefficie R^2 = 0.9881 (second-degree polynomial) (Figure 17); which shows the affinity betwe both techniques, and also corroborates the incidence of the use of CA in the behavior the CRMs.

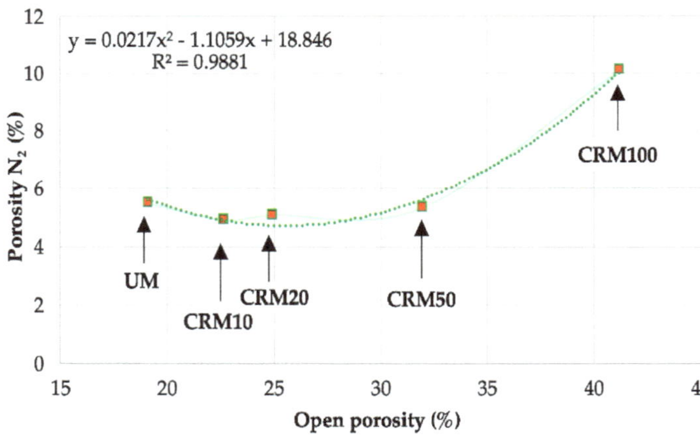

Figure 17. Correlation between open porosity and N_2 porosity (desorption phase).

In accordance with the above, and with the objective of establishing the optim technique to use—or in its case to establish the hierarchical order of importance in tl selection of the technique when porosity is to be established, it is determined that the op porosity technique is the one that provides more reliable data (in addition to doing so wi a minimum processing of the samples). Then, the N_2 adsorption analysis technique shou be used as a second alternative (verification of the first one, or with the contribution special and different porosity parameters, which can help to better explain the mechanic and physical behavior of CRMs); although it is true that its requirements for laborato equipment and its representativeness of the samples are more important and compl Finally, as a third option, porosity by SEM image analysis would be the last of the thr techniques; this limitation of choice is due to the fact that its numerical determination m involve factors such as: the choice of the image for its study, its representativeness of the complete mortar matrix, its study scale factor, its requirements of specialized laboratory equipment and its own complex sample preparation.

As additional information in this study, it is suggested that the optimum percentage of CA to be used in CRMs is 20%. Taking into account the porosity results obtained by the three techniques, it was observed that for each of them, the CRMs presented similar behavior —increased porosity, with increased CA content, particularly with the open porosity and N^2 porosity techniques; however, although this information is relevant, it prevents choosing the adequate percentage of CA. Therefore, with knowledge of the physical and mechanical properties of CRMs, the *fm* results (property considered important or representative in mortars and concretes) were taken as an indicator for such decision. According to the information presented in the present study, it is considered that the use of up to 20% of CA in CRMs will guarantee a similar behavior to UMs, having as an added value the environmental benefits that the use of CA entails (as previously mentioned in the introduction).

4. Conclusions

In this study, an experimental investigation was carried out to present the results referring to the porosity of CRMs, correlating these with the physical and mechanical properties of the mortars, as well as between the techniques used to study porosity. Based on the results obtained, the following conclusions can be drawn:

1. The incremental variability of the porosity in the CRMs (increase directly correlated with the increase in CA percentage content), is validated by the different test techniques used; in particular with that of open porosity and that of N_2 adsorption.
2. The results obtained from the open porosity calculation correlate best with the behavior of the CRMs—percentage of CA addition as well as with the results obtained from the physical and mechanical properties of the CRMs.
3. The N_2 adsorption test provides information related to the amount of adsorbed gas, pore size distribution (diameter/radius), BET surface area, diversity of pore radius types, etc., which can be used to analyze or compare with respect to the physical and mechanical properties of CRMs. The technique can be considered with a contribution of important diversity of parameters that can be used to understand the effect of the porous network of mortars.
4. The properties such as *fm*, *E* and *density* of the CRMs, presented the best fit with the variable of r_{ave} porosity, (in adsorption phase); while *absorption* and *drying shrinkage*, with respect to r_{max}. The former because it is a variable of general representation of the porous network with incidence on properties on which the whole mortar matrix has an impact; and the latter, because it is a parameter on which the related properties have an impact on the capacity of water mobility.
5. The BET surface area with respect to physical and mechanical properties, established acceptable coefficients R^2 ($R^2 \cong 1$), as well as equations that generate a trace of its curve coincident to the points of the related variables.
6. Porosity results obtained with SEM image analysis do not show an evident correlation with respect to physical and mechanical properties of CRMs; they are so for the extreme variables (UM and CRM100), but without evident order for the rest. This is attributed to the image analysis process itself.
7. It is determined that the open porosity technique is the one that provides the most reliable data, which can help to better explain the physical and mechanical behavior of the CRMs, followed by the N_2 adsorption analysis technique, and finally the SEM image analysis technique.
8. It is suggested that the optimal percentage of CA to be used in CRMs is 20%, which will guarantee similar properties to UMs.

Author Contributions: Conceptualization, J.M.G.-S. and F.G.C.-C.; methodology, J.M.G.-S., F.G.C.- and C.A.R.-C.; software, J.M.G.-S. and C.A.R.-C.; validation, J.M.G.-S., F.G.C.-C. and J.L.A.-S.; form analysis, F.G.C.-C. and J.M.G.-S.; investigation, F.G.C.-C. and J.M.G.-S.; resources, J.M.G.-S., J.L.A.- and J.M.B.-C.; data curation, J.M.G.-S., F.G.C.-C. and C.A.R.-C.; writing—original draft preparatio F.G.C.-C.; writing—review and editing, J.M.G.-S. and F.G.C.-C.; visualization, J.M.G.-S.; supervisio J.M.G.-S.; project administration, J.M.G.-S.; funding acquisition, J.M.G.-S., F.G.C.-C. and J.M.B.- Authorship must be limited to those who have contributed substantially to the work reported. A authors have read and agreed to the published version of the manuscript.

Funding: This research received no external funding.

Institutional Review Board Statement: Not applicable.

Informed Consent Statement: Not applicable.

Data Availability Statement: The data presented in this study are available on request from th corresponding author. The data are not publicly available due to the large amount and variety data that were processed.

Acknowledgments: The authors would like to thank to the Barcelona School of Building Constructio UPC, the Department of Architecture Technology-EPSEB-UPC, Faculty of Engineering Mochis-UA to CONACYT and Research and Postgraduate Coordination-School of Engineering Mazatlán-UA

Conflicts of Interest: The authors declare no conflict of interest.

Abbreviations

The following abbreviations are used in this manuscript:

r_{ave}	Average radius
BEI	Backscattered Electron Image
BJH	Barrett, Joyner and Halenda
BET	Brunauer, Emmett and Teller
c:a	Cement:aggregate ratio
R^2	Coefficient of determination R^2
fm	Compressive strength
C&D	Construction and Demolition
CDW	Construction and Demolition Waste
r_{cri}	Critical radius
A	Cumulative pore area
V	Cumulative pore volume
EDS	Energy Dispersive Spectrometer
ITZ	Interfacial Transition Zone
IUPAC	International Union of Pure and Applied Chemistry
r_{max}	Maximum radius
r_{med}	Medium radius
MIP	Mercury Intrusion Porosimetry
r_{min}	Minimum radius
E	Modulus of Elasticity
NA	Natural Aggregate
N_2	Nitrogen
OD	Oven-Dry condition
CA	Recycled Ceramic Aggregate
CRM	Recycled Ceramic Mortars
SSD	Saturation-Surface-Dry condition
SEM	Scanning Electron Microscopy
SEI	Secondary Electron Image
UM	Usual Mortar
V_{ads}	Volume adsorption phase
V_{des}	Volume desorption phase
w/c	Water:cement ratio
EDX	X-Ray Diffraction

eferences

Bektas, F.; Wang, K.; Ceylan, H. Effects of crushed clay brick aggregate on mortar durability. *Constr. Build. Mater.* **2009**, *23*, 1909–1914. [CrossRef]

Wang, G.; Tian, B. Effect of waste ceramic polishing powder on the properties of cement mortars. In Proceedings of the 2009 International Conference on Energy and Environment Technology, ICEET 2009, Guilin, China, 16–18 October 2009; IEEE Computer Society: Washington, DC, USA, 2009; pp. 101–104.

Silva, J.; De Brito, J.; Veiga, R. Recycled Red-Clay Ceramic Construction and Demolition Waste for Mortars Production. *J. Mater. Civ. Eng.* **2010**, *22*, 236–244. [CrossRef]

Lehmann, S. Optimizing Urban Material Flows and Waste Streams in Urban Development through Principles of Zero Waste and Sustainable Consumption. *Sustainability* **2011**, *3*, 155–183. [CrossRef]

Jiménez, J.; Ayuso, J.; López, M.; Fernández, J.; de Brito, J. Use of fine recycled aggregates from ceramic waste in masonry mortar manufacturing. *Constr. Build. Mater.* **2013**, *40*, 679–690. [CrossRef]

Lu, G.; Fan, Z.; Sun, Z.; Liu, P.; Leng, Z.; Wang, D.; Oeser, M. Improving the polishing resistance of cement mortar by using recycled ceramic. *Resour. Conserv. Recycl.* **2020**, *158*, 104796. [CrossRef]

Liu, Q.; Singh, A.; Xiao, J.; Li, B.; Tam, V.W. Workability and mechanical properties of mortar containing recycled sand from aerated concrete blocks and sintered clay bricks. *Resour. Conserv. Recycl.* **2020**, *157*, 104728. [CrossRef]

Corinaldesi, V.; Giuggiolini, M.; Moriconi, G. Use of rubble from building demolition in mortars. *Waste Manag.* **2002**, *22*, 893–899. [CrossRef]

Silva, J.; De Brito, J.; Veiga, R. Fine ceramics replacing cement in mortars Partial replacement of cement with fine ceramics in rendering mortars. *Mater. Struct.* **2007**, *41*, 1333–1344. [CrossRef]

Naceri, A.; Hamina, M.C. Use of waste brick as a partial replacement of cement in mortar. *Waste Manag.* **2009**, *29*, 2378–2384. [CrossRef]

Moriconi, G.; Corinaldesi, V.; Antonucci, R. Environmentally-friendly mortars: A way to improve bond between mortar and brick. *Mater. Struct.* **2003**, *36*, 702–708. [CrossRef]

Pereira-De-Oliveira, L.A.; Castro-Gomes, J.P.; Santos, P.M. The potential pozzolanic activity of glass and red-clay ceramic waste as cement mortars components. *Constr. Build. Mater.* **2012**, *31*, 197–203. [CrossRef]

Li, L.G.; Lin, Z.H.; Chen, G.M.; Kwan, A.K.H. Reutilizing clay brick dust as paste substitution to produce environment-friendly durable mortar. *J. Clean. Prod.* **2020**, *274*, 1–11. [CrossRef]

Ge, Z.; Feng, Y.; Zhang, H.; Xiao, J.; Sun, R.; Liu, X. Use of recycled fine clay brick aggregate as internal curing agent for low water to cement ratio mortar. *Constr. Build. Mater.* **2020**, *264*, 120280. [CrossRef]

Silva, J.; de Brito, J.; Veiga, R. Incorporation of fine ceramics in mortars. *Constr. Build. Mater.* **2009**, *23*, 556–564. [CrossRef]

Corinaldesi, V.; Moriconi, G. Behaviour of cementitious mortars containing different kinds of recycled aggregate. *Constr. Build. Mater.* **2009**, *23*, 289–294. [CrossRef]

Corinaldesi, V. Environmentally-friendly bedding mortars for repair of historical buildings. *Constr. Build. Mater.* **2012**, *35*, 778–784. [CrossRef]

Mohammadhosseini, H.; Lim, N.H.A.S.; Tahir, M.M.; Alyousef, R.; Alabduljabbar, H.; Samadi, M. Enhanced perfor-mance of green mortar comprising high volume of ceramic waste in aggressive environments. *Constr. Build. Mater.* **2019**, *212*, 607–617. [CrossRef]

López Gayarre, F.; López Boadella, Í.; López-Colina Pérez, C.; Serrano López, M.; Domingo Cabo, A. Influence of the ceramic recycled agreggates in the masonry mortars properties. *Constr. Build. Mater.* **2017**, *132*, 457–461. [CrossRef]

O'Farrell, M.; Wild, S.; Sabir, B. Pore size distribution and compressive strength of waste clay brick mortar. *Cem. Concr. Compos.* **2001**, *23*, 81–91. [CrossRef]

Higashiyama, H.; Yagishita, F.; Sano, M.; Takahashi, O. Compressive strength and resistance to chloride penetration of mortars using ceramic waste as fine aggregate. *Constr. Build. Mater.* **2012**, *26*, 96–101. [CrossRef]

Filho, R.T.; Gonçalves, J.; Americano, B.; Fairbairn, E. Potential for use of crushed waste calcined-clay brick as a supplementary cementitious material in Brazil. *Cem. Concr. Res.* **2007**, *37*, 1357–1365. [CrossRef]

AENOR. *UNE-EN 197-1. Cemento Parte 1: Composición, Especificaciones y Criterios de Conformidad de Los Cementos Communes*; Asociación Española de Normalización y Certificación: Madrid, Spain, 2011; p. 40.

ASTM International. *C144-99. Standard Specification for Aggregate for Masonry Mortar*; American Society for Testing and Materials: West Conshohocken, PA, USA, 1999; p. 2.

ASTM International. *C128-04a. Standard Test Method for Density, Relative Density (Specific Gravity), and Absorption*; American Society for Testing and Materials: West Conshohocken, PA, USA, 2004; p. 6.

ASTM International. *C29/C29M–97. Standard Test Method for Bulk Density ("Unit Weight") and Voids in Aggregate*; American Society for Testing and Materials: West Conshohocken, PA, USA, 1997; p. 4.

ASTM International. *C117-95. Standard Test Method for Materials Finer than 75-µm (No. 200) Sieve in Mineral*; American Society for Testing and Materials: West Con-shohocken, PA, USA, 1995; p. 4.

ASTM International. *C136-06. Standard Test Method for Sieve Analysis of Fine and Coarse Aggregates*; American Society for Testing and Materials: West Conshohocken, PA, USA, 2006; p. 5.

29. Cabrera-Covarrubias, F.G.; Arredondo-Rea, S.P.; Gómez-Soberón, J.M.; Almaral-Sánchez, J.L.; Corral-Higuera, R. Mechanic properties of mortars containing recycled ceramic as a fine aggregate replacement. *Rev. Constr.* **2015**, *14*, 22–29. [CrossRef]
30. ASTM International. *C109/C109M-05. Standard Test Method for Compressive Strength of Hydraulic Cement Mortars*; American Socie for Testing and Materials: West Conshohocken, PA, USA, 2005; p. 9.
31. ASTM International. *C230/C230M-03. Standard Specification for Flow Table for Use in Tests of Hydraulic Cement*; American Society f Testing and Materials: West Conshohocken, PA, USA, 2003; p. 7.
32. AENOR. *UNE-EN 1015-10. Parte 10: Determinación de la Densidad Aparente en Seco Del Mortero Endurecido*; Asociación Española Normalización y Certificación: Madrid, Spain, 2000; p. 8.
33. AENOR. *UNE-EN 1936: Determinación de la Densidad Real y Aparente y de la Porosidad Abierta y Total*; Asociación Española Normalización y Certificación: Madrid, Spain, 2007; p. 13.
34. Cabrera-Covarrubias, F.G.; Gómez-Soberón, J.M.; Almaral-Sánchez, J.L.; Arredondo-Rea, S.P.; Gómez-Soberón, M.C.; Corr Higuera, R. An Experimental Study of Mortars with Recycled Ceramic Aggregates: Deduction and Prediction of the Stress-Stra *Materials* **2016**, *9*, 1029. [CrossRef]
35. Mendivil-Escalante, J.M.; Gómez-Soberón, J.M.; Almaral-Sánchez, J.L.; Cabrera-Covarrubias, F.G. Metamorphosis in the Porosi of Recycled Concretes Through the Use of a Recycled Polyethylene Terephthalate (PET) Additive. Correlations between t Porous Network and Concrete Properties. *Materials* **2017**, *10*, 176. [CrossRef] [PubMed]
36. Sing, K.S. Reporting physisorption data for gas/solid systems with special reference to the determination of surface area ar porosity (Recommendations 1984). *Pure Appl. Chem.* **1985**, *57*, 603–619. [CrossRef]
37. De Belie, N.; Krátký, J.; Van Vlierberghe, S. Influence of pozzolans and slag on the microstructure of partially carbonated ceme paste by means of water vapour and nitrogen sorption experiments and BET calculations. *Cem. Concr. Res.* **2010**, *40*, 1723–17 [CrossRef]
38. Quantachrome Instruments. The Gas Sorption Process. 2018. Available online: http://www.quantachrome.com/pdf_brochure evo_07160_brochure.pdf (accessed on 10 December 2020).
39. Palos, A.; D'Souza, N.A.; Snively, C.; Reidy, R.F. Modification of cement mortar with recycled ABS. *Cem. Concr. Res.* **2001**, 1003–1007. [CrossRef]
40. Chinchillas-Chinchillas, M.J.; Rosas-Casarez, C.A.; Arredondo-Rea, S.P.; Gómez-Soberón, J.M.; Corral-Higuera, R. SEM Ima Analysis in Permeable Recycled Concretes with Silica Fume. A Quantitative Comparison of Porosity and the ITZ. *Materials* **20** *12*, 2201. [CrossRef] [PubMed]
41. Gómez-Soberón, J.M.V. Relationship between Gas Adsorption and the Shrinkage and Creep of Recycled Aggregate Concre *Cem. Concr. Aggreg.* **2003**, *25*, 1–7.
42. Santos, A.R.; Veiga, M.D.R.; Silva, A.S.; de Brito, J. Microstructure as a critical factor of cement mortars' behaviour: The effect aggregates' properties. *Cem. Concr. Compos.* **2020**, *111*, 103628. [CrossRef]
43. Santamarina, J.C.; Klein, K.A.; Wang, Y.H.; Prencke, E. Specific surface: Determination and relevance. *Can. Geotech. J.* **2002**, 233–241. [CrossRef]
44. Gómez-Soberón, J.M.V. Porosity of recycled concrete with substitution of recycled concrete aggregate: An experimental stu *Cem. Concr. Res.* **2002**, *32*, 1301–1311. [CrossRef]
45. Da Silva, P.; de Brito, J. Experimental study of the porosity and microstructure of self-compacting concrete (SCC) with binary ar ternary mixes of fly ash and limestone filler. *Constr. Build. Mater.* **2015**, *86*, 101–112. [CrossRef]
46. Atahan, H.N.; Oktar, O.N.; Taşdemir, M.A. Effects of water–cement ratio and curing time on the critical pore width of harden cement paste. *Constr. Build. Mater.* **2009**, *23*, 1196–1200. [CrossRef]
47. Liu, C.; Liu, H.; Xiao, J.; Bai, G. Effect of old mortar pore structure on relative humidity response of recycled aggregate concre *Constr. Build. Mater.* **2020**, *247*, 118600. [CrossRef]

materials

MDPI

ʳticle

nfluence of Cement Replacement with Fly Ash and Ground ʿand with Different Fineness on Alkali-Silica Reaction ɔf Mortar

uwat Ramjan [1], Weerachart Tangchirapat [2,*], Chai Jaturapitakkul [2], Cheah Chee Ban [3], Peerapong Jitsangiam [4] nd Teewara Suwan [4]

1 Building Control Bureau, Department of Public Works and Town & Country Planning, Bangkok 10140, Thailand; tience14@hotmail.com

2 Construction Innovations and Future Infrastructure Research Center (CIFIR), Department of Civil Engineering, Faculty of Engineering, King Mongkut's University of Technology Thonburi, Bangkok 10140, Thailand; Chai.jat@kmutt.ac.th

3 School of Housing, Building and Planning, Universiti Sains Malaysia, Penang 11800, Malaysia; cheahcheeban@usm.my

4 Center of Excellence in Natural Disaster Management, Department of Civil Engineering, Faculty of Engineering, Chiang Mai University, Chiang Mai 50200, Thailand; peerapong@eng.cmu.ac.th (P.J.); teewara.s@cmu.ac.th (T.S.)

* Correspondence: weerachart.tan@kmutt.ac.th

itation: Ramjan, S.; Tangchirapat, .; Jaturapitakkul, C.; Chee Ban, C.; sangiam, P.; Suwan, T. Influence of ement Replacement with Fly Ash d Ground Sand with Different neness on Alkali-Silica Reaction of ortar. *Materials* **2021**, *14*, 1528. tps://doi.org/10.3390/ma14061528

ademic Editor: Miguel Bravo

eceived: 4 March 2021
ccepted: 15 March 2021
ıblished: 20 March 2021

Abstract: The alkali-silica reaction (ASR) is an important consideration in ensuring the long-term durability of concrete materials, especially for those containing reactive aggregates. Although fly ash (FA) has proven to be useful in preventing ASR expansion, the filler effect and the effect of FA fineness on ASR expansion are not well defined in the present literature. Hence, this study aimed to examine the effects of the filler and fineness of FA on ASR mortar expansion. FAs with two different finenesses were used to substitute ordinary Portland cement (OPC) at 20% by weight of binder. River sand (RS) with the same fineness as the FA was also used to replace OPC at the same rate as FA. The replacement of OPC with RS (an inert material) was carried out to observe the filler effect of FA on ASR. The results showed that FA and RS provided lower ASR expansions compared with the control mortar. Fine and coarse fly ashes in this study had almost the same effectiveness in mitigating the ASR expansion of the mortars. For the filler effect, smaller particles of RS had more influence on the ASR reduction than RS with coarser particles. A significant mitigation of the ASR expansion was obtained by decreasing the OPC content in the mortar mixture through its partial substitution with FA and RS.

Keywords: alkali-silica reaction; blended cement; filler; cement-based materials; fly ash; mortar

1. Introduction

Fly ash (FA) is one of main by-products from coal combustion in thermal power plants. Currently, the Mae Moh power plant in Thailand produces about 2.1 million tons of fly ash, increasing annually. Fly ash (FA) is also a well-known mineral admixture and has been widely accepted for use as a pozzolanic material to partially replace ordinary Portland cement (OPC) in concrete. The use of fly ash as a pozzolanic material is constantly increasing because it improves the properties of mortar and concrete, namely the workability, durability, and strength in the long term. Generally, fly ash from pulverized combustion has a spherical particle shape and high fineness. However, some fly ashes also contain irregular or angular particles. The size and shape of fly ash particles vary depending on the source and combustion conditions. The chemical composition of fly ash also depends on the characteristics of the coal and burning condition. FA has been studied for a long time and has been continuously developed to improve many of the properties of concrete. In 2010, Sharma et al. [1] studied the influence of FA characteristics from 14 thermal power plants and found that calcium content and particle size are the main paramete affecting the compressive strength of concrete. Compressive strength is an importa factor to consider when selecting concrete mix proportions, and the use of FA can provid compressive strengths as high as OPC concrete [2–5]. Moreover, replacing OPC with F in the concrete mixture can produce high strength concretes, with compressive strengt of 77.3 to 82.5 MPa at an age of 28 days [6]. Once FA was accepted as a way to improv the compressive strength of concrete, many researchers studied FA for improving oth properties of concrete, such as workability and durability. The use of FA can significant improve the elementary properties of paste and mortar, such as normal consistency, settin time, and water requirements [7].

In terms of durability, the replacement of 20 to 50% of OPC with FA (by weight binder) has been shown to produce concrete with lower water permeability than concre without FA. In addition, similar FA contents have been used to produce concrete with higher resistance to sodium and magnesium sulfates than OPC concrete [8]. FA is a popul pozzolan for preventing chloride penetration into concrete in marine environments [The use of FA also contributes to a lower chloride penetration depth of recycled aggrega concrete than that of OPC concrete, by approximately 2–2.5 times for recycled aggrega concrete without FA [8]. Additionally, FA has been used to improve both the fresh an mechanical properties of concrete containing recycled aggregate [2]. FA has also been use as a major cementing material to produce calcium carbide residue-FA concrete with ze OPC [9] and alkali-activated materials [10].

Turk et al. [11] confirmed that FA is a good pozzolan for use in controlling alkali-sili reactions (ASRs). It had been established that the use of low-calcium Class F FA at 20 blended with limestone powder can mitigate ASR expansion. Saha et al. [12] stated that t use of Class F FA can reduce ASR expansion more significantly than Class C FA. Howev the findings of Moser et al. [13] showed that the use of FA does not always help to redu the ASR expansion of mortars with very highly reactive aggregates. In their study, it w reported that the control mortar had an ASR expansion of 0.70% at 14 days. Meanwhi replacing 20% of the OPC with Class C FA failed to reduce the ASR to below 0.20% 14 days.

Although FA has been proven by many researchers to be a useful pozzolanic materi for preventing ASR expansion, the primary reduction of the expansion may be due other factors, such as the filler effect or the reduction of the OPC content in the mixtu Additionally, the filler effect and the effects of FA fineness on ASR expansion have n been well defined in the present literature. Therefore, this investigation was aimed examining the influence of the replacement of OPC with FA with different levels of finene on the ASR expansion of mortar. Two fineness levels of FA were also considered in ord to observe the effect of fineness on ASR expansion. Ground river sand (RS) with the sam fineness as FA was used to replace OPC at the same rates of FA to study the filler effects FA and RS on the ASR expansion of mortar.

2. Research Significance

Alkali-silica reaction (ASR) is a dangerous phenomenon in concrete structures, becau ASR builds internal pressure causing the concrete to crack. After that, the concrete structu losses its compressive strength and durability, making it easy for moisture and oth chemical solutions to migrate into the internal concrete, and leading to the corrosion steel in reinforced concrete structures. Although, the replacement of OPC by fly ash ca reduce the expansion due to ASR, other factors are still not clear, such as the filler effect an the effects of the fineness of the materials. Therefore, the knowledge obtained from th study will help the civil engineer to understand the factors affecting the ASR expansion mortar, and lead to applying this knowledge in concrete structures in order to obtain t safety and sustainability of structures.

3. Experiment

3.1. Materials

A low alkali cement obtained from a cement manufacturer in Thailand, which was classified as type I according to ASTM C150 [14], was used to cast all mortars. Fly ash (FA) was collected from the Mae Moh electricity generating plant, Thailand, where lignite coal is burned using a pulverized burning system. The FA was obtained directly from the plant and was used as received. The fineness of the FA was measured by weighing particles of FA retained on a No. 325 sieve [15]. The original FA had 32.8% of its particles retained on the No. 325 sieve, which does not exceed the limit defined by ASTM C618 [16] of 34%. This FA was called "33FA". The FA was then ground by a ball mill to have 5.0% of its particles by weight retained on a No. 325 sieve (5FA). River sand was used as an inert material in order to study the filler effect on the ASR expansion of the mortar. River sand used in this study was collected from the central part of Thailand. To examine the effect of the fineness of FA on ASR, the RS was also ground to have the same fineness as the FA.

After completing the grinding processes, the crystallinity of the 5FA and 5RS were analyzed by X-ray diffraction (XRD), and the results are presented in Figures 1 and 2, respectively. The 5FA had crystals of quartz (SiO_2), calcium carbonate ($CaCO_3$), and mullite ($3Al_2O_3{\cdot}2SiO_2$), while the 5RS had mainly crystals of quartz. However, the quartz in RS does not react with $Ca(OH)_2$ and was treated as an inert material, since the area under the XRD graph was very low. The XRD results of 33FA and 33RS were not analyzed because the results of 33FA and 33RS were assumed to be similar to those of 5FA and 5RS, respectively. Chindaprasirt et al. [17] and Kroehong et al. [18] proved that fly ash and palm oil ash (with the same source) with different finenesses had only slight differences in their XRD results.

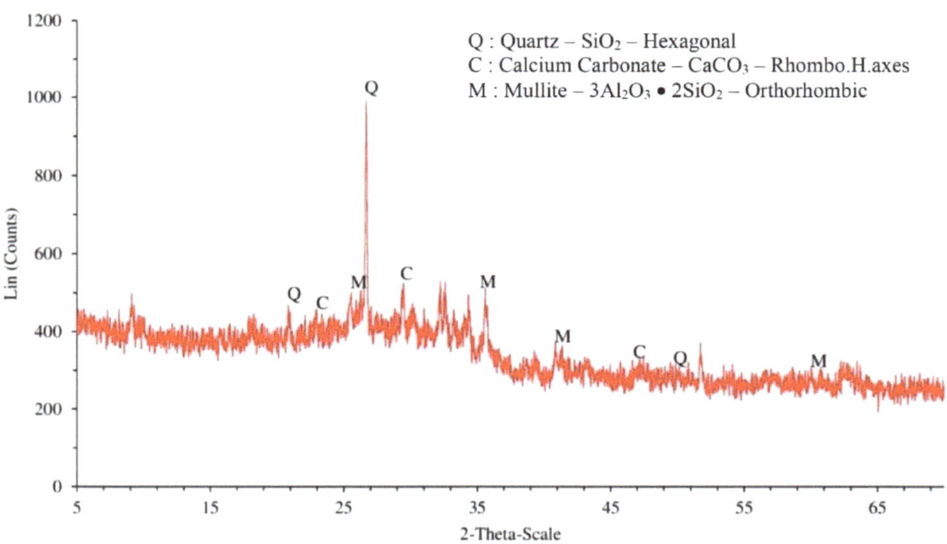

Figure 1. The qualitative results (XRD) of 5FA.

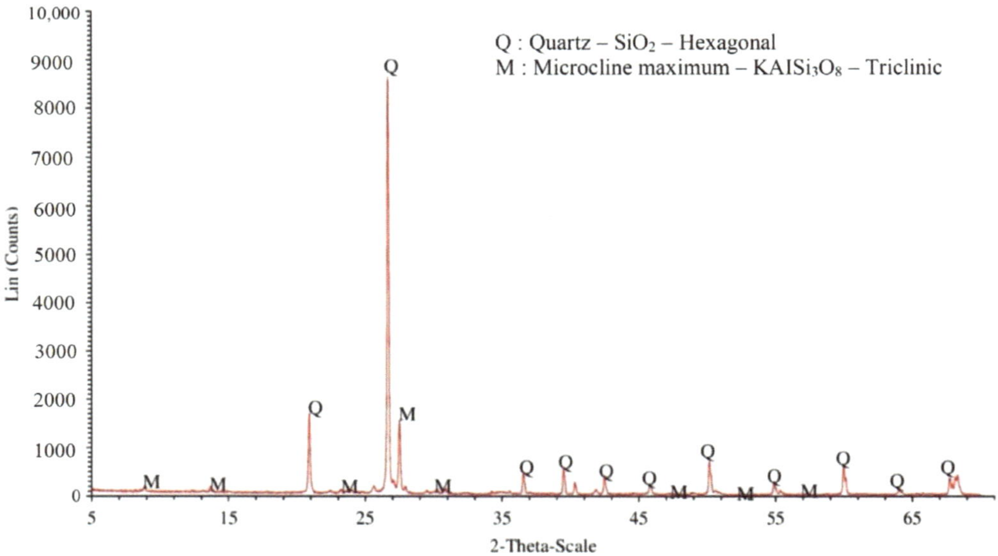

Figure 2. The qualitative results (XRD) of 5RS.

Many prior studies have established that the fineness of pozzolanic materials h little influence on their chemical compositions [19]. The chemical properties of 5FA m the requirements of Class C FA according to ASTM C618 [16], which requires that t summation of SiO_2 + Al_2O_3 + Fe_2O_3 is more than 50%, SO_3 is less than 5%, and loss ignition is less than 6%, while the alkali equivalent (Na_2O_{eq}) is 2.8% by weight. The maj oxide of 5RS is SiO_2, which makes up 92.5% of its content, and the rest are other oxid The chemical compositions of the materials are given in Table 1, and Table 2 shows t physical properties and strength activity index of materials.

Table 1. Chemical compositions of ordinary Portland cement (OPC), 5FA, and 5RS.

Sample	Chemical Composition (%)									
	SiO_2	Al_2O_3	Fe_2O_3	CaO	SO_3	MgO	Na_2O	K_2O	Na_2O_{eq}	LOI
OPC	20.8	4.7	3.3	65.4	2.7	1.3	0.3	0.4	0.6	1.0
5FA	33.2	17.1	12.8	25.2	3.2	1.9	1.2	2.5	2.8	2.9
5RS	92.0	5.0	0.6	0.2	0.2	0.0	0.4	0.8	0.9	0.1

Table 2. Physical properties of materials.

Sample	Specific Gravity	Retained on a No. 325 Sieve (%)	Strength Activity Index (%)	
			7 days	28 days
OPC	3.14	20.0	—	—
5FA	2.41	5.0	86	105
33FA	2.33	32.8	81	93
5RS	2.72	4.8	67	68
33RS	2.63	32.5	62	62

For mortars containing an ASR reactive aggregate, the coarse reactive aggregate w ground and sieved until the gradation met the requirements of ASTM C1567 [20] for AS investigation. The fine reactive aggregate had a specific gravity (saturated surface d SSD) and fineness modulus of 2.63 and 2.90, respectively. The gradation of the fine reacti aggregates in this study was compared with the required gradation of fine aggregates specified in ASTM C1260 [21], as shown in Figure 3.

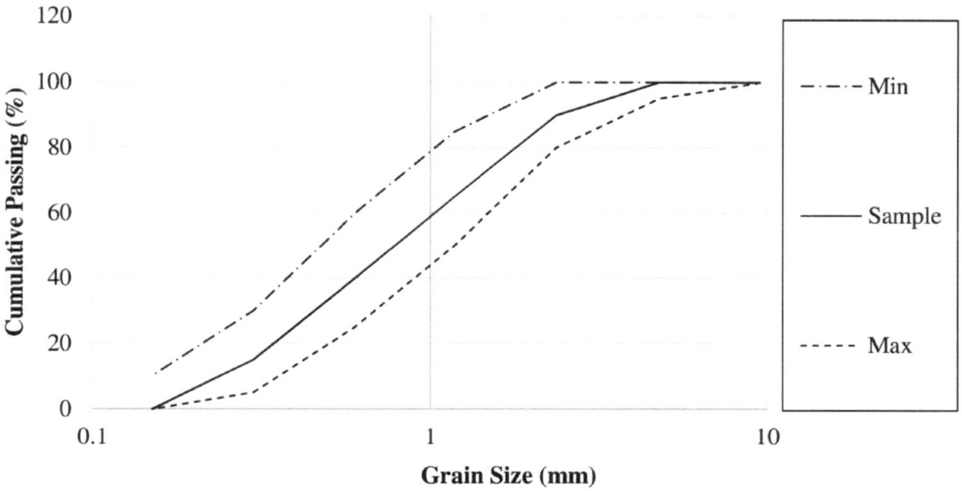

Figure 3. Particle size distributions of the reactive fine aggregate used in this study.

3.2. ASR Testing Procedure

The ASR of the mortar was investigated by an accelerated method, and according to ASTM C1567 [20]. For the first step, a binder to fine aggregate ratio of 2.25 was kept, while a water to binder (W/B) ratio of 0.47 was used to cast mortar bars. It is noted that the W/B ratio of the mortar bars was calculated with the effective water content, which already compensates for the water absorption by the fine reactive aggregate. For the binder in the mortar bar mixtures, FA and RS were used to partially replace OPC at 20% by weight of binder. The mix proportions of the mortar bars, including their flows, are presented in Table 3.

Table 3. Mix proportions of mortar bars in the alkali–silica reaction (ASR) investigation.

Mortar Bars	Mix Proportions (By Weight)					
	Cement	**FA**	**RS**	**Fine Aggregate**	**W/B**	**Flow (%)**
CT	1.00	—	—	2.25	0.47	100
5FA	0.80	0.20	—	2.25	0.47	101
33FA	0.80	0.20	—	2.25	0.47	99
5RS	0.80	—	0.20	2.25	0.47	92
33RS	0.80	—	0.20	2.25	0.47	88

Next, two stainless steel gauge studs were cast-in at both ends of the mortar bar mold for use as an indicator in the measuring process. A mortar mixture was filled and compacted in three molds with dimensions of 25 × 25 × 285 mm^3. Mortar bars were de-molded after 24 h of casting and were then cured in tap water at a temperature of 80 °C. The initial length of the mortar was measured 24 h after completion of the previous process. Next, the mortar bars were immersed in a 1 N sodium hydroxide (NaOH) solution at the temperature of 80 °C. The length of mortar bars was read for 28 days and the reading at each time was targeted to be the nearest 0.001% of the nominal gauge length. The reported ASR expansion was presented as the percent change in length, which was obtained from the average of three mortar specimens. The acceptable range of test results between samples at the same testing age and mix proportion was 8.3% of the average as specified by ASTM C1567 [20].

The effectiveness of using fly ash or ground river sand in controlling the ASR expansion of mortars was determined according to Equation (1), given by ASTM C441 [2 as follows:

$$R_e = \frac{E_{ct} - E_t}{E_{ct}} \times 100 \quad (1)$$

where R_e is the reduction of the expansion of mortar (%), E_t is the average length of mortar bars containing fly ash or ground river sand (%), and E_{ct} is the average length of the contr mortar bars (%).

4. Results and Discussion

4.1. ASR Expansion of Mortars Containing FA

As shown in Figure 4, the control mortar (CT), in which the crushed reactive coars aggregate was used as a fine reactive aggregate, had an ASR expansion of 0.234 ± 0.013% 14 days, exceeding 0.20% and suggesting that the fine aggregate used should be classified a highly reactive aggregate according to ASTM C1567 [20]. In addition, this highly reactiv aggregate had a high ASR mortar expansion similar to trachyte, trachyandesite, rhyolit and glass aggregates [23,24]. Various mechanisms for the occurrence of ASR in mortar concrete without pozzolan have been suggested by studies. Several researchers [25,2 have found that increasing the amount of $Ca(OH)_2$ promotes a high yield of ASR produ Additionally, Hou et al. [27] demonstrated a direct relationship between the additio of OPC and the development of ASR gels. The formation of ASR expansion produc was found in the CT mortar because the NaOH solution provided a high alkalinity lev while the primary calcium source was released by OPC during the hydration process [2 However, the reduction of the OPC content in the mixture through the partial replaceme of OPC by added minerals or pozzolanic materials could reduce the amount of Ca(OH thereby reducing the degree of ASR expansion.

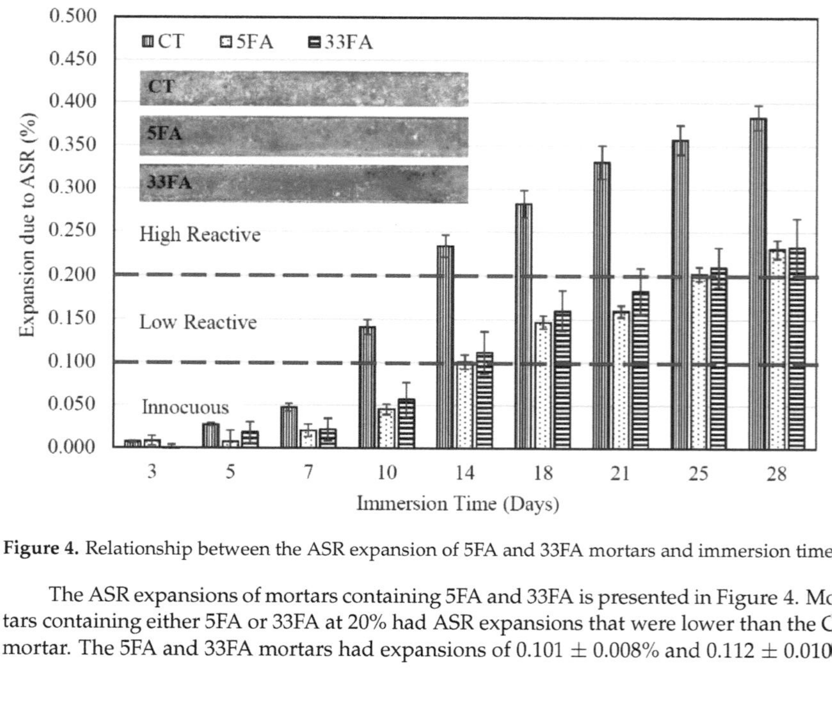

Figure 4. Relationship between the ASR expansion of 5FA and 33FA mortars and immersion time

The ASR expansions of mortars containing 5FA and 33FA is presented in Figure 4. Mo tars containing either 5FA or 33FA at 20% had ASR expansions that were lower than the C mortar. The 5FA and 33FA mortars had expansions of 0.101 ± 0.008% and 0.112 ± 0.010

at 14 days, which increased to 0.231 ± 0.010% and 0.233 ± 0.020% at 28 days, respectively. The partial substitution of OPC with FA or other pozzolans in the mortar mixture was shown to reduce the ASR expansion, which is consistent with the findings of a number of prior studies [26,29,30]. In this study, the degree of fineness of the FA used as a replacement for OPC had a marginal influence on the degree of ASR expansion of the mortars. This is consistent with the reported findings of Aydin et al. [31], who found that changing between the original and ground FA from the same source had no significant influence on the degree of ASR expansion. For example, mortars incorporating 5FA and 33FA had expansions at 14 days of 0.101 ± 0.008 and 0.112 ± 0.010%, respectively. While the results of Aydin et al. [31] showed that ASR expansions at 14 days of mortars containing original and ground fly ash were approximately 0.095 and 0.085%, respectively. It was clearly seen that a reduction in the fineness of fly ash could decrease the expansion of mortar due to an ASR of not more than 5%.

A mechanism for the reduction of the ASR expansion due the addition of pozzolan was described by Topçu et al. [30], who reasoned that the pozzolanic reaction between FA and $Ca(OH)_2$ (from the cement hydration reaction) leads to a decreased pH in the pore solution. Lindgård et al. [32] stated that supplementary cementitious materials (SCMs) made from low CaO and high SiO_2 pozzolans are the most critical factors in reducing the alkalinity of pore solutions. The corresponding reduction in the concentration of hydroxyl ions in the pore solutions could prevent ASR expansion or reduce its degree. Conversely, the FA used in the current study had a high amount of CaO (25.2 wt.%) and a low amount of SiO_2 (33.2 wt.%), but it also reduced the expansions of the mortars. Topçu et al. [30] explained that the pozzolanic reaction of FA could reduce the reactivity between the alkali and silica due to the formation of a dense paste structure. This also results in the subsequent reduction in the amount of water or solution which is able to penetrate the paste matrix. A higher proportion of OPC replacement by FA or other pozzolans has been shown to establish a lower $Ca(OH)_2$ content in the hardened paste. Moreover, it also decreases the availability of alkali reactants, due to the partial consumption of $Ca(OH)_2$ by pozzolanic reaction [13,33–35]. In addition, Hong and Glasser [36] suggested that the Al_2O_3 content might also have an important influence on the alkali capacity of an SCM binder.

The reduction of ASR expansion in mortar or concrete containing FA is attributed to the reduction of alkali in the binder. However, the main factor that limits the ASR expansion is the lower amount of OPC in the mixture, achieved by using FA as a partial substitute for OPC. To verify this claim, RS (inert material) was used to replace OPC instead of FA, to demonstrate the effect of the cement content on ASR expansion.

4.2. ASR Expansion of Mortars Containing RS

Figure 5 shows the relationship between the ASR expansions of RS mortars and immersion time. It is interesting to note that the replacement of OPC by 5RS and 33RS in mortars resulted in lower ASR expansions as compared to the CT mortar. The 5RS mortar had an ASR expansion below 0.10% at 14 days, indicating that a 20% 5RS content could reduce the degree of ASR expansion of mortar from a highly reactive level to an innocuous level (less than 0.10%). However, it should be noted that RS is ground river sand, which is an inert material due its high crystallinity, as shown in Figure 2. Thus, the mitigation of ASR expansion was a result of the decrease of the OPC content in the binder matrix.

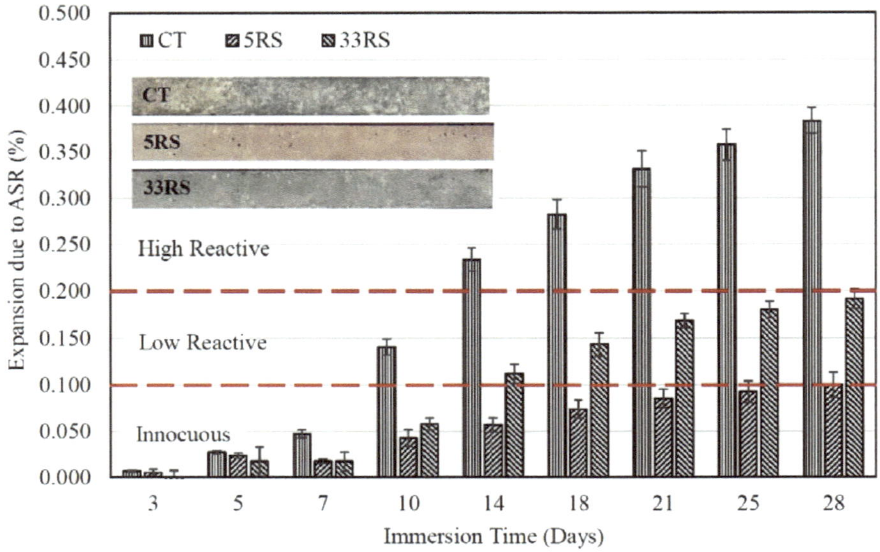

Figure 5. Relationship between the ASR expansion of 5RS and 33RS mortars and immersion time.

The replacement of OPC with 5RS (high fineness) reduced the ASR expansion of tl mortar more than replacement with 33RS (low fineness), because the 5RS particles cou better fill voids and resist the penetration of the 1 N NaOH solution into the mortar. F instance, the 5RS mortar had an ASR expansion of 0.061 ± 0.003% at 14 days, while th of the 33RS mortar was 0.112 ± 0.009%. The fineness of RS is one significant factor th influences the reduction of ASR in mortar. Finer RS also provided a higher compressi strength to the mortar than RS with lower fineness, largely due to the filler effect [37].

Although the replacement of OPC by RS can reduce ASR expansion, it can also resu in a lower compressive strength of mortar [37–39]. Thus, the utilization of RS in controlli the ASR expansion must also take into account the required compressive strength of mort

4.3. ASR Expansion of Mortars Containing FA and RS

Figure 6 shows the relationship between the ASR expansion of the 5FA and 5I mortars and immersion time. The expansion of the 5FA mortar was higher than for tl 5RS mortar at 28 days. However, the ASR expansion of the 5FA mortar was similar to th of the 5RS mortar at periods before 10 days. This is largely due to the slow pozzolar reaction of FA at early ages. After 10 days, the expansion of the 5FA mortar continuous increased until it reached 0.231 ± 0.010% at 28 days, while the 5RS mortar had an AS expansion of 0.093 ± 0.007% at 28 days. The increase in the ASR expansion of the 5F mortar was more than for the 5RS mortar because the CaO in 5FA increases the alkalini of the mixture. A similar behavior could be observed in the 33FA and 33RS mortars (s Figure 7), in which the non-reactive period of the 33FA mortar was marginally longer tha that of the 5FA mortar (before 14 days). On the other hand, the difference in the AS expansion between the 33FA and 33RS mortars was lower than the difference betwe the 5FA and 5RS mortars because the coarser CaO in 33FA reacted more slowly and le completely than the finer CaO in 5FA.

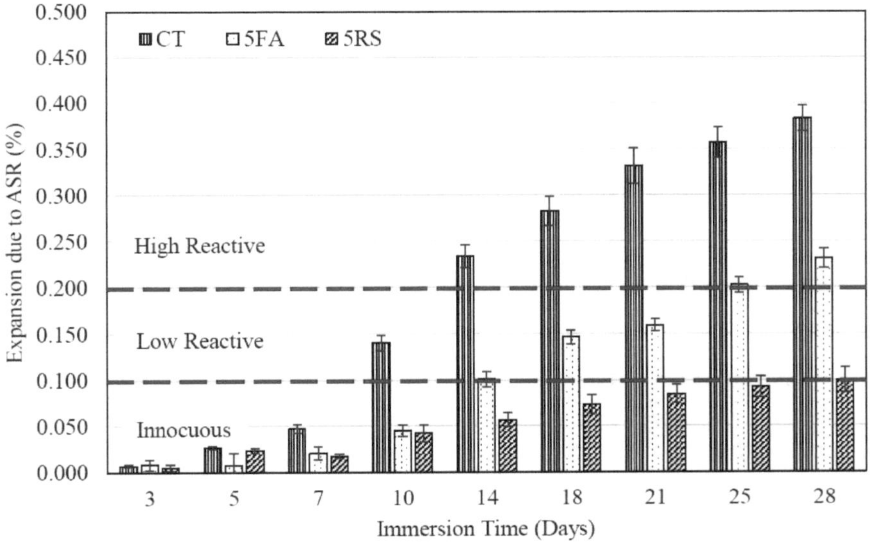

Figure 6. Relationship between the ASR expansion of 5FA and 5RS mortars and immersion time.

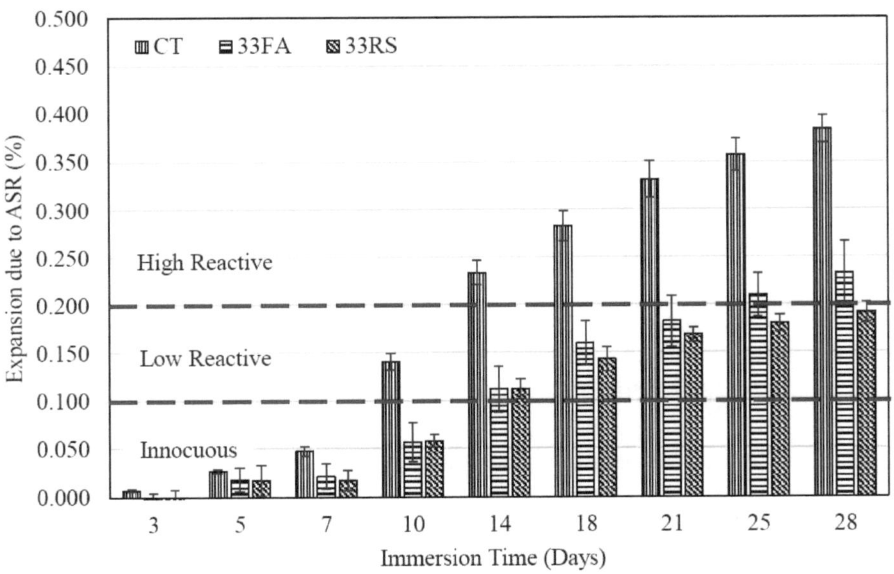

Figure 7. Relationship between the ASR expansion of 33FA and 33RS mortars and immersion time.

4.4. Filler Effect of Different Finenesses of RS on the ASR Expansion of Mortar

Figure 8 shows the relationship between the ASR expansions of RS mortar and immersion time. The filler effect of RS on the ASR expansion of mortar was determined by Equation (2) when the immersion times of all mortars were the same.

$$F_{Filler} = E_{RS} - E_{CT} \quad (2)$$

where F_{Filler} is the filler effect of RS on the ASR expansion of the mortar (%), E_{RS} is th ASR expansion of the RS-containing mortars (%), and E_{CT} is the ASR expansion of the C mortar (%).

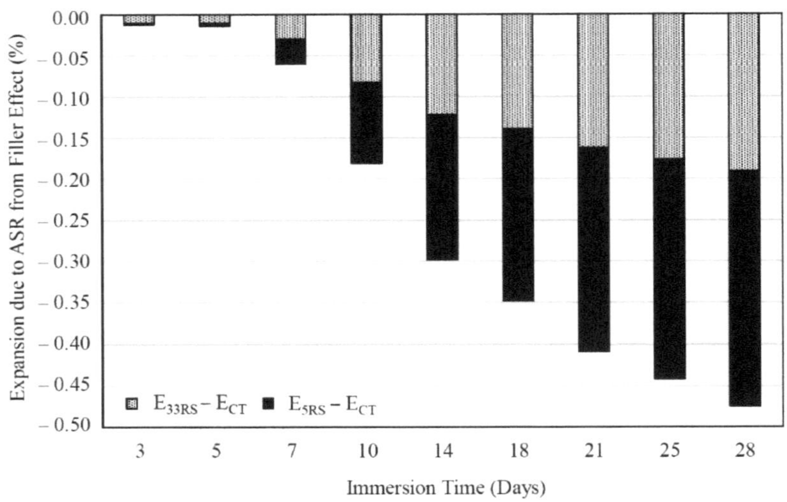

Figure 8. Filler effect of 5RS and 33RS particles on the ASR expansion of mortars.

The filler effect of finer particles improves the mechanical and durability properti of mortar or concrete. Since 33RS and 5RS were ground river sands and consisted main of quartz, which does not react with OPC hydration products, the different reductions expansion between 33RS and 5RS mortars were considered to be predominantly due to th filler effect of the RS material.

The filler effect due to RS was an important factor in lowering the ASR expansio and the filler effect had a higher degree of effectiveness at a later age. For instance, 5R exhibited changes in the ASR mortar expansions of −0.177 and −0.283% at ages of 14 an 28 days, respectively, due to the filler effect. It should be noted that the negative values the expansion indicate that there was a reduction of the ASR mortar expansion compare to the CT mortar. The reduction of the OPC content from 100 to 80% by weight meant the was a lower amount of $Ca(OH)_2$ available to react with SiO_2 in the reactive aggregate, an there was thus a lower content of ASR products. Moreover, $Ca(OH)_2$ continually reacte with SiO_2 in the reactive aggregate, and the filler effect could be observed clearly aft 5 days of immersion, with more apparent results after longer immersion durations.

When the OPC comprised only 80%, the ASR expansion of the mortars incorporatir 5RS and 33RS (with equal OPC contents in the mortar) was 0.177 and 0.122%, respective which were lower than the ASR expansion of the CT mortar. Meanwhile, increasing th fineness of RS, from 33RS to 5RS, produced an absolute increase in the ASR expansion about 0.055% at 14 days, and 0.092% at 28 days. It is noted that the filler effect of RS (33R and 5RS) was less effective than the 20% reduction of OPC in the binder, since the lowerir of OPC in the mortar by 20% reduced the ASR expansion to about 0.122% (when usir 33RS as a replacement for OPC) or 0.177% (when using 5RS as a replacement for OPC while the reduction of the ASR expansion by the filler effect of 33RS and 5RS was on 0.055% at 14 days.

4.5. *Effectiveness of FA in Controlling ASR Expansion of Mortar*

The effectiveness of FA in controlling expansion due to ASR depended on the amou of FA used to substitute OPC. It was observed in previous research [31] that the expansic due to ASR was reduced by 27 to 87% when ground FA was used to replace OPC at 10 to 40%. The 5FA and 33FA mortars in this study reduced the ASR expansion by 57 and 52% at 14 days, respectively, suggesting that the different particle sizes of FA had a marginal effect on its ability to reduce the ASR expansion. Similarly, the replacement of 20% of the OPC with Class C FA in previous research showed reductions in ASR expansion in the range of 17 to 65% [40,41]. Meanwhile, the replacement of 30 to 40% of the OPC with Class C FA could reduce the ASR expansion in the range of 27 to 96% [29,40,41].

The use of 5RS and 33RS to replace 20% of the OPC reduced the ASR expansion of mortars at 14 days by 76 and 52%, respectively. The use of 33FA and 33RS provided similar reductions in the ASR expansion of the mortar, while the use of 5RS was more effective than the use of 5FA for the reduction of the ASR expansion. However, the replacement of OPC with RS resulted in a lower compressive strength of the mortars, compared to those with FA. Mortars containing 20% of 5RS and 33RS had a strength activity index of 89% and 78%, respectively, at the age of 28 days. In contrast, the mortars containing 20% of 5FA and 33FA had a strength activity index of 108 and 93%, respectively, at the age of 28 days [37].

5. Conclusions

Based on the results from this study of the influence of cement replacement with fly ash and ground sand with different fineness on the alkali-silica reaction of mortar, the conclusions can be summarized as follows:

1. The use of FA with 5 and 33 wt.% of its particles retained on a No. 325 sieve to replace 20% of OPC reduced the ASR expansion of the mortar. The extent of the reduction depended on the replacement of OPC by FA rather than on the fineness of FA.
2. The use of RS to partially substitute OPC reduced ASR expansion of the mortar by decreasing the $Ca(OH)_2$ content available from OPC.
3. The filler effect of RS is a factor that contributed to the reduced ASR expansion of the mortar. Moreover, the filler effect of RS with a higher fineness was more effective than that of RS with a lower fineness.
4. The use of Class C FA with different levels of fineness (retained on No. 325 sieve at 5 and 33% by weight) to replace OPC at 20% reduced the ASR expansion by more than 50% at 14 days. In addition, it is suggested that reducing the CaO content (by reducing the OPC content) in mortar could be more effective in mitigating the ASR expansion than the filler effect of the cement replacement material.

Author Contributions: Conceptualization, S.R.; Methodology: W.T.; Data curation, S.R., W.T. and C.J.; Formal analysis: S.R., W.T. and C.J.; Visualization, C.C.B., P.J. and T.S.; Supervision, C.J.; Writing—original draft, S.R. and W.T.; Writing—review and editing, C.C.B., P.J. and T.S. All authors have read and agreed to the published version of the manuscript.

Funding: This research was funded by Thailand Research Fund (TRF) under TRF Research Career Development, Grant No. RSA6280032 and the Ministry of Science, Technology and Innovation (MOSTI) for funding the study under the International Collaboration Fund, Reference No. IF0420I1224. The APC was funded by Thailand Research Fund (TRF).

Institutional Review Board Statement: Not applicable.

Informed Consent Statement: Not applicable.

Data Availability Statement: The data presented in this study are available on request from the corresponding author.

Acknowledgments: The authors are grateful for financial supports from Thailand Research Fund (TRF) under TRF Research Career Development Grant No. RSA6280032. We would also like to thank the Ministry of Science, Technology and Innovation (MOSTI) for funding the study under the International Collaboration Fund (Reference No. IF0420I1224).

Conflicts of Interest: The authors declare no conflict of interest.

References

1. Sharma, V.; Chandra, S.; Choudhary, R. Characterization of fly ash bituminous concrete mixes. *J. Mater. Civ. Eng.* **2010**, 1666–1673. [CrossRef]
2. Tangchirapat, W.; Rattanashotinunt, C.; Buranasing, R.; Jaturapitakkul, C. Influence of fly ash on slump loss and strength concrete fully incorporating recycled concrete aggregates. *J. Mater. Civ. Eng.* **2013**, *25*, 243–251. [CrossRef]
3. Hsu, S.; Chi, M.; Huang, R. Effect of fineness and replacement ratio of ground fly ash on properties of blended cement mort *Construct. Build. Mater.* **2018**, *176*, 250–258. [CrossRef]
4. Choi, S.J.; Lee, S.S.; Monteiro-Paulo, J.M. Effect of fly ash fineness on temperature rise, setting, and strength development mortar. *J. Mater. Civ. Eng.* **2012**, *24*, 499–505. [CrossRef]
5. Chalee, W.; Jaturapitakkulm, C.; Chindaprasirt, P. Predicting the chloride penetration of fly ash concrete in seawater. *Mar. Stru* **2009**, *22*, 341–353. [CrossRef]
6. Jaturapitakkul, C.; Kiattikomol, K.; Sata, V.; Leekeeratikul, T. Use of ground coarse fly ash as a replacement of condensed sili fume in producing high-strength concrete. *Cem. Concr. Res.* **2004**, *34*, 549–555. [CrossRef]
7. Cheerarot, R.; Jaturapitakkul, C. A study of disposed fly ash from landfill to replace Portland cement. *Waste Manag.* **2004**, 701–709. [CrossRef]
8. Somna, R.; Jaturapitakkul, C.; Made, A.M. Effect of ground fly ash and ground bagasse ash on the durability of recycled aggreg concrete. *Cem. Concr. Compos.* **2012**, *34*, 848–854. [CrossRef]
9. Dueramae, S.; Tangchirapat, W.; Jaturapitakkul, C. Strength and heat generation of concrete using carbide lime and fly ash a new cementitious material without Portland cement. *Adv. Powder Techol.* **2018**, *29*, 672–681. [CrossRef]
10. Dueramae, S.; Tangchirapat, W.; Sukontasukkul, P.; Chindaprasirt, P.; Jaturapitakkul, C. Investigation of compressive streng and microstructures of activated cement free binder from fly ash–calcium carbide residue mixture. *J. Mater. Res. Technol.* **2019** 4757–4765. [CrossRef]
11. Turk, K.; Kina, C.; Bagdiken, M. Use of binary and ternary cementitious blends of F-Class fly-ash and limestone powder mitigate alkali-silica reaction risk. *Constr. Build. Mater.* **2017**, *151*, 422–427. [CrossRef]
12. Saha, A.K.; Khan, M.N.N.; Sarker, P.K.; Shaikh, F.A.; Pramanik, A. The ASR mechanism of reactive aggregates in concrete and mitigation by fly ash: A critical review. *Constr. Build. Mater.* **2018**, *171*, 743–758. [CrossRef]
13. Moser, R.D.; Jayapalan, A.R.; Garas, V.Y.; Kurtis, K.E. Assessment of binary and ternary blends of metakaolin and Class C fly a for alkali-silica reaction mitigation in concrete. *Cem. Concr. Res.* **2010**, *40*, 1664–1672. [CrossRef]
14. ASTM C150/150M. *Standard Specification for Portland Cement*; ASTM International: West Conshohocken, PA, USA, 2012.
15. ASTM C430. *Standard Test Method for Fineness of Hydraulic Cement by the 45-µm (No. 325) Sieve*; ASTM International: W Conshohocken, PA, USA, 2008.
16. ASTM C618. *Standard Specification for Coal Fly Ash and Raw or Calcined Natural Pozzolan for Use in Concrete*; ASTM Internation West Conshohocken, PA, USA, 2012.
17. Chindaprasirt, P.; Jaturapitakkul, C.; Sinsiri, T. Effect of fly ash fineness on microstructure of blended cement paste. *Constr. Bu Mater.* **2007**, *21*, 1534–1541. [CrossRef]
18. Kroehong, W.; Sinsiri, T.; Jaturapitakkul, C.; Chindaprasirt, P. Effect of palm oil fuel ash fineness on the microstructure of blend cement paste. *Constr. Build. Mater.* **2011**, *25*, 4095–4104. [CrossRef]
19. Paya, J.; Monzo, J.; Borrachero, M.V.; Peris, E. Mechanical treatments of fly ashes. Part I: Physico-chemical characterization ground fly ashes. *Cem. Concr. Res.* **1995**, *25*, 1469–1479. [CrossRef]
20. ASTM C1567. *Standard Test Method for Determining the Potential Alkali-Silica Reactivity of Combinations of Cementitious Materials a Aggregate (Accelerated Mortar-Bar Method)*; ASTM International: West Conshohocken, PA, USA, 2013.
21. ASTM C1260. *Standard Test Method for Potential Alkali Reactivity of Aggregates (Mortar-Bar Method)*; ASTM International: W Conshohocken, PA, USA, 2014.
22. ASTM C441. *Standard Test Method for Effectiveness of Pozzolans or Ground Blast-Furnace Slag in Preventing Excessive Expansion Concrete due to the Alkali-Silica Reaction*; ASTM International: West Conshohocken, PA, USA, 2011.
23. Tapan, M. Alkali–silica reactivity of alkali volcanic rocks. *Eur. J. Environ. Civ. Eng.* **2015**, *19*, 94–108. [CrossRef]
24. Penacho, P.; Brito, J.; Silva, A.S.; Veiga, M.R. Risk of ASR in coating mortars incorporating glass aggregates and a Portlan limestone cement. *Eur. J. Environ. Civ. Eng.* **2017**, *23*, 1–19. [CrossRef]
25. Bleszynski, R.F.; Thomas, M.D. Microstructural studies of alkali-silica reaction in fly ash concrete immersed in alkaline solutio *Adv. Cem. Based Mater.* **1998**, *7*, 66–78. [CrossRef]
26. Shehata, M.H.; Thomas, M.D. The effect of fly ash composition on the expansion of concrete due to alkali-silica reaction. *Ce Concr. Res.* **2000**, *30*, 1063–1072. [CrossRef]
27. Hou, X.; Struble, L.J.; Kirkpatrick, R.J. Formation of ASR gel and the roles of C-S-H and portlandite. *Cem. Concr. Res.* **2004**, 1683–1696. [CrossRef]
28. García-Lodeiro, I.; Palomo, A.; Fernández-Jiménez, A. Alkali–aggregate reaction in activated fly ash systems. *Cem. Concr. R* **2007**, *37*, 175–183. [CrossRef]
29. Ramyar, K.; Çopuroğlu, O.; Andiç, Ö.; Fraaij, A. Comparison of alkali-silica reaction products of fly-ash- or lithium-salt-beari mortar under long-term accelerated curing. *Cem. Concr. Res.* **2004**, *34*, 1179–1183. [CrossRef]

Topçu, İ.B.; Boğa, A.R.; Billir, T. Alkali–silica reactions of mortars produced by using waste glass as fine aggregate and admixtures such as fly ash and Li2CO3. *Waste Manag.* **2008**, *28*, 878–888. [CrossRef] [PubMed]

Aydın, S.; Karatay, Ç.; Baradan, B. The effect of grinding process on mechanical properties and alkali-silica reaction resistance of fly ash incorporated cement mortar. *Powder Technol.* **2010**, *197*, 68–72. [CrossRef]

Lindgård, J.; Andiç-Çakır, Ö.; Fernandes, I.; Rønning, T.F.; Thomas, M.D. Alkali–silica reactions (ASR): Literature review on parameters influencing laboratory performance testing. *Cem. Concr. Res.* **2012**, *42*, 223–243. [CrossRef]

Duchesne, J.; Bérubé, M.A. The effectiveness of supplementary cementing materials in suppressing expansion due to ASR: Another look at the reaction mechanisms part 1: Concrete expansion and portlandite depletion. *Cem. Concr. Res.* **1994**, *24*, 73–82. [CrossRef]

Ichikawa, T. Alkali–silica reaction, pessimum effects and pozzolanic effect. *Cem. Concr. Res.* **2009**, *39*, 716–726. [CrossRef]

Turanli, L.; Uzal, B.; Bektas, F. Effect of large amounts of natural pozzolan addition on properties of blended cements. *Cem. Concr. Res.* **2005**, *35*, 1106–1111. [CrossRef]

Hong, S.Y.; Glasser, F.P. Alkali sorption by C-S-H and C-A-S-H gels: Part II. Role of alumina. *Cem. Concr. Res.* **2002**, *32*, 1101–1111. [CrossRef]

Tangpagasit, J.; Cheerarot, R.; Jaturapitakkul, C.; Kiattikomol, K. Packing effect and pozzolanic reaction of fly ash in mortar. *Cem. Concr. Res.* **2005**, *35*, 1145–1151. [CrossRef]

Jaturapitakkul, C.; Tangpagasit, J.; Songmue, S.; Kiattikomol, K. Filler effect and pozzolanic reaction of ground palm oil fuel ash. *Constr. Build. Mater.* **2011**, *25*, 4287–4293. [CrossRef]

Kroehong, W.; Sinsiri, T.; Jaturapitakkul, C. Effect of palm oil fuel ash fineness on packing effect and pozzolanic reaction of blended cement paste. *Procedia Eng.* **2011**, *14*, 361–369. [CrossRef]

Esteves, T.C.; Rajamma, R.; Soares, D.; Silva, A.; Ferreira, V.M.; Labrincha, J.A. Use of biomass fly ash for mitigation of alkali-silica reaction of cement mortars. *Constr. Build. Mater.* **2012**, *26*, 687–693. [CrossRef]

Shon, C.S.; Zollinger, D.G.; Sarkar, S.L. Evaluation of modified ASTM C1260 accelerated mortar bar test for alkali-silica reactivity. *Cem. Concr. Res.* **2002**, *32*, 1981–1987. [CrossRef]

materials

MDPI

Article

Influence of Nano-SiO_2 on the Mechanical Properties of Recycled Aggregate Concrete with and without Polyvinyl Alcohol (PVA) Fiber

Shenglin Wang and Baolong Zhu *

Department of Civil Engineering and Architecture, Southwest University of Science and Technology, Mianyang 621000, China; wsltarget@163.com
* Correspondence: zhubaolong@swust.edu.cn

Abstract: In recent years, recycled aggregate concrete (RAC) has become a research hotspot the field of urban construction because of its resource utilization of construction waste. Howev compared with original concrete, its strength is still low, which requires additional nano-SiO_2 (N and fiber. In order to study the mechanism of strength improvement of RAC, this paper takes NS ar polyvinyl alcohol (PVA) fiber as variable parameters; uniaxial and triaxial compression tests we carried out on RAC with PVA fiber and NS, and the mechanical properties of RAC were investigat The result shows that within the range of 3% NS content, an increase in the NS substitution ra causes the mechanical properties of RAC to improve significantly. The compressive strength of RA increases again after adding PVA fiber; through a SEM (scanning electron microscopy) analysis the specimen, it was found that the NS filled the micro-pores and micro-cracks in the RAC, and t PVA fiber changed the contact range between recycled aggregate and mortar, so the microstructu of the material was more compact. The mechanism of RAC strength improvement is explained in t microcosmic view.

Keywords: nano-SiO_2; polyvinyl alcohol (PVA) fiber; SEM; recycled concrete; uniaxial compressic conventional triaxial compression

check for updates

Citation: Wang, S.; Zhu, B. Influence of Nano-SiO_2 on the Mechanical Properties of Recycled Aggregate Concrete with and without Polyvinyl Alcohol (PVA) Fiber. *Materials* **2021**, *14*, 1446. https://doi.org/10.3390/ma14061446

Academic Editors: Miguel Bravo, Rui Vasco Silva and António P.C. Duarte

Received: 29 January 2021
Accepted: 5 March 2021
Published: 16 March 2021

1. Introduction

As urbanization gains increasing momentum, many buildings are demolished, r sulting in substantial construction waste [1]. Meanwhile, construction of new buildin requires a lot of natural resources and consumes numerous natural aggregates. Encoura ing the use of recycled concrete can help conserve natural resources and achieve the goa of sustainable development and environmental protection.

However, compared with natural concrete, recycled concrete has a much more compl structure. As recycled coarse aggregate boasts lots of micro-cracks, with the pressure rolling, the old cement mortar and the aggregate will give rise to cracks. Influenced by t minor cracks in recycled aggregate, the interface transition coefficient of natural concre is greater than recycled concrete [2]. In addition, the residual mortar left on recycl coarse aggregate boasts a comparably lower density and higher porosity, which lead to t recycled coarse aggregate exhibiting disadvantages such as lower strength and brittlene compared with natural concrete [3]. Meanwhile, as the polyvinyl alcohol (PVA) fiber characterized with benefits such as environmental conservation, hydrophilicity, as w as corrosive and soluble base resistance, it possesses great adhesive strength with ceme At the same time, it serves as a concrete modification material, capable of improving t flexibility and impact resistance of concrete [4].

In recent years, surveys on recycled concrete mixing with PVA fiber have revealed th concrete with PVA fiber exhibits good ductility [5], inhibits the generation of micro-crac and crack propagation, and narrows crack width [6]. At the same time, PVA fiber can mi mize the stress concentration caused by the initial defects in recycled concrete materia thus improving strength [7]. In terms of PVA fiber reinforcing concrete performance, varying amounts of fiber (0.05–0.38%) were added in the crack tensile tests of a standard PVA fiber concrete sample, which improved the plain concrete under the same circumstances [8]. Having thoroughly investigated the mechanical performance parameters of PVA fiber with regard to reinforcing concrete, Lin et al. found that the ideal PVA fiber content enhances the coherence of cement mortar and the compressive performance of the mortar [9]. Haskett et al. studied concrete mixed with PVA fiber and found that PVA fiber effectively prevents cracks from appearing in the concrete compression zone [10]. Yu et al. concluded that PVA fiber improves the compressive bearing capacity and fracture bridging capacity of concrete by improving the tensile toughness of the internal connection in concrete [11]. Having conducted tensile and other tests by adding PVA fibers into concrete, Cadoni et al. concluded that PVA fibers can significantly enhance the compressive strength and minimize the fracture and strain rate [12]. Additionally, Atahan et al. discovered that the flexural performance and impact resistance of composite materials alter significantly as the PVA quantity of fiber (at a mixing ratio between 0.5–2.0%) changes [13]. Meanwhile, the addition of PVA fiber also improves the impact toughness and durability of composite materials [14–16]. At present, most of the mechanical performance tests of PVA fiber on cementitious composite materials have confirmed increases in the porosity of the cement paste and the flexural strength of concrete [17,18].

Additionally, abundant studies have shown the capacity of nano-SiO_2 (NS) in enhancing the mechanical properties and durability of concrete. The addition of NS into the concrete accelerates the hydration of cement, which reduces the consumption of cement, and improves the concrete strength from the micro perspective [19]. Generally, as the addition of NS is less than 3%, the compressive strength and flexural strength of concrete within 28 days can be increased by 10% and 25%, respectively [20].

Gonzalez observed that with a mixture of 2% NS, the compressive strength of the cement mortar and the bending stiffness were improved by 50% and 16%, respectively [21]. Gesoglu et al. [22] pointed out that the addition of 2% NS improves the tensile strength of concrete by 37.5%. Givi et al. found that the rise of NS enhances the flexural, compressive and tensile strength of the concrete [23]. By examining the effect of NS on the mechanical properties and porosity of concrete, Zhang et al. found that as the content of NS rises from 0.5 to 5%, the cracks of the samples gradually expand, and the internal NS generated by the unhydrated cement fills the pores [24]. Given the disadvantages of recycled coarse aggregate, such as low density, high porosity, low strength and brittle failure, it is of great significance to investigate the influence of PVA fiber and NS on the mechanical properties of recycled aggregate concrete. Previously, Mahmoud et al. proved that NS can improve the performance of martial fiber concrete and increase the compressive strength [25]. Sikora et al. studied the changes in the mechanical properties of acrylic polymer concrete following the addition of NS [26]. Ling et al. considered the influence of PVA fiber content and NS on the mechanical properties of recycled aggregate concrete (RAC). The results show that PVA fiber can enhance the flexural strength of concrete, and the addition of NS mainly affects the workability and tensile strength [27]. Mukharjee et al. have carried out experimental research on the replacement of Portland cement with nano colloidal silica. The results show that the compressive and tensile strength of the material are improved. When a small amount of NS is added, the performance of RAC is similar to that of original concrete [28]. However, few studies exist concerning the mechanical properties of PVA fiber and NS reinforcing recycled aggregate concrete.

Based on previous studies on recycled concrete, PVA fiber concrete and NS concrete, this paper makes the uniaxial and triaxial compression tests on NS recycled aggregate concrete with and without PVA fiber and studies its impact on the work performance, mechanical strength and ultimate stress of the concrete from a macro perspective. Utilizing scanning electron microscopy (SEM), we discuss the influence of PVA fiber and nano NS on the microstructure of recycled aggregate concrete, which shows that the microstructure performance is in good agreement with the macroscopic mechanical properties. This paper also illustrates how NS fills the micro-pores and micro-cracks of the recycled aggrega and how PVA fiber improves the contact surface of the new and old mortar. To sum up, t collaboration of PVA fiber and NS enhances the strength of recycled concrete, providing efficient and effective purpose for its application.

2. Materials and Methods

2.1. Test Raw Materials

The sample fibers used in our work are PVA fibers (10–20 mms in length, Jiang Capability Technology Co., Ltd., Nanjing, China), and Table 1 shows the physical a mechanical properties of the sample fibers. The cement is PO42.5R common Portla cement. The recycled aggregate comes from abandoned construction concrete that h been artificially broken, with an original strength of C30. The concrete was crushed the jaw crusher into 3–10 mm particles and then sieved. With a sand ratio of 50%, the fi aggregate consists of particles with an average diameter of 0.44 mm and a mud conte below 2.5%. NS is the solid white powder with particles 5 nm in diameter. This is dedicat for VK-SH30 concrete, with a purity of 99.5%, a specific surface area within 150 to 200 m^2, and a pH within 5 to 7.

Table 1. Basic parameters of polyvinyl alcohol (PVA) fibers used in this study.

Young's Modulus	Elongation at Break	Young's Modulus	Fiber Diameter	Density	Tensile Strength
(GPa)	(%)	(GPa)	(μm)	(g/cm^3)	(MPa)
36	6	36	21	1.30	1280

2.2. Experiment Mixture Proportions

The introduction has described the current research towards PVA fiber concrete a summarized that PVA fiber enhances concrete strength [8,9], as shown in Figure 1. Wi the design parameter of the concrete remaining unchanged, the rise in the quantity PVA fibers enhances the strength of the sample from Point B to Point A, and thus the B segment is regarded as the rising phase. As the fiber content surpasses Point A, the concre strength gradually decreases to C due to internal reasons. Therefore, the AC segment w called the descending phase. As a result, the fiber content in our test was set at 3.6 kg/m

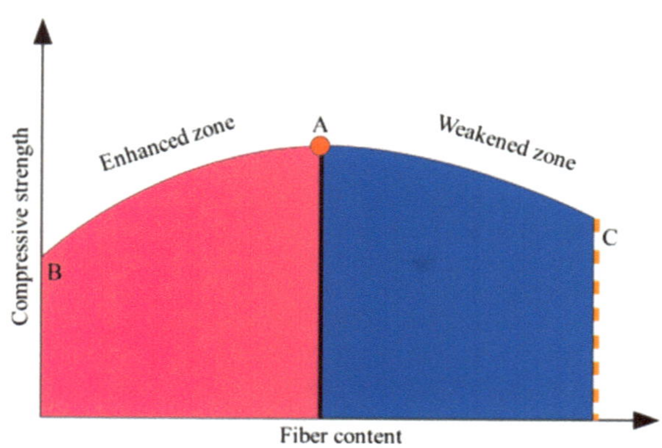

Figure 1. PVA fiber improves concrete strength [8].

The design of concrete mix ratio is shown in Table 2. In which the NS content 0%, 0.5%, 1.0%, 2.0%, 3.0% and 4.0% of the cement quality, respectively, RCN is recycl aggregate concrete mixed with NS, RCN-P is RCN mixed with PVA fiber, and the numb after the code N represents the percentage of NS content.

Table 2. Proportion of sample.

Numbering	Concrete Mix Design /(kg/m^3)					
	Nano -SiO_2 (kg/m^3)	PVA Fiber (kg/m^3)	Recycled Aggregate (kg/m^3)	Portland Cement (kg/m^3)	Water (kg/m^3)	Silica Sand (kg/m^3)
RCN0	0	0	1400	340	180	600
RCN0.5	1.5	0	1400	340	180	600
RCN1	3.	0	1400	340	180	600
RCN2	6	0	1400	340	180	600
RCN3	10	0	1400	340	180	600
RCN4	12	0	1400	340	180	600
RCN0-P	0	3.6	1400	340	180	600
RCN0.5-P	1.5	3.6	1400	340	180	600
RCN1-P	3.	3.6	1400	340	180	600
RCN2-P	6	3.6	1400	340	180	600
RCN3-P	10	3.6	1400	340	180	600
RCN4-P	12	3.6	1400	340	180	600

2.3. Test Methods

With the completion of curing, all samples went through uniaxial compression and traditional triaxial compression tests. Figure 2 illustrates the stress test model.

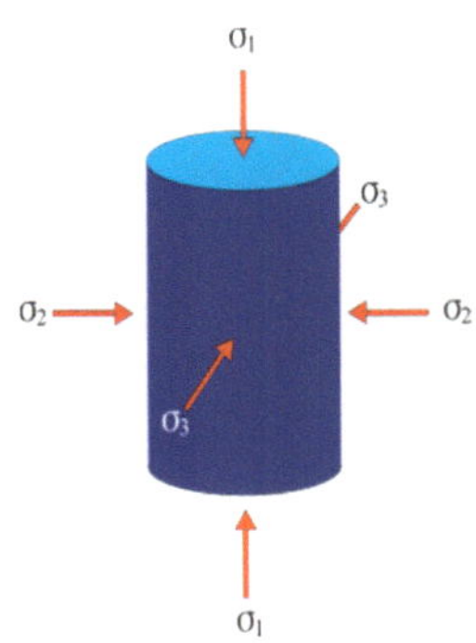

Figure 2. Stress test model.

After casting formation, the test piece was put into a standard curing room ((20 ± 2) °C, (95 ± 3)% RH) for 3 days, and then the mold was removed. As the test block without mold continued to stay in the standard curing room for a certain period of time, the test samples were examined on the DYS-2500 high temperature and pressure rock triaxial testing machine (Sichuan Dexiang Kechuang Instrument Co., Ltd., Chengdu, China), with 3 samples subjected to 3 values of confining pressures (0, 5, 10 MPa). Finally, the average of the test results was regarded as the final value. The loading path involves 2 stages: First, the confining pressure continued to increase at a speed of 0.25 MPa/s until the target confining pressure was reached; Second, under a constant confining pressure, the displacement rate was controlled at 0.06 mm/s.

3. Experiment Results

The Failure Form of the Sample

By comparing the failure modes of NS-recycled aggregate concrete with PVA fibers and the concrete without PVA fibers under the confining pressure, we found that the failure mode of the sample was related to the content of NS, the value of the confining pressure and whether it contained PVA fiber. Figure 3 shows the partial failure of the sample.

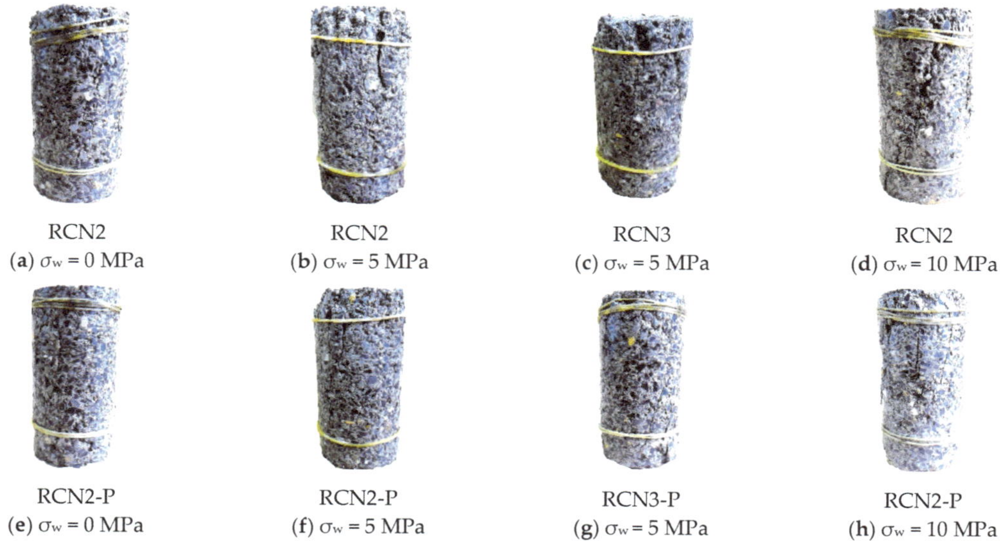

Figure 3. The failure mode of recycled concrete after compression.

Among them, the sample exhibited vertical splitting failure under uniaxial compre sion and oblique shear failure under triaxial compression. As there was no confinir pressure in the lateral direction, the concrete was under uniaxial compression. With tł content of NS within 3%, an increased content of NS gradually increased the elastic lim value of the recycled concrete and damaged the aggregate concrete. Furthermore, tł addition of PVA fiber further improved the elastic limit value of the recycled concrete. A the load exceeded the elastic limit value, vertical cracks appeared in the middle of the te piece and extended to both ends and, at the same time, new small transverse cracks we generated. Compared with NS-recycled aggregate, the one mixed with PVA fiber produce fewer small cracks, and when the load reached the maximum, the bearing capacity of tł sample lasted longer, but all the final samples showed split failure.

In triaxial compression, as the sample withstood the confining pressure from tł lateral restraint, the interior structure could endure greater pressure. In contrast wi uniaxial compression, the sample showed larger and more cracks. With the increase the NS, the sample exhibited large cracks, but the width of the small cracks decreased. the face of the corresponding confining pressure and NS replacement ratio, the PVA fib could shoulder the load with concrete as the PVA fibers were arbitrarily allocated in tł sample to construct a web system. Compared with the NS-recycled aggregate concre the concrete blended with PVA fiber showed higher durability and more compact crac Observation of the failure surface revealed that when the confining pressure value w 5 MPa, the interface between the recycled coarse aggregate and the cement paste w mainly sheared, but the coarse aggregate was hardly sheared. As the confining pressu value rose to 10 MPa, the recycled aggregate and the cement collision between the recycle aggregates were sheared and destroyed.

In conclusion, the failure mode of the specimen changes significantly with the increa of the confining pressure value. When the confining pressure value is 0, it is mainly tł splitting failure of the interface between the coarse aggregate and cementitious bod with the increase of the confining pressure value, the vertical cracks gradually chan into oblique cracks, and when the confining pressure value reaches 10 MPa, shear failu appears on the recycled aggregate. Compared with the failure mode of RAC without PV the aggregate at the failure surface of RAC with PVA is almost sheared into powder;

the failure surface of RAC with low nano silica content, the cement mortar wrapped with coarse aggregate is crushed into fine powder, and a small amount of coarse aggregate is sheared off.

4. Discussion

4.1. Analysis of the Compressive Strength under Uniaxial Compression

The compressive strength of the NS-recycled aggregate concrete with and without PVA fibers was compared, and the results are shown in Figure 4.

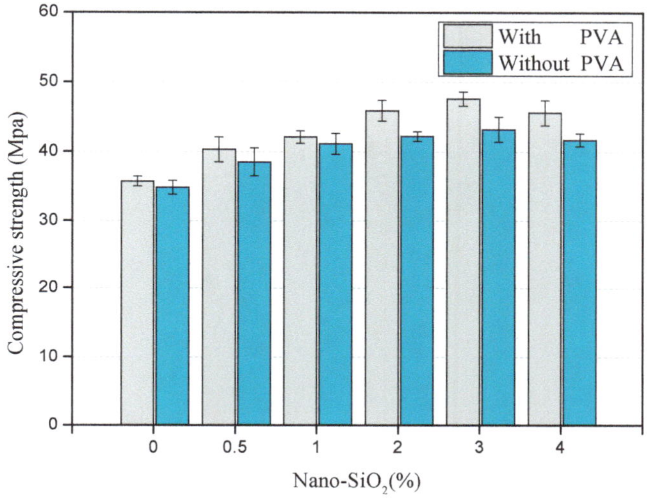

Figure 4. Compressive strength of nano-SiO_2 (NS)-recycled aggregate concrete with and without PVA fibers.

Figure 4 shows the uniaxial compressive strength of the concrete improves as the NS content increases within is within 3%. When the content of NS is 3%, the strength of the recycled aggregate concrete is 15.1% higher than that without NS, which indicates that NS can improve the compressive strength of the concrete.

When the content of NS is 3%, a 7.6% increase of the compressive strength is registered by the NS-recycled aggregate concrete containing PVA fibers. It shows that the addition of PVA fibers enhances the compressive strength of NS-recycled aggregate concrete. When the NS content remains low, PVA fibers exert minor impacts on the NS-recycled aggregate concrete. Meanwhile, with the rise of NS content, the compressive strength of NS-recycled aggregate concrete blended with PVA fiber is gradually improved, and NS is easily dispersed in the course of mixing the concrete. Thus, NS is prone to penetrating the interface of recycled aggregate, with the effective release of NS.

Figure 4 also shows that when the NS content climbs up to 4%, the compressive strength of NS-recycled aggregate concrete begins to decrease, which illustrates that 3% NS content is capable of repleting the micro-pores and cracks of the recycled aggregate. As the NS content exceeds 3%, excessive NS content brings more pores in the matrix, thereby reducing the compressive strength.

4.2. Conventional Triaxial Test Analysis

4.2.1. Stress-Strain Relationship Curve

Figure 5 presents the stress-strain relationship curve of the sample under different confining pressures. According to the data, as the lateral confining pressure value grows greater, the damage occurs later, and the strain during damage failure grows larger, which indicates that the lateral confining pressure value can virtually inhibit the occurrence ar development of internal damage in recycled concrete. Under the confining pressure, r obvious difference is observed in the occurrence of the initial strain between the NS-recycle aggregate concrete with PVA fiber and the one without. Under triaxial compression, at th initial stage of loading, all samples are in the linear elastic stage under triaxial confinir pressure. As they continue to load, the lateral confining pressure causes differences in t stress-strain process of the concrete samples, with the following main characteristics:

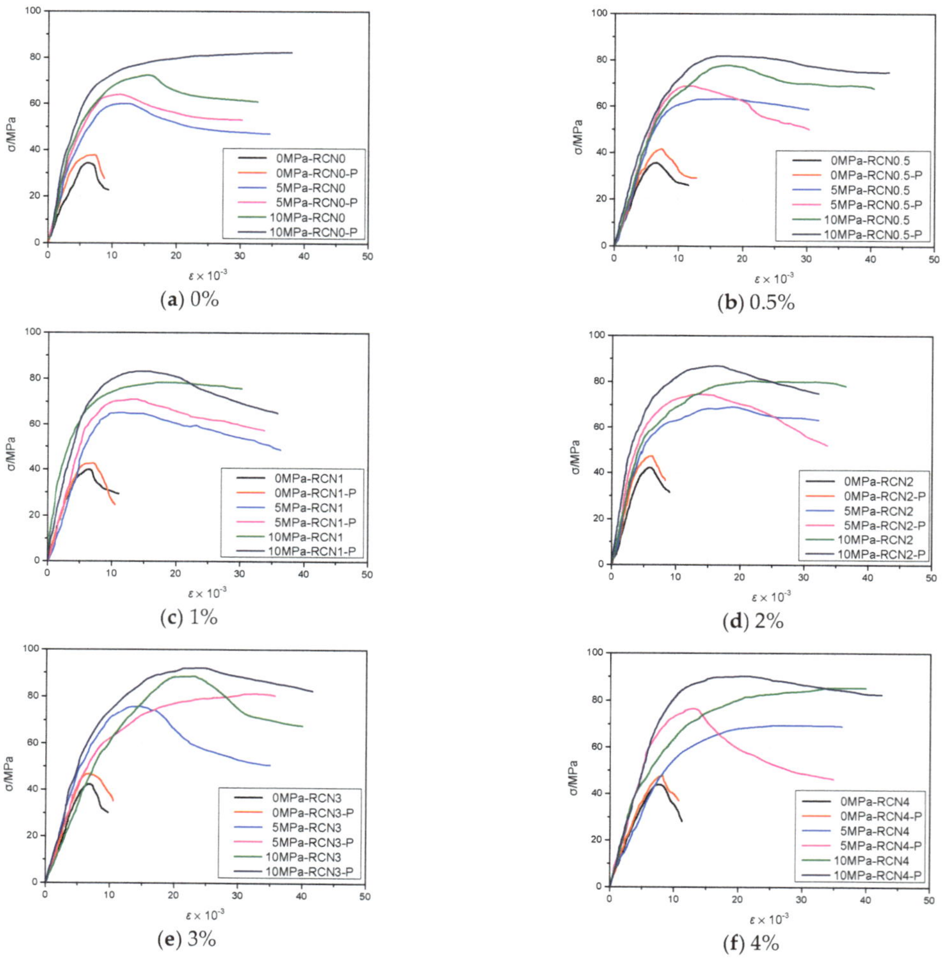

Figure 5. Changes of NS content under different confining pressures and stress-strain relationship curves of recycled aggregate concrete with and without PVA fibers.

(1) As the confining pressure exists, the stress-strain curve turns to develop smooth without an obvious peak, which shows that the deformation of the sample and the mic cracks in the section under the confining pressure can be virtually restrained. As t confining pressure rises from 0 MPa to 10 MPa, the ultimate strain range of the samp will be regulated. After the ultimate stress interval, the stress-strain curve presents plas features; however, as the applied load surpasses the appropriate axial stress and reach

the utmost strength of the concrete, the sample is instantly squashed. In other words, the stress-strain curve drops abruptly.

(2) With the confining pressure unchanged and the increase of the NS replacement ratio, the main compressive stress of the sample increases, along with a corresponding decrease in the axial strain. All in all, the curves of different NS replacement ratios show the same trend. When the NS replacement ratio remains constant, the rise of confining pressure leads to an increase in the main compressive stress and the axial strain. In contrast with the NS recycled aggregate concrete blended without PVA fiber, the concrete without fibers has a larger main compressive stress but a smaller axial strain. Owing to the confining pressure, the sample shows stronger plastic capacity and lasts longer in the triaxial test before it fails.

4.2.2. The Influence of Confining Pressure on Peak Stress

Figure 6 shows that the lateral confining pressure has a large impact on the peak stress of concrete. The peak stress increases in varying degrees with the rise of the confining pressure. With the NS content controlled at 3%, when the confining pressure value rises from 0 to 10 MPa, the peak stress of the NS-recycled aggregate concrete containing PVA fibers soars by 211.4%, and the one without fibers by 193.7%. Therefore, the restraining effect of confining pressure greatly improves the bearing capacity of the concrete.

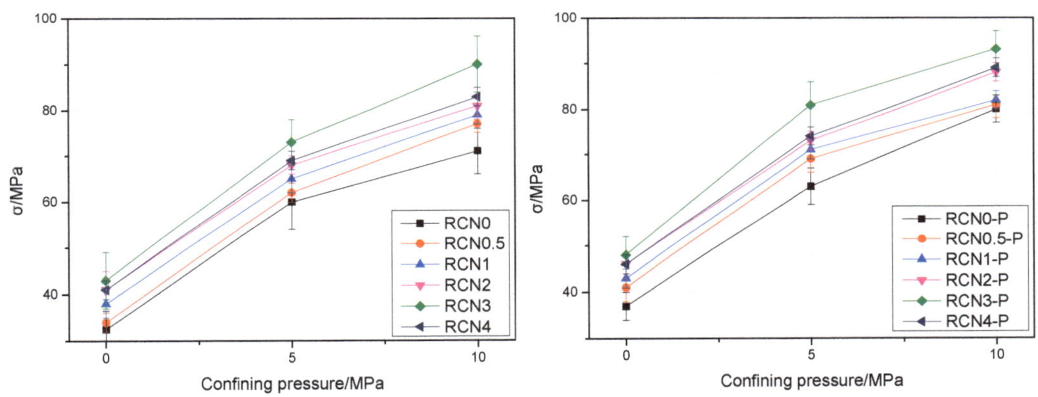

Figure 6. Relationship curve between confining pressure and peak stress.

Comparing the peak stress of NS-recycled aggregate concrete with and without PVA fiber, it is found that under the same loading conditions, the strength of the former is higher than the latter, which is attributed to the addition of PVA fibers. As micro-cracks are generated in the aggregate while crushing, the strength of the aggregate is reduced, but the PVA fibers restrain the development of internal cracks and weaken the impact of micro-cracks in the aggregate.

Figure 6 demonstrates that when the confining pressure remains constant, the peak stress of recycled aggregate concrete continues to change with the rise of the NS replacement ratio. As the NS content gradually reaches 3%, the overall peak stress witnesses an increase. As the NS replacement ratio remains unchanged, the concrete peak stress of the RCN group and RCN-P group shows a linear trend with the rise of the confining pressure. Under the condition of NS substitution, with the rate of substitution and confining pressure unchanged, the peak stress of the NS-recycled aggregate concrete with PVA fibers is 1.21–1.40 times that of the samples without PVA fibers.

4.2.3. SEM Microstructure Analysis

Figure 7 divides recycled aggregate concrete into three transition zones [29]: (1) the transition zone between the recycled aggregate and the old mortar stuck to the recycled aggregate; (2) the transition zone between the new and old mortar; (3) the transition zo between the recycled aggregate and new mortar.

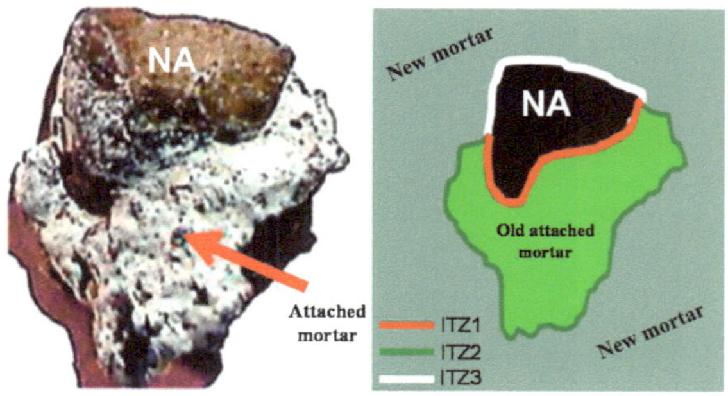

Figure 7. Schematic of interface transition zones [29].

Figure 8 shows the SEM images of the microstructure of recycled aggregate concre with different NS contents (0%, 1.0%, 2.0%, 3.0%). Figure 8a presents the microstructu without NS, and it can be seen that the strengthened result has formed a network structu of strengthened calcium silicate (C-S-H). Meanwhile, Figure 8b–d demonstrate that th process of cement hydration is promoted to varying degrees, and the density of ceme stone continues to rise with increased NS content. In this way, NS promotes the formatic of a C-S-H gel, which is consistent with the conclusion obtained from the test results Roncem [30]. Therefore, NS improves the strength of cement, cement mortar and concre

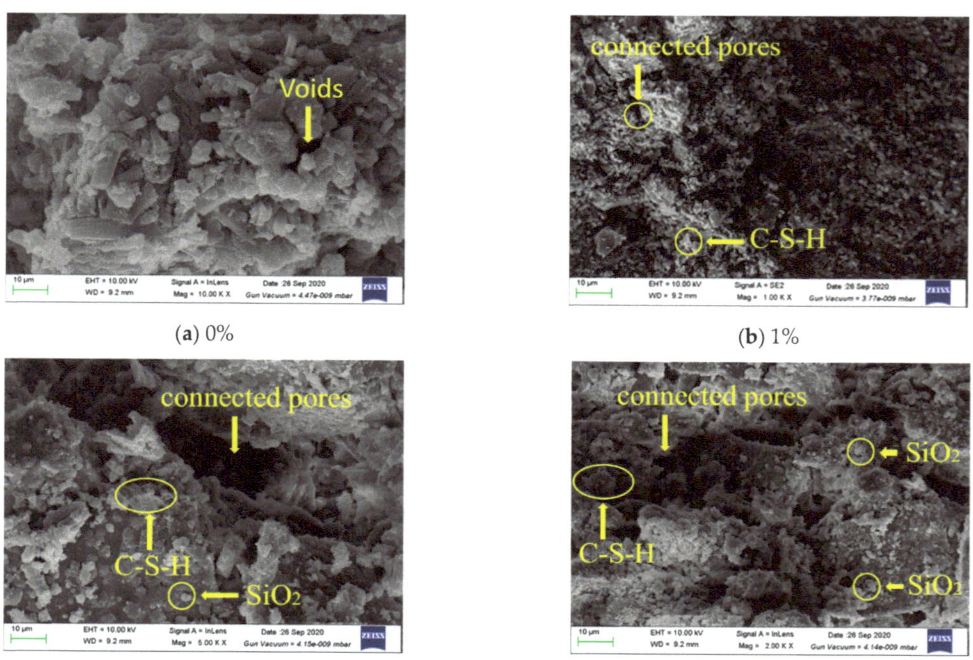

(**a**) 0% (**b**) 1%

(**c**) 2% (**d**) 3%

Figure 8. SEM images of the microstructure of recycled concrete with different NS contents.

Figure 9a,b are the SEM scanning results of the microstructure of NS-recycled concrete mixing with PVA fibers. With the incorporation of PVA fibers into the cementitious material, the microstructure of the material appears as a network structure. As the NS content increases, the network structure of reinforced C–S–H and PVA fibers tend to grow denser. When PVA fibers are incorporated, the PVA fibers and NS will work together, resulting in obvious ettringite (AFt) crystals. NS particles and PVA fibers can be found in the sample matrix, and significant amounts of reinforced C-S-H is found in the interface transition zone, which indicates that NS particles and PVA fibers can be distributed in the interface transition zone of recycled aggregate concrete. Obvious micro-cracks are observed in the transition zone of the concrete interface with a width of several nanometers, which may be related to the NS entering the recycled aggregate. Figure 9c,d illustrate that the mix of NS with recycled aggregate concrete has greatly improved the compactness of the interface transition zone, and few large micro-cracks are present in the interface transition zone. This proves that the incorporation of NS and PVA fibers can significantly enhance the interface transition zone of recycled aggregate concrete. To a certain extent, this phenomenon explains how the addition of NS and PVA fibers enhance the compressive strength of recycled aggregate concrete.

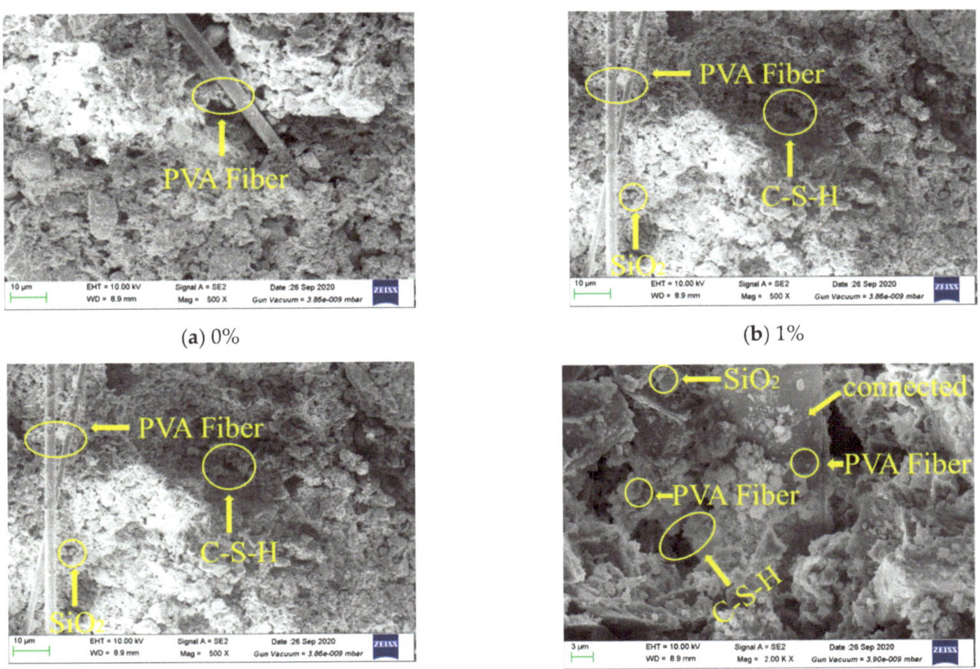

(**a**) 0%　(**b**) 1%

(**c**) 2%　(**d**) 3%

Figure 9. SEM images of the microstructure of recycled concrete with different NS contents and PVA fibers.

Figure 10 shows the collaboration mechanism of PVA fiber and NS. As PVA fibers are incorporated into the recycled aggregate, the collaboration of PVA fiber and NS will increase the adhesive strength of the new and old mortar as well as the aggregate in recycled concrete, thus increasing the compressive strength of recycled concrete.

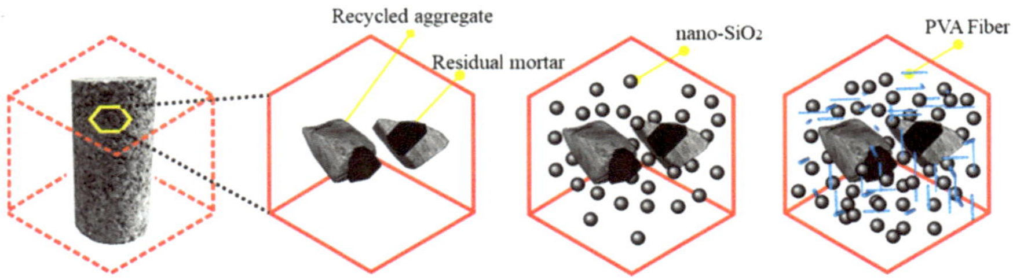

Figure 10. Synergistic mechanism of PVA fiber and NS.

5. Conclusions

By changing the replacement ratio of mixed NS (0%, 0.5%, 1%, 2%, 3%, 4%) mac scopically, we compared the NS recycled aggregate concrete with PVA fibers and the o without in terms of mechanical properties, the failure process, deformation and other ch acteristics; meanwhile, the microstructure changes were analyzed by the SEM techniq microscopically. In this way, we reached the following conclusions:

(1) The addition of NS within a certain range greatly improves the mechanical prop ties of the concrete. However, as the NS content exceeds a threshold, the performance concrete no longer improves for the dispersion of nanoparticles.

(2) The synergistic effect of PVA fibers and NS improves the recycled aggrega concrete, which enhances the adhesive strength between new and old mortar, as well aggregate in recycled concrete. Thus, the recycled concrete containing PVA fibers and N shows higher compressive strength than that mixed with NS alone.

(3) When the NS content is 3.0%, the compressive strength of concrete without a with PVA fiber reaches 43.2 MPa and 47.6 MPa, respectively. The addition of PVA fibe improves the compressive strength of recycled aggregate concrete containing differe contents of NS. As the NS replacement ratio remains constant and the confining pressu rises, NS-recycled aggregate concrete with and without PVA fibers will show a larg principal compressive stress but their axial strain is smaller. Comparison of uniax compression test and triaxial compression test, the triaxial compression test shows that N recycled aggre-gate concrete with PVA fiber improves its containing elastic characteris and the specimen will take longer to fail.

(4) As the replacement ratio of NS remains constant and confining pressure increas the peak stress of the concrete in the RCN group and the RCN-P group shows a line increase. Under the condition of NS substitution, with rate substitution and confini pressure unchanged, the peak stress of the NS recycled aggregate concrete with PVA fibe is 1.21–1.40 times that of the concrete without PVA fibers.

Author Contributions: S.W. conceived and designed the experiment; S.W. and B.Z. carried c experiments; S.W. analyzed the data and S.W. wrote the paper under the guidance of B.Z. All auth have read and agreed to the published version of the manuscript.

Funding: This research was financially supported by National Natural Science Foundation of Chi [Grant NO.: 41672342] and Sichuan Province Applied Basic Research Project (2019YJ0555).

Institutional Review Board Statement: Not applicable.

Informed Consent Statement: Not applicable.

Data Availability Statement: Data sharing is not applicable to this article.

Acknowledgments: This work was carried out at the Sichuan Provincial Key Laboratory of En neering Materials and Structural Shock and Vibration, School of Civil Engineering and Architectu Southwest University of Science and Technology.

Conflicts of Interest: The authors declare no conflict of interest.

eferences

Zhongfu, Q.; Nannan, S.; Haizhen, W. The application status research progress of recycled aggregate concrete. *Mater. Rep.* **2013**, *27*, 142–145.

Xianggang, Z.; Zongping, C.; Jianyang, X. Experimental study on the physical and mechanical properties of recycled concrete. *Bull. Chin. Ceram. Soc.* **2015**, *34*, 1684–1689.

Xiping, L.; Lijiao, X.; Keqiang, Z.; Dongliang, C. Basic performance test of recycled concrete with different replacement ratio. *J. Liaoning Tech. Univ.* **2014**, *33*, 1270–1274.

Yanan, Z. *Research on the Mechanical Properties of Nanoparticle PVA Fiber Cement-Based Composites*; Zhengzhou University: Zhengzhou, China, 2016.

Sun, C.; Jin, B.; Li, J. Orthogonal experimental study on the mechanical properties of PVA fiber recycled concrete. *J. Guangxi Univ. (Nat. Sci. Ed.)* **2018**, *43*, 1569–1575.

Irshidat, M.R.; Al-Shanna, A. Using textile reinforced mortar modified with carbon nano tubes to improve flexural performance of RC beams. *Compos. Struct.* **2018**, *200*, 127–134. [CrossRef]

Xiaoyun, S. Experimental Study on the Performance of PVA Fiber Reinforced Concrete Flexural Members. Master's Thesis, Tongji University, Shanghai, China, July 2001.

Zhaoqiang, Z. Experimental study on conventional triaxial compression of concrete mixed with PVA fiber. *Concr. Cem. Prod.* **2018**, *7*, 51–54.

Hui, L. Study on the Mechanical and Deformation Properties of PVA Fiber Concrete. Master's Thesis, Nanjing University, Nanjing, China, December 2006.

Haskett, M.; MSadakkathulla, M.; Oehlers, D.; Guest, G.; Pritchard, T.; Sedav, V.; Stapleton, B. Adelaide Research and Scholarship: Deflection of GFRP and PVA fibre reinforced concrete beams. In Proceedings of the 6th International Conference on FRP Composites in Civil Engineering, Rome, Italy, 13–15 June 2012.

Yu, K.; Wang, Y.; Yu, J.; Xu, S. A strain-hardening cementitious composites with the tensile capacity up to 8%. *Constr. Build. Mater.* **2017**, *137*, 410–419. [CrossRef]

Cadoni, E.; Meda, A.; Plizzari, G.A. Tensile behaviour of FRC under high strainrate. *Mater. Struct.* **2009**, *42*, 1283–1294. [CrossRef]

Atahan, H.N.; Pekmezci, B.Y.; Tuncel, E.Y. Behavior of PVA fiber-reinforced cementitious composites under static and impact flexural effects. *J. Mater. Civ. Eng.* **2013**, *25*, 1438–1445. [CrossRef]

Zhang, Y.; Sun, W.; Li, Z.; Zhou, X.; Chau, C. Impact properties of geopolymer based extrudates incorporated with fly ash and PVA short fiber. *Constr. Build. Mater.* **2008**, *22*, 370–383.

Viswanath, P.; Thachil, E.T. Properties of polyvinyl alcohol cement pastes. *Mater. Struct.* **2008**, *41*, 123–130. [CrossRef]

Ekaputri, J.J.; Limantono, H.; Susanto Susanto, T.E.; Abdullah, M.M.A. Effect of PVA fiber in increasing mechanical strength on paste containing glass powder. *Key Eng. Mater.* **2016**, *273*, 83–93.

Topic, J.; Proseka, Z.; Indrova, K.; Plachy, T.; Nezerka, V.; Kopecky, L.; Tesarek, P. Effect of pva modification on the properties of cement composites. *Acta Polytech.* **2015**, *55*, 64–75. [CrossRef]

Sobolev, I.F.K.; Torres-Martinez, L.M.; Valdez, P.L.; Zarazua, E.; Cuellar, E.L. Engineering of SiO2 nanoparticles for optimal performance in nano cementbased materials. *Nanotechnol. Constr.* **2009**, *3*, 139–148.

Singh, L.P.; Karade, S.R.; Bhattacharyya, S.K.; Yousuf, M.M.; Ahalawat, S. Beneficial role of nanosilica in cement based materials—A review. *Constr. Build. Mater.* **2013**, *47*, 1069–1077. [CrossRef]

Bastos, G.; Patino-Barbeito, F.; Patino-Cambeiro, F.; Armesto, J. Nano-inclusions applied in cement-matrix composites: A review. *Materials* **2016**, *9*, 1015. [CrossRef] [PubMed]

Gonzalez, M.; Tighe, S.; Hui, K.; Rahman, S.; Lima, A.O. Evaluation of freeze/thaw and scaling response of nanoconcrete for Portland cement concrete (PCC) pavements. *Constr. Build. Mater.* **2016**, *120*, 465–472. [CrossRef]

Gesoglu, M.; Güneyisi, E.; Asaad, S.D.; Muhyaddin, F.G. Properties of low binder ultra-high performance cementitious composites: Comparison of nanosilica and microsilica. *Constr. Build. Mater.* **2016**, *102*, 706–713. [CrossRef]

Givi, A.N.; Rashid, S.A.; Aziz, F.N.A.; Salleh, M.A.M. Experimental investigation of the size effects of SiO2 nano-particles on the mechanical properties of binary blended concrete. *Compos. Part B* **2010**, *41*, 673–677. [CrossRef]

Zhang, P.; Dai, X.; Gao, J.; Wang, P. Effect of nano-SiO2 particles on fracture properties of concrete composite containing fly ash. *Curr. Sci.* **2015**, *108*, 2035–2043.

Mahmoud, K.; Ghazy, A.; Bassuoni, M.T.; El-Salakawy, E. Properties of nanomodified fiber-reinforced cementitious composites. *J. Mater. Civ. Eng.* **2017**, *29*, 1–12. [CrossRef]

Sikora, P.; Łukowski, P.; Cendrowski, K.; Horszczaruk, E.; Mijowska, E. The effect of nanosilica on the mechanical properties of polymer cement composites (PCC). *Procedia Eng.* **2015**, *108*, 139–145. [CrossRef]

Ling, Y.; Zhang, P.; Wang, J.; Chen, Y. Effect of PVA fiber on mechanical properties of cementitious composite with and without nano-SiO2. *Constr. Build. Mater.* **2019**, *229*, 117068. [CrossRef]

Mukharjee, B.B.; Barai, S.V. Influence of Nano-Silica on the properties of recycled aggregate concrete. *Constr. Build. Mater.* **2014**, *55*, 29–37. [CrossRef]

29. Xie, J.; Zhang, H.; Duan, L.; Yang, Y.; Yan, J.; Shan, D.; Liu, X.; Pang, J.; Chen, Y.; Li, X.; et al. Effect of nano metakaolin on compressive strength of recycled concrete. *Constr. Build. Mater.* **2020**, *256*, 119393. [CrossRef]
30. Roncero, J.; Valls, S.; Gettu, R. Study of the influence of superplasticizers on the hydration of cement paste using nuclear magnetic resonance and X-ray diffraction techniques. *Cem. Concr. Res.* **2002**, *32*, 103–408. [CrossRef]

materials

rticle

[E]urocode Shear Design of Coarse Recycled Aggregate Concrete: [R]eliability Analysis and Partial Factor Calibration

[Jo]ão Pacheco [1], Jorge de Brito [1,*], Carlos Chastre [2] and Luís Evangelista [3]

[1] CERIS, IST, Universidade de Lisboa, 1049-001 Lisbon, Portugal; joaonpacheco@tecnico.ulisboa.pt
[2] CERIS, NOVA FCT, Universidade Nova de Lisboa, 2825-149 Caparica, Portugal; chastre@fct.unl.pt
[3] CERIS, ISEL, 1959-007 Lisbon, Portugal; luis.evangelista@isel.pt
* Correspondence: jb@civil.ist.utl.pt; Tel.: +351-21841-9709

Abstract: This paper contributes to the definition of design clauses for coarse recycled aggregate concrete. One of the main reasons for scepticism towards recycled aggregate concrete is the perceived notion that the heterogeneity of recycled aggregates may increase the uncertainty of the behaviour of concrete. Therefore, the paper uses structural reliability concepts to propose partial factors for recycled aggregate concrete's design for shear failure. The paper builds upon a previous publication by the authors, in which the model uncertainty of recycled aggregate concrete elements designed for shear, with and without shear reinforcement, was compared with that of natural aggregate concrete elements. In that paper, the statistics of the model uncertainty for recycled aggregate concrete shear design were indeed found to be less favourable than those of natural aggregate concrete. Therefore, a partial factor for recycled aggregate concrete design is needed to ensure safety. This paper presents partial factors calibrated with explicit reliability analyses for different cases of design concerning beams (in the case of shear design of elements with shear reinforcement) and slabs (for the design of elements without shear reinforcement). For full incorporation of coarse recycled concrete aggregates and the design of elements without shear reinforcement, the calibrated partial factor reduces the design value of shear resistance by 10% (design with EN1992) or 15% (design with prEN1992) in comparison to natural aggregate concrete's design. For the shear design of elements with shear reinforcement, the partial factor decreases resistance by 5% but a sensitivity analysis showed that the reduction might be, under pessimistic expectations, of up to 20%.

Keywords: EN1992; prEN1992; shear; structural concrete; recycled aggregate concrete; reliability analysis; coarse recycled concrete aggregates; partial factor calibration

[Cit]ation: Pacheco, J.; de Brito, J.; [Ch]astre, C.; Evangelista, L. Eurocode [She]ar Design of Coarse Recycled [Ag]gregate Concrete: Reliability [An]alysis and Partial Factor [Cal]ibration. *Materials* **2021**, *14*, 4081. [htt]ps://doi.org/10.3390/[ma]14154081

[Aca]demic Editor: Alessandro [P]antilli

[Rec]eived: 18 June 2021
[Acc]epted: 17 July 2021
[Pub]lished: 22 July 2021

1. Introduction

1.1. Shear Resistance of Recycled Aggregate Concrete Elements

Research on the structural behaviour of recycled aggregate concrete (RAC) demonstrates the feasibility of using coarse recycled aggregates (RAs) for structural concrete applications [1,2]. This is relevant since RAs are produced from construction and demolition waste (CDW), whose recycling is a key objective of the European Union's strategy for a circular economy [3]. Notwithstanding abundant research arguing in favour of RAC as a structural material [4,5], clear design guidelines are lacking. This is an obstacle for systematic applications of RAC, since without guidelines, designers, contractors and clients alike are reticent towards this structural material.

Guidelines for RAC are needed since RAs are different from coarse natural aggregates (NAs), changing the properties of concrete. The main differences between NAs and RAs are the following:

- NAs are composed of particles of a single type of stone (limestone, granite and basalt are the most common), while RAs are composed of a mix of several constituents (concrete, mortar, unbound stone, ceramics, glass, and other deleterious contaminants) of different quality and properties [6,7];

- This implies that, in general, RAs are weaker, more deformable, more porous a have larger water absorption than NAs [8,9];
- At the same time, the mechanical and durability properties of concrete are detrim tally affected by the incorporation of RAs: for the same compressive strength, RA is typically found to have a smaller Young's modulus, larger creep and shrinka and worse durability properties [9–11]. Fracture energy and tensile strength are al detrimentally affected, especially when the strength class of concrete is larger [12,1
- Regarding the structural behaviour of reinforced concrete, the use of RAs is found result in larger short- and long-term deflections [14,15]. Ductility and resistance a only relevantly affected when the resistance mechanism relies relevantly on concre rather than on reinforcement [1].

Research on RAC is mostly concerned with RAs produced from concrete was rather than RAs produced from mixed CDW, which includes concrete waste and the oth aforementioned constituents. RAs produced from concrete waste are better suited for t production of RAC [16], since the other constituents (such as gypsum-based materials a weak ceramics) strongly impair the properties of concrete [6]. As such, this paper concer the shear resistance of RAC elements made with coarse recycled concrete aggregates or

Research on the shear resistance of RAC elements is comprehensive and includes t following topics:

- Comparison of the shear resistance of NAC and RAC elements with full incorporati of RAs (RAC100), with or without shear reinforcement [17–20];
- The influence of the incorporation ratio of RAs on shear resistance [21];
- The shear resistance of prestressed RAC beams [22];
- The shear resistance of RAC elements made with RAs that are treated with benefic tion methods [23];
- Meta-analyses that compare the shear resistance of NAC and RAC based on seve investigations [24];
- The punching shear resistance of RAC elements [25–27].

The main findings of research on the shear resistance of RAC elements are that:

- The behavioural pattern of elements failing in shear is unaffected by the incorporati of RAs [17,20,21];
- The shear resistance of members without shear reinforcement decreases as the conte of RAs increases [17,21,28,29];
- The previous finding is validated by a meta-analysis [24] that compares the moc uncertainties (θ_R) [30] of Eurocode shear resistance models for NAC and RAC desi These θ_R show that the resistance models of EN1992 [31] and prEN1992 [32] for e ments without shear reinforcement overestimate the resistance of RAC in comparis to NAC elements;
- In most cases, the shear resistance of members with shear reinforcement is less affect by the incorporation of RAs [20,28].

The incorporation of RAs decreases shear resistance when shear reinforcement absent because the RAs decrease the shear strength of concrete, as found in an appraisal [1 of several push-off experiments that compared the shear strength of NAC and RAC [33–3 Shear strength decreases because:

- Aggregate interlock is a preponderant mechanism in shear strength mobilisation [3
- RAs are weaker than NAs [8,40], i.e., RAs break more easily and limit aggregate interlo

The decrease in shear strength of concrete caused by the use of RAs is followed b decrease in shear resistance in the RAC element.

When elements have shear reinforcement, resistance models based on struts and t are used in the design. In typical cases, these models are limited by the stress transfer of ti so the only material property included in the resistance model is the reinforcement's yie stress (f_y). This reflects the smaller relevance of concrete in the shear mechanism of comm beams subjected to shear and implies that the detrimental effects of the incorporation of RAs on these cases of design is small.

Notwithstanding the comprehensive state-of-the-art knowledge on the shear behaviour of RAC elements, conventional research programmes cannot fully assess the need to change the provisions of NAC codes for RAC design. Research programmes that investigate the structural behaviour of reinforced concrete elements are carried out with a small number of specimens (typically a single specimen for each reinforcement layout and prototype geometry). Therefore, the influence of RAs on the structural reliability of reinforced concrete cannot be assessed. This is discussed in the next section.

1.2. Codified Shear Design of Recycled Aggregate Concrete

Since RAC is a viable structural material, research should be focused on giving designers the necessary means for the codified design of RAC structures. The ultimate limit state design of current codes has an underlying probabilistic basis. In the case of the Eurocodes, reinforced concrete design is made through characteristic values of material properties and load-effects, which underestimate the former and overestimate the latter. Moreover, partial factors are used to further decrease resistance and increase load-effects, providing an additional margin of safety. The following partial factors are typically used in the design equations of Eurocode reinforced concrete design:

- Partial factors $\gamma_G = 1.35$ and $\gamma_Q = 1.50$ increase actions. Permanent loads are multiplied by γ_G and variable loads are multiplied by γ_Q;
- Partial factors $\gamma_S = 1.15$ and $\gamma_C = 1.50$ decrease material properties or resistance (depending on the resistance model). In most cases, the characteristic yield stress of the reinforcement is divided by γ_S, while the compressive strength of concrete is divided by γ_C;
- γ_C may be modelled as $\gamma_C = \gamma_c \times \gamma_{Rd}$, where γ_c is a partial factor for material variability and γ_{Rd} is a partial factor that accounts for the uncertainty in geometry and in resistance modelling.

The margins of safety provided by the partial factors of the Eurocodes are calibrated based on the variability involved in structural design and require that the safety of structural design complies with the expectations of society. Any new guideline for the ultimate limit state design of RAC (the case of this paper for shear design) needs to take this into account. Therefore, the paper calibrates a partial factor for shear design (γ_{RAC}).

Codes have a probabilistic basis because most of the variables that are relevant to structural design are subjected to uncertainty. Therefore, the calibration of codes requires that load-effects, geometry, material properties and the parameters related to modelling be modelled as stochastic variables. Then, reliability analyses are used to calibrate factors used in the design equations of codes. The calibration of these factors ensures that the reliability index (β) of codified design complies with target values based on the expectation of society regarding safety.

Since RAs are more heterogeneous [41] and weaker than NAs:

- The behaviour of RAC may be more variable than that of NAC;
- The resistance models used for NAC may not be as representative for RAC.

The latter statement is justified by the different stress paths, crack propagation and damage progression of RAC in comparison to NAC [42,43]. This influences the θ_R of the resistance models of codes. Differences in θ_R are relevant in the context of structural reliability, since they are accounted for in the definition of the partial factors of the Eurocodes [44]. If these uncertainties are affected by the incorporation of RAs, then the assumptions and partial factors of EN1992 [31] may be unsafe for RAC design.

In a recent publication [24], the authors proposed a preliminary partial factor γ_{RAC} that accounts for the θ_R of RAC in the safety of the shear design of elements without shear reinforcement when either EN1992 [31] or prEN1992 [32] is used. This preliminary partial factor γ_{RAC} was put forward based on a methodology that conforms with Annex C ISO2394 [45] and Annex D of EN1990 [44].

A database of validated experiments on the shear resistance of NAC and RAC w gathered and stochastic models for the model uncertainty (θ_R) of the shear resistance mo elling of EN1992 [31] and prEN1992 [32] were estimated for NAC and RAC. Afterwarc these θ_R were used to define the preliminary partial factor γ_{RAC} for the shear design elements without shear reinforcement. Due to specificities of the shear resistance models EN1992 [31] and prEN1992 [32], partial factors for NAC design (γ_{NAC}) were also propose

However, the methodology of Annex C of ISO2394 [45] and Annex D of EN1990 [4 is based on assumptions that may not be accurate for the reliability of elements designe for shear. This occurs because shear resistance models have higher scatter than most oth resistance models [46] and assumptions concerning the direction cosines (α) presente in ISO2394 [45] and EN1990 [44] may not be accurate. Therefore, this paper remov these assumptions from the estimates of γ_{NAC} and γ_{RAC} for elements without she reinforcement through calibration based on reliability analysis.

Concerning elements with shear reinforcement, no partial factor is proposed in [2 since research on RAC shear resistance of this type of element is not sufficient for a sour database. Thus, reliability methods based on sensitivity analyses are used to propose γ_{RA}

1.3. Objectives

The main contribution of this paper is the proposal of design guidelines for the she design of RAC elements with and without shear reinforcement. These guidelines are bas on the Eurocode format and include a partial factor γ_{RAC} for RAC design when eith EN1992 [31] or prEN1992 [32] is used. The latter document presents significant changes [4 in comparison with the shear strength model of the current version of EN1992 [31]. Tl paper has the following objectives:

- Proposal of a resistance format with a specific partial factor for RAC design;
- Reliability analyses for representative cases of design using the stochastic models f θ_R proposed in [24];
- Calibration of γ_{RAC} to be used in the resistance format proposed;
- Sensitivity analyses to understand the robustness of the calibrated γ_{RAC} for the she design of elements with shear reinforcement.

This paper is part of an effort by the authors towards the reliability-based structur design of RAC. So far, publications have addressed:

- The within-batch variability of the mechanical properties of RAC [48];
- A probabilistic factor for the conversion of the compressive strength tested in 150 m cubes to the compressive strength used in design codes (ϕ 150 mm × 300 mm cyli ders) [49];
- The scatter of the compressive strength vs. tensile strength and compressive strength Young's modulus relationships of EN1992 [31], Model Code 2010 [50] and prEN1992 [5 and ACI318 [52] when used for RAC [12];
- The model uncertainty of the flexural resistance models of EN1992 [31], prEN1992 [5 and ACI318 [52] for RAC [4];
- The model uncertainty of the shear resistance models of EN1992 [31], prEN1992 [5 and Model Code 2010 [53] for RAC [24];
- The model uncertainty of the bond strength and the reliability-based calibration the development length of ribbed steel reinforcement for RAC when fib Bulletin 72 prEN1992 [51] is used [40];
- The reliability-based calibration of concrete cover for EN1992 [31] design of RAC f chloride-prone environments [54].

2. Design Equations of the Eurocode Format

2.1. Design of Members without Shear Reinforcement

According to EN1992 [31], the design value for the shear resistance of elements without shear reinforcement is:

$$V_{Rd} = \left(\frac{0.18}{\gamma_C} \times k \times \sqrt[3]{100\, \rho_l\, f_{ck}} \right) \times d \times b \geq v_{min} \times d \times b \quad (1)$$

where:

- γ_C is the partial factor for concrete;
- b is the width of the web of the beam;
- d is the effective depth of the beam;
- $k = 1 + \sqrt{\frac{200}{d}} \leq 2.0$ accounts for size effects. In this equation, d is in mm;
- ρ_l is the geometric ratio of the longitudinal tensile reinforcement. In this equation, $\rho_l \leq 2\%$;
- f_{ck} is the characteristic compressive strength of concrete and is in MPa;
- $v_{min} = 0.035 \times k^{3/2} \times \sqrt[2]{f_{ck}}$ is the minimum shear stress. This condition is always complied with in this paper and is not mentioned from here on.

Equation (1) uncouples γ_C from f_{ck}. This is not a common option of the resistance models of EN1992 [31] but allows that:

- In the case of NAC, γ_C is replaced with γ_{NAC};
- In the case of RAC, γ_C is replaced with γ_{RAC}.

Partial factor γ_{NAC} differs from 1.50, the value of γ_C of EN1992 [31], because of the preliminary findings of [24]. These findings suggest that using $\gamma_C = 1.50$ may lead to overly safe (thus uneconomical) design in the case of the shear design of elements without reinforcement. If the reliability analysis developed in this paper finds that β is below expectations, γ_{NAC} will be calibrated. γ_{RAC} is calibrated using reliability analyses on representative cases of design. The calibration criterion is stated in Section 3.1.

In the case of prEN1992 [32], the design value for the shear resistance of elements without shear reinforcement is:

$$V_{Rd} = \left(\frac{0.6}{\gamma_C} \times \sqrt[3]{100\rho_l \times f_{ck} \times \frac{d_{dg}}{a_v}} \right) \times d \times b \geq V_{Rd,c,min} \quad (2)$$

where:

- $d_{dg} = 16 + d_{max} \leq 40$ mm if $f_c \leq 60$ MPa;
- $d_{dg} = 16 + (d_{max}/f_c)^2 \leq 40$ mm if $f_c > 60$ MPa;
- $a_v = \sqrt{\frac{a_{cs}}{4} \times d}$, where $a_{cs} = |M_{cs}/V_{cs}| (\geq d)$;
- $a_v = d$ if $a_{cs} > 4d$;
- $V_{R,c,min} = \left(\frac{10}{\gamma_C}\right) \times \sqrt[2]{\frac{f_{ck}}{f_{yk}/\gamma_s} \times \frac{d_{dg}}{d}}$;
- f_{yk} is the characteristic yielding stress of the longitudinal reinforcement;
- $\gamma_s = 1.15$ is the partial factor of steel reinforcement;
- M_{cs} and V_{cs} are the bending moment and shear stress at the control section;
- In this code, no limit on the ρ_l of EC2 (2004) is imposed.

Alternatively, this code allows Equation (2) to be replaced with a simplified and conservative expression. This paper calibrates partial factors for Equation (2) only, but these factors may be used in both expressions. The same option is taken in prEN1992 [32], where a single γ_C is used for NAC.

As for the design equation of EN1992 [31], partial factor γ_C is replaced with γ_{NAC}.

2.2. Design of Members with Shear Reinforcement

EN1992 [31] and pEN1992 [32] use the same resistance model for the design val of the resistance of elements with shear reinforcement. This resistance model is based struts and ties. The load-bearing capacity of ties is given by:

$$V_{Rd,tie} = \frac{A_{sw}}{s} \times 0.9d \times \frac{f_{yk}}{\gamma_S} \times \cot(\Omega) \leq V_{R,strut}$$

where:

- A_{sw} is the shear reinforcement area;
- s is the distance between shear reinforcement;
- f_{yk} is the characteristic yield stress of the shear reinforcement;
- $\gamma_S = 1.15$ is the partial factor for reinforcement strength, including geometric a modelling uncertainty;
- Ω is the angle of the strut with the longitudinal axis of the element. This angle may assumed as any value in the region of 21.8 to 45°.

The model assumes that the resistance of the compression struts is:

$$V_{Rd,strut} = \frac{b \times 0.9d}{\cot(\Omega) + \tan(\Omega)} \times \frac{\nu_1 \times f_{ck}}{\gamma_C}.$$

EN1992 [31] defines $\nu_1 = 0.6$ if $f_{ck} < 60$ MPa, while for a design with pEN1992 [3 $\nu_1 = 0.5$ may be used.

EN1992 [31] and pEN1992 [32] allow designers to assume a value for angle Ω. In th paper, the most common option of designers was used: Ω was assumed as 30°.

In the typical cases found in conventional reinforced concrete design, $V_{Rd,tie}$ limits t resistance mechanism. Hereafter, all mentions in the paper to V_{Rd} concern shear streng design limited by the design value of resistance of ties (Equation (3)).

Equation (3) does not include γ_C, since its underlying resistance model does n consider the influence of concrete on the resistance mechanism. Nevertheless, concre bears part of the shear force (this is acknowledged in refined formulae, such as the she resistance model of Model Code 2010 for Level of Approximation 3 [53]) and the inc poration of RAs will affect the resistance mechanism. Therefore, a partial factor for R is calibrated.

3. Reliability Analysis for Partial Factor Calibration

3.1. Calibration Procedure and Reliability Method

Equations (1)–(3) result in five expressions for the design value of resistance th need to be calibrated. Equations (5)–(8) concern the design of elements without she reinforcement and are adaptations of either Equation (1) for EN1992 [31], or Equation for prEN1992 [32]. Equation (9) is intended for the shear design of RAC elements wi shear reinforcement. In this case, EN1992 [31] and pEN1992 [32] use the same resistan model and design equation.

$$V_{Rd,EN1992,NAC} = \left(\frac{0.18}{\gamma_{NAC}} \times k \times \sqrt[3]{100\,\rho_l \times f_{ck}}\right) \times d \times b,$$

$$V_{Rd,EN1992,RAC} = \left(\frac{0.18}{\gamma_{RAC}} \times k \times \sqrt[3]{100\,\rho_l \times f_{ck}}\right) \times d \times b,$$

$$V_{Rd,prEN1992,NAC} = \left(\frac{0.6}{\gamma_{NAC}} \times \sqrt[3]{100\rho_l \times f_{ck} \times \frac{d_{dg}}{a_v}}\right) \times d \times b,$$

$$V_{Rd,prEN1992,RAC} = \left(\frac{0.6}{\gamma_{RAC}} \times \sqrt[3]{100\rho_l \times f_{ck} \times \frac{d_{dg}}{a_v}} \right) \times d \times b, \tag{8}$$

$$V_{Rd,RAC} = \frac{1}{\gamma_{RAC}} \times \left[\frac{A_{sw}}{s} \times 0.9d \times \frac{f_{yk}}{\gamma_S} \times \cot(\Omega) \right]. \tag{9}$$

The calibration concerns partial factors γ_{NAC} and γ_{RAC} only. All other partial factors (see Section 1.2) are kept as those of the Eurocodes, so that the design guidelines proposed in this paper are easily implemented. The calibration is made for concrete with 50% incorporation of RAs (RAC50) and for concrete with full incorporation of RAs (RAC100), but partial factors for intermediate incorporation ratios of RAs may be linearly interpolated.

In addition, to ensure an unbiased estimate of β, the reliability analyses are made for the exact correspondence of the design value of load-effects (E_d) with the design value of resistance (V_{Rd}).

The load combination presented in Equation (6.10) of EN1990 [44] is used to calculate E_d. A permanent load and a single variable load are assumed:

$$E_d = \gamma_G \times P_k + \gamma_Q \times Q_k \tag{10}$$

where $\gamma_G = 1.35$ is the partial factor for permanent loads and $\gamma_Q = 1.50$ is the partial factor for variable loads, both including modelling uncertainty; P_k is the characteristic value of permanent loads; and Q_k is the characteristic value of the variable load.

The calibration of these expressions is made through reliability analyses using an iterative procedure. For each of Equations (5)–(9):

1. Cases of design are defined;
2. The load combination presented in Equation (10) is used to determine E_d;
3. An iterative process takes place, beginning with the preliminary proposal of the authors [24] for γ_{NAC} and γ_{RAC}. For a given E_d, the equation for the design value of resistance is used to design the structural element;
4. A reliability analysis takes place, in which a limit state function of the type $g_x = R - E$ is used to determine β. R is the random outcome of resistance and E is the random outcome of load-effects. The limit state functions used are presented in Sections 3.2 and 3.3;
5. When β is below expectations, a new partial factor is checked and a new iteration (starting at step 3 of this bullet list) takes place;
    a. The calibration criteria for shear design of elements without shear reinforcement are that:
        i. In the case of NAC, γ_{NAC} ensures that the target reliability index (β_{target} = 3.8, since reliability class 2 and a 50-year reference period are considered [44]) is complied with in the majority of cases of design. Moreover, the β value should be similar to those obtained in seminal reliability assessments of the Eurocodes [55,56];
        ii. In the case of RAC, the criterion is that γ_{RAC} results in a similar β value to that obtained when γ_{NAC} is used for NAC design;
    b. In the case of the shear design of elements with shear reinforcement, no partial factor for NAC is used and the criterion is that the calibrated γ_{RAC} leads to a similar β value to that when NAC elements are designed.

The reliability analyses use a Rosenblatt transform to convert stochastic variables to the standard normal space and the reliability algorithm used is a First Order Reliability Method [57]. The discussion of the results of the reliability analyses (Section 5) is mostly focused on β and on α^2, the square of the direction cosines of the stochastic variables. Since all stochastic variables are independent and uncorrelated, the α^2 of a stochastic variable is the percentage of the total uncertainty that is due to the uncertainty in that stochastic variable [58].

To ensure that the partial factors are representative of common design, the reliabili analyses concern cases of design that are common for either resistance mechanism. Therefo

- The partial factors for shear resistance of elements without shear reinforcement a calibrated for the shear design of slabs;
- The partial factor for shear resistance of elements with shear reinforcement is calibrat for the shear design of beams.

The cases of design are presented in Section 4.1.

3.2. Limit State Function for Slabs without Shear Reinforcement

The limit state functions compare the random outcome of resistance and load-effec Thus, all parameters of design that are subjected to uncertainty are modelled as stochast variables. These include the magnitude and modelling of loads, material and geometr uncertainty, and θ_R. Equations (11) and (12) are the limit state functions used for th reliability analysis of design using EN1992 [31] and prEN1992 [32] for RAC. In the expressions, parameters that include stochastic modelling are presented in bold.

$$g(x)_{,EN1992} = \theta_R \times \left[\left(0.18 \times k \times \sqrt[3]{100\, \rho_{Asl} \times f_c}\right) \times d \times b\right] - \theta_E \times (P+Q) \quad (1$$

where $k = 1 + \sqrt{\frac{200}{d}} \leq 2$ and $\rho_{Asl} = \frac{A_s}{d \times b}$.

$$g(x)_{,prEN1992} = \theta_R \times \left[\left(0.6 \times \sqrt[3]{100\rho_{Asl} \times f_c \cdot \frac{d_{dg}}{a_v}}\right) \times d \times b\right] - \theta_E \times (P+Q). \quad (1$$

θ_E is the model uncertainty of load-effect modelling, f_c is the random outcome of co pressive strength, and P and Q are the random outcomes of permanent and variable loadir

The stochastic models are presented in Section 4.2.

3.3. Limit State Function for Beams with Shear Reinforcement

In the case of elements with shear reinforcement, the same equation is used f EN1992 [31] and prEN1992 [32]. This $g(x)$ is meant for reliability analyses when the desig is made with Equation (3) for NAC or Equation (9) for RAC.

$$g(x) = \theta_R \times \left(\frac{A_{sw}}{s} \times 0.9d \times f_y \times \cot(\Omega)\right) - \theta_E \times (P+Q). \quad (1$$

4. Cases of Design and Modelling

4.1. Cases of Design

Since the results of reliability analyses depend on the specific cases of design [5 several cases of design covering conventional reinforced concrete design were used in t calibration procedure.

In the case of slabs, the cases of the design were based on Figure 1 and included differe cross-sectional height and longitudinal reinforcement detailing. Two compressive streng classes were analysed: C25/30 and C40/50 [60]. RAC applications of higher strength class than C40/50 are not expected in the near future, due to the reservations of construction ager and because of the restrictions of standards [32] and national specifications [61].

Table 1 presents the main parameters of design of the five slabs analysed wh prEN1992 [32] was used. In this table, Φ_{Asl} is the diameter of longitudinal reinforceme s is its spacing, H is the height of the beam and cy is the concrete cover in the vertic direction. Despite the reliability analyses concerning three incorporation ratios of R (NAC, RAC50, and RAC100), only NAC slabs are presented, since the design of RA depends on the iterative calibration of γ_{RAC}.

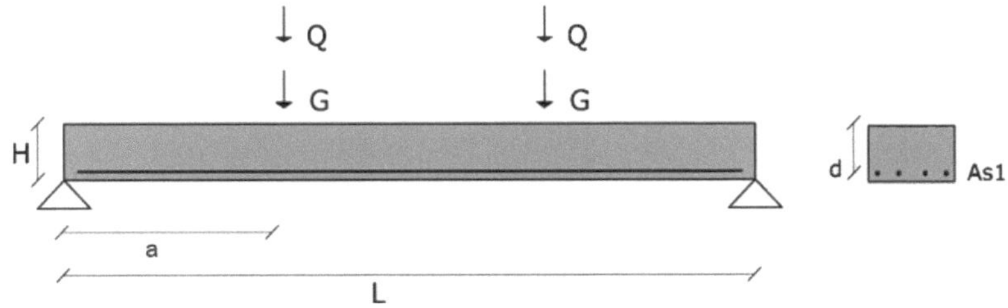

Figure 1. Case of design of the reliability analysis. Slab without shear reinforcement.

Table 1. Cases of design for shear design of NAC slabs without shear reinforcement using prEN1992 [32].

Slab	f_{ck} (MPa)	d_{max} (mm)	H (mm)	c (mm)	L (m)	a/d	Φ_{Asl} (mm)	s (mm)	ρ_l	V_{Rd} (kN)
1	25	20	170	25	5.0	2.5	12	100	0.81%	122.1
2	25	20	140	25	5.0	2.5	16	100	1.88%	135.5
3	25	20	210	25	5.0	2.5	16	150	0.76%	140.0
4	40	20	170	25	6.5	2.5	16	150	0.97%	150.4
5	40	20	145	25	6.5	2.5	16	75	2.39%	177.1

The γ_{NAC} and γ_{RAC} used in the first iteration of the calibration procedure are presented in Table 2 and were derived in [24].

Table 2. Partial factors for elements without shear reinforcement used in the first iteration of the calibration procedures [24].

Code	γ_{NAC}	γ_{RAC50}	γ_{RAC100}
EN1992 [32]	1.45	1.55	1.60
prEN1992 [31]	1.40	1.60	1.70

Table 3 shows the NAC cases when the design was made with EN1992 [31]. In these cases, the slabs were thicker than those of prEN1992 [32] to prevent parameter k from being limited to 2.0—see Equation (1). The parameters of design were adjusted so that the V_{Rd} values of both prEN1992 [32] and EN1992 [31] were the same.

Table 3. Cases of design for shear design of NAC slabs without shear reinforcement using EN1992 [31].

Slab	f_{ck} (MPa)	H (mm)	c (mm)	L (m)	a/d	Φ_{Asl} (mm)	s (mm)	ρ_l	V_{Rd} (kN)
1	25	190	25	5.0	2.5	16	104.1	1.23%	122.1
2	25	190	25	5.0	2.5	16	86.2	1.49%	135.5
3	25	215	25	5.0	2.5	16	100	1.10%	140.0
4	40	175	25	6.5	2.5	16	73	1.94%	150.4
5	40	200	25	6.5	2.5	20	94.2	2.02%	177.1

In the case of beams, a single case of design was studied because:

- The uncertainty in the outcome of resistance of this type of design is virtually lognormally distributed and reliability is predominantly dependent on the moments of θ_R and f_y, which do not depend on the case of design. This occurs because Equation (13) has a multiplicative nature, is mainly composed of lognormal distributions, and depends mostly on θ_R and f_y;
- The uncertainty in the outcome of load-effects depends on θ_E, P, and Q only. Since the statistics of θ_E are fixed for all cases of design, loads are given by P+Q and the

variability of P and Q are defined in terms of their coefficient of variation (Co different cases of design lead to similar uncertainty in the outcome of load-effects.

Thus, in the case of beams, different cases of design result in the same outputs reliability analysis, including β as long as V_{Rd} is equal to the design value of load-effec (which is already a condition of the reliability analyses performed). Table 4 presents t case of design analysed. As in the case of Table 1, only NAC is presented.

Table 4. Case of design for reliability analysis of shear design of NAC beam with shear reinforcement.

f_{ck} (MPa)	b (mm)	H (mm)	c (mm)	d (mm)	Φ_{Asl} (mm)	Φ_{Asw} (mm)	s (mm)	A_{sw}/s (cm²/mm)	ρ_w	$\rho_w \cdot f_{yk}/\gamma_S$ (MPa)	Ω	V_{Rd} (kN)
25	250	450	25	407	16	8	150	6.7	0.27%	1.17	30°	185.9

The design complied with the maximum spacing between shear reinforcement a with the minimum geometric ratio of shear reinforcement (ρ_w) of either code [31,32]. stated in Section 2.2, V_{Rd} is given by $V_{Rd,tie}$ in all cases. $\gamma_{RAC} = 1.0$ in the first iteration the calibration stage.

Each case of design of both beams and slabs included seven separate reliability an yses, since different ratios of variable to total load were analysed. For that purpo parameter χ was defined as shown in Equation (14). The analysis of χ vs. β relationshi was made because other code calibration procedures have shown that the β values Eurocode design depend on χ [55,56].

$$\chi = \frac{\gamma_G \times G_k}{\gamma_G \times P_k + \gamma_Q \times Q_k}. \quad (1$$

4.2. Deterministic and Stochastic Modelling

The deterministic and stochastic modelling is shown in Tables 5–7. These models a needed for the design equations, which were presented in Equations (3) and (5)–(9), a limit state functions, presented in Equations (11)–(13).

The incorporation of RAs does not affect reinforcement, loads, load-effect modelling, geometry; therefore, the criterion was to use the stochastic models of seminal publicatio (those of the Joint Committee for Structural Safety [62] and of other relevant document e.g., concerning the calibration and partial factor assessment of the Eurocodes). Tabl shows the references of each stochastic model. In the case of the compressive strength concrete, stochastic modelling was made after the proposal of Bartlett and McGregor [6 Parameter λ of this proposal was defined based on the quality control records of Portugue ready-mixed concrete production for the year 2017 [64].

No modelling for b (the width of the web of the cross-section) is presented in Tabl since this parameter is used for the shear resistance of elements without shear—s Equations (5)–(8)—but the cases of design concerned slabs (and geometric uncertair in this case was negligible). In addition, no modelling for variability of height is present since vertical deviations concerned the effective depth (given by the effective depth, whi accounts for the uncertainty in cover).

Table 5. Deterministic and stochastic modelling of load-effects, geometry, and material properties.

Parameter	Deterministic Model		Stochastic Model				Reference
	Symbol	Fractile	Symbol	Mean	Standard Deviation (σ)	Probability Distribution	
Permanent load	P_k	50 %	P	P_k	$0.10 \times P$	Normal	[44]
Maximum variable load (50 years)	Q_k	*	Q	$0.60 \times Q_k$	$0.35 \times Q$	Gumbel	[65–70]
Model uncertainty of load-effects	Absent from deterministic modelling		θ_E	1.00	0.05	Lognormal	[55,71]
Compressive strength	f_{ck}	5% fractile	f_c	$f_c = \lambda \times F_2 \times f_{ck}$, with f_{ck} assumed deterministic			[63,64]
Specified to delivered strength	Absent from deterministic modelling	λ	1.20		0.17		Lognormal
Standard to strength-in-structures	Absent from deterministic modelling	F_2	0.95		0.13		Lognormal
Yield stress of the reinforcement	f_{yd}	5% fractile	f_y	$f_{yd} + 2\sigma$	30 MPa	Lognormal	[55,56,62,72];
Young's modulus of the reinforcement			E_s = 200 GPa assumed as deterministic				
Cross-sectional area of the reinforcement			A_s assumed as deterministic				
Height of the cross-section			H assumed as deterministic				
Concrete cover (vertical)	cy	Nominal value	$cy + \Delta cy$	Δcy: 5 mm (slabs) Δcy: −5 mm (beams)	Δcy: 5 mm (slabs) Δcy: −5 mm (beams)	Δcy: Normal cy is deterministic	[13,62]

* Q_k is conceptualized as the 98% fractile of the one-year maximum load, but this concept is not used for loads in buildings due to scarce statistical information [73]. The Q_k of loads on buildings of EN1991 [74] are based on relevant reports on the subject and on engineering judgment [75–77].

The θ_R of the elements without shear reinforcement is presented in Table 6 and its derivation may be consulted in [24]. No deterministic model is presented for θ_R because model uncertainties were omitted from design equations—see Equations (3) and (5)–(9).

Table 6. Stochastic modelling of the model uncertainty of elements without shear reinforcement.

θ_R	Incorporation Ratio of RAs	Mean Value	Standard Deviation (σ)	Probability Distribution
EN 1992 [31]	NAC	1.03	0.113	Lognormal
	RAC50	1.00	0.120	Lognormal
	RAC100	0.95	0.114	Lognormal
prEN 1992 [32]	NAC	0.98	0.088	Lognormal
	RAC50	0.93	0.102	Lognormal
	RAC100	0.93	0.121	Lognormal

Since research on the shear resistance of elements with shear reinforcement that comply with the criteria typically used in the assessment of θ_R [78] is not abundant, the stochastic models presented in [24] for the design of elements with shear reinforcement are preliminary, and additional research is needed prior to the definition of statistics for θ_R. Therefore, the approach taken in this paper was to perform a sensitivity analysis for two assumptions:

- Assumption 1, in which the statistics of NAC are those presented in *fib* Bulletin 80 [79] for $\rho_w \cdot f_{yd}$ between 1 and 2 MPa. Concerning RAC, this case assumes that the mean value of θ_R is unaffected by the incorporation of RAs, but the standard deviatic increases as the RA incorporation ratio increases;

- Assumption 2, in which the statistics of NAC are those presented in *fib* Bulletin 80 [7 for $\rho_w \cdot f_{yd}$ between 1 and 2 MPa and pessimistic expectations for the influence of R on the mean value and standard deviation of θ_R are assumed.

Table 7. Stochastic modelling of the model uncertainty of elements with shear reinforcement.

θ_R	Incorporation Ratio of RA	Mean	Standard Deviation (σ)	Probability Distribution	Source
Assumption 1	NAC RAC50 RAC100	1.25 1.25 1.25	0.312 (CoV = 25.0%) 0.343 (CoV = 27.5%) 0.375 (CoV = 30.0%)	Lognormal Lognormal Lognormal	NAC: *fib* Bulletin 80 [79] RAC: Same mean value as NAC pessimistic expectation of the Co
Assumption 2	NAC RAC50 RAC100	1.25 1.21 1.17	0.312 (CoV = 25.0%) 0.333 (CoV = 27.5%) 0.351 (CoV = 30.0%)	Lognormal Lognormal Lognormal	NAC: *fib* Bulletin 80 [79] RAC: Pessimistic expectation of the mean value and CoV

Table 7 summarises these assumptions. The statistics of this θ_R were characterized high CoV for both NAC and RAC. This was expected to lead to small β values.

5. Results

5.1. Slabs without Shear Reinforcement

5.1.1. Design with prEN1992

The results of the reliability analyses are presented in Figure 2 for all NAC slabs. shown there, the β values of the slabs are similar. The same β vs. χ trend reported in oth documents concerning Eurocode design [55,56] was observed. Moreover, the 50-year β th resulted from a design with $\gamma_{NAC} = 1.40$ was an analogue to that of Eurocode reliabili assessments and calibration efforts [67,68,73], and no calibration of γ_{NAC} was needed.

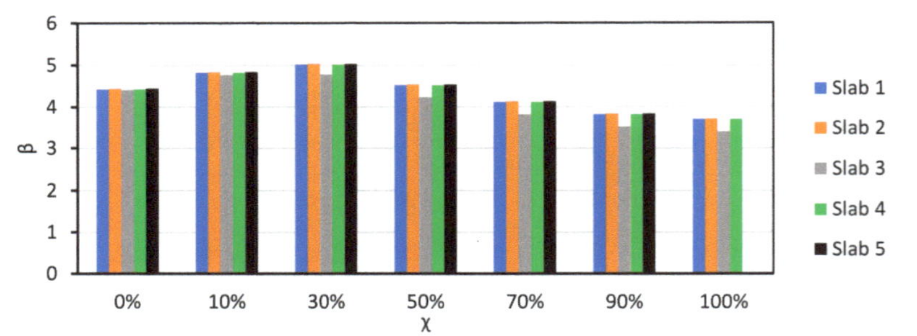

Figure 2. χ vs. β for shear resistance design of NAC slabs without shear reinforcement. prEN1992 [3 $\gamma_{NAC} = 1.40$.

Since the trends observed for all slabs were similar, the discussion of results is focus on Slab 1. Figure 3 compares the β vs. χ relationship of NAC, RAC50 and RAC100. It shown that: (i) β was above $\beta_{target} = 3.80$ in all cases except for NAC and $\chi = 100\%$; (the β of RAC was above that of NAC, which means that the partial factor γ_{RAC} present in Table 2 for RAC50 and RAC100 may be decreased.

To better understand the relative importance of all sources of uncertainty, Figure shows the α^2 values of Slab 1 for $\chi = 0$ and NAC, RAC50, and RAC100 design. The figu shows that the uncertainties in θ_R and P are the most relevant for the overall uncertair Moreover, as the incorporation ratio of RA increased, the α^2 of θ_R increased. This was d to the detrimental influence of RAs on the statistics of θ_R (see Table 6) and suggested th notwithstanding the preliminary partial factor γ_{RAC} leading to a larger β value than th of NAC, a specific partial factor for γ_{RAC} is still needed.

Findings for other χ values were similar to those reported in Figure 3. The only noteworthy differences are that as χ increased: (i) the α^2 of Q increased; (ii) the α^2 of P decreased relevantly; and (iii) the α^2 of all other stochastic variables decreased proportionally to the α^2 reported in Figure 3.

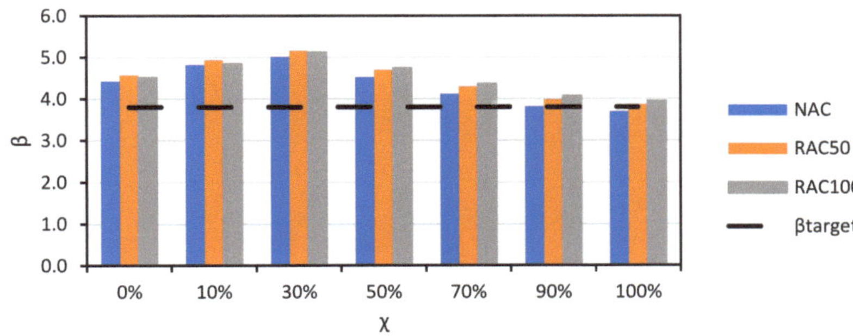

Figure 3. β vs. χ of Slab 1 for NAC and RAC. prEN1992 [32]. γ_{RAC} of Table 2 (not calibrated).

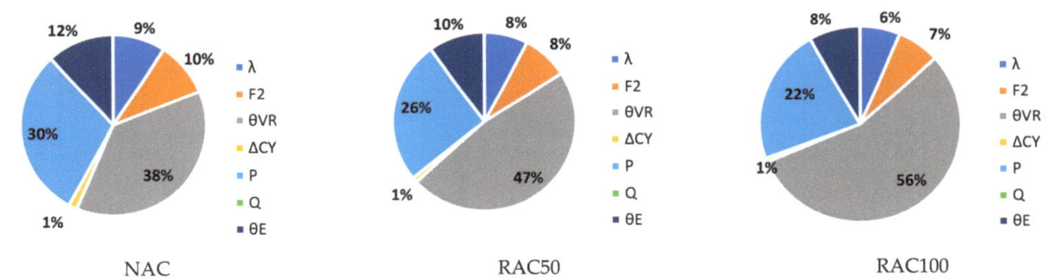

Figure 4. α^2 for $\chi = 0$ and Slab 1 and the preliminary partial factors for shear design of elements without stirrups presented in Part 3. prEN1992 [32]. $\gamma_{NAC} = 1.40$; $\gamma_{C,V,RAC50} = 1.60$; $\gamma_{C,V,RAC50} = 1.70$.

The calibration procedure presented in Section 3.1 was used. Figure 5 and Table 8 show that the β value of the RAC design became similar to that of NAC when $\gamma_{RAC50} = 1.50$ and $\gamma_{RAC100} = 1.60$.

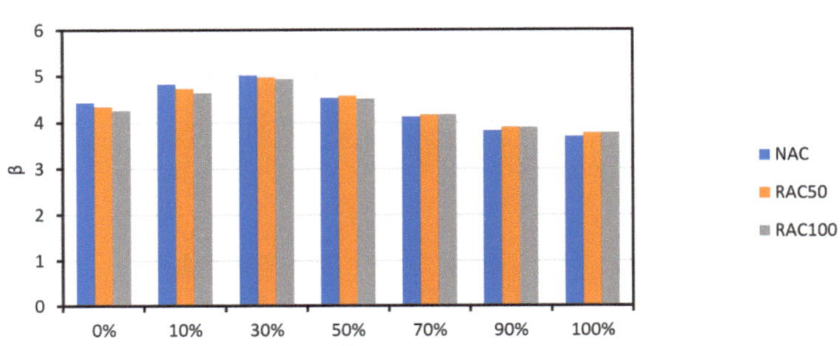

Figure 5. β vs. χ of Slab 1 for NAC and RAC. prEN1992 [32]. $\gamma_{NAC} = 1.40$ and $\gamma_{CRAC50} = 1.50$, $\gamma_{RAC100} = 1.60$ (after calibration).

Table 8. Ratio β RAC/β NAC of slabs. prEN1992 [32]. $\gamma_{NAC} = 1.40$ and $\gamma_{CRAC50} = 1.50$; $\gamma_{RAC100} = 1.60$ (after calibration).

RA	Slab	χ						
		0%	10%	30%	50%	70%	90%	100%
RAC50	Slab 1	98%	98%	99%	101%	101%	102%	102%
	Slab 2	98%	98%	99%	101%	101%	102%	102%
	Slab 3	98%	98%	100%	101%	102%	102%	102%
	Slab 4	98%	98%	99%	101%	103%	103%	102%
	Slab 5	100%	102%	103%	104%	104%	104%	104%
RAC100	Slab 1	96%	96%	98%	100%	101%	102%	102%
	Slab 2	93%	92%	96%	99%	101%	102%	102%
	Slab 3	94%	93%	100%	97%	102%	102%	103%
	Slab 4	94%	93%	96%	100%	101%	102%	102%
	Slab 5	94%	93%	96%	99%	101%	102%	102%

5.1.2. Design with EN1992

Figure 6 presents the relationship between β and χ of all NAC slabs designed wi EN1992 [31]. As in the case of prEN1992 [32], the results of all slabs were similar. Moreov the comparison between Figures 4 and 6 showed that the design with either code result in similar β values, notwithstanding the smaller value γ_{NAC} and the smaller mean θ_R of t resistance model of prEN1992 [32] in comparison to those of EN1992 [31]. This means th shear design of NAC members without shear reinforcement tends to be more economi when prEN1992 [32] is used, due to the better precision of its resistance model.

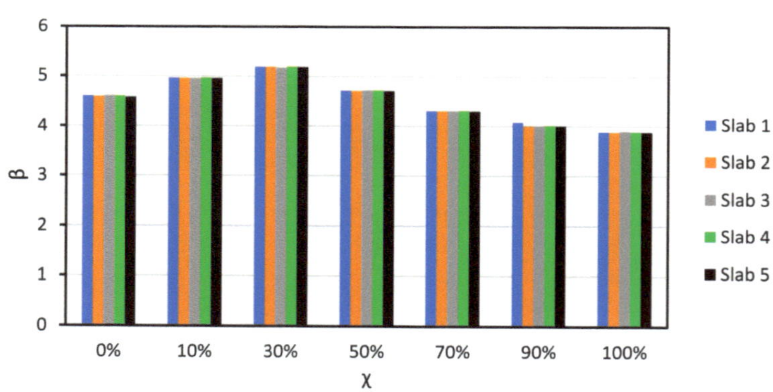

Figure 6. χ vs. β for shear resistance design of NAC slabs without shear reinforcement. EN1992 [$\gamma_{NAC} = 1.45$; $\gamma_{RAC50} = 1.55$; $\gamma_{RAC100} = 1.60$ (no calibration needed).

The reliability analyses also found that the preliminary partial factors proposed in [and presented in Table 2 resulted in β values that were similar to those of the NAC desi Table 9 shows these results.

Figure 7 is an example (Slab 1 and χ = 0) of the α^2 of all stochastic variables. It sho that the trends reported for designs with prEN1992 [32] were also observed for EN1992 [

Table 9. Ratio β RAC/β NAC of slabs. EN1992 [31]. $\gamma_{NAC} = 1.45$; $\gamma_{RAC50} = 1.55$; $\gamma_{RAC100} = 1.60$ (no calibration needed).

RA	Slab	χ						
		0%	10%	30%	50%	70%	90%	100%
RAC50	Slab 1	100%	100%	100%	101%	101%	100%	102%
	Slab 2	100%	100%	100%	101%	101%	102%	102%
	Slab 3	102%	100%	101%	101%	101%	101%	102%
	Slab 4	100%	101%	100%	101%	101%	102%	102%
	Slab 5	101%	100%	100%	101%	101%	102%	102%
RAC100	Slab 1	98%	98%	99%	100%	100%	98%	100%
	Slab 2	98%	97%	98%	99%	100%	100%	100%
	Slab 3	101%	101%	98%	97%	101%	100%	99%
	Slab 4	98%	98%	99%	99%	100%	100%	100%
	Slab 5	98%	97%	98%	99%	100%	100%	100%

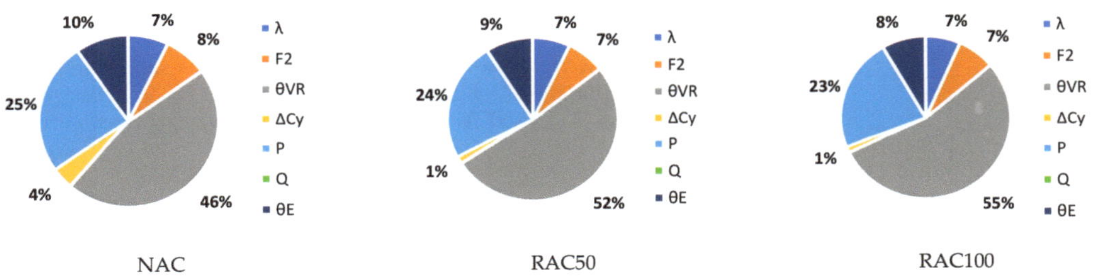

Figure 7. α^2 for $\chi = 0$ and Slab 1. Slabs without shear reinforcement. EN1992 [31]. $\gamma_{NAC} = 1.45$ and $\gamma_{CRAC50} = 1.55$, $\gamma_{RAC100} = 1.60$ (no calibration needed).

5.2. Beams with Shear Reinforcement

Figure 8 shows the β vs. χ relationship of NAC and RAC beams. Partial factor γ_{RAC} is yet to be calibrated ($\gamma_{RAC} = 1.0$ at this stage). As expected, the incorporation of RAs resulted in a relevant decrease of reliability, particularly when Assumption 2 for θ_R (Table 7) was analysed. Moreover, the β value achieved for the shear design of elements with shear reinforcement was below β_{target} (even in the case of NAC); for χ it was in the region of 0 to 10% and for χ above 70%, which are uncommon cases of reinforced concrete design [55].

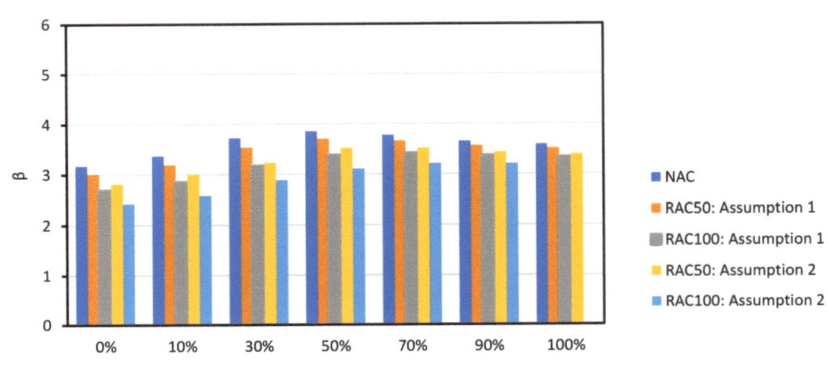

Figure 8. χ vs. β for shear resistance design of NAC and RAQ beams with shear reinforcement. EN1992 [31] and prEN1992 [32]. $\gamma_{RAC} = 1.0$ (not calibrated).

Figure 9 shows large values of the α^2 of θ_R, which demonstrated that θ_R was the cau of most of the uncertainty in this type of design. Figure 9 only concerns Assumption 1 Table 7 since findings for Assumption 2 followed the same rationale.

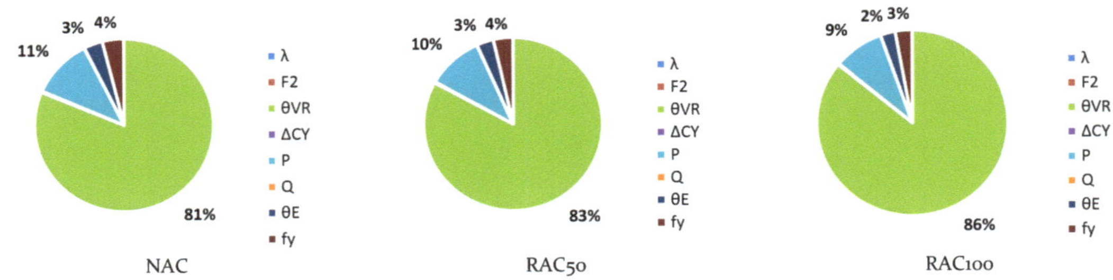

Figure 9. α^2 for $\chi = 0$ and θ_R Assumption 1. EN1992 [31] and prEN1992 [32]. $\gamma_{RAC} = 1.0$ (not calibrated).

Partial factor γ_{RAC} was calibrated and Figure 10 shows the results of calibration. observed, the β value of the RAC design was equivalent to that of the NAC design when t calibrated partial factors were used. The next section presents all calibrated partial facto

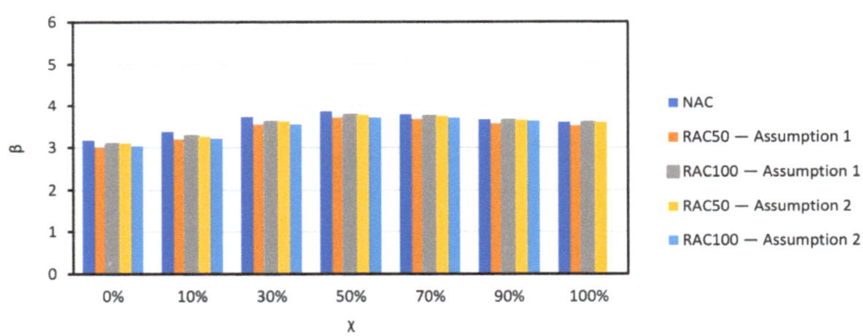

Figure 10. β vs. χ for both assumptions for θ_R. EN1992 [31] and prEN1992 [32]. Results the calibration: Assumption 1: $\gamma_{RAC50} = 1.00$; $\gamma_{RAC100} = 1.05$; Assumption 2: $\gamma_{RAC50} = 1.$ $\gamma_{RAC100} = 1.20$ (after calibration).

Since the θ_R for elements with shear reinforcement was based on scarce data a assumptions, the elasticities of the mean and of the standard deviation [58] of this stochas variable are studied and presented in Table 10.

Table 10. Elasticities of $\theta_{VR,with\ shear}$ for $\chi = 0$. EN1992 [31] and prEN1992 [32]. Assumption 1: $\gamma_{RAC50} = 1.00$; $\gamma_{RAC100} = 1.05$; Assumption 2: $\gamma_{RAC50} = 1.10$; $\gamma_{RAC100} = 1.20$ (after calibration).

Elasticity (%)	NAC	RAC50 Assumption 1	RAC100 Assumption 1	RAC50 Assumption 2	RAC100 Assumption 2
Mean	2.01	2.04	1.99	1.97	1.94
Standard deviation	−0.86	−0.88	−0.89	−0.89	−0.91

These elasticities showed that, if the statistics of the θ_R for elements with shear re forcement differed from those assumed, the influence on β would be relevant. For instan

- For Assumption 1 and χ = 50%, the beam made with RAC100 has a 50-year β of 3. Assumption 1 models θ_R with a mean of 1.25;

- If the mean of θ_R is 1.20 instead of 1.25, the actual 50-year β would correspond to roughly:

$$\beta = 3.75 \cdot \left[1 + \left(\frac{1.20 - 1.25}{1.25}\right) \times 1.99\right] = 3.45. \quad (15)$$

This corresponds to a decrease in β of 8%. This decrease emphasized that further experiments on the shear resistance of beams with stirrups should be performed prior to definite proposals for θ_R and partial factors for RAC design in order to base the statistics of this θ_R on a comprehensive set of data.

5.3. Recommendations for Design

Partial factors for shear design of elements with and without shear reinforcement were calibrated for the design of:

- Elements without shear reinforcement using EN1992 [31]—Equations (5) and (6);
- Elements without shear reinforcement using prEN1992 [32]—Equations (7) and (8);
- Elements with shear reinforcement using either EN1992 [31] or prEN1992 [32]—Equations (3) and (9).

The recommended partial factors, which were discussed in Sections 5.1 and 5.2, are presented in Table 11.

Table 11. Recommended partial factors for shear design of NAC and RAC elements.

Type of Concrete	Design			
	Shear without Shear Reinforcement		Shear with Shear Reinforcement—EN1992 [31] and prEN1992 [32]	
	EN1992 [31]	prEN1992 [32]	Assumption 1 (Moderate)	Assumption 2 (Pessimistic)
NAC	1.45	1.40	1.00	1.00
RAC50	1.55	1.50	1.00	1.10
RAC100	1.60	1.60	1.05	1.20

These partial factors ensure that the β values of reinforced concrete elements made with incorporation of RAs are similar to those of conventional reinforced elements designed for analogue conditions. Partial factors for other incorporation ratios may be determined by linear interpolation.

The partial factor calibrated for Assumption 2 of the shear design of elements with shear reinforcement is a conservative upper bound of the implications of RAs for shear design, and its main purpose is to show that additional experiments on the shear resistance of elements with shear reinforcement are needed prior to a definite calibration of a partial factor for this resistance model.

6. Conclusions

This paper provided partial factors for the shear design of reinforced concrete elements made with the incorporation of coarse recycled aggregates produced from concrete waste. The paper addressed the design of elements with and without shear reinforcement. The design equations concerned two codes: the current version of EN1992 and prEN1992 (the next generation of EN1992, under approval by CEN).

An overview of research on the shear resistance of recycled aggregate concrete elements was provided. The fundamental reason for recycled aggregates affecting the uncertainty in shear resistance modelling was stated: since recycled aggregates are weaker than natural aggregates, aggregate interlock decreases without this being accounted for by the shear resistance model. The probabilistic basis of structural codes was discussed and the development of design guidelines for Eurocode recycled aggregate concrete design was contextualised with the partial factor format of the Eurocodes.

A partial factor was then added to the design equations of EN1992 and prEN1992 and calibrated using a procedure that follows the general rules and recommendations of

ISO2394 and EN1990. Relevant increases in the partial factor were found for the design elements without shear reinforcement and, without the calibrated partial factor for recycl aggregate concrete design, structural safety was compromised. In the case of elements wi shear reinforcement, since research is not as comprehensive, the calibrated partial fact was preliminary and defined based on a sensitivity analysis and assumptions. It was fou that the partial factor was sensitive to deviations in the statistics of the model uncertair and that additional research is recommended prior to the definite calibration of a part factor. In the meantime, a partial factor calibrated based on engineering judgement a either moderate or fairly pessimistic assumptions was calibrated and proposed.

The authors recommend future research on the punching shear resistance of recycl aggregate concrete; research and partial factor calibration concerning the shear resistance elements made with recycled aggregates produced from construction and demolition was and research in which the shear behaviour of concrete and the properties of the recycl aggregates are thoroughly characterised, so that the resistance models for shear design a appropriately changed with physically based coefficients rather than partial factors.

Author Contributions: Conceptualisation, J.P.; methodology, J.P.; software, J.P.; validation, J.d C.C., L.E.; formal analysis, J.P., J.d.B., C.C., L.E.; investigation, J.P.; resources, J.d.B.; data curation, J writing—original draft preparation, J.P.; writing—review and editing, J.d.B., C.C., L.E.; visualisati J.P.; supervision, J.d.B., C.C., L.E.; project administration, J.d.B.; funding acquisition, J.d.B. All auth have read and agreed to the published version of the manuscript.

Funding: The scholarship of the first author was funded by the Portuguese Foundation for Scier and Technology (FCT), grant number PD/BD/113643/2015.

Institutional Review Board Statement: Not applicable.

Informed Consent Statement: Not applicable.

Data Availability Statement: The data presented in this study are available on reasonable requ from the corresponding author.

Acknowledgments: The support of IST, University of Lisbon, the CERIS research centre, and EcoCoRe doctoral programme of FCT is acknowledged.

Conflicts of Interest: The authors declare no conflict of interest. The funders had no role in the desi of the study; in the collection, analyses, or interpretation of data; in the writing of the manuscript, in the decision to publish the results.

List of Acronyms and Symbols

CDW	construction and demolition waste
CoV	coefficient of variation
NA	coarse natural aggregate
RA	coarse recycled aggregate
RAC	concrete with partial or total incorporation of coarse recycled aggregate concrete
RAC50	recycled aggregate concrete elements with 50% incorporation of coarse recycled aggregates
RAC100	recycled aggregate concrete elements with full incorporation of coarse recycled aggregates
A_s	cross-sectional area of the reinforcement
A_{sw}	area of shear reinforcement
E	random outcome of load-effects
E_d	design value of load-effects
E_s	Young's modulus of reinforcement

F_2	conversion of delivered strength measured on standard specimens to the strength within structural elements
H	height of the beam
M_{cs}	bending moment at the control section
P	random outcome of permanent loading
P_k	characteristic value of permanent loading
Q	random outcome of variable loading
Q_k	characteristic value of variable loading
R	random outcome of resistance
V_{cs}	shear stress at the control section
V_{Rd}	design value of shear resistance
$V_{Rd,cmin}$	design value of the minimum shear resistance of elements without shear reinforcement
$V_{Rd,strut}$	design value of the shear resistance of the compression struts of the resistance model of beams with shear reinforcement
$V_{Rd,tie}$	design value of the shear resistance of the ties of the resistance model of beams with shear reinforcement
b	width of the web of the beam
c	concrete cover
d	effective depth of the beam
d_{max}	maximum aggregate diameter
f_c	random outcome of the compressive strength of concrete
f_{ck}	characteristic value of the compressive strength of concrete
f_y	random outcome of the yield stress of the reinforcement
f_{yd}	design value of the yield stress of the reinforcement
f_{yk}	characteristic yield stress of the reinforcement
g_x	limit state function
s	distance between reinforcement
ΔB	uncertainty in the width of the beam
Φ_{Asl}	diameter of longitudinal reinforcement
Ω	angle of the compression strut with the longitudinal axis used in the resistance model of elements with shear reinforcement
α	direction cosine of a stochastic variable
β	reliability index
β_{target}	target reliability index
γ_C	partial factor for the strength of concrete
γ_c	partial factor for the variability of the strength of concrete
γ_G	partial factor for permanent loads
γ_Q	partial factor for variable loads
γ_{NAC}	partial factor for shear design of natural aggregate concrete elements without shear reinforcement
γ_{RAC}	partial factor for shear design of recycled aggregate concrete elements
γ_{RAC}	partial factor for shear design of recycled aggregate concrete elements with 50% incorporation of coarse recycled aggregates
γ_{RAC100}	partial factor for shear design of recycled aggregate concrete elements with full incorporation of coarse recycled aggregates
γ_{Rd}	partial factor for the uncertainty in geometry and in resistance modelling
γ_S	partial factor for the yield stress of the reinforcement
θ_E	model uncertainty of load-effect modelling
θ_R	model uncertainty of the resistance model
λ	stochastic model for the conversion of specified to delivered strength
ρ_l	geometric ratio of longitudinal tensile reinforcement
ρ_w	geometric ratio of shear reinforcement
v_{min}	minimum shear stress of the resistance model of elements without shear reinforcement
χ	ratio of the design value of the variable loading to the total design value of loading

References

1. Pacheco, J.; De Brito, J.; Ferreira, J.; Soares, D. Destructive horizontal load tests of full-scale recycled aggregate concrete structur *Aci Struct. J.* **2015**, *112*, 815–826. [CrossRef]
2. Rangel, C.S.; Amario, M.; Pepe, M.; Yao, Y.; Mobasher, B.; Filho, R.T. Tension stiffening approach for interface characterization recycled aggregate concrete. *Cem. Concr. Compos.* **2017**, *82*, 176–189. [CrossRef]
3. European Environment Agency. *Construction and Demolition Waste: Challenges and Opportunities in a Circular Economy*; Briefing 14/2019; European Environment Agency: Copenhagen, Denmark, 2020.
4. Pacheco, J.; De Brito, J.; Chastre, C.; Evangelista, L. Uncertainty models of reinforced concrete beams in bending: Code comparis and recycled aggregate incorporation. *J. Struct. Eng.* **2019**, *145*, 04019013. [CrossRef]
5. Silva, R.; De Brito, J. Reinforced recycled aggregate concrete slabs: Structural design based on Eurocode 2. *Eng. Struct.* **2020**, 2 110047. [CrossRef]
6. Pacheco, J.; De Brito, J. Recycled aggregate concrete: Properties and behaviour, applications and production challenges. Proceedings of the fib Symposium 2021: Concrete structures: New Trends for Eco-Efficiency and Performance, Lisbon, Portug 13–16 June 2021.
7. Del Bosque, I.S.; Zhu, W.; Howind, T.; Matías, A.; de Rojas, M.S.; Medina, C. Properties of interfacial transition zones (ITZs) concrete containing recycled mixed aggregate. *Cem. Concr. Compos.* **2017**, *81*, 25–34. [CrossRef]
8. Pepe, M.; Grabois, T.M.; Silva, M.A.; Tavares, L.M.; Filho, R.D.T. Mechanical behaviour of coarse lightweight, recycled and natu aggregates for concrete. *Proc. Inst. Civ. Eng. Constr. Mater.* **2018**, *173*, 1–9. [CrossRef]
9. Bravo, M.; de Brito, J.; Pontes, J.; Evangelista, L. Mechanical performance of concrete made with aggregates from constructi and demolition waste recycling plants. *J. Clean. Prod.* **2015**, *99*, 59–74. [CrossRef]
10. Xiao, J.; Li, W.; Fan, Y.; Huang, X. An overview of study on recycled aggregate concrete in China (1996–2011). *Constr. Build. Ma* **2012**, *31*, 364–383. [CrossRef]
11. Tošić, N.; Torrenti, J.M.; Sedran, T.; Ignjatović, I. Toward a codified design of recycled aggregate concrete structures: Backgrou for the new fib Model Code 2020 and Eurocode 2. *Struct. Concr.* **2021**. [CrossRef]
12. Pacheco, J.; de Brito, J.; Chastre, C.; Evangelista, L. Scatter of constitutive models of the mechanical properties of concre Comparison of major international codes. *J. Adv. Concr. Technol.* **2019**, *17*, 102–125. [CrossRef]
13. Pacheco, J. Reliability Analysis of Eco-Concrete. Ph.D. Thesis, Civil Engineering, Department of Civil Engineering, Architectu and Gerorresources, Instituto Superior Técnico, University of Lisbon, Lisbon, Portugal, 2020.
14. Pacheco, J.; de Brito, J.; Ferreira, J.G.; Soares, D. Flexural load tests of full-scale recycled aggregates concrete structures. *Con Build. Mater.* **2015**, *101*, 65–71. [CrossRef]
15. Tošić, N.; Marinković, S.; De Brito, J. Deflection control for reinforced recycled aggregate concrete beams: Experimental datab and extension of the fib Model Code 2010 model. *Struct. Concr.* **2019**, *20*, 2015–2029. [CrossRef]
16. Silva, R.; de Brito, J.; Dhir, R. Properties and composition of recycled aggregates from construction and demolition waste suita for concrete production. *Constr. Build. Mater.* **2014**, *65*, 201–217. [CrossRef]
17. Choi, H.; Yi, C.; Cho, H.; Kang, K.; Choi, H.; Yi, C.; Cho, H.; Kang, K. Experimental study on the shear strength of recycl aggregate concrete beams. *Mag. Concr. Res.* **2010**, *62*, 103–114. [CrossRef]
18. Yun, H.-D.; Choi, W.-C. Shear Strength of Reinforced Recycled Aggregate Concrete Beams Without Shear Reinforcements. *J. C Eng. Manag.* **2016**, *23*, 76–84. [CrossRef]
19. Schubert, S.; Hoffmann, C.; Leemann, A.; Moser, K.; Motavalli, M. Recycled aggregate concrete: Experimental shear resistance slabs without shear reinforcement. *Eng. Struct.* **2012**, *41*, 490–497. [CrossRef]
20. Ignjatović, I.S.; Marinković, S.B.; Tošić, N. Shear behaviour of recycled aggregate concrete beams with and without she reinforcement. *Eng. Struct.* **2017**, *141*, 386–401. [CrossRef]
21. Arezoumandi, M.; Drury, J.; Volz, J.S.; Khayat, K. Effect of Recycled Concrete Aggregate Replacement Level on Shear Strength Reinforced Concrete Beams. *ACI Mater. J.* **2015**, *112*, 559–568. [CrossRef]
22. Brandes, M.R.; Kurama, Y.C. Behavior of shear-critical prestressed concrete beams with recycled concrete aggregates unc ultimate loads. *Eng. Struct.* **2018**, *165*, 237–246. [CrossRef]
23. Katkhuda, H.; Shatarat, N. Shear behavior of reinforced concrete beams using treated recycled concrete aggregate. *Constr. Bu Mater.* **2016**, *125*, 63–71. [CrossRef]
24. Pacheco, J.; de Brito, J.; Chastre, C.; Evangelista, L. Uncertainty of shear resistance models: Influence of recycled concr aggregate on beams with and without shear reinforcement. *Eng. Struct.* **2020**, *204*, 109905. [CrossRef]
25. Francesconi, L.; Pani, L.; Stochino, F. Punching shear strength of reinforced recycled concrete slabs. *Constr. Build. Mater.* **2016**, 1 248–263. [CrossRef]
26. Reis, N.; de Brito, J.; Correia, J.R.; Arruda, M.R.T. Punching behaviour of concrete slabs incorporating coarse recycled concr aggregates. *Eng. Struct.* **2015**, *100*, 238–248. [CrossRef]
27. Sahoo, S.; Singh, B. Punching shear capacity of recycled-aggregate concrete slab-column connections. *J. Build. Eng.* **20** *41*, 102430. [CrossRef]
28. Etxeberria, M.; Mari, A.; Vázquez, E. Recycled aggregate concrete as structural material. *Mater. Struct.* **2006**, *40*, 529–5 [CrossRef]

Han, B.C.; Yun, H.D.; Chung, S.Y. Shear capacity of reinforced concrete beams made with recycled-aggregate. *ACI Spec. Publ.* **2001**, *200*, 503–516.

Sýkora, M.; Holicky, M. Assessment of Uncertainties in Mechanical Models. *Appl. Mech. Mater.* **2013**, *378*, 13–18. [CrossRef]

EN1992-1-1. *Eurocode 2-Design of Concrete Structures: Part 1-1: General Rules and Rules for Buildings*; Comité Européen de Normalisation (CEN): Brussels, Belgium, 2008.

CEN/TC-250/SC-2. *prEN 1992-1-1 D6 Working File (2020-10-05 Rev. 7)*; CEN: Brussels, Belgium, 2020.

Waseem, S.; Singh, B. An experimental study on shear capacity of interfaces in recycled aggregate concrete. *Struct. Concr.* **2018**, *19*, 230–245. [CrossRef]

Khaldoun, N.R.; Abdul-Lateef, A.-K. Shear-friction behavior of recycled and natural aggregate concrete—an experimental investigation. *ACI Struct. J.* **2015**, *112*, 725–734.

Fakitsas, C.G.; Papakonstantinou, P.E.A.; Kiousis, P.D.; Savva, A. Effects of recycled concrete agregates on the compressive and shear strength of high-strength self-consolidating concrete. *J. Mater. Civ. Eng.* **2012**, *24*, 356–361. [CrossRef]

Xiao, J.; Sun, C.; Lange, D.A. Effect of joint interface conditions on shear transfer behavior of recycled aggregate concrete. *Constr. Build. Mater.* **2016**, *105*, 343–355. [CrossRef]

Waseem, S.A.; Singh, B. Shear transfer strength of normal and high-strength recycled aggregate concrete—An experimental investigation. *Constr. Build. Mater.* **2016**, *125*, 29–40. [CrossRef]

Xiao, J.; Xie, H.; Yang, Z. Shear transfer across a crack in recycled aggregate concrete. *Cem. Concr. Res.* **2012**, *42*, 700–709. [CrossRef]

Walraven, J. Aggregate Interlock: A Theoretical and Experimental Analysis. Ph.D. Thesis, TU Delft, Delft, The Netherlands, 1980.

Pacheco, J.; de Brito, J.; Chastre, C.; Evangelista, L. Bond of recycled coarse aggregate concrete: Model uncertainty and reliability-based calibration of design equations. *Eng. Struct.* **2021**, *239*, 112290. [CrossRef]

Khoury, E.; Ambrós, W.; Cazacliu, B.; Sampaio, C.H.; Remond, S. Heterogeneity of recycled concrete aggregates, an intrinsic variability. *Constr. Build. Mater.* **2018**, *175*, 705–713. [CrossRef]

Xiao, J.; Li, W.; Sun, Z.; Shah, P.S. Crack propagation in recycled aggregate concrete under uniaxial compressive loading. *Mater. J.* **2012**, *109*, 451–461.

Guo, M.; Alam, S.Y.; Bendimerad, A.Z.; Grondin, F.; Rozière, E.; Loukili, A. Fracture process zone characteristics and identification of the micro-fracture phases in recycled concrete. *Eng. Fract. Mech.* **2017**, *181*, 101–115. [CrossRef]

EN1990. *Eurocode: Basis of Structural Design*; Comité Européen de Normalisation (CEN): Brussels, Belgium, 2002.

ISO2394. *General Principles on Reliability for Structures*; ISO: Geneve, Switzerland, 2014.

Yousefpour, H.; Bayrak, O. *Shear design of beams without stirrups: Survey of participants in the fib workshop on beam shear. Fib Bulletin 85: Towards a Rational Understanding of Shear in Beams and Slabs*; Fib: Zurich, Switzerland, 2016.

PT-SC2-T1. *D3BG-Background Documents to the Final PT1 Draft prEN 1992-1-1:2018*; CEN: Brussels, Belgium, 2018.

Pacheco, J.; de Brito, J.; Chastre, C.; Evangelista, L. Experimental investigation on the variability of the main mechanical properties of recycled aggregate concrete. *Constr. Build. Mater.* **2019**, *201*, 110–120. [CrossRef]

Pacheco, J.N.; De Brito, J.; Chastre, C.; Evangelista, L. Probabilistic Conversion of the Compressive Strength of Cubes to Cylinders of Natural and Recycled Aggregate Concrete Specimens. *Materials* **2019**, *12*, 280. [CrossRef]

Fib. *Bulletin 65. Model Code 2010 Final Draft*; Fib: Lausanne, Switzerland, 2010; Volume 1.

CEN/TC-250/SC-2. *Updated Draft by SC2/WG1/CDG prEN 1992-1-1-D5 Working File (Rev. 11) 2020-05-2*; CEN: Brussels, Belgium, 2020.

ACI318-14. *Building Code Requirements for Structural Concrete (ACI 318-14) and Commentary*; The American Concrete Institute: Farmington Hills, MI, USA, 2014.

Fib. *Bulletin 66. Model Code 2010 Final Draft*; Fib: Lausanne, Switzerland, 2010; Volume 2.

Albuquerque, A.; Pacheco, J.; Brito, J. Eurocode Design of Recycled Aggregate Concrete for Chloride Environments: Stochastic Modeling of Chloride Migration and Reliability-Based Calibration of Cover. *Crystals* **2021**, *11*, 284. [CrossRef]

BRE. *Client Report 210297. An Independent Technical Expert Review of the SAKO Report*; Gulvanessian, H., Calgaro, J.A., Spehl, P., Jensen, B., Eds.; BRE: Watford, UK, 2003.

NKB/SAKO. NKB1999:01E Basis of design of structures. In *Proposals for Modification of Partial Safety Factors in Eurocodes, I.-B.*; Nordic Committee on Building Regulations (NBK): Oslo, Norway, 1999.

Rackwitz, R.; Fiessler, B. Structural reliability under combined random load sequences. *Comput. Struct.* **1978**, *9*, 489–494. [CrossRef]

Lemaire, M. *Structural Reliability*; John Wiley & Sons: Hoboken, NJ, USA, 2009.

Ditlevsen, O.; Madsen, H.O. *Structural Reliability Methods*; Department of Mechanical Engineering, Technical University of Denmark: Lyngby, Denmark, 2005.

EN206. Concrete: Specification, Performance, Production and Conformity. In *Incorporating Corrigendum May 2014*; CEN: Brussels, Belgium, 2013.

LNEC E471. *Guide for the Use of Coarse Recycled Aggregates in Concret-Guia Para a Utilização de Agregados Reciclados Grossos em Betões de Ligantes Hidráulicos*; LNEC: Portugal, Spain, 2009.

JCSS. Probabilistic Model Code. In *Part 3: Material Properties*; Danish Technical University: Lyngby, Denmark, 2001.

Bartlett, F.M.; MacGregor, J.G. Statistical analysis of the compressive strength of concrete in structures. *Mater. J.* **1996**, *93*, 158–168.

64. Pacheco, J.; de Brito, J.; Chastre, C.; Evangelista, L. Statistical analysis of Portuguese ready-mixed concrete production. *Con. Build. Mater.* **2019**, *209*, 283–294. [CrossRef]
65. Gulvanessian, H.; Holicky, M. Eurocodes: Using reliability analysis to combine action effects. *Proc. Inst. Civ. Eng. Struct. Bu* **2005**, *158*, 243–252. [CrossRef]
66. Holicky, M. Reliability-based analysis of codified design allowing for production quality. In Proceedings of the 4th Internatio ASRANet Colloquium, Athens, Greece, 25–27 June 2008.
67. Gulvanessian, H.; Holický, M. Reliability based calibration of Eurocodes considering a steel member. In Proceedings of the JC Workshop on Reliability Based Code Calibration, Zurich, Switzerland, 21–22 March 2002.
68. Holicky, M.; Markova, J. Calibration of reliability elements for a column. In Proceedings of the JCSS Workshop on Reliabil Based Code Calibration, Joint Committee for Structural Safety, Zurich, Switzerland, 21–22 March 2002.
69. Caspeele, R.S.; Miroslav, S.; Taerwe, L. Influence of quality control of concrete on structural reliability: Assessment usin Bayesian approach. *Mater. Struct.* **2014**, *47*, 105–116. [CrossRef]
70. Caspeele, R.; Taerwe, L. Influence of concrete strength estimation on the structural safety assessment of existing structures. *Con Build. Mater.* **2014**, *62*, 77–84. [CrossRef]
71. Gulvanessian, H.; Holický, M. Annex C-Calibration procedure. In *Implementation of Eurocodes: Development of Skills Facilitat Implementation of Eurocodes. Handbook 2: Reliability Backgrounds*; Leonardo da Vinci Pilot Project CZ/02/B/F/PP-134007; Prag Czech Republic, 2005.
72. Holický, M.; Retief, J.; Wium, J. Partial factors for selected reinforced concrete members: Background to a revision of SA 10100-1. *J. S. Afr. Inst. Civ. Eng.* **2010**, *52*, 36–44.
73. Gulvanessian, H.; Calgaro, J.A.; Holicky, M. Designers' Guide to EN1990 Eurocode: Basis of Structural Design. In *Designers' Gu to the Eurocodes*; Gulvanessian, H., Ed.; Thomas Telford: London, UK, 2002.
74. EN1991-1. *Eurocode 1: Actions on Structures-Part 1-1: General actions-Densities, Self-Weight, Imposed Loads for Buildings*; CE Brussels, Belgium, 2002.
75. Gulvanessian, H.; Formichi, P.; Calgaro, J.A. Designers' guide to Eurocode 1: Actions on buildings. In *Designers' Guide to Eurocodes*; Harding, G., Ed.; Thomas Telford: London, UK, 2009; Volume 266.
76. Sedlacek, G.; Gulvanessian, H. *Eurocode 1: Basis of Design and Actions on Structures: Part 2.1: Densities, Self Weight, Imposed Lo IABSE Rep*; IABSE: Zurich, Switzerland, 1996; Volume 74.
77. Holický, M.A. Chapter 1-Self-weight and imposed loads on buildings. In *Implementation of Eurocodes: Development of Sk Facilitating Implementation of Eurocodes. Handbook 3: Action Effects for Buildings*; Leonardo da Vinci Pilot Project CZ/02/B/F/ 134007; Aachen, Germany, 2005.
78. Holický, M.; Retief, J.V.; Sýkora, M. Assessment of model uncertainties for structural resistance. *Probabilistic Eng. Mech.* **2016**, 188–197. [CrossRef]
79. Fib. *Bulletin 80: Partial Factor Methods for Existing Concrete Structures*; Fib: Lausanne, Switzerland, 2016.

MDPI AG
Grosspeteranlage 5
4052 Basel
Switzerland
Tel.: +41 61 683 77 34

Materials Editorial Office
E-mail: materials@mdpi.com
www.mdpi.com/journal/materials

www.ingramcontent.com/pod-product-compliance
Lightning Source LLC
Chambersburg PA
CBHW041421210926
51850CB00013B/3327
* 9 7 8 3 7 2 5 8 2 3 2 8 4 *